Chemistry 11 In Focus

Abdul Jalil Shakur

The New Awakening Books—Ontario, Toronto CA
ISBN: 979-8-9885555-2-0
Library of Congress Control Number: 2021916065
Title: *Chemistry 11 In Focus*
Author: Abdul Jalil Shakur
Digital distribution | 2021
Paperback | 2021

Editor in Chief:
Mohamad Kazim
C.P.A., M.B.A., C.G.M.A.
Assistants:
Haimnauth Ramkirath
C.P.A., C.M.A.
Clive Sankardyal
BA. Dip.ED.
Dr. Yaseer Shakur
BSc., MSc., PhD., MD.
Graphics and Design:
Abdul J. Shakur
Assistant:
Nafeesa Shakur
BA. MA. OTC
Cover page:
Nafeesa Shakur
Cover Molecule:
Abdul J. Shakur
BSc., B.Ed, OTC
Significant Digit Consultant:
Jaspal Singh Ughra
BSc. MSc., B.Ed, OTC
Technical Assistant:
El Houssine Chouyak
BSc., Dip. Comp. Sc.

Introduction

This is a grade 11 Chemistry textbook that is suitable for students in any part of the world who desire to master the conceptual framework of Chemistry at this level. The goal of this project is to produce a book that is essentially a comprehensive guide and almost like a personal tutor to all students studying grade 11 Chemistry.

After teaching grade 11 Chemistry for over 40 years, I have gained invaluable insights into how students learn grade 11 Chemistry successfully. For years, I have experimented with many innovative and creative teaching strategies that were successful in enabling students to master Chemistry in an effective manner. However, teaching this course made me realized that the mole is a daunting concept for students to understand, particularly from the backdrop that this concept permeates so many aspects of the curriculum. I feel it is imperative that students have a firm grasp of this concept to succeed.

One of the problems students face today is a limited understanding of grade 11 Chemistry. Unlike other science courses, it is a highly integrated one, and it is critical that students learn all aspects of it thoroughly. This is a course that is pivoted on the "step" approach to learning where the mastery of each ensuing concept is dependent on a thorough understanding of that which preceded it. Having a superficial knowledge of some crucial topics such as the mole concept and its applications, will affect students' overall performance since they encompass a major part of the curriculum. This text provides an unprecedented and thorough exploration of this concept and its applications which enabled my former students to master it. I am confident that it will empower other students to do likewise. Furthermore, while the other texts in use today are useful in their own ways, I know first-hand their inadequacies in addressing this concern and other prevailing problems. As such, I was impelled to develop a repertoire of teaching skills that were very successful in helping students master not only this concept, but others that posed challenges to learning. I have dedicated several years to writing and perfecting this text to preserve a rich teaching legacy that has been very successful.

The pedagogical approach that I used in creating this text is one that took nothing for granted in its endeavor to maximize learning. Concepts are introduced from first principles, and are then meticulously and thoroughly explained with the aid of suitable illustrations, demonstrations, simulations, and other diagrams. For greater clarity, and to emphasize points of importance, color codes are used throughout the text. Every effort is made to identify and further clarify concepts with which students normally encounter difficulties and have misconceptions. To reinforce and enhance understanding, at the end of every concept taught, numerous examples of problems and their solutions are provided. Wet labs are included and the results of these are presented in the form of pictures, tables, graphs, and charts with follow-up questions. Assessment and evaluation resources are included throughout the text. The quality of content, and profound approach used in this text, will enable students to acquire a most solid knowledge of grade 11 Chemistry, and even beyond the requirements at this level.

This book consists of five units: **matter, quantities in chemical reactions, solutions, and gases.** The unit in **matter** begins by exploring the various theories of the atom, and trends in the periodic table. This is followed by the chemical bonding, molecular geometry, chemical reactions and ends with nomenclature of binary compounds.

The unit in **quantities in chemical reactions** deals comprehensively with the mole concept and stoichiometry. After establishing a profound understanding of the mole concept, calculations relating to mass, moles and number of entities are introduced. Percentage composition, empirical and molecular formulas of compounds follow these. Finally, stoichiometry is taught where the concepts of excess and limiting reagents are fully dealt with.

The unit in **solutions** begins with properties of the water molecule with the objective of understanding its ability to dissolve various substances, and the solution process. This is followed by the concept of concentration of solutions; the various factors that determine it, and solution preparation. The effect of temperature on solubility is then investigated and solubility graphs are analysed. The actions of soaps and detergents in the cleaning process are described, and the effects of hard water on these are investigated. Dissolved chemicals in water are studied with the objective of maintaining safe levels in drinking water. Sewage and water treatments are described and the effects of organic wasted are investigated for their effects on dissolved gases in aquatic ecosystems. Theories of acids and bases, acid-base neutralization, pH and pH changes during titration, and the nomenclature of binary and polyatomic acids are taught with the aid of with suitable demonstrations and experiments.

The unit on **gases** begins with an explanation and demonstrations of atmospheric pressure. The various gas laws and their applications are then taught. Gas mixtures, Dalton's law of partial pressure, Avogadro's hypothesis, molar volumes, and gas stoichiometry follow. Finally, the Ideal gas equation is introduced to allow calculation of gas volumes at various temperature and pressure conditions.

I attribute nothing in this book to myself, but I thank and praise God for granting me the intelligence, perseverance, guidance, and inspiration, in allowing me to put together the invaluable works of present and past erudite scholars, authors and scientists, in ways to maximize learning. I am also very grateful to all those who have encouraged and assisted me in any way to make this textbook a reality. I encourage the purchase of this book, as the proceeds earned will be used for charitable and developmental projects in poor nations, such as my native land of Guyana.

Abdul Jalil Shakur
14/07/2021

Dedication

I would like to dedicate this book to my beloved parents, the late Abdul Shakur and Jafiran Shakur, my loving in-laws, Haji Rustum Ali and Hajin Rafeekhan Ali, my loving wife, Lilatool Shakur, my four wonderful children: Shazeeda, Yaseer, Nazeera and Nafeesa and their spouses and my most wonderful grandchildren: Saleh, Shuayb, Munira, Abdullah, Nusaybah, Muhammad, Salma, Ali, Fatima, Maryam and Anwar. Also, my twelve siblings: Sidique, Nazeera, Farida, Aisha, Ryhan, Shaheed, Arman, Majeed, Safoora, Khadija, Aleem and Shaharazaad, and all other family members too numerous to mention. I would like to pay special tribute to my sister, Aisha and my brothers; Arman and Sidique for their unstinted support in helping our parents provide the basic amenities to sustain our family.

Table of Contents

Acknowledgement

I would like to acknowledge the following people for their unstinted support and encouragement:
Lilatool Shakur, Shazeeda Shakur, Dr. Yaseer Shakur, Nazeera Shakur, Nafeesa Shakur, Safraz Shakur, Arif Assim, El Houssine Chouyakh, Mohamad Kazim Yusuff, Jaspal Singh Ughra, Clive Sankardayal, Dr. Monday Gala, Haimnauth Ramkirath, Derrick Mach, Janice Hew, Salini Sornalingam, Dr. Elaine Sinclair, Nicloe Cheung-Seekit, Ranbir Dhoot, Inga Teper, Bishnauth Shewprashad, Sabrina Azeez, Nazra Hussain, Dr. Mirza Kamaludeen, Indrani Sankardayal, Marina M., Rukhsana Khan, Zeenatul Shadick, Fizal Sattaur, Shaeeza Khan and Yusuf Ali.

Accuracy Reviewers

Monaday Gala, BSc. MSc., PhD., B.Ed, OTC
Principal Westview S.S. School
Toronto District School Board

Derrick Mach, **BSc., B.Ed, OTC**
C.W.Jefferys C.I
Toronto District School Board
Mirza Kamaludeen, **PhD., P.Eng, C. Chem**
Chief Scientist, iG2 Group Inc.
Toronto Canada
(Former Assistant Professor at Ryerson University)
"Beautifully tailored to meet the needs of grade eleven students by providing a unique blend of science and learning instructions. This text will definitely stimulate students' curiosity in chemistry, while meticulously walking them through the foundations of chemistry. An excellent contribution to STEM.,."

Abdul Yaseer Shakur, BSc., MSc., PhD., Md.
Trillium and Sunnybrook Hospitals

Elaine Sinclair, BSc., PhD., B.Ed, OTC
William Lyon Mackenzie C. I.
Toronto District School Board

Nicole Cheung-Seekit, BSc., B.Ed, OTC
William Lyon Mackenzie C. I.
Toronto District School Board

Inga Teper, BSc. MSc., B.Ed, OTC
William Lyon MacKenzie C. I.
Toronto District School Board

Jaspal Singh Ughra, BSc. MSc., B.Ed, OTC
C.W.Jefferys C.I.
Toronto District School Board

UNIT 1
Matter and Chemical Bonding

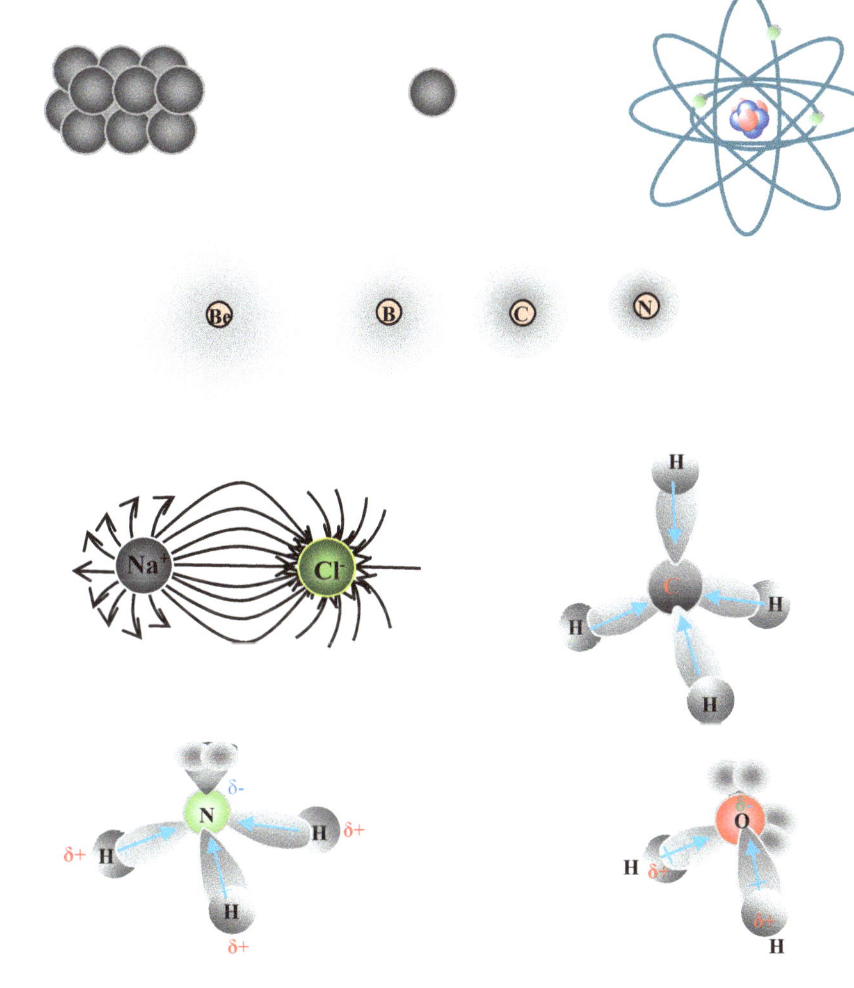

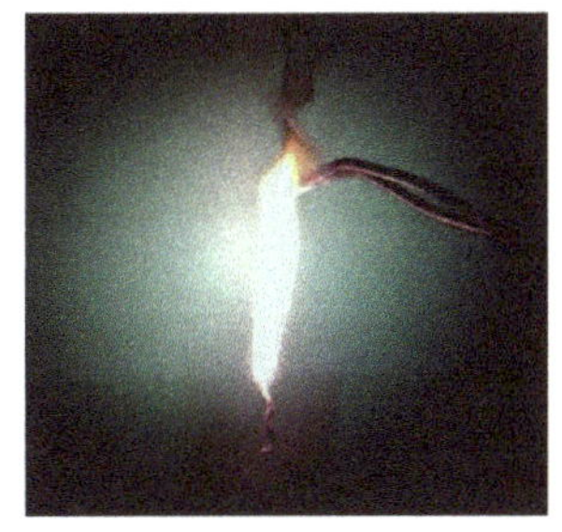

CHAPTER 1
Classification of Matter

The various states of matter are the result of the relative magnitude of the forces of attraction that exists between the constituent particles of substances. In the solid state, the particles are held together in relatively 'fixed' positions because of strong forces of attraction between the particles. In the liquid state, the forces of attraction between the particles are weak just enough to allow the particles to slip and slide over each other. In the gaseous state, the forces of attraction between the particles are so weak that they are free to escape from each other and diffuse to infinite distances. Matter can change states progressively from solid to liquid to gas by the addition of energy in the form of heat. As the particles of matter gain heat energy, their average kinetic energy increases causing them to vibrate to greater distances from each other. As a result of this, the forces of attraction between the particles gets progressively weaker, allowing for an eventual change of state. The opposite happens if energy is removed; the particles come closer to each other and attract each other more strongly. Knowledge of the physical properties of matter is essential for their effective and safe use. Iron, for example, is used to build important infrastructures such as buildings, and bridges. The tensile strength of iron is therefore important to calculate its weight-bearing capacity when it is used in these structures. The high electrical conductivity of copper and aluminum makes them excellent for use in electrical cables.

Knowledge of the chemical properties of matter is also essential for their safe handling, use and storage. Yellow phosphorus, for example, ignites in air spontaneously at room temperature. It is therefore stored in water, with which it does not react, to keep it cool. Potassium, on the other hand, if left exposed to air would quickly oxidize forming potassium oxide or would react explosively in direct contact with water. To prevent these from happening, potassium is stored in paraffin oil with which it does not react. Oxidizing materials such as nitrates and chlorates can easily decompose to release oxygen which supports combustion; they are thus stored separately from other chemicals.

1.1 **Matter:** This is defined as anything that has mass and occupies space.

Element: A substance that cannot be separated into other simpler substances. Examples of these are gold, silver, copper, zinc and lead. The rest can be seen on the periodic table.

Compound: A substance that is made up of two or more elements that are chemically combined. Examples of compounds are water (H_2O), sodium chloride (NaCl), sodium hydroxide (NaOH) and hydrochloric acid (HCl).

Pure substance: A substance in which all the particles are of one type only.

Elements are pure substances because their atoms are the same.

Atoms:

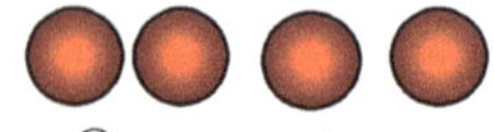

Hydrogen atoms   Oxygen atoms

Figure 1.1. The elements hydrogen and oxygen with their identical atoms.

Like all other elements, hydrogen and oxygen are pure substances since each contains only one type of atoms, as indicated in the drawings before. However, if the elements hydrogen and oxygen are chemically combined, the compound, water is formed as shown below.

Molecules:

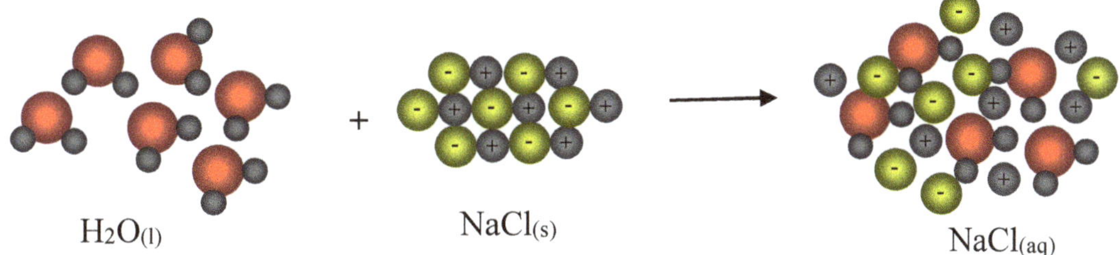

Figure 1.2. The compound, water, having identical molecules.

In the case of water, the particles are called molecules. It can be seen from the drawings above, that all the molecules are identical. Even though the molecules are made up of two different atoms, the compound is still pure as there is only one type of particles (*molecules*). *All compounds are pure substances since the particles they contain; molecules or formula units are identical.*

Mixtures: A mixture is formed when two or more pure substances are **physically** combined. For example, when some sodium chloride is added to water, a mixture is formed.

Figure 1.3. The formation of a salt water mixture.

Mixtures can further be classified as either *homogeneous* (mechanical mixtures) or *heterogeneous (solutions).* In a homogeneous mixture, the various components mix in such a way that everything appears as only one phase. Examples of these are salt solutions, sugar solutions and alcohol beverages. In contrast to this, in a heterogeneous mixture, the various components mix in such a way that the various components can be seen as separate entities. Examples of these are salad dressings, pizza and ketchup.

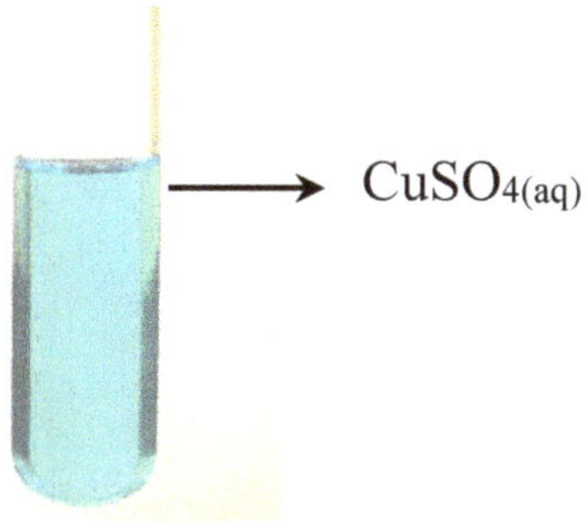

Figure 1.4(a). Homogeneous mixture.    **Figure 1.4(b).** Heterogeneous mixture.

1.2 States of Matter

Matter exists in three physical states:
- Solid
- Liquid
- Gas

1.3 Physical Property

The physical properties of any substance are those that are unique only to it and can be used to identify it. For example, a liquid which is colourless, transparent and having a boiling point of 100 °C and density of 1.0 g/mL, can be identified as water, since these physical properties together are unique only to water.
Some physical properties of matter are as follows:

• Boiling point	• Transparency	• Ductility	• Solubility
• Melting point	• Shape	• Malleability	• Conductivity
• Colour	• Hardness	• Viscosity	• Dimension
• Density	•Texture	• State	

Physical properties may be categorized as either **qualitative** or **quantitative.** For example, colour and density can be categorized as being qualitative and quantitative respectively. As an exercise, try to categorize those listed above.

1.4 Physical Change

This is defined as a change in a physical property of matter, without it being changed into a new substance (the original composition of the substance stays the same). For example, a change of state is a physical change. The following figure depicts some physical changes

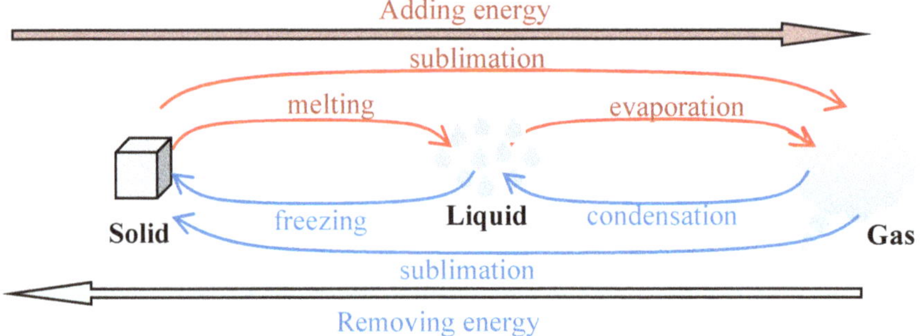

Figure 1.5. The changes of state of water.

As illustrated above, water can undergo various changes of state by the removal or addition of energy. However, each state has its own physical properties, but the original substance remains the same, namely water molecules; no new substance is formed.

1.5 Chemical Property

A chemical property of a substance describes what can be observed when it is changed to a different substance.

An example:

One of the chemical properties of the element potassium, for example, is that it reacts explosively with water to produce hydrogen gas and a solution of potassium hydroxide.

Figure 1.6. Potassium metal reacting explosively with water.

1.6 Chemical Change

During a chemical change, the original substance(s) is changed to one or more new substances which have different chemical and physical properties than the original substance(s). For example, when chlorine gas reacts with sodium metal, the compound sodium chloride is formed. Chlorine is a greenish-yellow poisonous gaseous element and sodium is a very reactive corrosive metal. But the sodium chloride produced, is a white crystalline solid compound which is edible.

Figure 1.7. The chemical change involving the formation of sodium chloride.

1.7 Concept Summary:

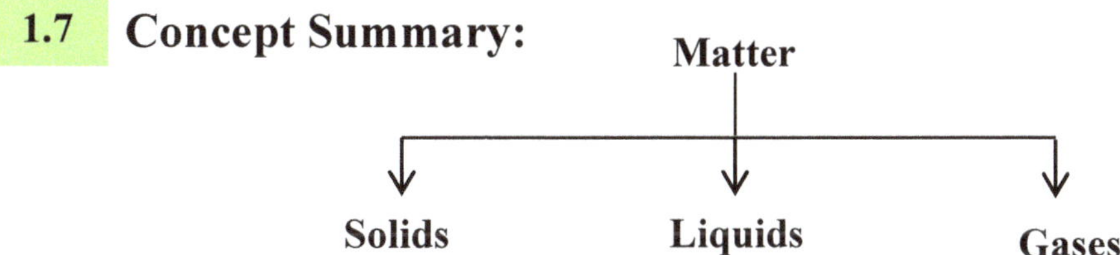

Table 1.1. Classification of matter summary.

Substance	Particle type	Pure or Impure?
Elements	**Atoms**	Pure
Compounds	**Molecules** (molecular compounds) **Formula Units** (ionic compounds)	Pure
Mixture	•**Molecules and Molecules** eg. water + alcohol •**Molecules and Ions** eg. water + salts •**Atoms and Atoms** etc. eg. metal alloys	Impure

1. Identify each of the following as *solid, liquid or gas:* K/U

   (a) Honey in a bottle

   (b) Crystals of sugar in a jar

   (c) The air in your lungs

   (d) Mercury in a bottle

   (e) Lead pellets in a box

2. Identify each of the following *as melting, freezing, evaporation, condensation or sublimation.* K/U

   (a) Dry ice on an exhibition stage disappearing

   (b) Fog disappearing

   (c) Snow on your roof turns into water

   (d) Your bathroom mirror gets cloudy after a long hot shower

   (e) Water in the lakes turn to ice during winter

3. Categorise the following as *physical change, physical property, chemical property or chemical change:* K/U

   (a) Sugar dissolves in water

   (b) Copper can be easily streched into wires

   (c) Magnesium reacts slowly with cold water

   (d) Frying an egg

   (e) Hydrogen reacts explosively with oxygen

   (f) Your glass window is shattered

   (g) Honey flows slower than water

4. Categorise the following substances as *pure, heterogeneous mixture or homogeneous mixture:* K/U

   (a) A spoon full of sugar

   (b) A glass of milk

   (c) Tea sweetened with honey

   (d) Salad dressing made of water, venegar, olive oil and honey

   (e) Salt water

   (f) A beaker of distilled water

5. For each of the following changes decide whether heat is *absorbed or released:* K/U

   (a) Freezing of water

   (b) Condensation of steam on your hand

   (c) The evaporation of freon gas in the freezing chamber of your refrigerator

(d) Carbon dioxide gas is changed to dry ice

(e) The evaporation of sweat on your skin

(f) Your bare hand becomes burnt by holding dry ice

6. Use the following words to fill in the blanks. One word may be used more than once. K/U

pure   impure   qualititative   quantitative   weakest

(a) Sodium is a __________ substance because it is made up of only atoms.

(b) Density is a __________ physical property.

(c) In the gaseous state the force of attraction between particles is ________.

(d) Texture is of a substance is a __________ physical property.

(e) Sugar water is a(n) ________ substance since it is made of two different pure substances.

(f) All mixture are __________.

(g) Distilled water is __________ since it is made up of one type of molecules.

7. Decide whether or not the following statements are *true or false?* K/U

(a) Heterogeneous mixtures has one phase. ________

(b) Tap water is a pure substance.________

(c) Substances that dissolve well in each other form homogeneous mixtures.________

(d) Sodium oxide must be impure since it is made up of sodium and oxygen.___

(e) The melting of ice is endothermic since it feels cold.____

(f) Light is scattered as it passes through an aqueous heterogeneous mixture.___

(g) The removal of energy from gaseous carbon dioxide changes it to dry ice.

CHAPTER 2
Atomic Theories

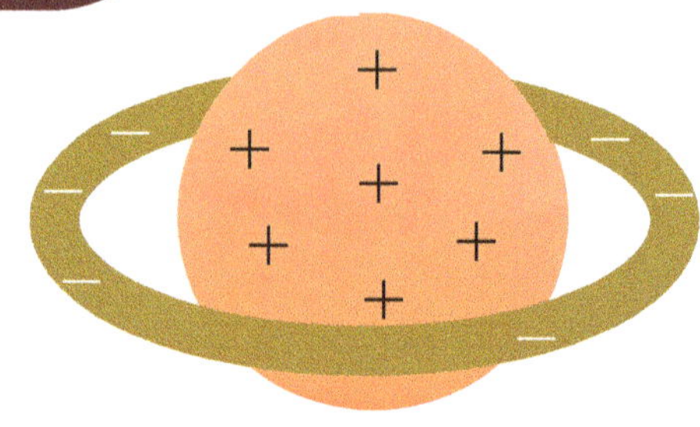

Chapter Content:

2.1 How could one describe what one could not see? This was the difficulty that countless number of scientists encountered in their endeavor to decipher the atom and its structure. Over thousands of years, many theories of atoms were proposed, only to be subsequently discarded for others that were deemed to be more scientifically sound. The advancement of technology has greatly revolutionized our understanding of the atom and its structure. The following is a timeline of some atomic theories and their proponents:

Democritus – 460– 370 BC	He believed that all matter is made up of indestructible units called atoms (400 BC).
Aristotle – 384 – 322 BC	He believed that there are five types of matter: earth, air water, fire and ether (300 BC).
John Dalton – 1766 – 1844	He stated that all matter is made up of atoms. The atoms of one element are identical and they cannot be changed to others nor be destroyed. During chemical reactions atoms are not destroyed but are merely rearranged. The atoms of elements react in specific ratios to form compounds (1803).
Eugen Goldstein – 1850 – 1930	He discovered the proton and that it was positively charged (1886).
J.J. Thomson – 1856 – 1940	He discovered the electron and calculated its charge to mass ratio. He proposed the raisin-bun model of the atom.
Hantaro Nagaoka – 1865 -1950	He proposed a Saturnian model of the atom. He was first to propose a spherical positive nucleus with electrons around it in the form of a ring. The arrangement of the electrons to the sphere is similar to the way the rings are in the planet, Saturn.
Ernest Rutherford – 1871 – 1937	He proposed a model of the atom that is composed of a large, massive, positively large sphere surrounded by revolving electrons.
Niels Bohr – 1885 – 1962	He stated that electrons move around the nucleus of the atom in discrete regions called energy levels. Electrons can absorb energy and jump to higher levels. As they fall back they emit radiant energy (1922).
James Chadwick –1891 – 1974	He discovered the neutron and found that it was neutral (1932).

Atomic Theories:

Some of the earliest theories pertaining to the atom were proposed by the following two Greek philosophers: *Empedocle, Democritus and Aristotle.*

Democritus: He theorized that if all matter is divided into smaller and smaller pieces, then there comes a point at which it could not be further divided. He called this smallest piece of any matter, "atomos," the Greek word for indivisible.

Empedocle: He believed in a four element composition of matter: fire, air, water and earth.

Aristotle: He opposed the theory of matter proposed by Democritus. Instead, he adopted **Empedocle's** four element theory of fire, air, water and earth, adding his fifth element of divinity, ether. Being brilliant and very influential, Aristotle's theory stood its ground for over two thousand years until it was proven wrong by more scientifically sound theories.

2.2 Dalton's Atomic Theory:

In 1809, the English scientist and teacher, John Dalton used the atomic theory proposed by Democritus as the basis for his own description of atoms and matter. Dalton's atomic theory is as follows:

- All matter **is made of very tiny particles called atoms, too small to be divided.**

Atoms →

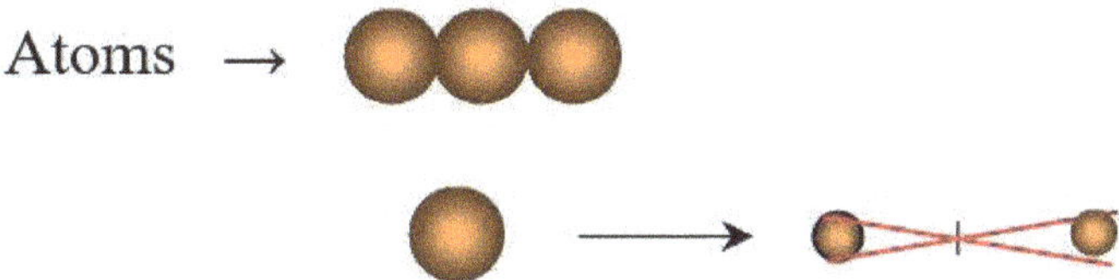

Figure 2.1. Atoms found in a sample of matter.

- **The atoms of one element have identical properties, in terms of size and mass.**

Figure 2.2. Atoms found in a sample of matter with identical

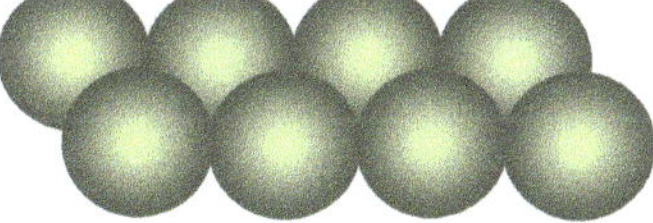

Figure 2.3. Atoms in another element having different properties from the example above.

- **The atoms of one element cannot be changed into atoms of another element, nor can they be created or destroyed.**

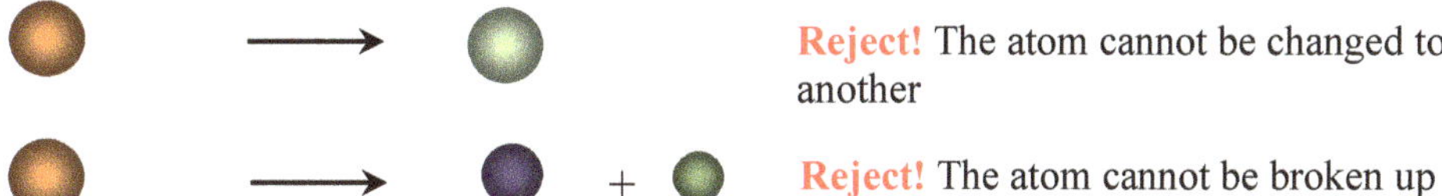

Figure 2.4. The atoms of one element cannot be changed into atoms of another element nor be broken up or destroyed

- **Atoms of different elements can combine in specific ratios to form compounds.**

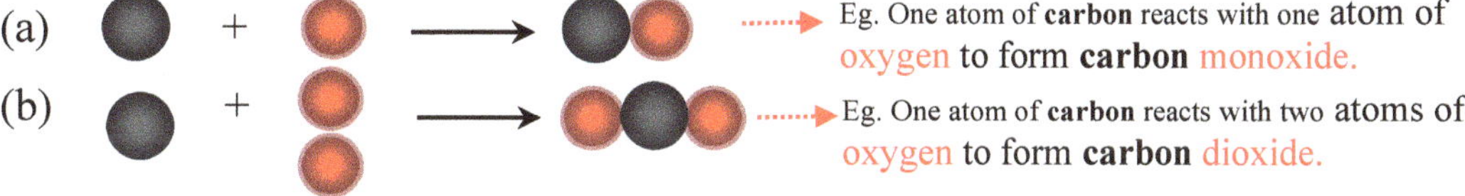

Figures 2.5 (a) and (b). The formation of compounds with different atom- ratios.

2.3   J.J. Thomson: Discovery of the Electron

Using a modified cathode ray tube, on April 30, 1897, J.J. Thomson was able to show that **the electron is a fundamental part of all atoms.**

The cathode ray tube is an evacuated glass tube in which two electrodes; a cathode and an anode are inserted at its opposite ends. The following drawing depicts this.

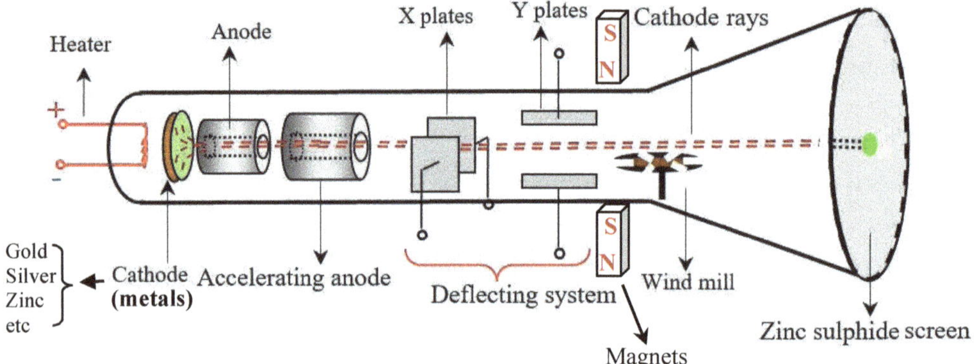

Figure 2.6. Diagram of a cathode ray tube.

When a high voltage is applied to the electrodes, a bright glow was observed on the screen, coated with a layer of zinc sulphide (**a phosphor).** Whatever caused the screen to glow was also found to have the following characteristics:

- It emanated from the cathode and travelled towards the anode because an object (mica cross) placed in its path casted a shadow on the screen.
- It was deflected by an electric field from pairs of electrodes placed perpendicular and horizontal to its path. The rays were always repelled by the negative electrode and attracted by the positive electrode.
- It was deflected by a magnetic field placed perpendicular to its path; it deflected in the opposite direction when the poles were reversed
- It was able to spin a tiny windmill placed in its path.

Based on the above observations, it was concluded that the *cathode rays were negatively charged*. The cathode rays, being negatively charged, were attracted and accelerated towards the positive anode placed infront of the cathode. The momentum of the cathode rays allowed some of them to pass through a hole placed in the center of the positively charged anode and subsequently impinged onto the phosphorescent screen behind. As these rays impinge on the phosphorescent screen they cause it to glow at that spot The effects of the electric and magnetic fields on the cathode rays were noted by the movement of the spot on the screen.

Thomson was also able to calculate the charge- to- mass ratio (e/m) of the particles that were streaming off the cathode. *When he changed the cathode using different metals, the nature of the particles emanating from the cathode remained identical.* Thomson thus hypothesized **that cathode rays were fundamental particles of all elements** and that **they were sub-atomic particles.** He renamed these particles **electrons.**

2.4   Eugen Goldstein: Discovery of the Proton

In 1886, the German physicist Eugen Goldstein observed that if **a gas** was placed in a cathode ray tube and *a hole was made on the cathode, instead of the anode,* **a faint glow** was observed on the end of the tube **next to the cathode** and **a bright glow** on the end of the tube **next to the anode**, Figure 2.7 (next page). Both ends were phosphor coated.

Goldstein gave the following explanation for his observations:

As the cathode rays sped towards the anode, they collided with the neutral gas particles in the tube and knocked electrons out of them. The cathode rays, along with the ejected electrons, produced the bright glow on the anode, where they ended up.. The faint glow at the cathode end must have been produced by particles that were positively charged, since they moved towards the negatively charged cathode. These positive charges must have been created as electrons were knocked out of the neutral gas particles. As these positive charges moved towards the cathode, their momentum must have allowed some of them to escape through the hole in the cathode and impinged on its phosphor coated walls. The process by which these events happen are illustrated below.

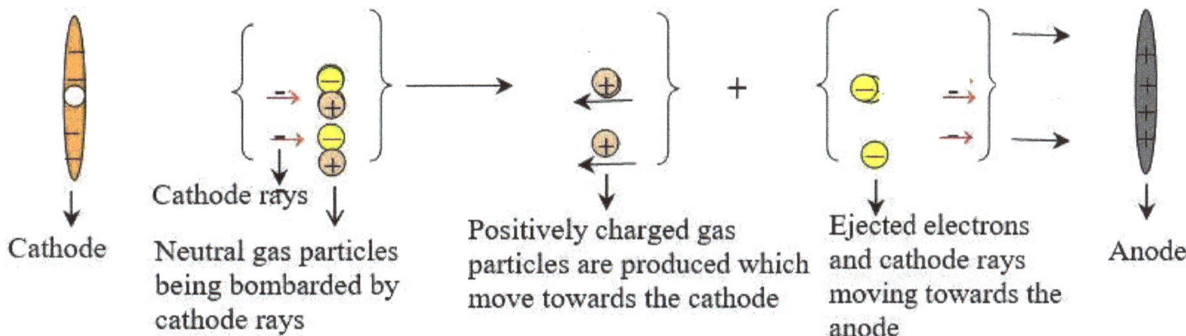

Figure 2.7(a). The formation of positive particles as happens in a gas-filled cathode ray tube due to the bombardment of cathode rays with hydrogen gas.

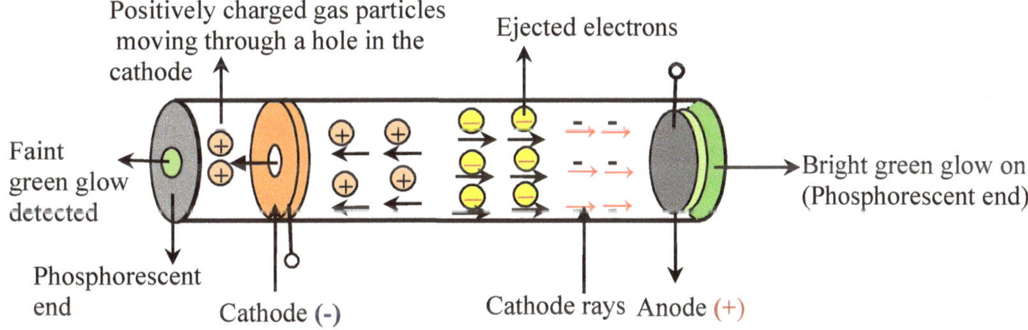

Figure 2.7(b). The detection of positive particles in a **gas-filled** cathode ray tube.

By experimentation outside the tube, Goldstein also able to confirm that the particles that produced the faint glow on the cathode end of the tube were indeed positively charged. Subsequently, scientists were also able to calculate the charge- to- mass ratios for these positively charged gas particles. Approximately 20 years later, they found that values obtained, varied depending on the nature of the gas that was placed in the tube, and that these were much smaller than that of the electron. Finally, they were able to calculate the charge- to- mass ratios for the gas particle with the smallest quantity of positive charge, after numerous trials. This smallest positive charge, (1+) was equal and opposite to the negative charge on the electron and had a mass of 1U or *one atomic mass unit*. It is now called the proton, **and the gas that produced this value on a consistent basis was hydrogen.** Different gas particles produced different charge- to- mass ratios because they had different number of protons in their molecules or atoms.

2.5 The J.J. Thomson's and H. Nagaoka's models of the atom

Based on those findings outlined above, Thomson proposed his own model of the atom. He suggested that the atom was a positive sphere with the electrons embedded in it like the way raisins are in a bun. According to this model, every atom of an element would have a uniform composition, in terms of its charge and density.

In the year, 1903 Japanese scientist, Hantaro Nagaoka proposed a model of the atom composing of a spherical positive centre surrounded by a ring of electrons that mimics the planet, Saturn. **He was the first scientist to propose a model of the atom with a nucleus**. After closer scrutiny, his model did not confirm to later more scientifically sound discoveries; it was thus shelved.

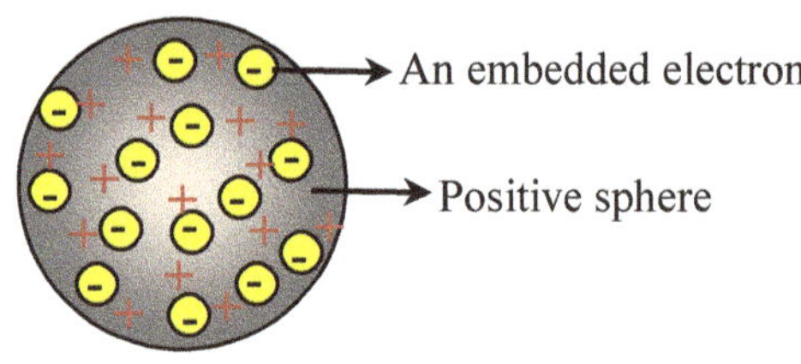

Figure 2.8 (a). J.J. Thomson's raisin-bun model of the atom.

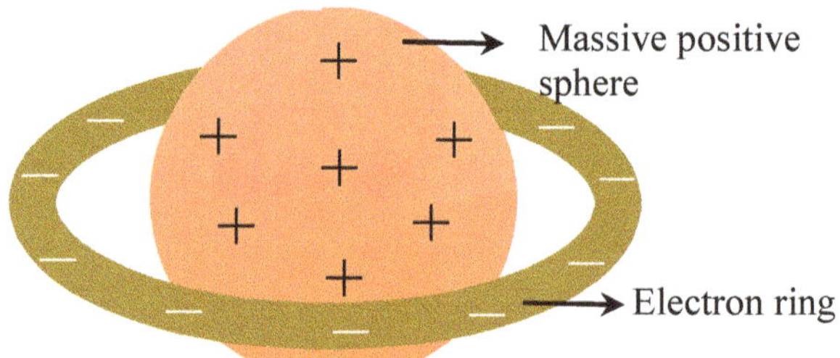

Figure 2.8 (b) Hantaro Nagaoka's Saturnian model of the atom.

2.6   Ernest Rutherford's Nuclear Model

With the discovery of **alpha particles,** emitted from radioactive substances, Ernest Rutherford in 1911, found it to be the ideal tool with which to put the J.J. Thomson's model of the atom to the test. Alpha particles are helium nuclei which consist of two protons and two neutrons. Rutherford reasoned that since they have mass and considerable energy with which they are ejected during radioactive decay, they could be used to penetrate a gold foil of few atoms of thickness.

During his experiment, Ernest Rutherford used a point source of alpha particles from a radioactive source and directed them at a gold foil. He placed zinc sulphide screens behind the gold foil and also at different positions in front to detect the pattern with which the alpha particles would exit the gold foil. When alpha particles strike a zinc sulphide screen, they produce an instantaneous glow on it. Based on J.J. Thomson's model of the atom (uniform density), *Rutherford expected that if alpha particles are shot at the gold foil at one point and they pass through freely there, then the same should happen if they are shot elsewhere on the foil.* The following figures illustrate Rutherford's predictions and what actually occurred in his investigations.

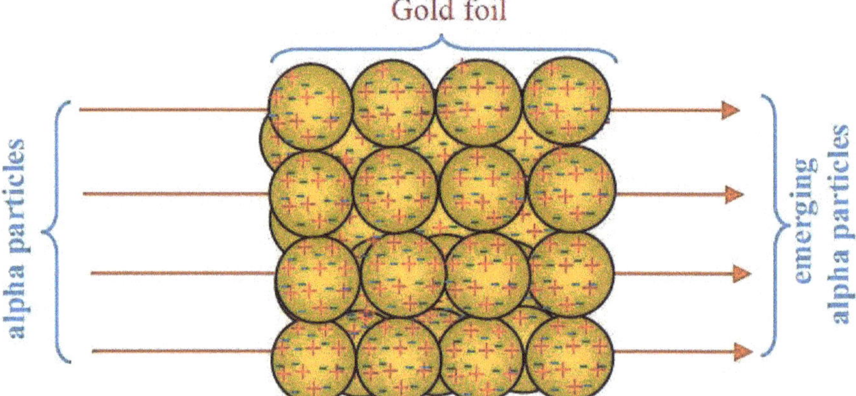

Figure 2.9. Rutherford's prediction of the movement of alpha particles through the gold foil.

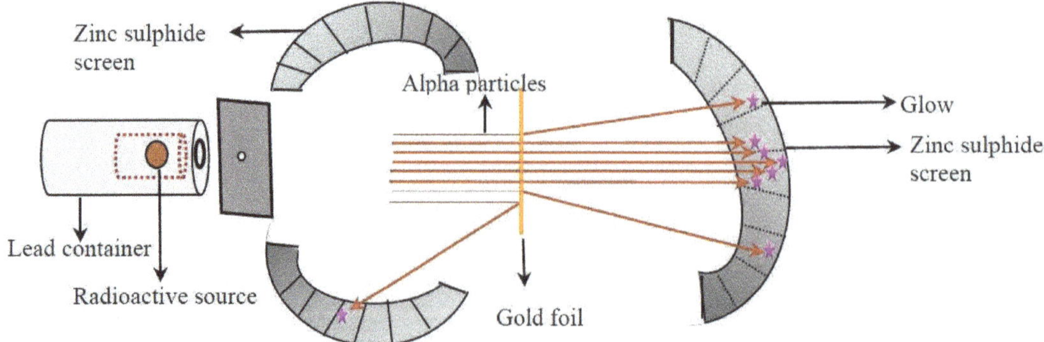

Figure 2.10. Actual movement of the alpha particles through the gold foil.

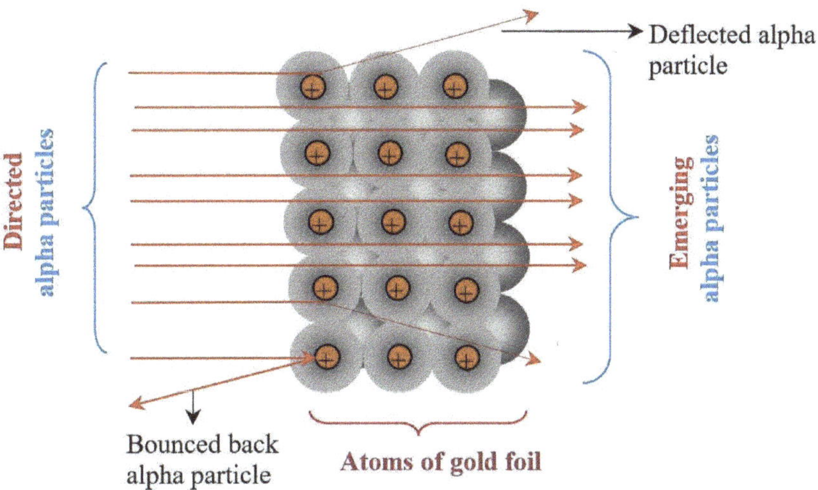

Figure 2.11. The actual passage of alpha particles through the atoms in the gold foil.

Rutherford observed that most of the alpha particles passed through the foil, but a few of them were deflected sideways and a few even bounced back directly. But he expected all the alpha particles to pass through the foil. To explain his observation, Rutherford made the following deductions:

- Since the positively charged alpha particles were deflected sideways and even rebounded, they must have encountered something massive in the atoms that was also positively charged. He called this mass the nucleus.
- Most of the alpha particles passed through the foil because most of the atom is empty space.
- The atom is made up of a massive nucleus which is positively charged and the rest of the atom is made up of empty space where the electrons are found.

Scientists now believe that the size of the nucleus with respect to the rest of the atom is analogous to placing a tennis ball in the centre of a soccer stadium; the nucleus is extremely small compared to the rest of the atom.

2.7 James Chadwick: Discovery of the Neutrons

In 1932, British physicist James Chadwick discovered that when atoms of beryllium were bombarded with high-speed alpha particles, beams of particles were emitted from the beryllium atoms. Analysis of these particles revealed that they were neither deflected by a magnetic or electric field, nor did they discharge charged electroscopes. The particles were thus neutral in nature. They were also found to have a mass of one atomic mass unit, **1u.** These newly discovered sub-atomic particles were called **neutrons.**

From the information gathered so far, it can be summarized that the atom has a massive nucleus which is composed of neutral particles called *neutrons* and positively charged particles called *protons*. The nucleus is surrounded by negatively charged particles called *electrons.* The negative electrons are attracted by the positively charged nucleus and they move around it like the way the planets orbit the sun.

2.8 Niels Bohr

To understand the work of Niels Bohr, it first necessary to study the visible spectrum since it plays an integral part of our understanding of the quantum theory that he used to explain his findings.

The Visible Spectrum
When white light is passed through a glass prism, it is resolved into what is called the continuous spectrum. This spectrum constitutes only a small part of the wider *electromagnetic spectrum,* as indicated next.

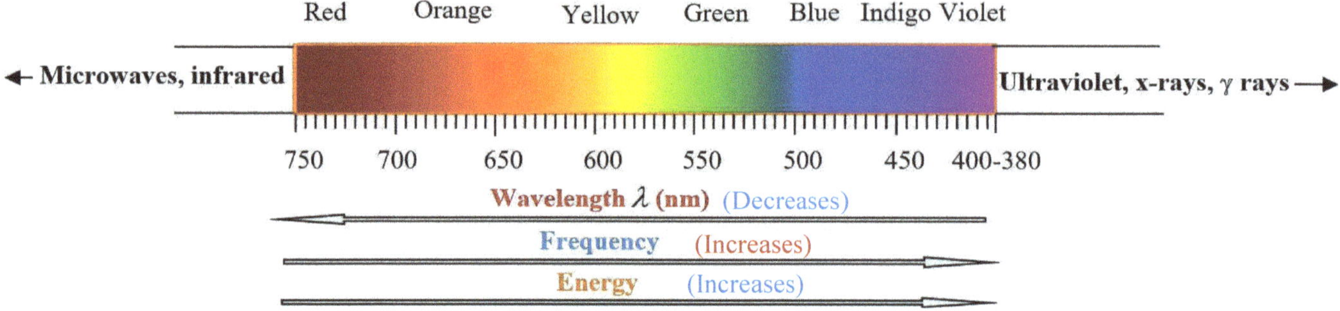

Figure 2.12. The visible spectrum.

The following figure shows what happens when gaseous hydrogen is placed in a glass discharge tube whose electrodes are connected to a power source of adjustable voltage. Starting with the power at zero, the voltage was raised gradually and *observations were made of the discharge tube through a diffracting grating*. At some point, **a line of red colour is first observed which was subsequently followed by three other lines of the different colours** as is indicated in the figure below. This collection of lines was called the *emission spectrum* of the element hydrogen. When samples of other elements in their gaseous state were placed separately in the discharge tube and the same was done to them, lines were observed that were characteristic for each element; *each element having its own unique emission spectrum.*

The German physicist, Max Planck believed that light is composed of photons or packages of energy, and that each **photon** has its own fixed amount of energy called a **quantum.** The visible spectrum is composed of a series of different photons jam-packed together. The amount of energy of a photon is given by the equation: $E = h\nu$, where **h** is *Planck's constant* and **ν** is the *frequency of the photon.* Based on this equation, it can be deduced that *as the frequency of a photon increases the more energy it would have*; a photon of violet light would therefore have more energy of than a photon of red light.

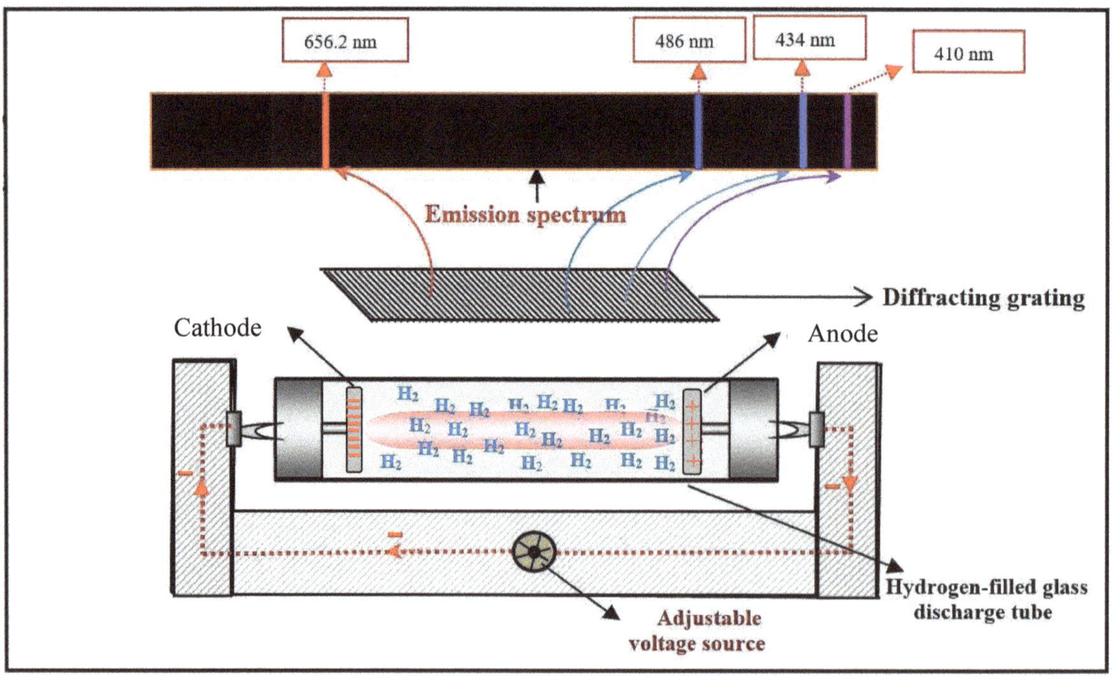

Figure 2.13. The emission spectrum for hydrogen as seen through a diffracting grating.

The Emission Spectrum of Hydrogen

To explain the emission spectrum of hydrogen, Niels Bohr proposed that electrons are found outside the nucleus of the atom in discrete regions with fixed energies. He called these regions of space, *energy levels* or *shells*; electrons cannot be located anywhere else except in these regions. The farther away the electrons are from the nucleus the more energy they would have. Niels Bohr *theorized that as the voltage of the discharge tube was increased, the energy with which the electrons from the cathode were ejected increased proportionately.* According to him, these electrons collided with the hydrogen atoms in the tube and bumped the atoms' *orbiting electrons* *from lower energy levels to* *higher energy levels*. When this happened, the electrons were said to have undergone **transitions** and the atoms became **excited.** The atoms in their unstable excited state, quickly reverted **to their more stable ground states,** by allowing the electrons to fall back to their original energy levels. As a result of these transitions; electrons falling from higher energy levels back to a lower energy levels, *quanta (*plural for quantum) *of energy were released.* The figures below depict how this happens in an atom of hydrogen.

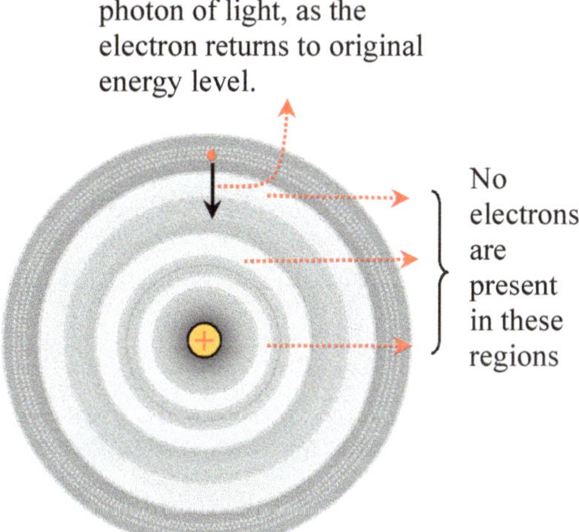

Figure 2.14(a). An electron absorbing energy and bumping up into a higher energy level.

Figure 2.14(b). An electron in an excited atom falling back to its original energy level and releasing energy.

Since a certain specific low voltage, a photon of light in the **red region** of the visible spectrum was first observed, Bohr reasoned that, at these, low voltages, the cathode rays are ejected with relatively low energy. These low-energy ejected catohe rays were only capable of knocking orbiting electrons of the hydrogen atom to a slightly higher energy level. At higher voltages, the cathode rays, having more energy, bumped *orbiting electrons* to progressively higher energy levels, so that when they transitioned back to their ground states, they released photons of higher energies, as indicated by the different colour of lines. After further experimentations, the following other observations were made:

Only when transitions happen from the second energy level upwards and back to it, are photons of light observed. These constitute the **Jacob Balmer series.** Radiations are observed only in the **visible portion** of the electromagnetic spectrum. Emission lines for hydrogen are shown below.

If transitions happen from the third energy level upwards and back to it, photons are detected in the IR region of the electromagnetic spectrum. These constitute the **Pachen** series.

If transitions happen from the first energy level upwards and back to it, photons are detected in the *UV region of the electromagnetic spectrum.* These constitute the **Lyman series**.
These series are named after those scientists who made these observations.

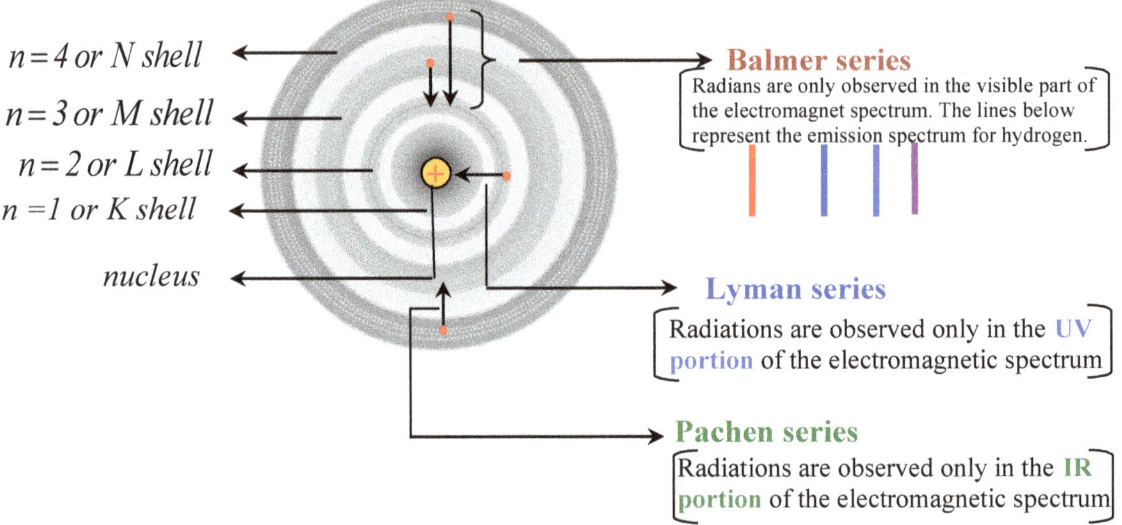

Figure 2.14c. The various transitions in the hydrogen atom.

In 1922, Bohr was awarded the Noble prize for his work. However, Bohr's theory was limited only to the hydrogen atom, as it could not be used to explain the emission spectra of other elements having more than one electron. **Bohr's theory that electrons orbit the nucleus like the planets orbiting the Sun, had to be modified to suit subsequent ones that were able to account for the emission spectra of elements other than hydrogen.** These other different theories, envision electron shells as composing of **orbitals** which are **regions of space around or about the nucleus where there are possibilities of finding the electrons**. **These orbitals are characterized by being negative and having different shapes**.

Flame Tests

The following are the flames obtained when salts of compounds containing the elements calcium, copper, sodium and lithium were heated in a Bunsen flame separately. As these flames appear one would be made to think that only on type of photon is emitted for each element. But, if each flame is viewed through a diffracting grating, it is resolved into a line spectrum that is characteristic of each element. The figures below show the line spectrum for each of these elements.

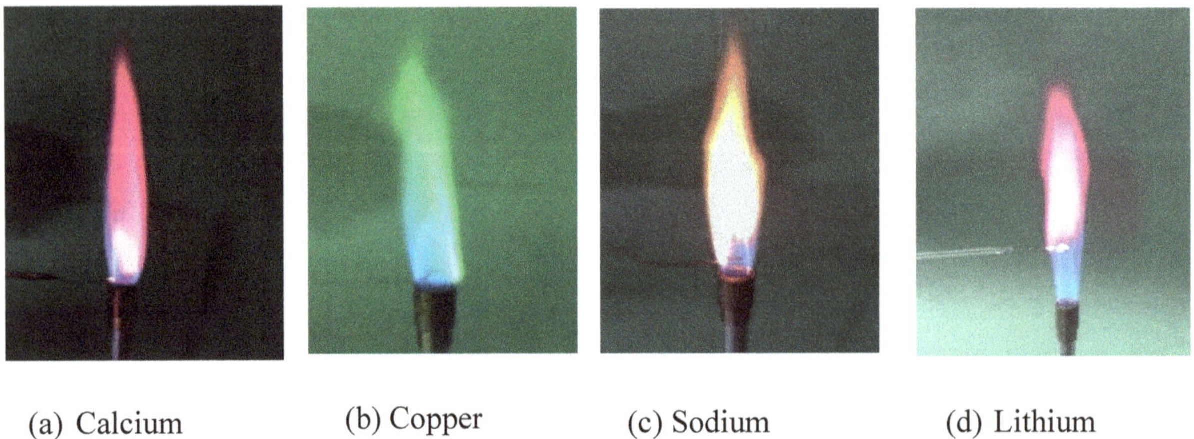

(a) Calcium   (b) Copper   (c) Sodium   (d) Lithium

Figure 2.15. The flames for the elements, calcium, copper, sodium and lithium respectively.

Like the discharge tube where the orbiting electrons are energized by the cathode rays (electrons) in a Bunsen flame, a similar thing happens. Only in this case, heat excites the atoms. The following are spectral lines produced for the above elements when viewed through a diffracting grating.

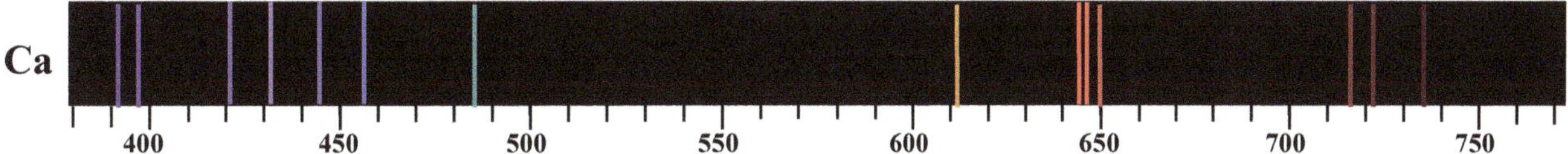

Figure 2.16(a). The spectral lines for the element, calcium.

Figure 2.16(b). The spectral lines for the element, copper

Figure 2.16(c). The spectral lines for the element, sodium

Figure 2.16(d). The spectral lines for the element, lithium

Note that in all these cases, the wave lengths of the emitted photons are measured in **nanometers.**

Exercise 2.1

1. **Match the following scientists with the statements that follow.** K/U

> *Dalton   Thomson   Rutherford   Bohr   Goldstein   Chadwick   Planck   Nagaoka*

(a) He discovered the sub-atomic particles with no charge.

(b) He discovered the electron and calculated its charge to mass ratio.

(c) He discovered the sub-atomic particles with a positive charge.

(d) He proposed the nuclear model of the atom.

(e) He stated that electrons move around the nucleus of the atom in discrete regions called energy levels.

(f) He stated that that light is composed of photons or packages of energy and that each **photon** has its own fixed amount of energy called a **quantum.**

(g) He stated that all matter is made of tiny, indivisible particles called atoms.

(h) He was first to propose a positively charged nucleus surrounded by an electron ring.

2. **Fill in the blanks in the following paragraph, as they relate to atomic structure, using the following words:** K/U

> *released   central   force   neutrons   positively   protons   shells*
> *excited   light   jump   electrons   quantum   line*
> *energy levels   ground*

An atom is composed of a ______ charge (a) nucleus surrounded by negatively charged ______ (b). The nucleus is made up of two sub-atomic particles; the _____ (c) and the ______(d). The electrons orbit the nucleus in regions called ______ (e) or ______ (f). Electrons are prevented from leaving the atom due to a ______ (g) of attraction from the ______ (h) charged nucleus. When the electrons are all in their lowest possible energy levels, the atom is said to be in its _______ (i) state. An electron can absorb a ______ (j) of energy and ______ (k) to a higher energy level. When this happens, the atom is said to be ______ (l). As the electron returns to its original state, a quantum of energy is ______ (m). This energy may be in the form of _______ (n). The spectrum resulting from electrons returning to their original lowest energy levels after being bumped to higher energy levels is called an _______ (o) spectrum.

3. In fireworks, metals or metal salts are heated up by igniting gunpowder (a mixture of potassium nitrate, sulphur and charcoal in 75: 15: 10 by mass). Use this information to answer the following: K/U C T/I

   a) What are the fuels in the gunpowdwer?

   b) What is the use of the potassium nitrate in the mixture?

   c) Explain why colours are produced when metals or their salts are heated to high temperatures.

   d) Referring to table 7.6 of flame test on p. 139, what combination of salts must be used to produce red, blue and green flames?

4. To block UV from affecting our skin sunscreens are used, and to block X-rays from producing mutation in our gonads lead screens are used. Use your knowledge of energy and penetrating powers of these two types of radiation to explain the choice of protections used. K/U C

5. Photosynthesis depends on energy from sunlight. The energy is absorbed by electrons in the chlorophyll molecule and get promoted across a certain membrane. As the electrons fall through this membrane back to their original position, energy is released that is used to drive the process of photosynthesis. Research the process of photosynthesis and graph the wavelengths of light that are absorbed during this process. T/I MC

CHAPTER 3
Isotopes

Chapter Content:
3.1 Relative Atomic Mass
3.2 Isotopes
3.3 Nuclear Reactions
3.4 Half-life

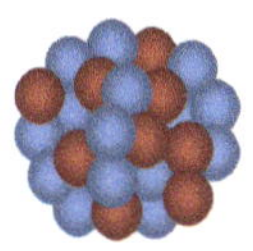

3.1 Relative Atomic Mass

Up to the latter part of the 18 th century century, hydrogen, being the lightest known element to chemists, was assigned an atomic mass of 1. Every other element was then assigned an atomic mass relative to that of hydrogen. This served as the basis for which **relative atomic mass** was ascribed to any newly discovered element, until 1961 when C-12 was chosen to as the new reference.

3.2 Isotopes

Pierre Curie and Marie Curie were first to discover that radiations emanated from the elements, polonium and radium. The fact that atoms of these elements changed their masses as they produced radiations until they became stable, inspired British chemist Frederick Soddy in 1906 to propose that atoms of an element can have more than one mass. His proposal was proven to be true when lead extracted from different sources contained atoms with two different masses, 206 and 208. He called these atoms **isotopes** of lead. Isotopes are now defined as atoms of the same element having the same atomic number (Z), but different mass numbers (A). This concept exemplified using the element, carbon in the following table.

Table 3.1. Isotopes of carbon.

Isotopes	$^{12}_{6}C$	$^{13}_{6}C$	$^{14}_{6}C$
# of protons	6	6	6
# of neutrons	6	7	8

As can be seen in the table above, the difference between these atoms is due to the difference in the number of neutrons they have. In the $^{12}_{6}C$, $^{13}_{6}C$ and $^{14}_{6}C$ atoms, the number of neutrons are **6, 7** and **8,** respectively.

Normally, the heavier isotopes are less abundant in nature and can be either stable or radioactive. Some radioactive isotopes (**radioisotopes**) can now be created and used in various ways as outlined below.

i) Isotopic Labelling.

Heavier isotopes such as ^{15}N and ^{13}C can be incorporated into a molecule to act as tracers to determine how that particular molecule is absorbed, metabolized or excreted by living organisms. Depending on whether the isotope is stable or radioactive, different technological means are used to trace their concentrations and locations in the organisms. The Geiger-Muller tube is one device that is used to trace and measure radioactivity.

ii) Nuclear Medicine.

Solutions of radioactive isotopes are introduced into body fluids such as the blood and traced through imagining techniques to determine how well blood flows through organs such as the heart and liver or to make the outlines of organs to be imaged more distinctly. The isotopes used in these techniques normally have very short half-lives. Radioisotopes can also be used to destroy cancer cells directly if they are localized in some tissue or organ of the body. Prostate cancer, for example, is treated by injecting radioactive pellets into it.

iii) Other uses.
Listed below are a number of ways that radioisotopes are also used; research these.

- Smoke detectors
- Sterilization of foods
- Archaeological dating
- Production of nuclear energy
- Production of nuclear weapons

3.3 Nuclear Reactions

The protons, being positively charged and very close together are supposed to repel each other within their nuclei, according to the law of electrostatics. If this were so, the question that could be asked is, "why don't the nuclei of atoms spontaneously break up because of the protons' mutual repulsions?" The answer to this question lies in the existence of very strong short-range forces of attraction between the nucleons (neutrons and protons) which are responsible for maintaining the integrity of the nucleus. However, it is observed that for the nuclear stability to endure, there must be a certain fixed ratio of the number of neutrons to the number of protons in the nucleus. This stabilizing ratio of protons to neutrons depends on the size of the nucleus. The table below shows this.

Table 3.2. Proton-neutron ratio for different sizes of nuclei.

Size of Nucleus	Small	Medium
Protons:Neutron ratio	1:1	$1:1\frac{1}{2}$

If the ratio of neutrons to protons in the nucleus of an atom is increased considerably, it may affect the nuclear stability to such a degree that it results in **nuclear reactions**. These reactions are described below.

Radioactive Decay

When an unstable atom decomposes spontaneously, it is said to undergo **radioactive decay.** Radioactive decay produces three types of rays. These are described next.

> **Alpha Decay**

This involves the loss of an alpha particle (α) from the nucleus of an atom. An alpha particle is a helium nucleus, $^{4}_{2}He$. It is composed of two protons and two neutrons. The following nuclear reaction is an example of alpha decay.

$$^{238}_{92}U \longrightarrow {}^{234}_{90}Th \; + \; {}^{4}_{2}He$$

# protons = 92	# protons = 90	# protons = 2
# neutrons = 146	# neutrons = 144	# neutrons = 2

Uranium-238 $\longrightarrow$ Thorium-234 + Helium nucleus (α particle)

In alpha decay, it can be seen that original nucleus is changed to another one that is slightly different; the original atom is changed to one that is different from it. This change in the nucleus composition is called **transmutation**. In this case, its mass number is reduced by four and its atomic number by two. Therefore, to know the atomic notation of the new element formed, reduce the mass number of the original atom by four and its atomic number by two.

➢ Beta Decay

Beta decay happens when *an electron* is emitted from *the nucleus* of an isotope. The electron having negligible mass and a negative charge is represented as $^{0}_{-1}e$. How can an electron be produced by the nucleus of an atom? Scientists believe that a neutron is composed of a proton and an electron. Therefore, to produce a beta particle all that has to happen is for a neutron to change into a proton and an energised electron. The latter is then ejected as a beta particle.

Figure 3.1. The transformation of a neutron into a proton and an electron.

The following nuclear reaction is an example in which a beta particle is emitted. The isotope of hydrogen, tritium decays by producing a beta particle.

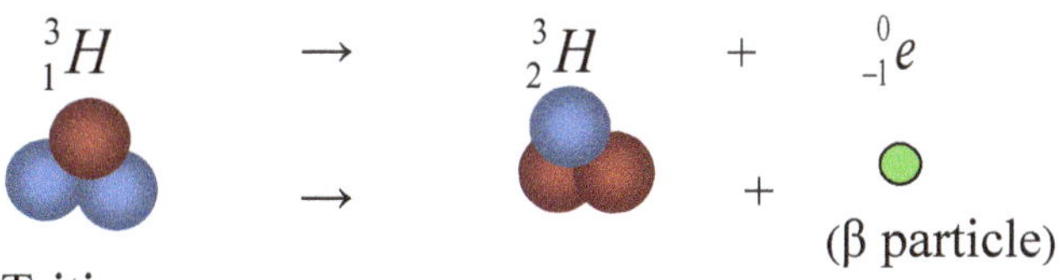

Figure 3.2. Beta decay of tritium.

In beta decay, ***the resulting nucleus*** *retains the same mass number,* but ***its atomic number increases by one***. This is so because one neutron is transformed into a proton; no mass is lost but the proton number increases.

➢ Gamma Radiation

Whenever an isotope decays by emitting either an alpha particle or beta radiation, its nucleus stays temporarily excited. Very shortly after, the protons and neutrons (nucleons) assume a more stable arrangement. As the nucleus attains this stability, it loses energy in the form of gamma (γ) rays. Gamma rays are high-energy electromagnetic radiation. Since they have no mass or charge, they are represented as $^{0}_{0}\gamma$.

Nuclear Fission

When a highly unstable isotope of a heavy nucleus is allowed to absorb a neutron, it may become unstable and split into two smaller nuclei. For example, when uranium-235 absorbs a neutron the following the following nuclear reaction happens:

$$^{235}_{92}U + \ ^{1}_{0}n \rightarrow \ ^{141}_{56}Ba + ^{92}_{36}Kr + 3\,^{1}_{0}n + \text{Energy}$$

This reaction can also proceed in a number of ways producing different products:

$$^{235}_{92}U + \ ^{1}_{0}n \rightarrow \ ^{87}_{35}Br + ^{146}_{57}La + 3\,^{1}_{0}n + \text{Energy}$$

$$^{235}_{92}U + \ ^{1}_{0}n \rightarrow \ ^{137}_{52}Te + ^{97}_{40}Zr + 2\,^{1}_{0}n + \text{Energy}$$

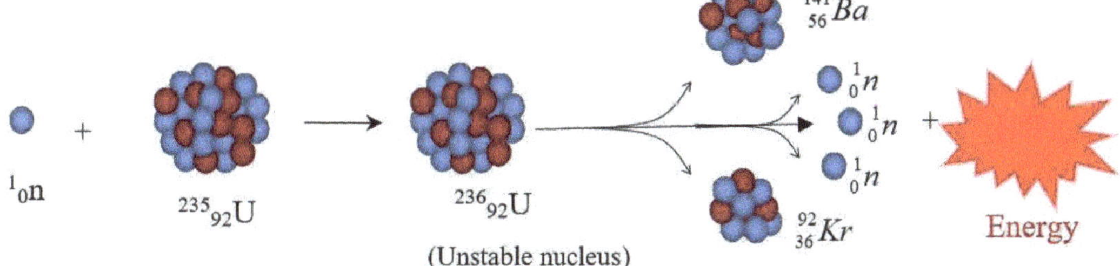

Figure 3.3. Nuclear Fission of U-235.

It should be noted that 3 neutrons are also produced during the above reaction, as depicted in Figure 3.3. If these neutrons are absorbed by a certain minimum mass of other uranium-235 isotopes, also called *critical mass*, then a chain reaction may proceed releasing a tremendous amount of energy. Nuclear reactors use fission in a controlled way to obtain useful energy to generate electricity, etc. The atomic bomb is designed to produce uncontrolled fission reaction to cause massive destruction.

Nuclear Fusion

***Nuclear fusion occurs** when smaller nuclei combine to form a larger nucleus*. An example of this reaction is what is taking place in the Sun and other stars. The following nuclear equation illustrates the fusion reaction happening in the Sun.

$$4\,{}^{1}_{1}H \;+\; 2\,{}^{0}_{-1}e \;\longrightarrow\; {}^{4}_{2}He \;+\; \text{Energy}$$

Fusion reactions produce far more energy than fission reactions. However, to initiate these reactions a large amount of energy is required. This is so because during fusion, the nuclei which are positively charged, must be brought together. They thus need a very high temperature to provide the adequate kinetic energy for them to collide with, in order to overcome the mutual repulsion between their positive nuclei. Other examples of fusion reactions are given by the following equations:

$$\triangleright \quad {}^{2}_{1}H \;+\; {}^{3}_{2}H \;\longrightarrow\; {}^{4}_{3}H \;+\; {}^{1}_{0}n \;+\; \text{Energy}$$

$$\triangleright \quad {}^{2}_{1}H \;+\; {}^{2}_{1}H \;\longrightarrow\; {}^{3}_{2}He \;+\; {}^{1}_{0}n \;+\; \text{Energy}$$

3.4 Half-life

Radioisotopes, being unstable are constantly undergoing decay, producing beta and /or alpha particles along with gamma rays. Depending on the nature of the radioisotope, the rate of decay varies. The concept of **half-life** is used to indicate the rate at which a radioisotope decays. **The half-life of a radioisotope is the time it takes for one half of the number of nuclei in a sample to decay.** For example, iodine-131 which is used to treat thyroid cancer has a half-life of 8.07 days. This means that if we start with 100 g of pure iodine-131 at zero time, after 8.07days, 50 g of the radioactive nuclei will remain along with the other stable nuclei. The following table shows the pattern of decay of radioactive iodine-131 with time.

Table 3.3. The rate of decay of radioactive iodine-131.

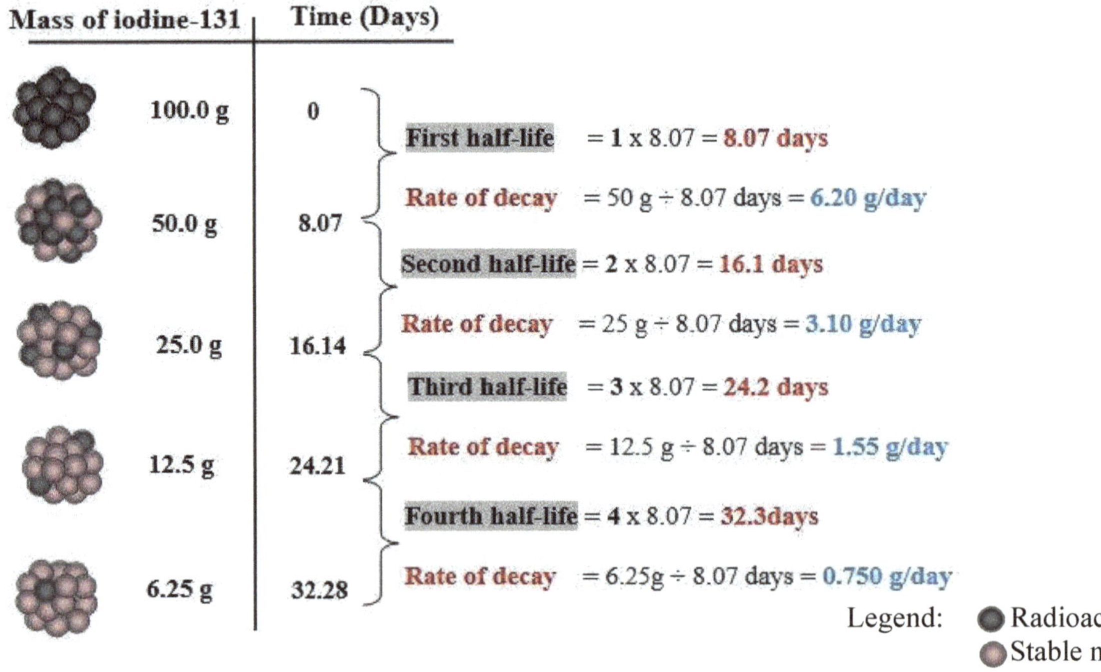

Common Student Misconceptions

When solving problems involving half-life, students make a number of mistakes:
- Taking the previous problem as an example, some students **would not start** a table of values with zero time but with the value for the half-life as illustrated below.

Table 3.4.   Incorrect starting time for decay table

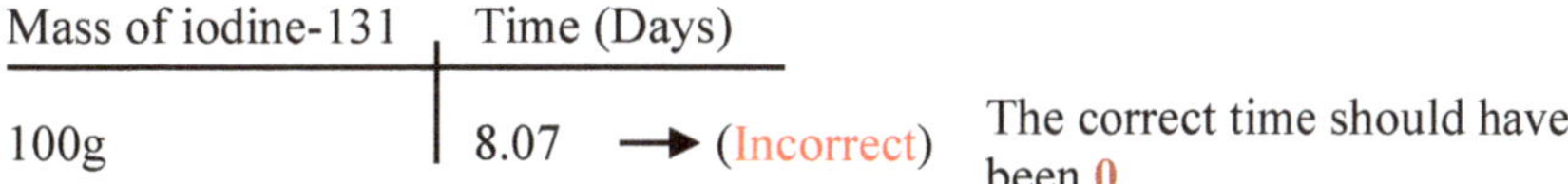

Mass of iodine-131	Time (Days)		
100g	8.07	→ (Incorrect)	The correct time should have been **0.**

- As calculated in the table of half-life, it can be seen that during the first half-life, **50 g** of the iodine-131 **decayed in 8.07 days** at an average rate of 6.2 g/day, and for the second half-life **25 g** decayed in the same **8.07 days** at an average rate of 3.1 g/day. Most students would be made to believe that the second half-life should be 4.035 days instead of 8.07 days. They should know that the reason why different masses took the same time (half-life) to decay is because the concentration of the original radioactive nuclei keeps on decreasing so the rate of decay slows down accordingly.

The following graph shows the decay of pure sample of radioactive iodine-131 with time.

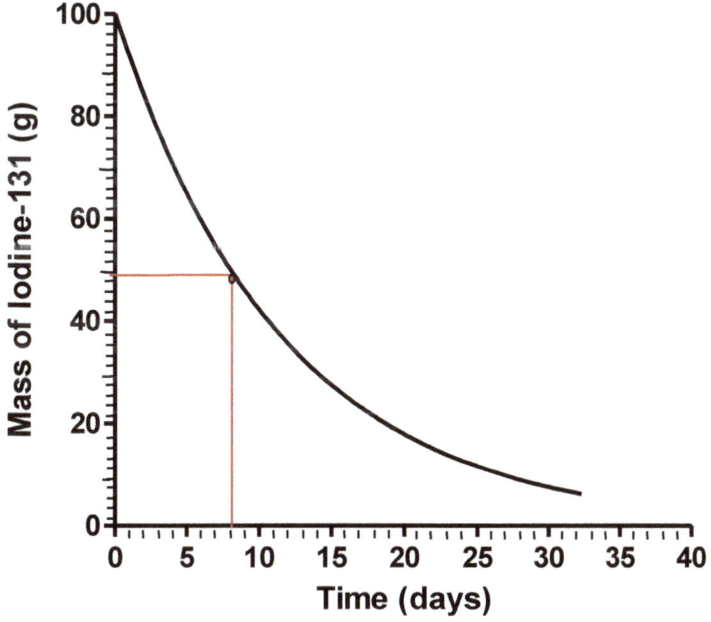

re 3.4. The half-life curve for the decay of iodine-131.

Exercise 3.1

Use the graph provided to answer the following questions. A T/I

1. What is the half-life for radioactive iodine-131?

2. At what time would 12.5 g of iodine-131 to still remain?

3. How long would it take for 93.75 g of iodine-131 to decay?

4. During which half-life period was the rate of decay the fastest?

5. If a cancer patient must have at least 25% of the initial dose of iodine-131 in the body, at what time after the first treatment must the second dose be given?

1. With the use of a periodic table complete the following nuclear reaction equations and state their type. K/U

(a)	$^{14}_{6}C$	$\rightarrow$	?	$+$	$^{0}_{-1}e$
(b)	$^{31}_{15}P$	$\rightarrow$	$^{28}_{14}Si$	$+$	?
(c)	$^{90}_{38}Sr$	$\rightarrow$	?	$+$	$^{0}_{-1}e$
(d)	$^{73}_{31}Ga$	$\rightarrow$	$^{0}_{-1}e$	$+$	?
(e)	$^{231}_{90}Th + ?$	$\rightarrow$	$^{232}_{90}Th$		
(f)	$^{12}_{5}B$	$\rightarrow$	?	$+$	$^{4}_{2}He$
(g)	$^{226}_{88}Ra$	$\rightarrow$	?	$+$	$^{4}_{2}He$

1. A certain antibiotic has a half-life of 40 h. If a patient is given a 160 mg of this drug, complete a table below to show how the amount of it changes with time in his / her body. Use the grid beside to plot a graph of the data. A T/I

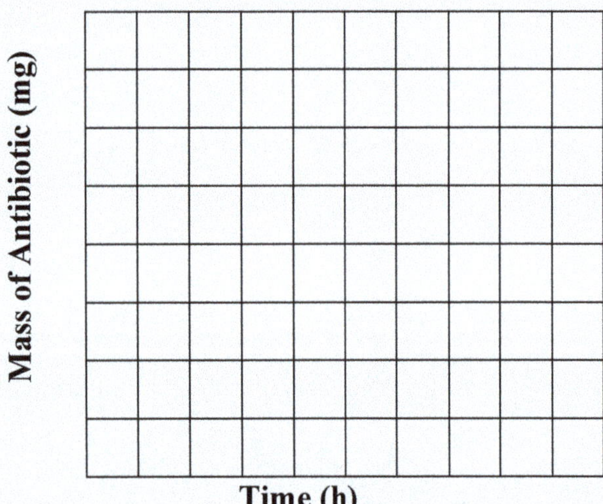

   (a) At what time would 10 mg of the drug still be in the body?
   (b) What time would it take for 120 mg of the drug to decay?
   (c) How much of the drug will disappear after 120 h?

2. A patient has an acute illness that lasts for only a few days. If you were a physician, what type of antibiotic would you prescribe, one with a short or long half-life? Justify your choice. A T/I

3. The following table shows the three isotopes of the element silicon and their percentage abundance. Calculate the average atomic mass for silicon. T/I

Isotopes	Si-28	Si-29	Si-30
% abundance	92.23	4.67	3.10

4. The average atomic mass for the element potassium is 39.10. Complete the following table for the missing isotope. T/I

Isotopes	K-39	?
% abundance	93.1	6.90

5. Many isotopes can be created by neutron radiation of select metals in a reactor depending on their needs. For example, Tc-97 is made from Ru-96. These isotopes have very short half-lives and they produce gamma radiation. Research how these man-made isotopes are used to make medical diagnosis or to treat some types of diseases. Discuss the pros and cons of nuclear medication. T/I C A

6. Radioactive carbon, C-14, is formed when $^{14}_{7}N$ is bombarded with cosmic neutrons in the atmosphere according to the following equation: $^{1}_{0}n + ^{14}_{7}N \longrightarrow ^{14}_{6}C$

    When any C-14 atom formed reacts with $O_{2(g)}$ to forms $CO_{2(g)}$, it gets incorporated into plants tissues by photosynthesis. Inevitably, a small percentage of this radioactive carbon enters the tissues of animals by eating plant tissues. In the body of the plant, the carbon-14 decays by beta emission producing N-14. Emission by a radioisotope is measured in units of bequerel, (Bq), which is one nuclear decay/second. To determine the average age of a fossilized plant, a certain mass of the plant material is first obtained and the rate of emission is measured. This rate is then compared to what its original rate would have been, and calculation is made using the half-life of carbon. For example, a plant material with C-14 of half-life 5730 years and 0.225Bq/g with rate of beta emission of 0.028Bq would be approximately 3x5730 years old = 17,190 years, since 3 half-lives present: $0.225 \longrightarrow 0.1125 \longrightarrow 0.0562 \longrightarrow 0.0281$. What would be the emission of a one gram of plant material with C-14 that is 22,920 years old? A T/I

CHAPTER 4
Trends in the Periodic Table

Chapter Content:

4.1 Atomic Radius (definition) 4.6 Electron Affinity
4.2 Trends Across a Period
4.3 Atomic and Ionic radius
4.4 Electron affinity
4.5 Trends Down a group

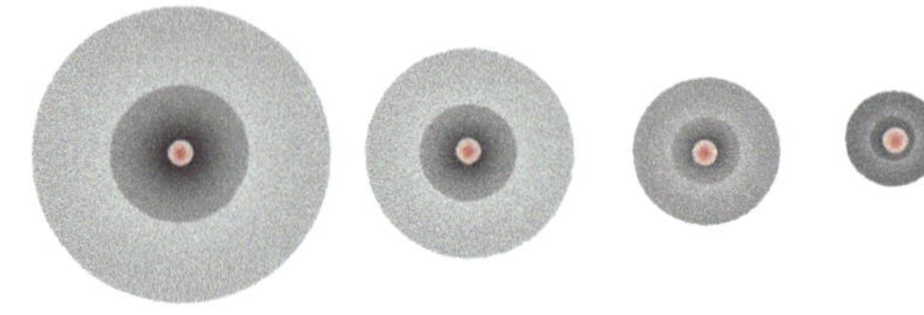

Systematically arranged by the great Russian scientist, Dmitri Mendeleev (1834-1907), the periodic table is a vast and intriguing storehouse of knowledge for scientists.

A sound knowledge of atomic structure and the systematic arrangement of atoms in the periodic table is essential for any student who aspires to excel in chemistry. Atomic structure, as it relates to electron configuration, provides useful information about an atom's radius, its ability to gain or lose electrons during chemical reactions and the energy changes accompanying these processes. Comparison of atoms and their radii can be used to predict their relative reactivity and give information of the amount of energy absorbed or liberated during chemical changes.

4.1 **Atomic radius,** (r) of an element can be defined in two ways: The distance from the centre of an atom to its outer edge, or the half the distance between the centres of two identical atoms, just touching each other.

Figure 4.1(a). Finding atomic radius using a single atom.   **Figure 4.1(b). Finding atomic radius using two atoms.**

The following figure depicts the relative atomic radii of the common group elements of the periodic table (not to scale).

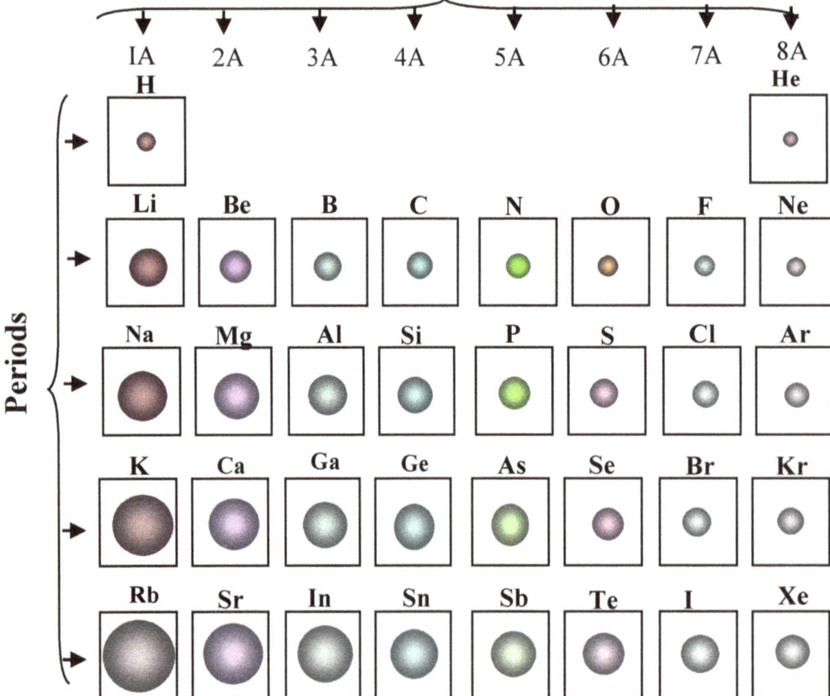

Figure 4.1(c). Relative atomic radii for some elements in the periodic table.

Trends in the Periodic Table

| 4.2 | **Trends across a period** |

In the following illustrations, the first four elements in period two of the Periodic Table are used to deduce the trends in atomic radius **across** that period. These trends are consistent with the trends across all other periods of the Periodic Table. The elements are shown with their respective number of protons and electrons.

$$^{9}_{4}Be \qquad\qquad ^{11}_{5}B \qquad\qquad ^{12}_{6}C \qquad\qquad ^{14}_{7}N$$

| # protons = **4** | # protons = **5** | # protons = **6** | # protons = **7** |
| # electrons = **4** | # electrons = **5** | # electrons = **6** | # electrons = **7** |

Using the plus sign, $(+)$ to represent each proton and the dot symbol (•) to represent each electron, we obtain the following Bohr-Rutherford diagrams for the atoms above. *Note that the neutrons are not shown in the nuclei as they do not affect atomic radius.*

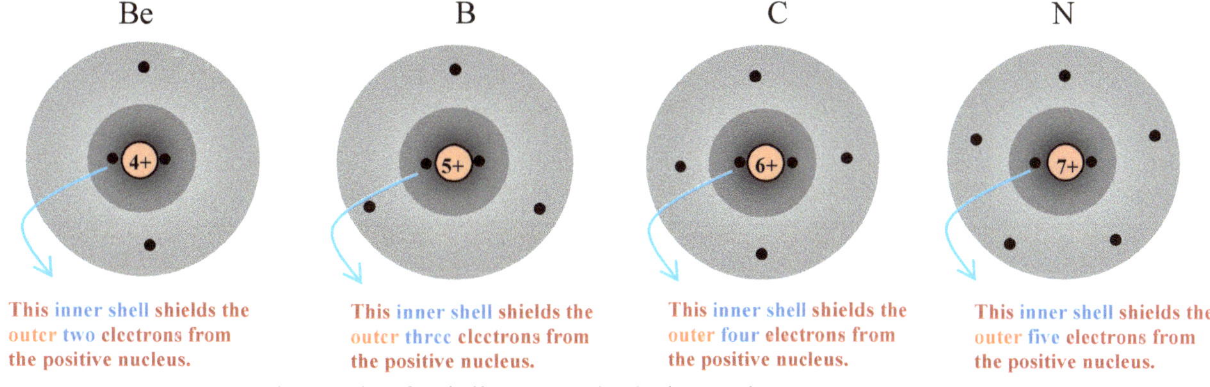

Figure 4.2. Bohr-Rutherford diagrams depicting only protons and electrons.

As these diagrams appear, it could be mistakenly concluded that each atom has the same atomic radius. However, this is not the case, as the following explanations and illustrations would show. The electron cloud of the two inner electrons of the K shell *shields* the outer electrons from the attractive force of the protons in the nucleus of each atom. Also, as the diagrams above show, the inner shell of each atom with their two electrons are very close to the protons. Because of their close proximity, the two inner shell electrons neutralize two protons in the nucleus of each atom. However, since the two electrons are not touching the two protons, there is only partial neutralization of charges. But for the sake of simplicity, we will assume that these *two inner shell electrons and two nuclear protons* **mutually neutralize each other.** The nucleus is thus left with a smaller net positive charge, called its **effective nuclear charge,** Z_{eff}. The following **new** nuclear compositions and electron configurations are obtained:

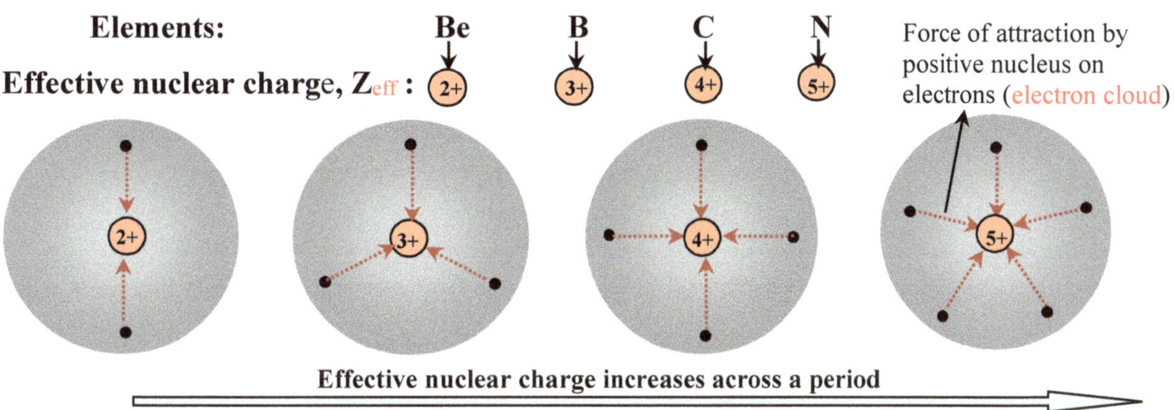

Figure 4.3. Bohr-Rutherford diagrams indicating how the nuclei are attracting the electrons.

Since two protons and the two inner core electrons have mutually neutralized each other, they are not shown in the previuos diagrams; only the **effective nuclear charges** and **the outer shell electrons** are shown, for simplicity. From the illustrations before, it can be concluded that *the effective nuclear charge of the atoms increases from left to right across a period.*

Since it is the nuclear charge of each atom that attracts the electrons, it can be concluded according to Coulomb's law that the attractive force on the outer electrons by its nucleus increases from left to right in a period. It should also be noted that *the electronic shells can contract or expand depending on the force of attraction from the nucleus on them. How close to the nucleus the electronic shells are attracted, will ultimately determine the size of the atomic radius, which is indicative of the size of the atom.* After analyzing the above structures, *it can be inferred that moving from left to right of the period, the outer electronic shells of atoms would be pulled progressively closer to their nuclei, rendering their atomic radii smaller and smaller.* The following figure shows how the atoms with different radii are obtained as a result of the effects different nuclear charges on their outer electrons.

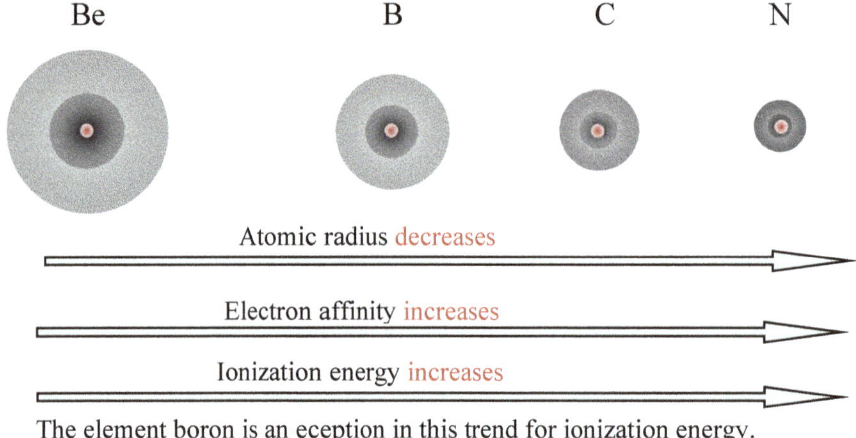

The element boron is an eception in this trend for ionization energy.
Use figures 4.8(a), 4.8(b) and 4.8(c) on pages 36 and 37 to figure out
this deviation.

Figure 4.4.   The relative atomic radii of atoms across a period.

Note how the atomic radius of nitrogen is much smaller than that of beryllium even though the nitrogen atom is heavier than that of beryllium. Because of this, *the electron density* of the atoms increases from left to right of the period.

<h2>4.3 Ionization Energy</h2>

If an atom or an ion of an element is in its gaseous state and it is given enough energy, one of its outermost electrons may be knocked out of its orbit. This minimum amount of energy that is required to do this, is called **ionization energy**.

This can be represented as follows using the element iron in the following example:

$$Fe_{(g)} + energy \ (\textbf{I.E}_1) \rightarrow Fe^+_{(g)} + e^-$$
$$Fe^+_{(g)} + energy \ (\textbf{I.E}_2) \rightarrow Fe^{2+}_{(g)} + e^-$$

In this example, the quantity of energy required to remove the first electron is called the **first ionization energy**, **I.E**$_1$, and the quantity of energy required to remove the second electron is called the **second ionization energy**, **I.E**$_2$. There can also be third and fourth ionization energies etc. The table on the following page shows the various ionization energies for a few elements.

Trends in Ionization Energies for Some Elements

Table 4.1. Ionization Energies (kJ/mol of electrons)

Elements	First $I.E_1$	Second $I.E_2$	Third $I.E_3$	Fourth $I.E_4$	Fifth $I.E_5$	Sixth $I.E_6$	Seventh $I.E_7$	Eight $I.E_8$
H	1 312							
He	2 372	5 220						
Li	520	7 298	11 815					
Be	899	1 757	14 849	21 006				
B	801	2 427	3 660	25 026	32 827			
C	1 087	2 353	4 621	6 223	37 831	47 277		
N	1 402	2 856	4 578	7 475	9 445	53 267	64 360	
O	1 314	3 388	5 3001	7 469	10 990	13 327	71 330	84 078
F	1 681	3 374	6 050	8 408	11 023	15 164	17 868	92 038
Ne	2 081	3 952	6 122	9 371	12 177	15 238	1999	23070
Na	496	4 562	6 910	9 543	13 354	16 613	20 117	25 496
Mg	738	1 451	7733	10 542	13 630	18 020	21 711	25 661

Analysis:

The data in the table above shows that there are patterns in ionization energy which are indicative of the number of electrons in the valence shell of an atom. The following is a summary of the findings:

An atom with one valence electron has a relatively small first ionization energy followed by a significantly larger one. This is exemplified by the elements, sodium and lithium. The values provided for sodium are as follow: $I.E_1$ = 497 kJ/mol, $I.E_2$ = 4562 kJ/mol.

- An atom with two valence electrons has a relatively small first ionization energy followed by a slightly larger one and then by a significantly larger one. This is exemplified by the elements, beryllium and magnesium. The values provided for magnesium are as follow: $I.E_1$ = 738 kJ/mol, $I.E_2$ = 1451 kJ/mol, $I.E_3$ = 7733 kJ/mol.
- An atom with three valence electrons has a relatively small first ionization energy followed by two slightly larger ones and then by a significantly larger one. This would be the case with aluminum and boron. The values provided for boron are as follow: $I.E_1$ = 801 kJ/mol, $I.E_2$ = 2427 kJ/mol, $I.E_3$ = 3660 kJ/mol, $I.E_4$ = 25026 kJ/mol.

- An atom with four valence electrons has a relatively small first ionization energy followed by three slightly larger ones and then by a significantly larger one. This is exemplified by the element carbon.
- The values provided for carbon are as follow: $I.E_1$ = 1087 kJ/mol, $I.E_2$ = 2353 kJ/mol, $I.E_3$ = 4621 kJ/mol, $I.E_4$ = 6223 kJ/mol, $I.E_5$ = 3783 kJ/mol.
- The same pattern continues for atoms having five, six and seven valence electrons.
- If the element is a noble gas, it starts with a relatively high first ionization energy. This is the case with the elements Helium and Neon.
- The high values highlighted in red is due fact that the noble gases have completely filled valences shells that impart tremendous stability to their atoms and to break this arrangement uses lots of energy.

Positive Ions

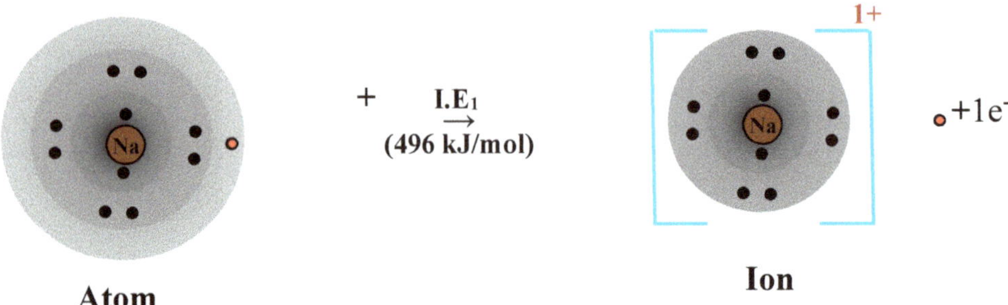

Figure 4.5(a). The **I.E₁** for the Na atom   **Figure 4.5(b).** The Bohr-Rutherford diagram of Na^{1+} ion.

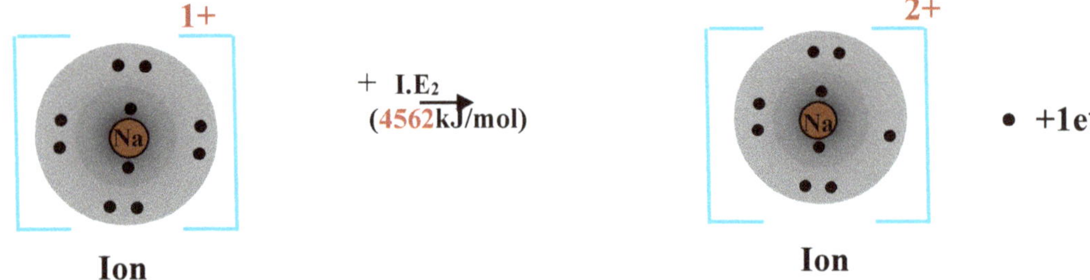

Figure 4.5(c). The **I.E₂** for the Na$^+$ ion.   **Figure 4.5(d).** The Bohr-Rutherford diagram of Na^{2+} ion.

Figure **4.5(a).** shows how a relatively moderate quantity of energy is required to remove the valence electron from a sodium atom. *As this electron is removed, an ion is formed whose radius is much smaller than that of the original atom*. This is because an entire energy level is now absent from the atom. Also, *the ion that is formed now has a valence shell with an octet*. **To remove an electron from an octet is very difficult**. For this reason, the second ionization energy for sodium is very high compared to the first. This concept is illustrated in figure 4.5(b).

The Electron-electron repulsion factor

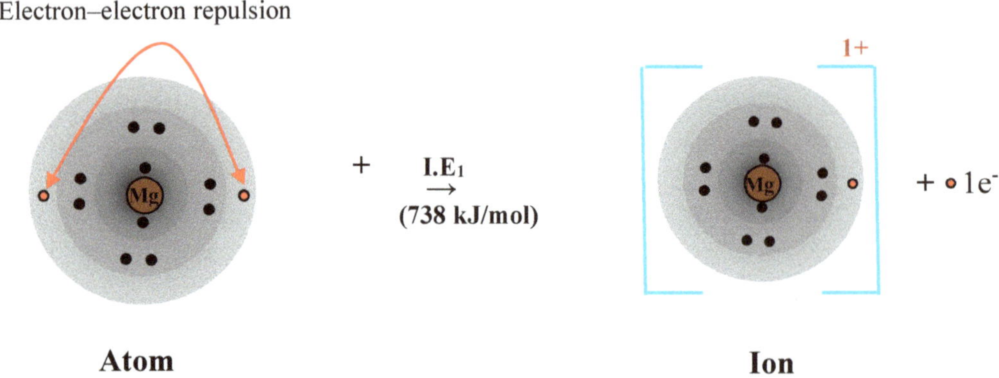

Figure 4.6(a). The **I.E₁** for Mg atom.   **Figure 4.6 (b).** The Bohr-Rutherford diagram of Mg^{1+} ion.

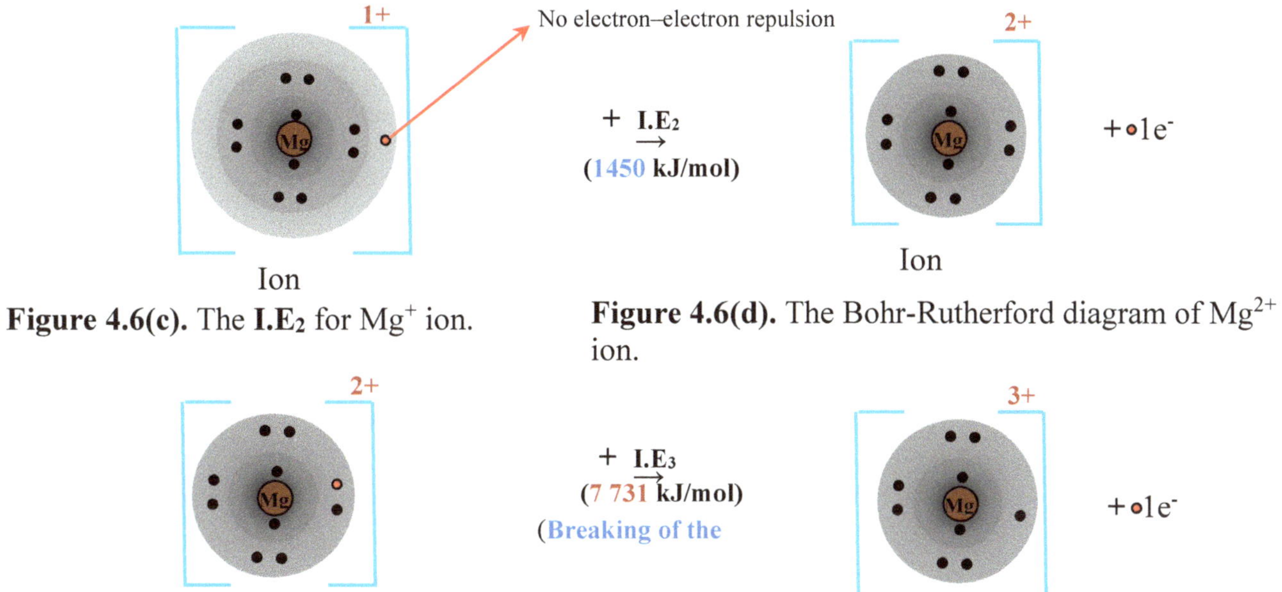

Figure 4.6(c). The **I.E₂** for Mg⁺ ion.

Figure 4.6(d). The Bohr-Rutherford diagram of Mg²⁺ ion.

Figure 4.6(e). The **I.E₃** for Mg²⁺ ion.

Figure 4.6(f). The Bohr-Rutherford diagram of Mg³⁺ ion.

The first ionization of an atom of Mg is 738 kJ/mol while the second one is 1450 kJ/mol. The question that arises here is why is this so, when both valence electrons are of equal distance from the nucleus? The answer to this has to do with the Mg atom having its two valence shell electrons **repelling** each other. Because of their mutual repulsion, **less energy is needed** to remove the first electron. **More energy** is needed to remove the second valence electron from the Mg⁺ ion because **there is no electron-electron repulsion** to aid the second ionization process, as was present before. Also, the radius of the Mg⁺ ion is smaller than the Mg atom due to the absence of electron-electron repulsion. The resulting Mg⁺ ion with a smaller ionic radius, as depicted by Figure 4.6(c), is the second factor responsible for a larger second ionization energy; the outer electron is now closer to the nuclear charge, needing more energy.

Similarly, as in the case of the Na⁺ ion, the Mg²⁺ ion has a much smaller ionic radius than the Mg atoms because one complete energy level is absent, as illustrated in Figure 4.6(f). The third ionization energy is thus considerably higher than the first two because of the presence of the octet.

Negative Ions

$^{35}_{17}Cl$

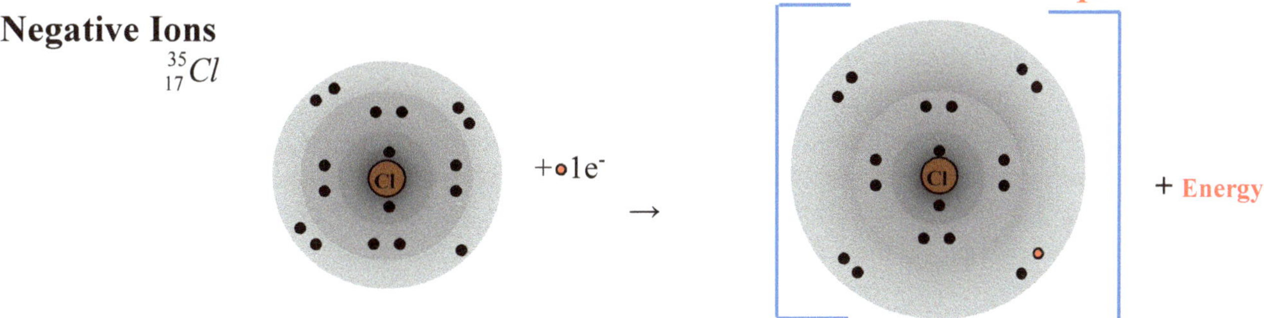

Figure 4.7(a). The Bohr-Rutherford diagram depicting chlorine's atomic radius.

Figure 4.7(b). The Bohr-Rutherford diagram depicting chlorine's **much larger** ionic radius.

As depicted in Figures 4.7a and 4.7b, after the Chlorine atom accepts an electron, it is changed to a chloride ion. Note that the Cl⁻ ion has a greater radius than the Cl atom. The reason for this is due to the fact that the Cl⁻ ion is now having one more electron in its valence shell that causes greater electron-electron repulsion.

Based on what has been explained so far, the following summary can be made with respect to atomic and ionic radius:

- As an atom loses electron(s) it forms an ion whose radius is smaller than its atomic radius. An example of this is a magnesium atom losing one valence electron. If the atom loses an entire energy level, its ionic radius is very much smaller than its atomic radius.
- When an atom or ion gains electron(s), the radius of the resulting species is much larger than what it had before.

4.5 Trends down a Group

The following Bohr-Rutherford diagrams are used to depict how the atomic radii of atoms change down the groups in the periodic table.

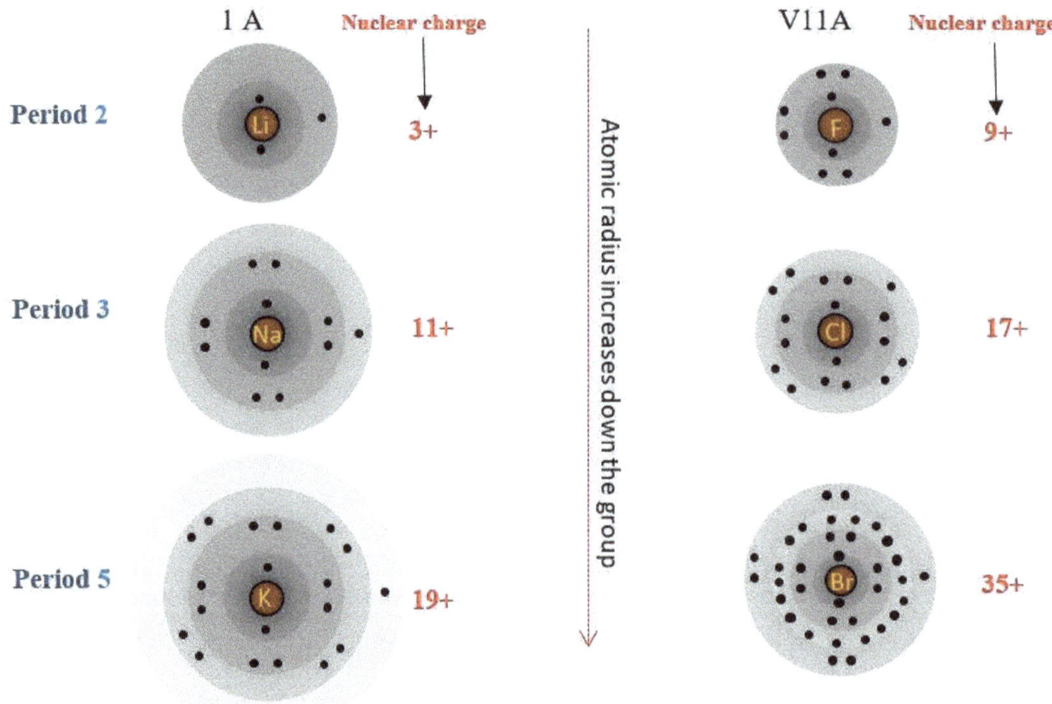

From these depictions, it can be deduced that the atomic radius increases down a group with an increase in periodic number. Two of the major factors responsible for this trend are: the increase in the number of electron shells down the group and the accompanying increase in nuclear charge of the atoms. An increase in the number of electron shells causes an overwhelming increase in atomic radius, while an increase in nuclear charge causes a decrease. The interplay between these two factors is solely responsible for the relative increase in atomic radius. Sodium for example, has a larger atomic radius that lithium, in spite of the fact that it has a greater nuclear charge than lithium. The reason for this is because sodium has one more electron shell than lithium, and this negates any significant decrease in atomic radius due to the increased nuclear charge. Both potassium and bromine have the same number of electron shells and their outer electrons are shielded from their positive nuclei by the same number of intervening shells (3). Yet, the atomic radius of bromine is much smaller that of potassium. This due to the greater nuclear charge of the bromine atom, compared to that of potassium. The attractive effect of bromine's nuclear charge on its outer electrons is greater that of potassium's nuclear charge on its outer single electron. The nucleus of bromine pulls in its outer electrons closer to itself, than the nucleus of potassium does its outer electron. The atomic radius of bromine is thus smaller than that of potassium.

Ionization energy is dependent on **the effective nuclear charge** on the outermost electron(s) of an atom. **The effective nuclear charge depends on how much shielding there is from intervening electrons shells**

between the outermost electron(s) and the charge on the nucleus, and the distance the outer electron (s) is away from the nucleus. **Lithium,** for example, has an effective nuclear charge of **1+** while that of **fluorine** is **5+.** In addition, the outer electrons of fluorine are much closer to its nucleus than the single electron of lithium is to its nucleus. For these two reasons; its greater effective nuclear charge, and its electrons being closer to its nucleus, the first ionization energy of fluorine is greater than that of lithium.

First Ionization Energy Trend

The following figure shows the graph of a plot of the first ionization energies of the elements in the first five periods of the periodic table.

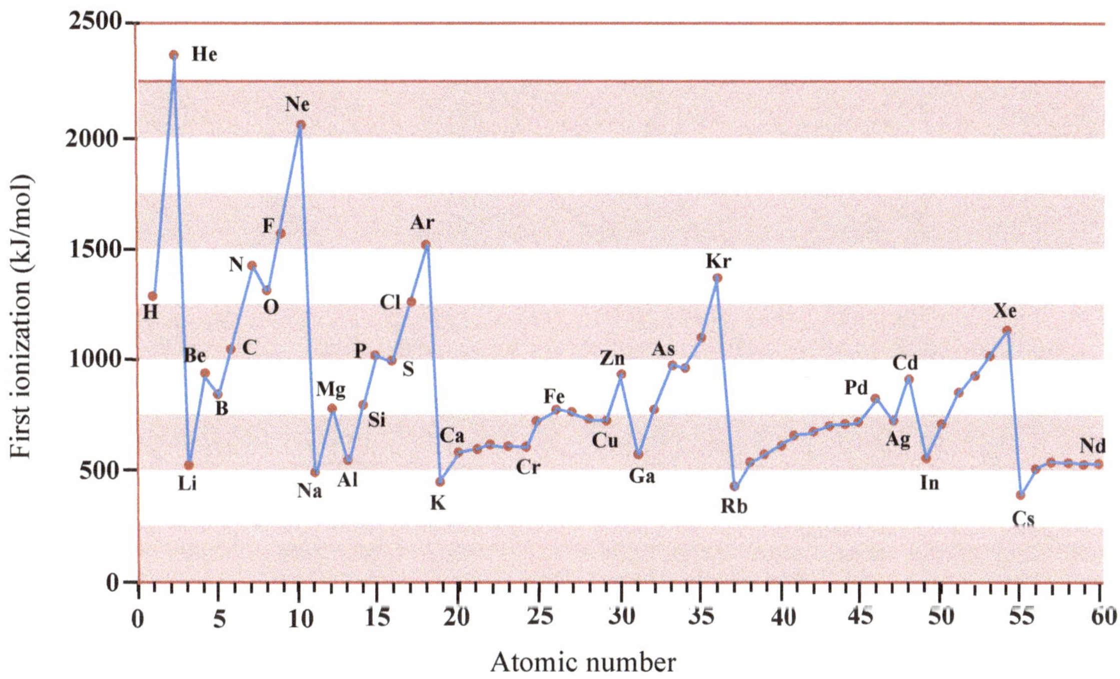

Figure 4.8 Graph of the first ionization energies for the first 60 elements

1. As you go down any group of the periodic table, the atomic radius increases. What are the factors that contribute to this trend? K/U

2. What trends exists in for first ionization energy down a group? Explain your answer using group 1 elements. In your answer you should use terms such as electron shells, atomic radius, inner-shell electron shielding, and nuclear charge. K/U C

3. Explain why the second ionization energy for beryllium is greater than its first. K/U C

4. Why would the radius of a fluoride ion be much bigger than the fluoride atom? K/U T/I

5. Using suitable diagrams, explain the differences in atomic radii and ionization energies for the elements, oxygen and fluorine. K/U C

6. Sodium and magnesium are in the same period yet magnesium has a larger first ionization than sodium. Explain. K/U C

4.6   Electron Affinity

If an electron is added to a neutral atom in its gaseous state and a stable anion is formed, energy is released. For example, when a bromine atom picks up an electron it forms a stable bromide ion, Br^- with the release of 325 kJ/mol. $Br_{(g)} + e^- \rightarrow Br^-_{(g)} +$ 325 kJ/mol (E.A. = -325 kJ/mol)

This energy change that occurs when an electron is added to atom in its gaseous state to form a negative ion is called **electron affinity (E.A); this sometimes called the first electron affinity.** The more stable the negative ion that is formed the more is the energy liberated. *Energy released is indicated by a negative number.* **The greater the negative number, the more stable is the anion formed.** The closer an electron comes to the positive nucleus of a particular atom, it results in greater stability of the anion and thus more energy is liberated. This, of course, depends on the unique electron configuration of the atom and its nuclear charge. The following figure depicts how energy is realesed when an electron's negative electric field enters that of the positive nucleus. The closer they come together, more energy is released.

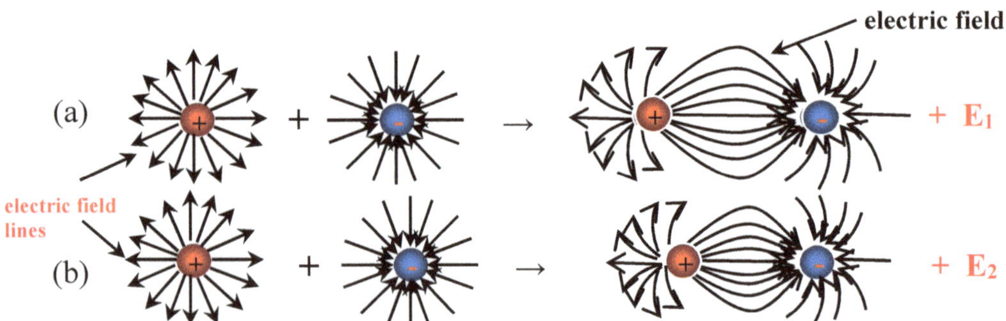

More energy will be released in case (b) than in case (a), therefore E_2 is greater than E_1.

In some atoms the negative ion formed is unstable, and it spontaneously breaks apart to produce a neutral atom and an electron. In these cases, to keep the ion intact, energy must be supplied, and this is indicated by *a positive number.* The greater the positive number, the more unstable is the anion formed.

All **second electron affinity** will have **positive numbers** because another electron added, would experience more electron repulsion from an aready negative anion making it more unstable and requiring more energy to keep it intact.

Table 4.2. Electron affinity values for some atoms

IA	IIA	IIIA	IVA	VA	VIA	VIIA	VIIIA
H -72.8							He +21
Li -59.6	Be +241	B -26.7	C -122	N 0	O -141	F -328	Ne +29
Na -52.9	Mg +230	Al -42.5	Si -134	P -72.0	S -200	Cl -349	Ar +34
K - 48.4	Ca +156					k Br -325	Kr +39

From the values provided in the table above a few patterns can be observed:

The atoms of groups IIA and VIIIA have positive numbers for electron affinity. This is so because for any of the group IIA atoms, when more electron(s) are added, it goes into a **p sub-shell** that is significantly farther away from their nucleus than the filled **s sub-shell**. For this reason, the anions formed are unstable. Similarly, the addition of one more electron to a group VIIIA atom means that it will be in a new energy level that is very much farther away from the nucleus. The following energy level diagrams illustrate the concept of orbital energy diagrams.

So far, our understanding of electrons arrangement has been that they are in shells or energy levels around the nucleus. However, scientists have discovered the shells can be further divided into sub-shells.

Within each sub-shell, the electrons there would have the same amount of energy but different from those in other sub-shells.

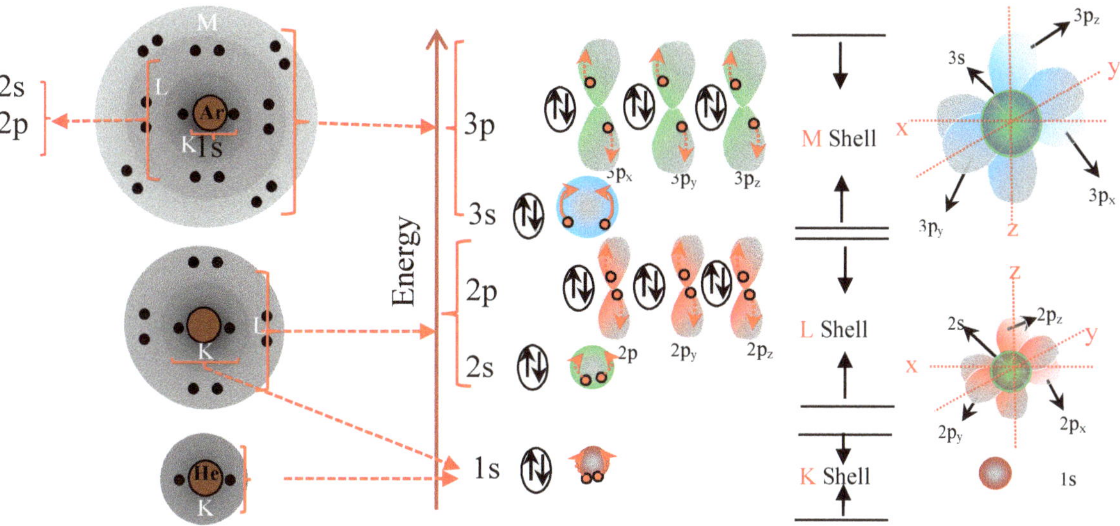

Figure 4.8(a). Orbital energy level diagrams and orbitals shapes and arrangements for the K, L and some M shell electrons and their spin orientations.

In the figures on the left it is shown how, with the exception of the K shell, the other shells are rearranged into their respective sub-shells. The higher up the sub-shells are, the more energy would the electrons located there would have. In these examples, the 2p electrons would be more energetic than the 2s electrons even though they are in the same L shell. This due to the difference in their shapes.

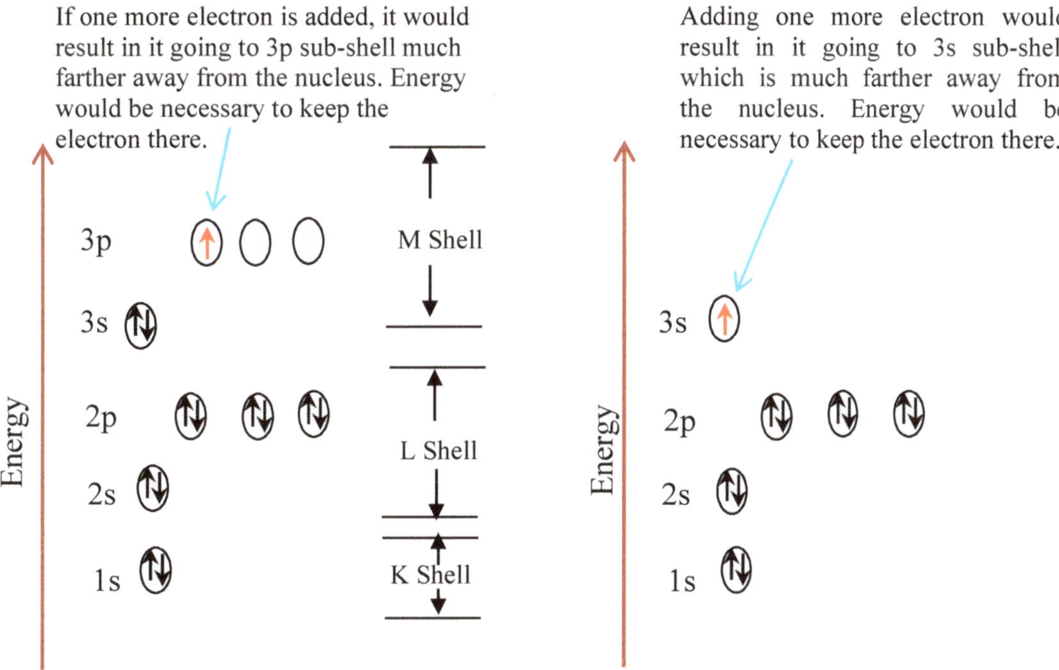

Figure 4.8(b). Orbital energy level diagram for the Mg atom accepting an electron.

Figure 4.8(c). Orbital energy level diagram for the Ne atom accepting an electron.

The concepts of atomic orbitals and energy level diagrams will be dealt with in more details in higher level chemistry courses than what is offered here.

Electron affinity increases across a period because the atomic radius gets smaller. Because of greater attraction between the nucleus and the outer electrons in an atom of small atomic radius, any addition of an electron to it would result in more energy being released. However, nitrogen is an exception to this trend as no energy is released when it accepts an electron. The following figure illustrates why.

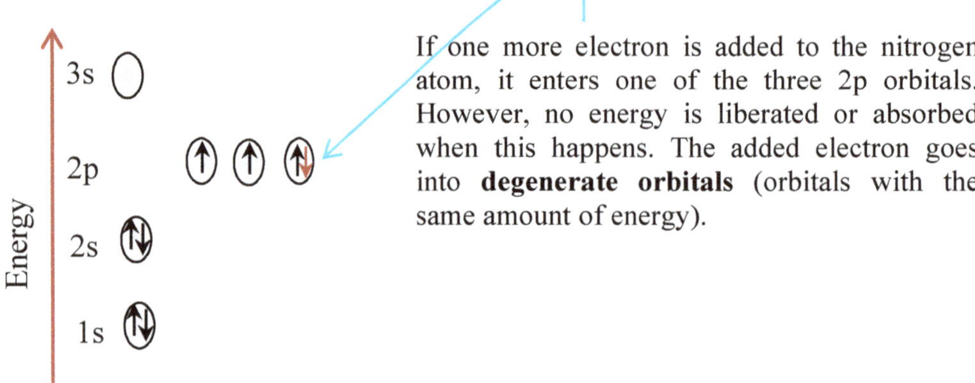

Figure 4.9. Orbital energy level diagram for the N atom accepting an electron.

CHAPTER REVIEW MATTER (1-4)

Matching

Match each term in the table below with the correct statements that follow:

A	Half-life	G	Isotopes
C	Thomson	H	Effective nuclear charge
B	Ionization energy	I	Alpha particles
D	Bohr	J	Chadwick
E	Transition	L	$E = h\nu$
F	Electron affinity	M	$4\,{}^{1}_{1}H + 2\,{}^{0}_{-1}e \rightarrow {}^{4}_{2}He + Energy$

1. As this increases the atomic radius decreases.

2. The amount of energy required to the outermost electrons from an atom.

3. The energy change associated when an electron is added to an atom in its gaseous state.

4. He calculated the charge-to-mass ratio of the electron.

5. These were used by Rutherford to probe the atoms of a gold foil.

6. He discovered the neutron.

7. He proposed that electrons are found in energy levels around the nucleus of atoms.

8. The equation for the amount of energy of a photon.

9. This describes the movement of an electron from a lower energy level to a higher energy level.

10. This equation represents a nuclear fusion reaction.

11. The time it takes for half the amount of radioactive atoms in a sample to decay.

12. The nuclei of these have the same number of protons but different number of neutrons.

True or False

Read each of the following statements and then decide if it is *True* or *False*.

1. All compounds are pure substances.

2. Milk is a homogeneous mixture.

3. The sublimation of $CO_{2(s)}$ is chemical change.

4. Viscosity is an example of a physical property.

5. The second ionization energy is always greater than the first.

6. The electron affinity of chlorine is greater than that of fluorine.

7. Cathode rays are negatively charged.

8. Beta radiation is more energetic than gamma radiation.

9. Gamma radiation is electromagnetic in nature.

10. An atom in its excited is more stable than in its ground state.

11. Carbon has a smaller atomic radius than boron.

12. To break the octet requires a large amount of energy.

13. When an atom absorbs an electron, energy is always released.

14. The mass of one proton has the exact mass as that of a neutron.

Multiple Choice

Choose the letter that best answers the questions:

1. Both oxygen and hydrogen are classified as pure substances because they each contain only one type atoms. Water is also classified as a pure substance even though it's made up of both hydrogen and oxygen atoms. This so because

a	both oxygen and hydrogen are pure substances
b	water is composed of only one type of particles, which in this case are molecules
c	oxygen and water molecules are chemically combined to form only water molecules
d	hydrogen and water are physically combined to form water molecules
e	b and c

2. Potassium explodes when added to water. This change can be classified as

a	physical	*d*	evaporation
b	sublimation	*e*	none of the above
c	chemical		

3. Cathode rays were reclassified as electrons because

a	they originated from the cathode and travelled towards the anode
b	they were repelled by the negative electrode and attracted by the positive electrode
c	they were deflected by a magnetic field placed perpendicular to their path.
d	they had the same charge to mass ratio when produced from different cathode materials
e	all of the above

4. Isotopes of an element are due to atoms of that element having

a	the same number of protons but different number of neutrons
b	the same number of protons but different number of electrons
c	the same number of protons and neutrons
d	the same number of neutrons, but a different number of protons
e	different number of protons and neutrons

5. Since the relative atomic mass of Zinc is (65.41 u), it means that

a	the nuclei of zinc have fractions of nucleons (protons and neutron)
b	the element zinc is composed of a mixture of isotopes.
c	the nuclei zinc atoms are broken
d	zinc atoms have unstable nuclei
e	zinc atoms are radioactive.

6. For an emission spectrum to be **visible** in any excited atom, electrons must fall back to an energy level of

a	principal quantum number. n = 1	*d*	principal quantum number. n = 4
b	principal quantum number. n = 2	*e*	principal quantum number. n = 5
c	principal quantum number. n = 3		

7. Nitrogen has a smaller atomic radius than beryllium because

a	a nitrogen atom is lighter than a beryllium atom
b	a nitrogen atom has lesser number of electronic shells than beryllium atom
c	a nitrogen atom has greater effective nuclear charge than a beryllium atom
d	a nitrogen atom is in a different period than a beryllium

8. The half-life of plutonium-239 is 24 100 years. If 100 g sample of it undergoes radioactive decay for 48 200 years, what mass of it would remain?

a	25 g	*d*	12.5 g
b	50 g	*e*	None of the above
c	75 g		

9. Potassium has a bigger atomic radius than lithium because

a	as you go down a group the atomic radius increases
b	potassium has more protons in its nucleus than beryllium
c	potassium weighs more than lithium
d	potassium has more electronic shells than lithium
e	a and d

10. For the following nuclear equation: $^{226}_{88}Ra \longrightarrow \ ? \ + \ ^{4}_{2}He$

The equation above can be completed by which of the following elements:

a	$^{222}_{86}Rn$	*d*	$^{230}_{90}Ra$
b	$^{209}_{84}Po$	*e*	$^{227}_{89}Ac$
c	$^{210}_{85}At$		

11. The atomic mass of any atom is primarily due to its number of

a	protons and electrons in the atom	*d*	neutrons and electrons in the atom
b	electrons in the atom	*e*	protons in the nucleus
c	protons and neutrons in the nucleus		

12. The ionic radius of a cation is smaller than its corresponding atomic radius because

a	the cation has a greater effective nuclear charge than the corresponding atom
b	the atom has a greater effective nuclear charge than the corresponding cation
c	the cation may have less energy levels than the corresponding atom
d	the cation has less electron-electron repulsion in its valence shell than the corresponding atom
e	both c and d

13. The ionic radius of an anion is greater than its corresponding atomic radius because

   a   the anion has a greater effective nuclear charge than the corresponding atom

   b   the atom has a greater effective nuclear charge than the corresponding anion

   c   the anion has more energy levels than the corresponding atom

   d   the atom has more energy levels than the corresponding anion

   e   the anion has a greater electron-electron repulsion in its valence shell than the corresponding atom

14. The second ionization energy of magnesium is greater than its first because

   a   the ionic radius is bigger than the atomic radius

   b   there is now a greater effective nuclear charge

   c   there is now less electron-electron repulsion in its valence shell

   d   there is now more electron-electron repulsion in its valence shell

   e   none of the above

15. Groups IIA elements have positive values for electron affinity because

   a   the change is endothermic

   b   the added electron(s) goes to a subshell that is farther away from the nucleus

   c   energy is liberated

   d   no energy change happens

   e   a and b

16. Nitrogen has a greater first ionization energy than carbon because

   a   they are both in the same period but nitrogen has a smaller atomic radius than carbon

   b   carbon has a smaller effective nuclear charge than nitrogen

   c   nitrogen needs more energy to remove its electron

   d   all of the above

   e   none of the above

17. The energy of a photon of light is dependent on

a	its frequency		d	its source
b	its intensity		e	a and c
c	wavelength			

18. Hydrogen ions from dilute acids accepts electrons from dilute acids and liberates hydrogen according to the following equation: $2H^+_{(aq)} + 2e^- \longrightarrow H_{2(g)}$. Three metals X, Y and Z have the following first ionization energies respectively: $E_1 = 906, 8991$ and 738 (kJ/mol). Which of these two metals will liberate hydrogen at the fastest rate from dilute acids?

a	X		d	not enough information given
b	Y		e	None of the above
c	Z			

19. An element has the following four successive ionization energies in kJ/mol.

1st	2nd	3rd	4th
801	2 427	3 660	25 026

The element is most likely to be a

a	an inert gas	d	group 4 element
b	group 2 element	e	group 1 element
c	group 3 element		

20. The atomic mass unit (amu) can be which of the following?

a	the mass of a proton	d	all of the above
b	the mass of a neutron	e	the mass of an electron
c	The mass of one twelfth the mass of a C-12 atom		

21. Bromine has a bigger atomic radius than chlorine because

a	bromine is less reactive than chlorine
b	bromine has more protons in its nucleus than beryllium
c	bromine weighs more than chlorine
d	bromine has more electronic shells than chlorine
e	none of the above

22. Which of the following sub-shells does the K shell have?

a	s alone
b	p alone
c	both s and p
d	s, p and d
e	none of the above

CHAPTER 5
Chemical Bonding

Chapter Content:

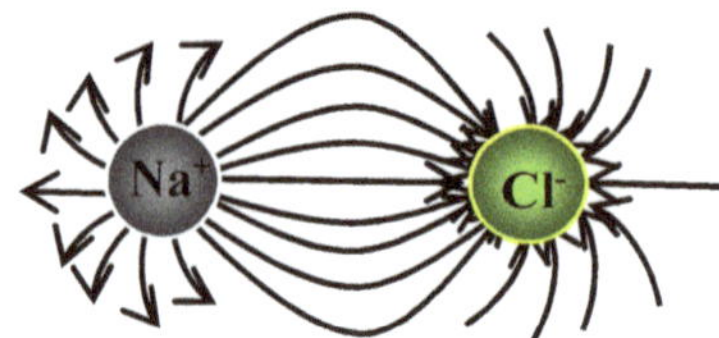

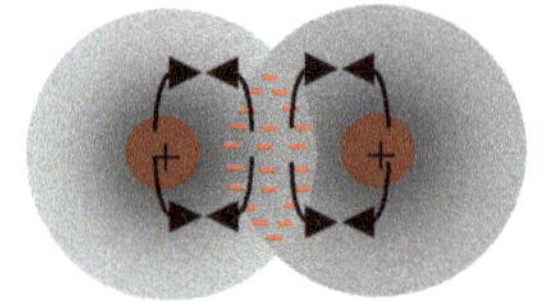

What would life be without the various types of chemical bonding? From the simplest of molecules to the most complex ones, bonding is essential. Molecules of proteins, fats and carbohydrates, for example, along with water, and vitamins are critically involved in building and maintaining bodily structures, and in carrying out various metabolic processes in living organisms. These molecules are all held together by special types of bonding. The replication of the DNA molecule is made possible because of the weak hydrogen bonding that exists between its two strands that can easily break and reform. Water is also a liquid at room temperature because of the intermolecular forces between its molecules.

In our non-living world, all structural entities, natural or man-made, are maintained by some type of bonding. The softness of graphite to the remarkable hardness of diamond is attributed to the difference in which the carbon atoms are naturally bonded within these two substances. The softness to the firmness of our foam mattresses is achieved by deliberately manipulating the type of bonding within the foam. The same can be said of the nature of adhesives that are made to bond firmly to specific substrates, or of fast setting dental pastes and alginates. Every day, marvelous compounds are created for household and industrial uses. These include the drugs in the medicines we use to the compounds needed to make concrete roads, buildings, bridges, aircrafts and countless number of other man-made structures around us. Without bonding to hold atoms and molecules together, everything including our bodies would spontaneously break apart into their respective atoms.

It is therefore important study how bonding takes place such that we become aware of naturally occurring compounds and their potential uses in society and in living things and also to create newer more beneficial ones.

Chemical Bonding

The relative stability of the noble gases is attributed to their unique electron configurations. As can be recalled, these elements have full outer electron shells. With the exception of helium, which has two electrons in its outermost shell, all the others have eight electrons in their outermost shell. This arrangement of two or eight electrons in the outermost shell of atoms is called the **duplex and octet** arrangements respectively. The following Bohr-Rutherford diagrams show these arrangements for some of the inert gases.

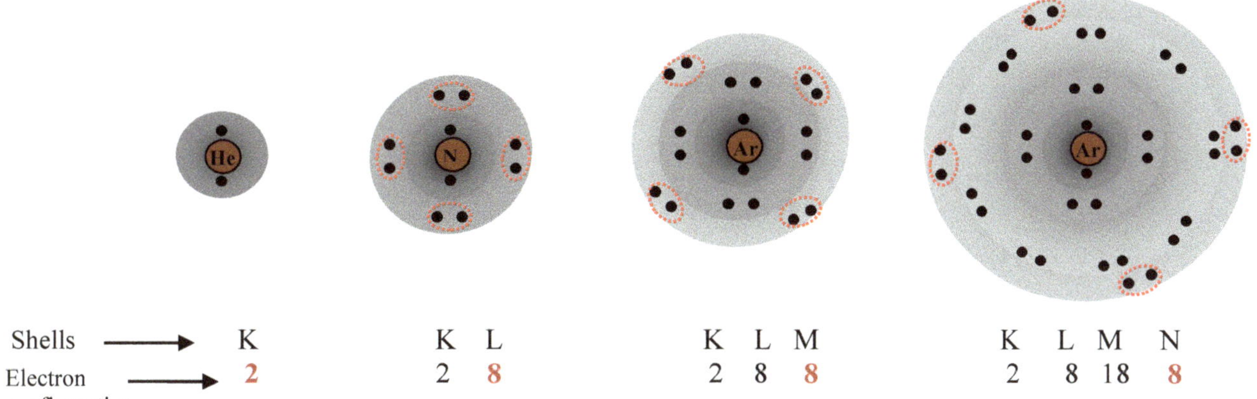

Figure 5.1. Bohr-Rutherford diagrams for the first four inert gases.

The above numbers (indicated by letters) summarize the number of electrons in each electron shell of the inert gas atoms above.

5.1 The Octet Rule

Since these electron arrangements (**duplex** or **octet)** bring stability to atoms, the atoms of other elements try to gain an electron configuration that is same as the electron configuration of an inert gas that most closely resembles it. Atoms of elements do this in one of **three ways**:

a) By gaining electron(s),
b) By losing electron(s),
c) By sharing electrons(s).

As can be recalled from the periodic table, elements in their respective groups have the same number of electrons in their outer shell, more commonly referred to as *valence electrons*. If the following groups of elements are examined, we would have the electron arrangements as shown in the following table.

5.2 Ionic Bondingwq

Table 5.1. Electron configurations for elements in groups IA, IIA, IIIA, IVA and VIIIA of the periodic table.

Groups	Group IA	Group IIA	Group IIIA	Group IVA	Group VIIIA
Elements	Li	Be	B	C	He
# electrons	3	4	5	6	2
Electron arrangement	2, 1	2, 2	2, 3	2, 4	2
Elements	Na	Mg	Al	Si	Ne
# electrons	11	12	13	14	10
Electron arrangement	2, 8, 1	2, 8, 2	2, 8, 3	2, 8, 4	2, 8
Elements	K	Ca			Ar
# electrons	19	20			
Electron arrangement	2, 8, 8, 1	2, 8, 8, 2			2, 8, 8

Analysis of Group IA elements shows that they look similar to the inert gases except for the **one electron** in their valence shell. To gain an inert gas electron configuration, each of these atoms will lose this single valence shell electron. This can be illustrated as follows:

<table>
<tr><td colspan="3">Group IA</td><td>Group VIIIA</td></tr>
<tr><td>Li</td><td>Li$^+$</td><td></td><td>He</td></tr>
<tr><td>2, 1</td><td>$\rightarrow$ 2</td><td>$+ 1e^-$</td><td>2</td></tr>
<tr><td>Na</td><td>Na$^+$</td><td></td><td>Ne</td></tr>
<tr><td>2, 8, 1</td><td>$\rightarrow$ 2, 8</td><td>$+ 1e^-$</td><td>2, 8</td></tr>
<tr><td>K</td><td>K$^+$</td><td></td><td>Ar</td></tr>
<tr><td>2, 8, 8, 1</td><td>$\rightarrow$ 2, 8, 8</td><td>$+ 1e^-$</td><td>2, 8, 8</td></tr>
<tr><td>Atoms</td><td colspan="2">Ions with inert gas
electron configurations</td><td>Inert gases and their
electron configurations</td></tr>
</table>

Doing the same for groups IIA and IIIA elements we have:

<table>
<tr><td colspan="3">Group IIA</td><td>Group VIIIA</td></tr>
<tr><td>Be</td><td>Be^{2+}</td><td>$+ 2e^-$</td><td>He</td></tr>
<tr><td>2, 2</td><td>$\rightarrow$ 2</td><td></td><td>2</td></tr>
<tr><td>Mg</td><td>Mg^{2+}</td><td>$+ 2e^-$</td><td>Ne</td></tr>
<tr><td>2, 8, 2</td><td>$\rightarrow$ 2, 8</td><td></td><td>2, 8</td></tr>
<tr><td>Ca</td><td>Ca^{2+}</td><td>$+ 2e^-$</td><td>Ar</td></tr>
<tr><td>2, 8, 8, 2</td><td>$\rightarrow$ 2, 8, 8</td><td></td><td>2, 8, 8</td></tr>
<tr><td>Atoms</td><td colspan="2">Ions with inert gas
electron configurations</td><td>Inert gases and their
electron configurations</td></tr>
</table>

<table>
<tr><td colspan="3">Group IIIA</td><td>Group VIIIA</td></tr>
<tr><td>B</td><td>B^{3+}</td><td>$+ 3e^-$</td><td>He</td></tr>
<tr><td>2, 3</td><td>$\rightarrow$ 2</td><td></td><td>2</td></tr>
<tr><td>Al</td><td>Al^{3+}</td><td>$+ 3e^-$</td><td>Ne</td></tr>
<tr><td>2, 8, 3</td><td>$\rightarrow$ 2, 8</td><td></td><td>2, 8</td></tr>
<tr><td>Atoms</td><td colspan="2">Ions with inert gas
electron configurations</td><td>Inert gases and their
electron configurations</td></tr>
</table>

Note that the element, boron has both metal and non-metal properties and is thus described as a **metalloid** element. It is also an electron deficient element that also forms covalent bonds.

As can be observed, atoms of group IA elements lose 1 electron each, group IIA elements lose 2 electrons each and group IIIA elements lose 3 electrons each to gain inert gas electron configurations. As can be reasoned, it is easier for these elements to lose rather than to gain electrons for this achievement. These elements that lose electrons to achieve the inert gas electron configurations are called **electropositive** elements. Note that as the atoms of these elements lose their valence electrons, they become positively charged ions. The charge that the ions have, sometimes referred to as (**valence**), is determined by the number of electrons that are lost by the respective atoms.

It is easier for other elements to gain rather than losing electrons to achieve the inert gas electron configurations. These elements are called **electronegative** elements. The following illustrations show how groups VA, VIA and VIIA elements achieve the inert gas electron configurations.

Group IVA

$$C \qquad C^{4-}$$

$$2,\ 4 + 4e^- \rightarrow 2, 8$$

Atom Ion with inert gas
    electron configuration

Group VIIA

$$Ne$$

$$2, 8$$

Neon's electron configuration

Note that the element, carbon, will only form ions when reacting with very electropositive elements such as calcium. When it reacts with calcium, for example, it forms the ionic compound, calcium carbide. Otherwise, carbon will share electrons with other non-metals to form covalent compounds. This is so because it is difficult for carbon to lose or gain four electrons to gain an inert gas electron configuration. It will only do this in extreme conditions. The element, silicon, another element with four valence electrons, will form only covalent bonds.

Table 5.2. Electron configurations for elements in groups VA, VIA, VIIA and VIIIA of the periodic table.

Groups	Group VA	Group VIA	Group VIIA	Group VIIIA
Elements # electrons ⟶	N 7	O 8	F 9	Ne 8
Electron arrangement	2, 5	2, 6	2, 7	2, 8
Elements # electrons ⟶	P 15	S 16	Cl 17	Ar 18
Electron arrangement	2, 8, 5	2, 8, 6	2, 8, 7	2, 8, 8
Elements # electrons ⟶			Br 37	Kr 38
Electron arrangement			2, 8, 18, 7	2, 8, 18, 8

Group VA

$$N \qquad\qquad N^{3-}$$
$$2, 5 \ + 3e^- \rightarrow \ 2, 8$$

$$P \qquad\qquad P^{3-}$$

$$2, 8, 5 + 3e^- \rightarrow \ 2, 8 , 8$$

Atoms Ions with inert gas
    electron configurations

Group VIIIA

$$Ne$$
$$2, 8$$

$$Ar$$

$$2, 8, 8$$

Inert gases and their
electron configurations

Group VIA		Group VIIIA	
O	O^{2-}	Ne	
2, 6 + 2e⁻ →	2, **8**	2, **8**	
S	S^{2-}	Ar	
2, 8, 6 + 2e⁻ →	2, 8, **8**	2, 8, **8**	
Atoms	Ions with inert gas electron configurations	Inert gases and their electron configurations	

Group VIIA		Group VIIIA	
F	F^-	Ne	
2, 7 + 1e⁻ →	2, **8**	2, **8**	
Cl	Cl^-	Ar	
2, 8, 7 + 1e⁻ →	2, 8, **8**	2, 8, **8**	
Br	Br^-	Kr	
2, 8, 18, 7 + 1e⁻ →	2, 8, 18, **8**	2, 8, 18, **8**	
Atoms	Ions with inert gas electron configurations	Inert gases and their electron configurations	

From the illustrations above, it can be seen that group VA elements need to gain 3 electrons, group VIA need to gain 2 electrons and group VIIA need to gain 1 electron respectively to gain an inert gas electron configuration. Note that as the atoms of these elements gain electrons, they become negatively charged ions. The charge that the ions have (**valence**), is determined by the number of electrons that are gained by the respective atoms.

5.3   Bohr-Rutherford diagrams for Positive Ions

The following Bohr-Rutherford diagrams are used to illustrate how a positive ion is formed when an atom loses an electron. The atom sodium will be used in this example.

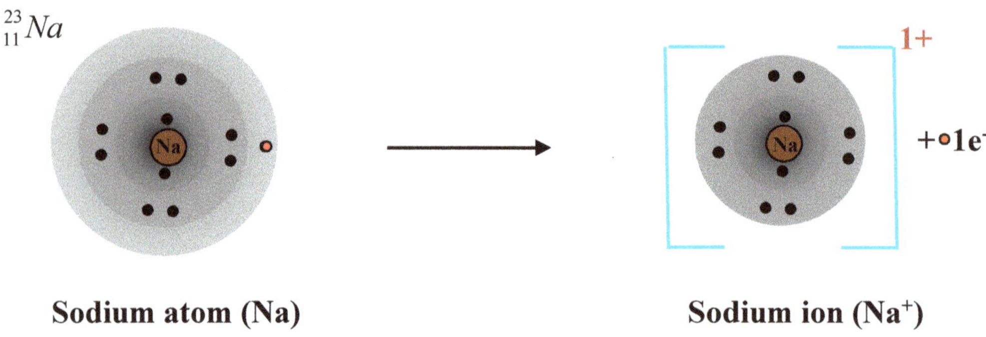

Figure 5.2. Bohr-Rutherford diagrams showing the formation of the sodium ion.

The following tally of protons versus electrons in these two figures above illustrates the difference between the atom from the ion in terms of net charge.

Atom	Ion
# protons = 11 = 11+	# protons = 11 = 11+
# electrons = 11 = <u>11-</u>	# electrons = 10 = <u>10-</u>
Net charge = 0	Net charge = 1+

Since all atoms are electrically neutral, then their number of protons equals their number of electrons. When an atom loses an electron(s) it would have more protons than electrons. As a result, it would be changed to an ion that is positively charged. The number of positive charges on the ion will be determined by the number of electrons lost, as that will determine how many more protons than electrons the ion will have.

Show how Bohr-Rutherford diagrams for the ions for the following elements are formed. K/U T/I

    (a) potassium

    (b) calcium

    (c) aluminium

    (d) magnesium

    (e) lithium

5.4   Bohr-Rutherford Diagrams for Negative Ions.

The following Bohr-Rutherford diagrams are used to illustrate how a negative ion is formed when an atom gains an electron(s). The chlorine atom will be used.

Chloride atom   Chloride ion (Cl⁻)

Figure 5.3. Bohr-Rutherford diagram showing the formation of the chloride ion.

The following tally of protons versus electrons in these two species illustrates the difference between the atom from the ion in terms of net charge.

Atom	Ion
# protons = 17 = 17+	# protons = 17 = 17+
# electrons = 17 = <u>17-</u>	# electrons = 18 = <u>18-</u>
Net charge = 0	Net charge = 1-

When an atom gains an electron(s), it will have more electrons than protons. As a result, it would be changed to an ion that is negatively charged. The number of negative charges on the ion will be determined by the number of electrons the atom gains, as that will determine how many more electrons than protons the ion will have.

Note that *when an atom gains an electron(s), its ionic radius becomes greater than its atomic radius.*

Exercise 5.2

Show how Bohr-Rutherford diagrams for the ions of the following elements are formed. K/UT/I

   (a) nitrogen

   (b) sulphur

   (c) phosphorus

   (d) fluorine

5.5   Bohr-Rutherford Diagrams for Ionic Compounds

From the preceding information, we have learned that atoms of elements need to gain or lose electron(s) to gain an inert gas electron configuration. The question that can be asked is, how do they do this? The answer to this question would be that atoms of the electropositive elements react with atoms of electronegative elements in fixed ratios, depending on the number of electrons needed to be lost or gained. The following figure illustrates how sodium reacts with oxygen.

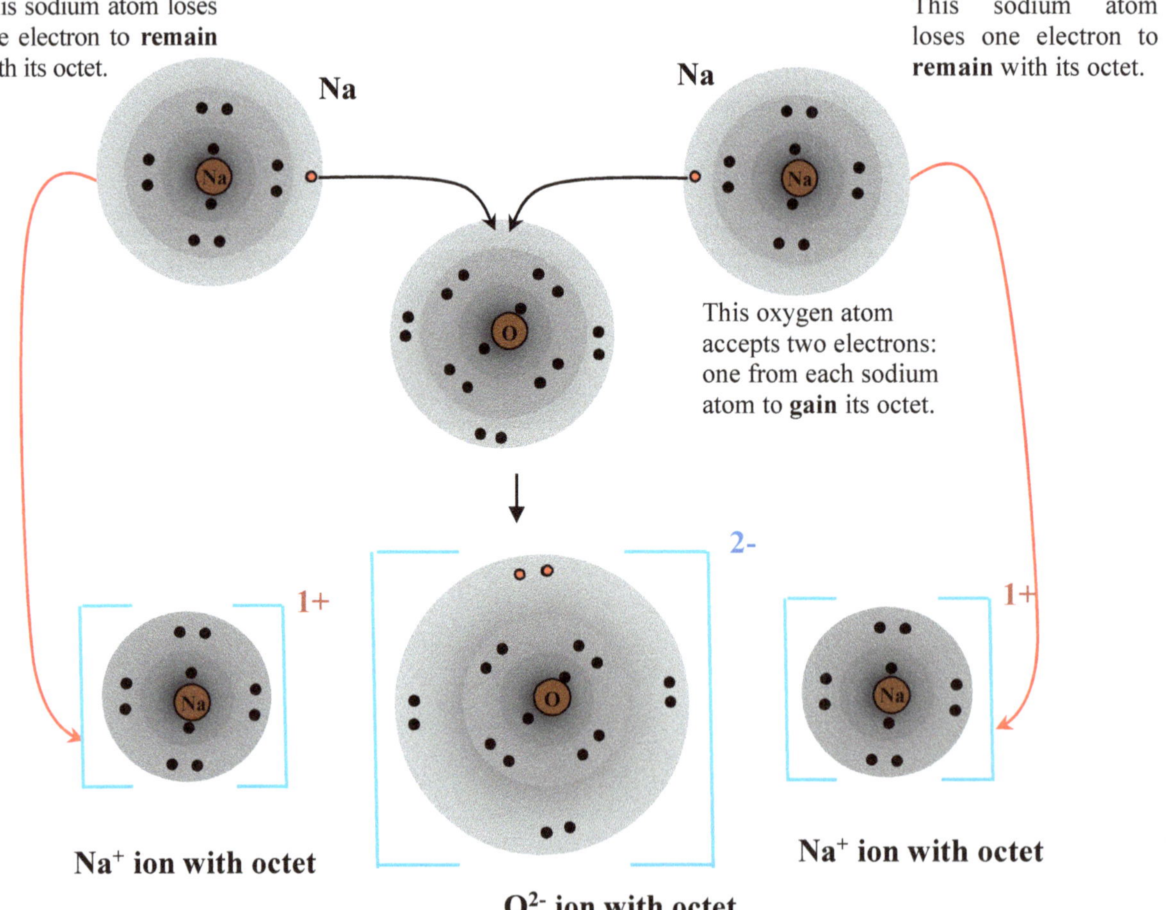

Figure 5.4. Bohr-Rutherford diagrams showing the acquisition of the octet by the oxygen and sodium atoms for the compound sodium oxide, Na_2O.

In the preceeding example, since each oxygen atom needs two electrons to complete its octet, and each sodium atom needs to lose only one electron to remain with its octet, then two sodium atoms are required for reaction with one atom of oxygen. The result is the formation one oxide ion and two sodium ions in one formula unit of sodium oxide.

Exercise 5.3

Draw Bohr-Rutherford diagrams showing how compounds are formed between the following pairs of elements. K/U T/I

    (a) lithium and fluorine

    (b) magnesium and chlorine

    (c) calcium and oxygen

    (d) aluminum and oxygen

Electrostatic Force of Attraction and the Ionic Bond

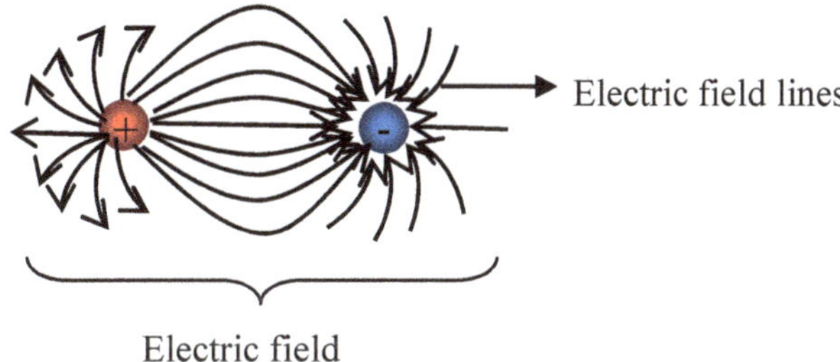

Figure 5.5. Electric field between a positive and negative charge.

Every positive and negative charge has its own electric field surrounding it. Each consists of electric field lines. **When two opposite charges are in each other's electric field, their electric field lines link and shorten, causing the two charges to attract each other.** Similarly, when **oppositely charged ions** are close to each other, the same thing happens; they bond with each other because of this electrostatic force of attraction. This bond is now called the **ionic bond.**

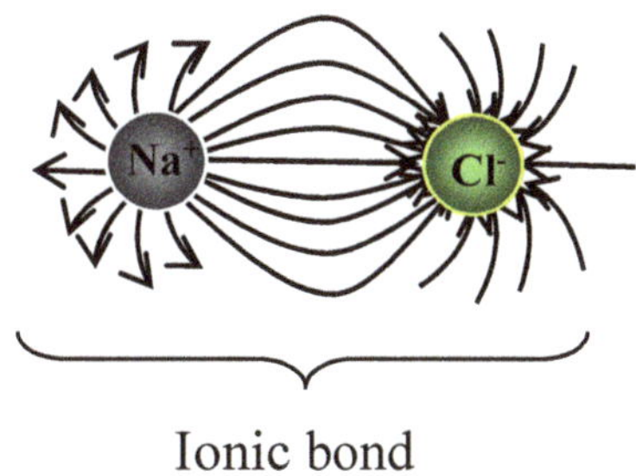

Figure 5.6. The ionic bond between a sodium ion and a chloride ion.

5.6   Electron Dot Diagrams for Ionic Compounds

To simplify the way that ionic bonds are represented, G.N. Lewis (1875-1946) used the **valence electrons** of the reacting atoms to show how electrons are lost and gained in the formation of ionic compounds. For example, if potassium were to react with fluorine we would have the following situation.

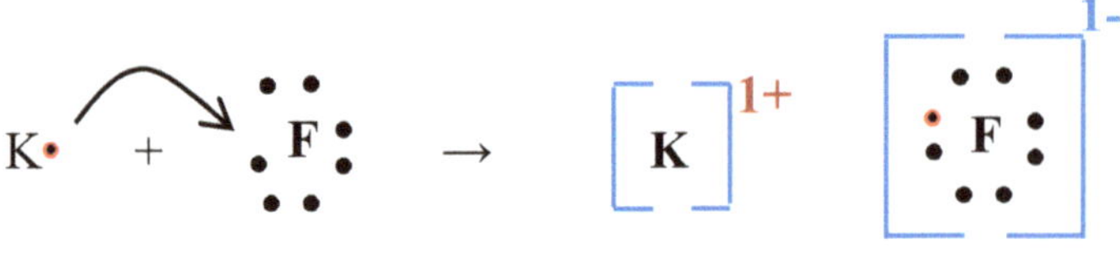

Chemical formula = KF

Figure 5.6. Lewis dot diagram for the reaction between potassium and fluorine.

In the example above, one atom of potassium donates a single electron to an atom of fluorine to form the compound potassium fluoride. Both elements have gained the octet.

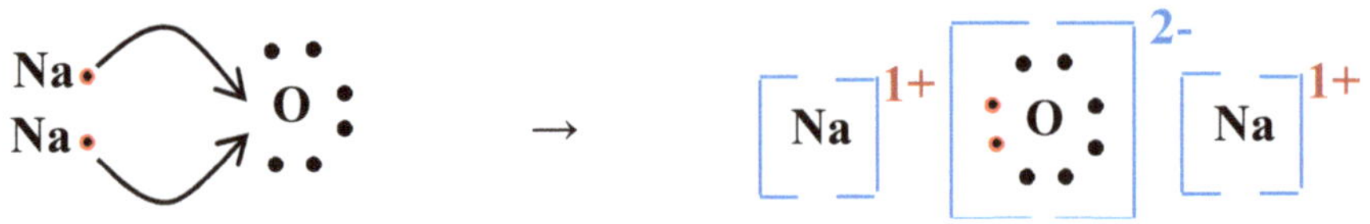

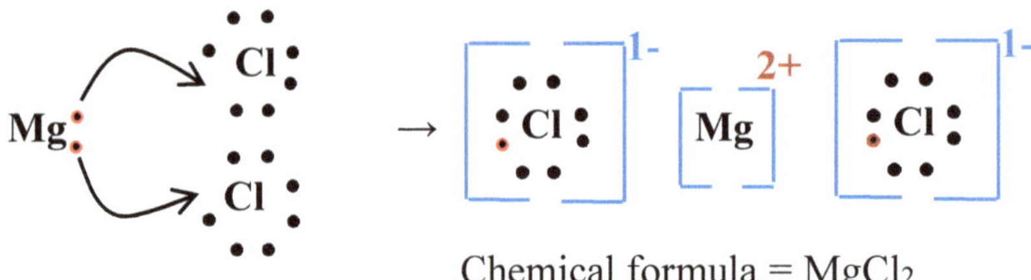

Chemical formula = Na$_2$O

Figure 5.7. Lewis dot diagram for the reaction between sodium and oxygen.

In the example above, two atoms of sodium donate one electron each to one atom of oxygen to form the compound, sodium oxide. Both elements have gained the octet.

Chemical formula = MgCl$_2$

Figure 5.8. Lewis dot diagram for the reaction between magnesium and chlorine.

In this example, one atom of magnesium donates two electrons; one to each of the two chlorine atoms to form the compound, magnesium chloride.

Note *that when Lewis dot diagrams are used to represent the formation of ionic compounds, the octet is shown **only** on the negative ion(s) **and not** on the positive ion(s).*

Use Lewis dot diagrams to show the formation of ionic compounds from the reaction between the following pairs of elements. K/UT/I

    (a) lithium and bromine

    (b) sodium and sulphur

    (c) calcium and iodine

    (d) magnesium and oxygen

    (e) aluminum and oxygen

    (f) aluminum and fluorine

5.7 Covalent Bonding

Covalent bonds are formed only from the reaction between non-metal elements. Take the non-metal element, chlorine for example. One atom of this element, as can be recalled, needs one electron to form the octet. If there is a willing donor such as sodium, chlorine will accept an electron to form a chloride ion. The resulting compound, sodium chloride will be ionic in nature. However, in the absence of electron donors, chlorine may react with fellow non-metals by sharing their valence shell electrons to gain an inert gas electron configuration. For example, it reacts with fellow non-metal hydrogen to form the molecular compound, hydrogen chloride. The compounds formed by covalent bonding are called **covalent compounds.** These can exist as diatomic molecules; ($Cl_{2(g)}$, $Br_{(l)}$ and $I_{2(s)}$), simple polyatomic molecular compounds ($CH_{4(g)}$, $CO_{2(g)}$, $C_6H_{12}O_{6(s)}$), giant molecular compounds such as DNA, and covalent network crystal such as graphite, diamond and quartz.

Non-metal such as carbon and silicon which have four electrons in their valence shells, find it most easy to react by sharing electrons, it would be difficult for these to loose or gain four electrons to form ionic compounds. The following figures illustrate how covalent bonds are formed by the reaction between the elements, carbon and hydrogen.

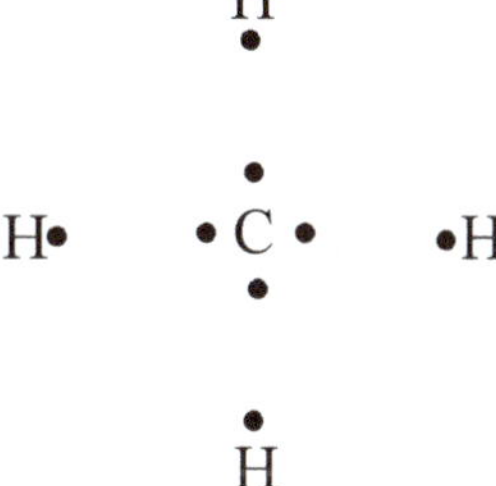

Figure 5.9(a). One carbon atom and four hydrogen atoms with their valence electrons.

Figure 5.9 (a) shows how carbon with its **four** un-bonded electrons is about to bond with four hydrogen atoms with their single electrons.

From a theoretical point of view, carbon can gain the octet by gaining one electron from each of the four hydrogen atoms. If this happened the following situation would result:

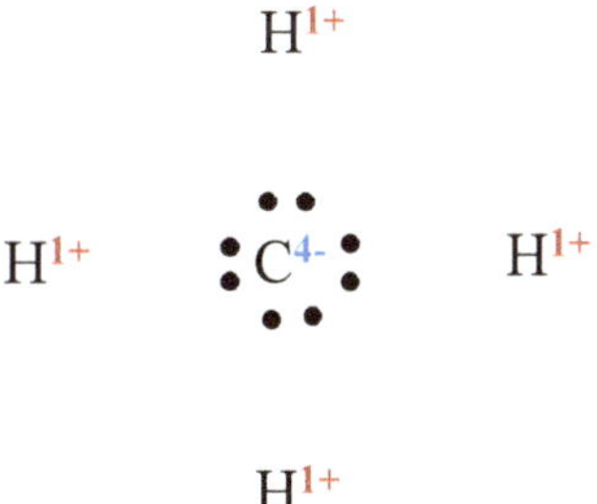

Figure 5.9(b). Carbon atom accepting one electron from each of the four hydrogen atoms.

In this case, carbon would not only gain the octet but also four negative charges. Each hydrogen atom would become positively charged.

If this change were to be reversed, we would end up with the original situation.

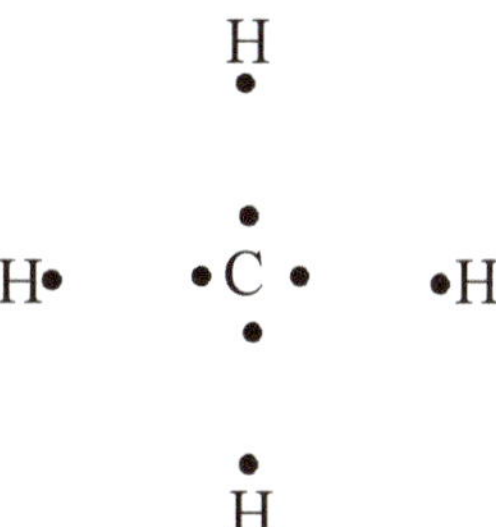

Figure 5.9(c). One carbon atom and four hydrogen atoms with their original valence electrons.

If each hydrogen atom now decides to take one electron from the carbon atom, the following new situation would result.

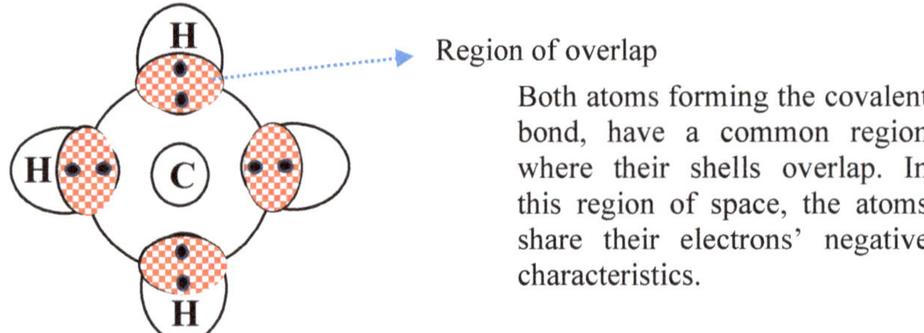

Figure 5.9(d). Carbon atom donating one electron to each of the four hydrogen atoms.

In this case, the hydrogen atoms would achieve their duplex electron configurations, becoming negatively charged with the carbon becoming positively charged. If these exchanges of electrons between carbon and hydrogen happened for only one second, then every other second carbon and hydrogen would become alternately positive and negative. *If the time for the exchange of electrons can be imagined to be reduced to one millionth of a second, then it would be difficult to conceive of any charges remaining on either the carbon or hydrogen atoms.* The pairs of electrons would be thought to have been positioned between the two atoms without any movement. In this case, the atoms would be said to **share the electrons** and the bonds formed between them would be called **covalent bonds.** The following figures represent some of the ways covalent compounds are drawn.

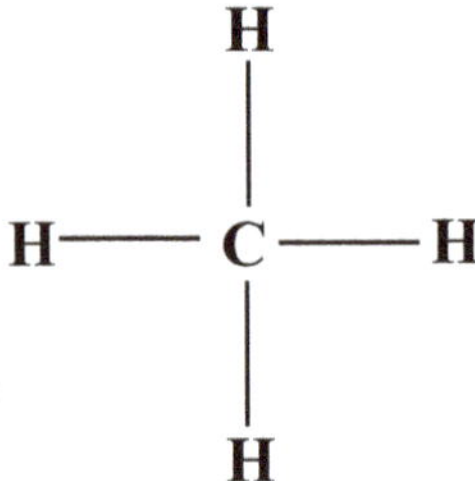

Both atoms forming the covalent bond, have a common region where their shells overlap. In this region of space, the atoms share their electrons' negative characteristics.

Figure 5.10. Sharing of electrons as represented by the overlapping of electron shells.

H
|
H — C — H
|
H

Figure 5.11. Sharing of electrons as represented by the dashes between bonded atoms.

In Figure 5.10., a pair of electrons placed between the two atoms is used to represent a single covalent bond. In Figure 5.11., a dash placed between two atoms is used to represent a single covalent bond.

In the figures above, it can be seen that both the carbon and hydrogen atoms have attained their inert gas electron configurations; carbon having the octet and each hydrogen having the duplex.
Note that in assigning electrons to atoms, to ascertain if they have achieved the inert gas electron configuration, the bonded electron pairs (bond pairs) are counted twice; once on each of the two bonded atoms. In this case, the **four bond pairs** are assigned to the carbon atom giving it the octet, and then each of the four **bond pairs** is assigned separately to each of the four hydrogen atoms, giving each of them the duplex. **In other words, it is assumed that** the shared pair of electrons belongs to both atoms.

5.8 The Nature of the Covalent Bond

The following example uses the wave mechanical model to illustrate how a covalent bond is formed between two hydrogen atoms. In this model, each hydrogen atom is thought to have a positively charged nucleus surrounded by a spherical electron cloud. According to wave mechanical model, the electrons are delocalized and are in constant motion around or about the nucleus. Since electrons are negatively charged and are constantly moving around or about the nucleus, the volume of space they occupy is negative in nature. This volume of space about or around the nucleus of an atom, where there is a probability of finding an electron is called an orbital and it is characterized by a **negative electron cloud**.

Using the two hydrogen atoms, we shall start by showing the electron clouds before and after they overlap with each other.

Figure 5.12. Two separate hydrogen atoms with their negative electron clouds and positive nuclei.

In the case of hydrogen atoms, the shape of the electron cloud (**orbital**) is that of a **sphere**. In other atoms having multiple electrons, there are in addition to spherical orbitals, orbital of others shapes (page 36).

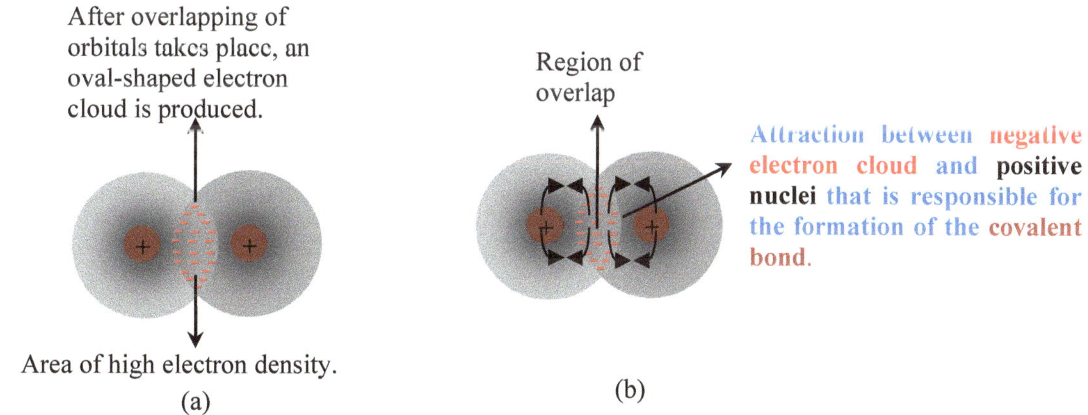

Figure 5.13. The overlapping of the electron clouds of the two hydrogen atoms (a) and subsequent bonding (b).

Where the electron clouds of the atoms overlap, an area of high electron density is created between the two nuclei. Since this area is highly negative, it attracts the positively charged nuclei bonding the two atoms, as depicted by Figure 5.13. **For any covalent bond, the greater the degree of overlap, the greater is the bond strength.** *Note that in the formation of a single covalent bond,* **no more than two electrons** *are used by the two bonding atoms;* **one from each.**

5.9 Bonding capacity

Bonding capacity is the number of bonds that an atom must form to achieve an inert gas electron configuration. To do this, atoms can lose, gain or share electrons. In the formation of methane, for example, carbon which had four electrons in its valence shell, needed to form four covalent bonds; it thus reacted with the four hydrogen atoms, forming four single bonds. The following table shows the bonding capacity of some common non-metal elements.

Table 5.3. Determining the bonding capacity of atoms.

Atom	Number of valence electrons	Number of bonding electrons	Bonding capacity
Carbon	4	4	4
Nitrogen	5	3	3
Oxygen	6	2	2
Halogens	7	1	1
Hydrogen	1	1	1

Analysis of the data in the table above shows that the number of valence electrons in any atom is found by knowing what group of the periodic table the element is in. Nitrogen for example, is in Group VA and thus has the same number of electrons as the group number. Since nitrogen has five electrons in its valence shell, it only needs three more to gain the octet. This is achieved by forming **three covalent bonds** with other suitable atoms. The following figure illustrates how this happens between nitrogen and hydrogen to form ammonia:

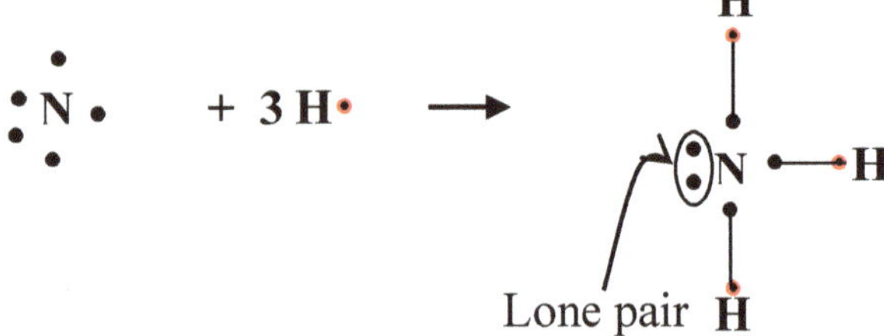

Figure 5.14. The formation of ammonia molecule by covalent bonds.

Three of nitrogen's five electrons form covalent bonds that account for six electrons. The remaining two electrons form a lone pair. The total number of electrons around the nitrogen atom is now eight. At the same time, each of the three hydrogen atoms has two electrons. Because nitrogen forms three covalent bonds, it has a bonding capacity of three. Note that **to *know the bonding capacity of any atom, subtract its number of valence electrons from the number, eight.***

5.10 Lewis Dot Structures for Covalent Compounds

Like ionic compounds, electron-dots can be used to represent bonding in molecular compounds also. As in ionic bonding, only the valence electrons of the atoms are involved in covalent bonding. In Lewis structures, a **dash** drawn between two bonded atoms is used to represent a single covalent bond. In these structures, **all the electrons, bonding and non-bonding ones, must be shown.**

To arrive at the Lewis structure for any species, a number of rules must be followed. Using the following examples, these rules will be progressively established:

Example 1:
Drawing the Lewis dot structure for the methane molecule having molecular formula, **CH$_4$**.

Step 1. Draw the skeletal structure of the molecule, using the atom with the highest bonding capacity as the central atom.
In this case the carbon atom has the highest bonding capacity, so it becomes the *central atom* and the hydrogen atoms become the *peripheral ones.*

H
|
H — C — H
|
H

54

Step 2. **Calculate the total number of valence electrons for all the atoms in the molecule.**

$$1 \text{ C atom} = \frac{1 \, atom \times 4 \, electrons}{atom} = 4 \, electrons$$

$$4 \text{ H atoms} = \frac{4 \, atoms \times 1 \, electron}{atom} = 4 \, electrons$$

Total number of valence electrons = *8*

In drawing Lewis dot structures, it must be shown how all the valence electrons are used; each electron must be accounted for.

Step 3. Since every covalent bond uses two electrons, **to know how many electrons are used in bonding, the number of covalent bonds formed are noted and then multiplied by the number 2.**

$$\textbf{Number of electrons used} = \frac{2 \, electrons}{bond} \times 4 \, bonds$$

$$= 8$$

Step 4. Subtract the total number of bonding electrons from the total number of valence electrons to find the number of remaining electrons.

$$\text{Total number of valence electrons} = 8 \, electrons$$
$$\text{Total number of bonded electrons} = \underline{-8 \, electrons}$$
$$\textbf{Number of electrons remaining} = 0$$

Step 5. Inspect the molecule to see if the various atoms have an inert gas electron configuration.
In this case, all the atoms have an inert gas electron configuration. The carbon atom has eight electrons and each of the four hydrogens atom has two electrons. The skeletal structure is the same as the required Lewis dot structured.

Example 2:
Drawing the Lewis dot structure for the ammonia molecule having molecular formula NH_3.
Step 1. To draw this structure, we shall follow all the rules outlined previously.

Step 2.
$$1 \text{ N atom} = \frac{1 \, atom \times 5 \, electrons}{atom} = 5 \, electrons$$

$$3 \text{ H atoms} = \frac{3 \, atoms \times 1 \, electron}{atom} = 3 \, electrons$$

Total number of valence electrons = *8*

Step 3.
$$\textbf{Number of electrons used} = \frac{2 \, electrons}{bond} \times 3 \, bonds$$
$$= 6$$

Step 4.
$$\text{Total number of valence electrons} = 8 \, electrons$$
$$\text{Total number of bonded electrons} = \underline{- 6 \, electrons}$$
$$\textbf{Number of electrons remaining} = 2$$

Step 5. Inspection of the skeletal structure (next page)shows that the nitrogen atom has only six electrons, but each hydrogen atom has two.

To successfully complete the Lewis dot structure for the NH_3 molecule, an additional rule must be followed, (step 6)

Step 6. If any atom has not achieved the inert gas electron configuration, use any remaining electrons to complete the octet.

In this case, there are two remaining electrons and these are placed on the nitrogen atom, in the form of a lone pair, to complete its octet. This is shown as follows:

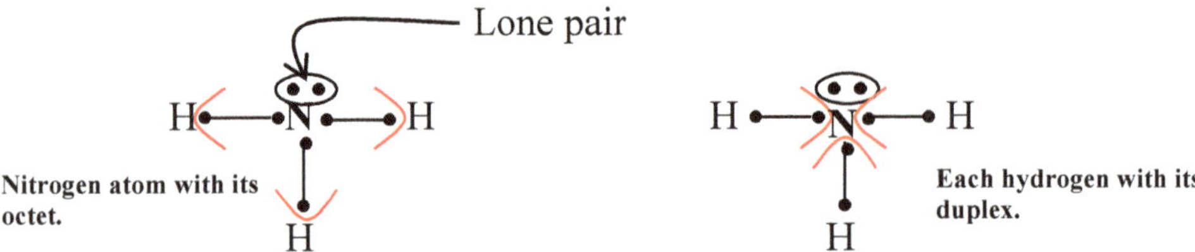

In the previous structures, it can be seen that both nitrogen and hydrogen have their own inert gas electron configurations.

Example 3.

Drawing the Lewis dot structure for the oxygen dichloride molecule having molecular formula, OCl_2.

Step 1.

$$O$$
$$Cl \quad Cl$$

Step 2. 1 O atom $= \dfrac{1\ atom \times 6\ electrons}{atom} = 6\ electrons$

2 Cl atoms $= \dfrac{2\ atoms \times 7\ electrons}{atom} = 14\ electrons$

Total number of valence electrons = *20*

Step 3. **Number of electrons used** $= \dfrac{2\ electrons}{bond} \times 2\ bonds$

$= 4$

Step 4. Total number of valence electrons = *20 electrons*
Total number of bonded electrons = *-4 electrons*
Number of electrons remaining = *16*

Step 5. Inspection of the skeletal structure (figures below) shows that neither oxygen nor chlorine has the octet; the oxygen atom has four electrons, and each chlorine atom has two electrons.

Step 6. Using the remaining 16 electrons we have the following:

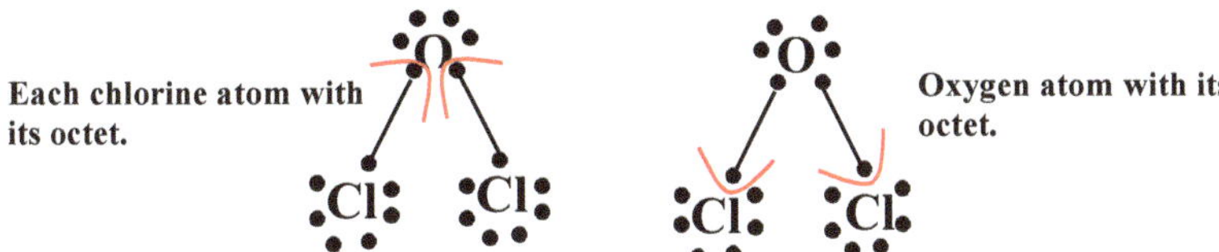

In the above drawings, it can be seen that all the atoms now have the inert gas electron configurations.

5.11   Molecules with Double Bonds

Example 4. Drawing the Lewis dot structure for the cabon dioxide molecule having molecular formula, CO_2.

Step1.

$$O\!-\!C\!-\!O$$

Step 2. 2 O atoms $= \dfrac{2\,atoms \times 6\,electrons}{atom} = 12\,electrons$

$\quad\quad\quad\quad$ 1 C atom $= \dfrac{1\,atoms \times 4\,electrons}{atom} = 4\,electrons$

Total number of valence electrons $= 16$

Step 3. **Number of electrons used** $= \dfrac{2\,electrons}{bond} \times 2\,bonds$

$$= 4$$

Step 4. Total number of valence electrons $= 16\,electrons$

Total number of bonded electrons $= \underline{-4\,electrons}$

Number of electrons remaining $= 12$

Step 5. Inspection of the skeletal structure shows that neither oxygen nor carbon has the octet; the carbon atom has four electrons and each oxygen atom has two electrons.

Step 6. Using the remaining twelve electrons, we have the following:

Further inspection of this new structure shows that the oxygen atoms have the octet, but not the carbon atom, which has only four electrons. To complete the Lewis structure, another step must be followed.

Step 7.

If all the remaining electrons are used up and still any atom has not achieved the octet, then use lone pairs of electrons from atoms with the octet to make multiple bonds.

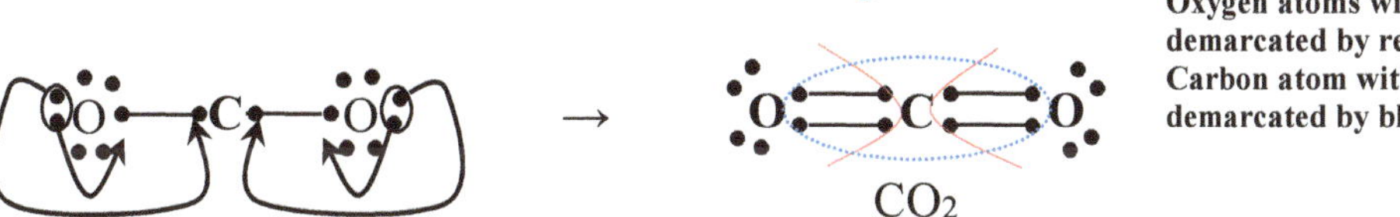

In the preceeding example, one pair of electrons was shifted from each oxygen atom to create an additional bond between the carbon and oxygen. In this case all the atoms have achieved the octet. ***Note that the oxygen atoms, from which the electrons were used, still retained their octets, since the electrons in the multiple bonds are also counted twice.***

5.12   Molecules with Triple Bonds

Example 5. Drawing the Lewis dot structure for the hydrogen cyanide molecule having molecular formula, HCN.

Step 1.   H——C——N

Step 2. $\cdot$N: 1 N atom $= \dfrac{1\ atom \times 5\ electrons}{atom} = 5\ electrons$

   $\cdot$C$\cdot$ 1 C atom $= \dfrac{1\ atoms \times 4\ electron}{atom} = 4\ electrons$

   H$\cdot$ 1 H atom $= \dfrac{1\ atom \times 1\ electron}{atom} = 1\ electron$

Total number of valence electrons $= 10$

Step 3.   **Number of electrons used** $= \dfrac{2\ electrons}{bond} \times 2\ bonds$

$$= 4$$

Step 4.   Total number of valence electrons $= 10\ electrons$

          Total number of bonded electrons $= -4\ electrons$

Number of electrons remaining $= 6$

Step 5.
In the skeletal structure shown in Step 1, neither the carbon nor the nitrogen has the octet; the carbon atom has four electrons while the nitrogen atom has two electrons.

Step 6. Using the remaining six electrons we have the following structure:

*Further inspection of this new structure shows that both **the nitrogen** and **hydrogen** atoms **have their inert configurations**, but the **carbon** atom atom has only **four** electrons.*

Step 7.

In the last step, two pairs of electrons were shifted from the nitrogen atom to create two additional bonds between the carbon and nitrogen atoms. The result is that all the atoms have now achieved their inert gas electron configurations. Please note the following:
   - If an atom is short of two electrons after all the remaining electrons have been assigned, then only one lone pair of electrons is used, so a double bond results.
   - If an atom is short of four electrons after all the remaining electrons have been assigned, then two lone pairs of electrons are used, so a triple bond results.

A. Negatively charged polyatomic ions

Example 6. Drawing the Lewis dot structure for the nitrate ion having formula, NO_3^-.

Because this species has one negative charge, it means that one more electron must be added to the total number of valence electrons of the nitrogen and oxygen atoms. All the other steps are to be followed, except that the ion is placed in square parentheses.

Step1.

$$O - N - O$$
$$|$$
$$O$$

Step 2.

$$3 \text{ O atoms} = \frac{3\,atoms \times 6\,electrons}{atom} = 18\,electrons$$

$$1 \text{ N atom} = \frac{1\,atom \times 5\,electrons}{atom} = 5\,electrons$$

Total number of valence electrons = *23 electrons*

1 electron added because of negative charge = _1 electron_

Total number of electrons = *24*

Step 3.   **Number of electrons used** $= \dfrac{2\,electrons}{bond} \times 3\,bonds$

$$= 6$$

Step 4.   Total number of valence electrons = *24 electrons*

Total number of bonded electrons = *-6 electrons*

Number of electrons remaining = *18*

Step 5. Inspection of the above structure shows that none of the atoms have the octet.

Step 6. Using the remaining eighteen electrons we get the following structure.

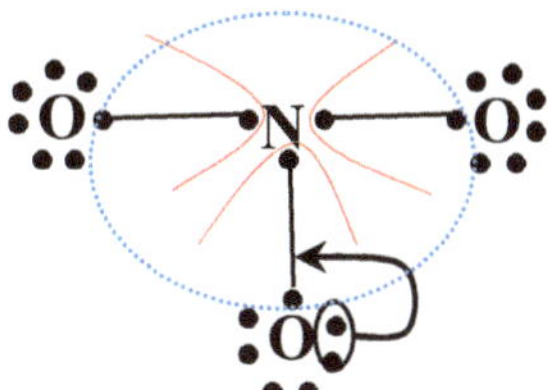

Further inspection of the new structure shows that only the oxygen atoms have the octet while the nitrogen atom does not because it has only six electrons.

Step 7. Since nitrogen needs only two more electrons, only one lone pair is shifted from any one of the oxygen atoms.

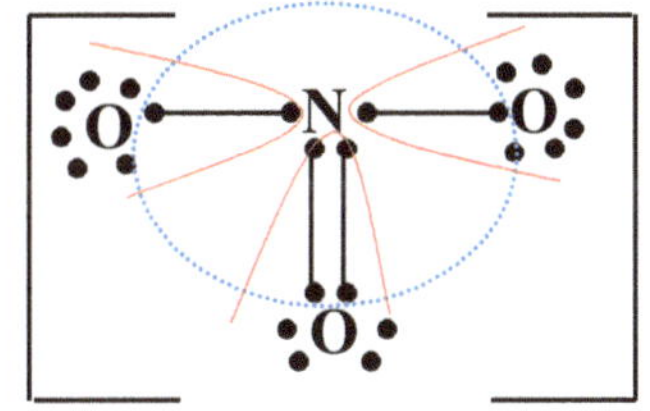

Oxygen atoms with their octets; demarcated by red lines.
Nitrogen atom with its octet; demarcated by blue oval.

Positively Charged Polyatomic Ions

Example 7. Drawing the Lewis dot structure for the NO^+ ion

*Because this species is positively charged, **it would have one less electron** than if it were neutral. One electron is thus subtracted from its total number of valence electrons.*

Step 1. $N \longrightarrow O$

Step 2.   1 O atom $= \dfrac{1\ atom \times 6\ electrons}{atom} = 6\ electrons$

   1 N atom $= \dfrac{1\ atom \times 5\ electrons}{atom} = 5\ electrons$

Total number of valence electrons $= 11\ electrons$

1 electron is subtracted because of positive charge $= -1$

Total number of electrons $= 10$

Step 3.   **Number of electrons used** $= \dfrac{2\ electrons}{bond} \times 1\ bond$

$= 2$

Step 4.   Total number of valence electrons $= 10\ electrons$

Total number of bonded electrons $= -2\ electrons$

Number of electrons remaining $= 8$

Step 5. Inspection of the above structure shows that none of the atoms has the octet.

Step 6. Using the remaining eight electrons we have the following:

Further inspection of the new structure shows that only the oxygen atom has the octet while the nitrogen atom does not because it has only four electrons.

Step 7. Since nitrogen needs only four more electrons, two lone pairs are shifted from the oxygen atom to create a triple bond between the two atoms.

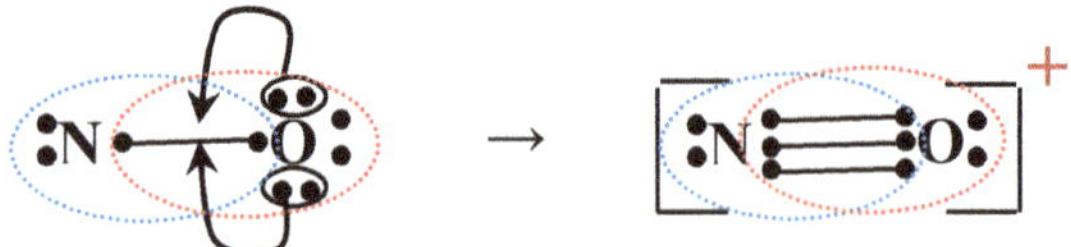

Oxygen atoms with their octets; demarcated by the red oval
Nitrogen atom with its octet; demarcated by blue oval.

*Note that in all of the examples done so far, dots were placed at the ends of the covalent bonds. This was only done to make it easy for the students to count all the electrons involved. **The final structure should not be drawn with the dots at the ends of the bonds.** The following drawings are what should be expected from the above examples.*

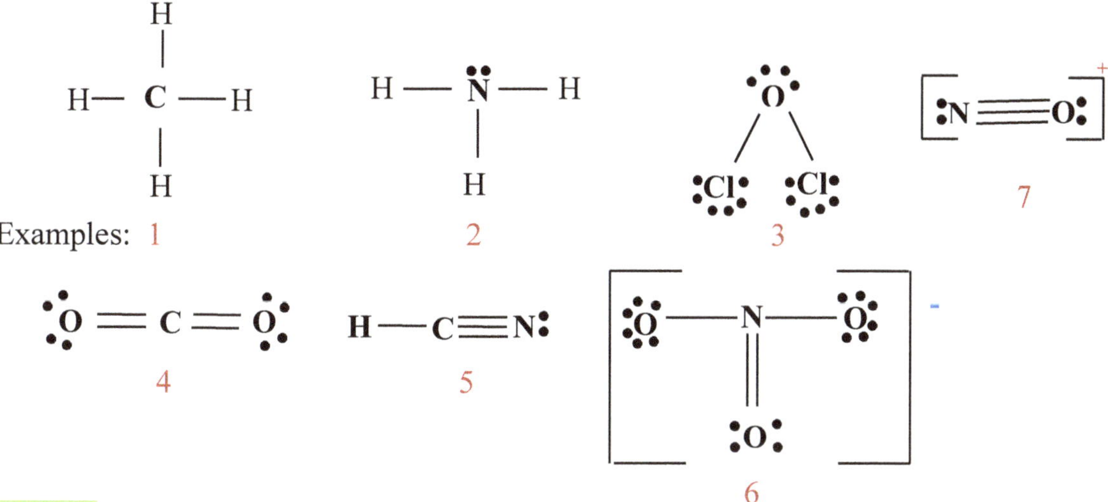

Examples: 1   2   3   7   4   5   6

5.14   The Coordinate Covalent Bond

Normally, when a single covalent bond is formed each of the two bonding atoms provides an electron. However, in some cases, one atom may lack a bonding electron but it may find another atom that has a lone pair with which it can bond. In these cases, the one with the lone pair provides both electrons for bonding. The bond is then called **a coordinate covalent bond.** The following reaction between $H^+_{(aq)}$ ion and $H_2O_{(l)}$ to form the hydronium ion ($H_3O_{(aq)}$) is used to illustrate this concept.

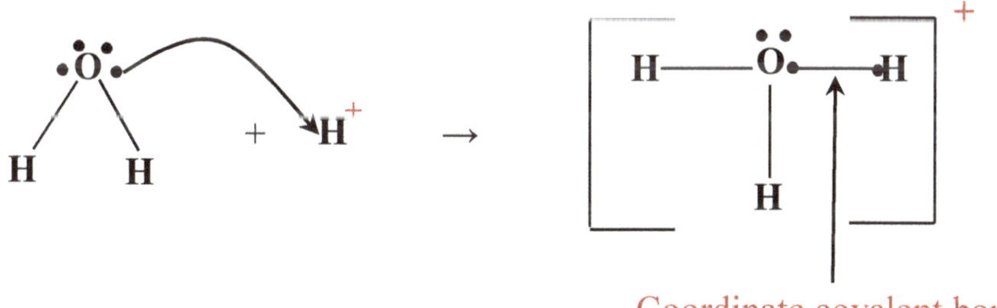

Coordinate covalent bond

In this case, the hydrogen ion has no electron for bonding, so one lone pair of electrons on the oxygen atom is used to create a coordinate covalent bond between the two atoms.

Exercise 5.5

Draw Lewis dot structures for the following: K/UT/I

(a) Cl_2 (b) PCl_3 (c) N_2 (d) PH_3 (e) H_2S

(f) H_2O_2 (g) CH_3OH (h) CN^- (i) CO_3^{2-}

Chapter Content:

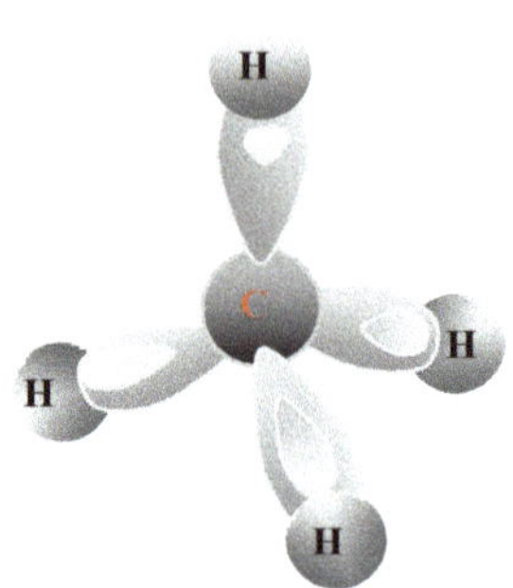

Knowledge of molecular geometry (a three dimensional structure of the atoms in a molecule) is important since it helps to predict the polarity of molecules, the type of intermolecular force that exists between molecules, the boiling and melting points of compounds, the solubility of compounds in other compounds and their reactivity. New molecules can be created with specific shapes and polarity to achieve some desired effects. The antibiotic penicillin, for example, has been modified to become ampicillin, a more effective broad-base antibiotic, by the addition of an amino acid chain. Many diseases are treated with drugs whose molecules have been specifically designed to destroy viruses and bacteria, or to influence the metabolism of cells in specific ways. Many drugs are designed to act as inhibitors of enzymes by blocking their active sites to achieve some desired effects; drugs for treating HIV infection, hypertension and erectile dysfunction in men, work on this basis. Knowledge enzymes' molecules and the nature of their active sites are important in this respect. In protein synthesis tRNA molecules choose the correct amino acids because of their specific shapes. Any mistake in choosing the correct amino acid could lead to the formation of a dysfunctional protein.

Intermolecular Force of Attraction

6.1   **Electronegativity:** *A measure of the ability of an atom in a molecule to draw the bonding pair electrons to itself.* The bonding electrons are in the form of an electron cloud of specific shape.

In 1992, Linus Pauling developed an electronegativity scale for the elements. The table below gives the electronegativity values he assigned to some of the common elements.

Table 6.1. Electronegativity values for some elements.

Elements	F	O	Cl	Br	I	S	C	Si	B	H	N	Be	Na	Mg	K
Electronegativity values	4	3.5	3	2.8	2.5	2.5	2.5	1.8	2	2.1	3.0	1.5	0.9	1.2	0.8

6.2   Bond Polarity

For diatomic molecules, whether a covalent bond is polar or not, depends solely on the difference in electronegativity of the two bonded atoms. In **homonuclear** diatomic molecules, (molecules containing two identical atoms), since there is no electronegativity difference between the bonded atoms, the bond is non-polar. Hydrogen and chlorine molecules are used to illustrate this concept as follows:

$$H_{2(g)} \qquad\qquad Cl_{2(g)}$$
$$H\text{———}H \qquad\qquad Cl\text{———}Cl$$

Electronegativity values   =   2.1   2.1   3.5   3.5

Electronegativity difference: =   0   0

 Both of these molecules are non-polar, because in each, there is even distribution of the shared electron clouds between the two bonded atoms. In these homonuclear molecules, the nuclei being identical, exert equal pulls on the shared electron cloud, and this is what is responsible for the even distribution of the electron clouds. The following figures illustrate how the electron clouds are distributed in these molecules.

H ●● H   Cl ●● Cl

(a)   (b)

Figures 6.1. The even electron cloud distributions for the hydrogen and chlorine molecules respectively.

6.3   Polar Bond in Diatomic Molecules

If the electronegativity difference between two bonded atoms is **greater than 0.4,** then the covalent bond becomes polar.

The bonding between chlorine and hydrogen is used to illustrate this concept.

$$\overset{\delta+}{H}\text{———}\overset{\delta-}{Cl}$$

Electronegativity values   =   2.1   3.5

Electronegativity difference =   1.4

The bond between the hydrogen and chlorine atoms would be a polar, because *the chlorine atom which is significantly more electronegative than the hydrogen atom,* exerts *a greater pull on the shared electron cloud. Because of this stronger pull, more of the electron cloud stays with the chlorine atom,* causing it to acquire a partial negative charge, symbolized as ($\delta-$) and the hydrogen atom acquiring a partial positive charge symbolized as ($\delta+$). This is usually what happens in **heteronuclear** diatomic molecules, (molecules containing two different atoms). The greater the electronegativity difference between two atoms, the more polar the bond becomes.

The following figure shows how the electron cloud between the two atoms is pulled more towards the chlorine atom.

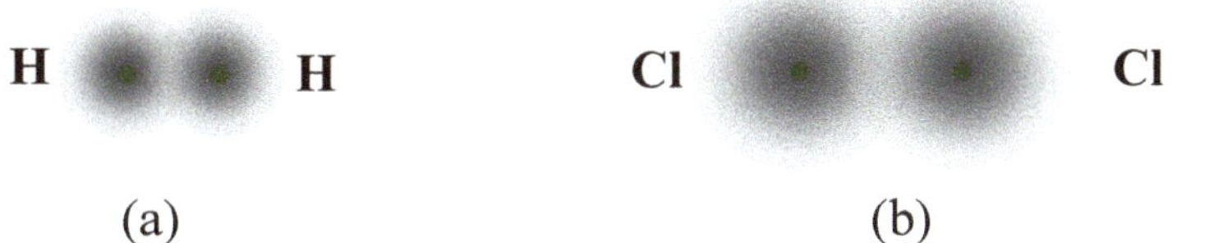

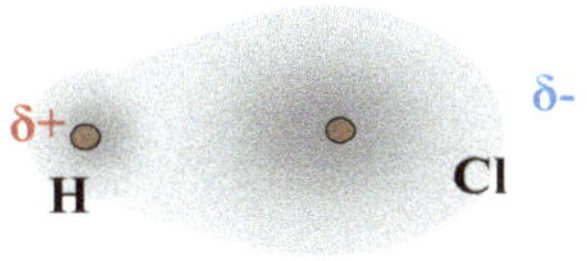

Figure 6.2. The uneven electron cloud distribution for the hydrogen chloride molecule.

If the electronegativity difference between the two bonded atoms reaches a certain high value, the more electronegative atom may take complete possession of the shared pair of electron and an ionic bond may result. Chemists generally agree that if this value is **greater than 1.7**, **the bond is ionic**, and if less, down to a value of **0.4,** the bond will be **polar covalent**.

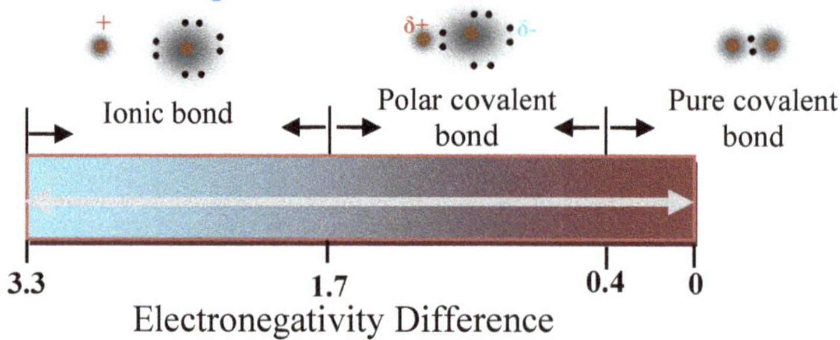

Figure 6.3. Electronegativity difference and bond polarity scale.

Decide if the following bonds are polar. K/U T/I

(a) H—F   (d) H—Br
(b) C—O   (e) H—I
(c) N—H   (f) Cl—F

Which of these would be the most polar molecule?

Diatomic Molecules
To determine whether a diatomic molecule is polar, the following three points must be considered:
1. **If the molecule is homonuclear and the bond is non-polar, the molecule is also non-polar.**
2. **If the molecule is heteronuclear and the bond is polar, the molecule is also polar**.
3. **If the molecule is heteronuclear and the bond is non-polar, the molecule is also non-polar.**

Determine whether the following diatomic molecules are polar or non-polar. K/U T/I
(a) H—F
(b) H—Br
(c) H—I
(d) Cl—F

6.4 Polyatomic Molecules
To know whether a polyatomic molecule is polar or non-polar, it is important to know the following:
1. Whether the bonds within the molecule are polar.
2. The geometry of the molecule. *In polyatomic molecules, the bonds within the molecule may be polar, but the molecule may still turn out to be non-polar.* The reason for this is due to the molecule's unique geometry.

Molecular Geometry

6.5 The VSEPR (Valence Shell Electron Pair Repulsion) Theory
According to this theory, the valence shell electron pairs on the **central atom** would repel each other as far away as possible to minimize the electrostatic force of repulsion between them.

What are valence shell electron pairs?

As can be recalled from your study of covalent bonding, it is the valence shell electrons on the central atom and those of the valence shells of peripheral atoms that are involved in bonding. Between the central atom and every peripheral atom, a pair of electrons is involved in bonding. **These bonding pairs of electrons** (bond pairs) along with any *lone pair of electrons on the central atom,* are referred to as the *valence shell electron pairs.*

Why do valence electron pairs repel each other?

As can be recalled from your study of covalent bonding, *covalent bonds are highly negative and would thus repel each other sideways,* **if placed close enough to one another other**. Lone pairs of electrons are also highly negative and would **repel each other,** as well as the other covalent bonds present next to them. This repulsion between covalent bonds and lone pairs of electrons would determine the directions in which the peripheral atoms in a molecule will be **with respect** to the central atom. This spatial orientation of the atoms would thus determine the geometry of the molecule. The methane molecule will be used to illustrate this concept.

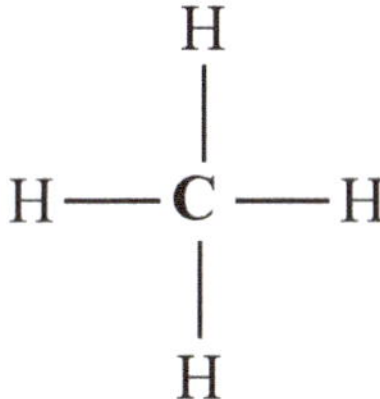

Figure 6.4. The structural formula for the methane molecule.

This molecule has four covalent bonds, or four valence shell electron pairs around the central carbon atom. Since each covalent bond is highly negative, we shall draw an electron cloud for each of the four bonds separately as follows.

Figure 6.5(a). The four covalent bonds that are formed in methane. **Figure 6.5(b).** Orbital of H and C

6.6 Molecular Geometry of Polyatomic Molecules

The AX₄ Type of Molecules

Since each of the four covalent bonds formed between hydrogen and carbon is highly negative, if they were to be placed around the central carbon atom, they would repel each other mutually until the electrostatic force of repulsion between each is as least as possible. The following figures show the arrangement of the atoms in a CH_4 molecule

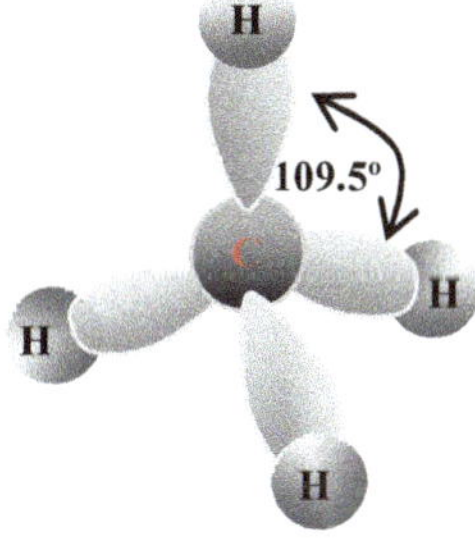

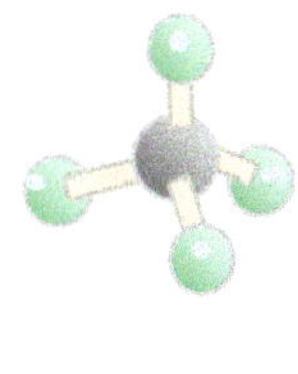

Figure 6.6(a). The space filling
model of the CH_4 molecule.

Figure 6.6(b). The ball and stick
model of the CH_4 molecule.

The geometry of this molecule will be that of a tetrahedron. The reason is because the four bonds formed, along with the four hydrogen atoms, exhibit tetrahedral arrangement around the central carbon atom. The hydrogen atoms, now called the **peripheral atoms**, are separated from each other by a bond angle of 109.5º.

When describing any molecule, the letter **A** is used to denote the **central atom** and the letter, **X** is used to indicate the **peripheral atom(s)**. Subscripts are used to indicate the relative number of atoms of A and X, respectively. The above molecule would thus be described as **AX₄**. *All AX₄ type of molecules would have tetrahedral arrangement of the electron pairs* ***and would have the geometry of a tetrahedron.***

The AX₃E Type of Molecules.

The ammonia molecule (NH₃) is an example of this type of molecule and its molecular geometry can be found as follows:

Step 1. Draw the Lewis dot diagram of molecule.

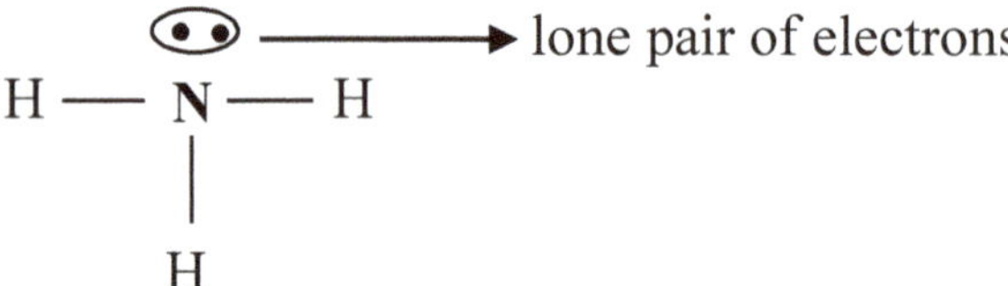

Figure 6.7. The Lewis dot diagram of the ammonia molecule (NH₃).

This molecule would be classified as **AX₃E**. The letter **E** is used to indicate a **lone pair** of electrons. Subscripts are also used to indicate the number of lone pairs.

Step 2. Use the geometry of an **AX₄** molecule that most closely resembles the NH₃ molecule and then modify it in a stepwise fashion to arrive at the **AX₃E** molecular geometry. The molecule used will be CH₄.

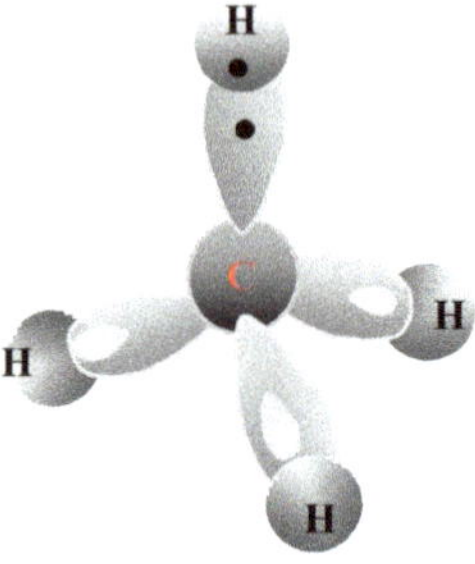

Figure 6.8. The methane molecule, showing one pair of electrons in one of its C-H bonds.

If one of the hydrogen atoms from this molecule is removed, it would have three hydrogen atoms like the NH₃ molecule. The bond would disappear but the pair of electrons forming it would now become a lone pair on the central nitrogen atom.

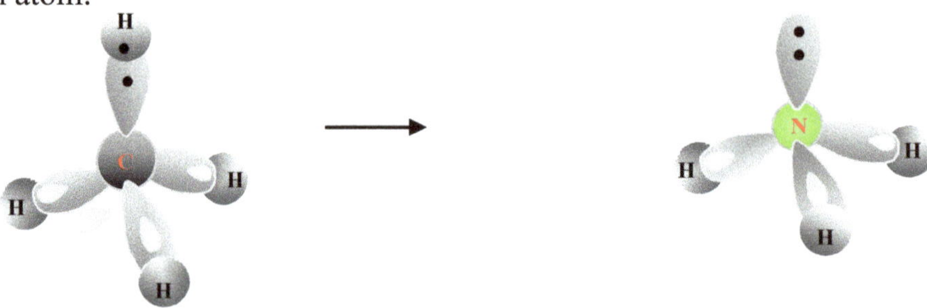

Figure 6.9. The formation of a lone pair of electrons on the central nitrogen atom.

If the bond finally collapses, we will end up with the molecular geometry of the NH_3 molecule as the following figure depicts:

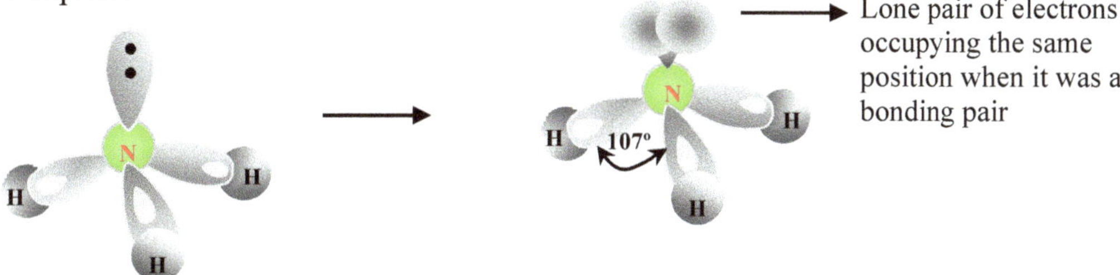

Figure 6.10. The final transformation of the bond into a lone pair of electrons.

It should be noted that the ***lone pair of electrons on the central nitrogen atom occupies the same position where a fourth bond would have been***, if the molecule were an AX_4 type. The electron pairs; bond pairs and lone pairs, would also have tetrahedral arrangement. However, the bond angles are less than 109.5° because the lone pair of electrons takes up more space, and also they are now closer to the three N—H bonds, pushing on them with a greater force of repulsion. Based on the appearance of space-filling model and other empirical evidence, the molecular geometry of NH_3 is found to have a **trigonal pyramid** conformation.

 A general rule is that in any AX_3E type of molecule, since the electron pairs *(bond pairs* **and** *lone pair)* *have tetrahedral arrangement around the central atom*, **the molecule will have the geometry of a trigonal pyramid.**

Figure. 6.11(a). The space-filling model of NH_3 with one lone pair'

Figure 6.11(b). The ball and stick model of the NH_3 molecule.

The AX_2E_2 Type of Molecules
The water molecule (H_2O) is an example of this type of molecule and its molecular geometry can be found as follows:

Step 1. Draw the Lewis dot diagram for the molecule.

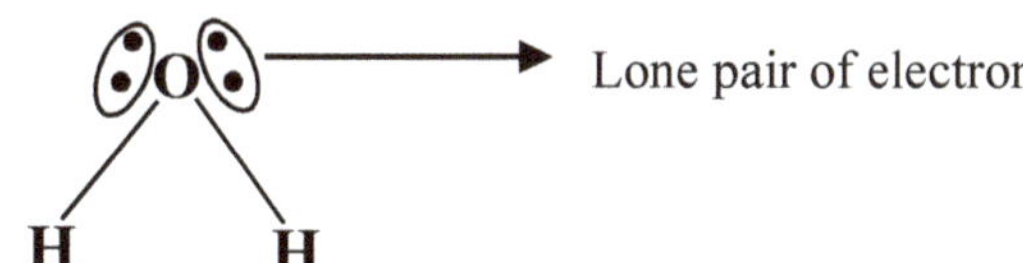

Step 2. Use the geometry of an **AX_3E** molecule that most closely resembles the H_2O molecule and then modify it in a stepwise fashion to arrive at the **AX_2E_2** molecular geometry. The molecule used will be NH_3.

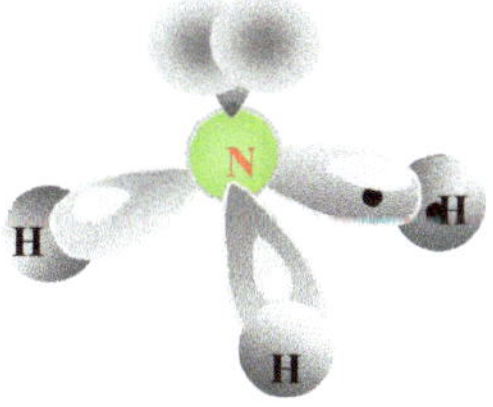

Figure 6.11. The ammonia molecule showing a pair of electrons in one of its bonds.

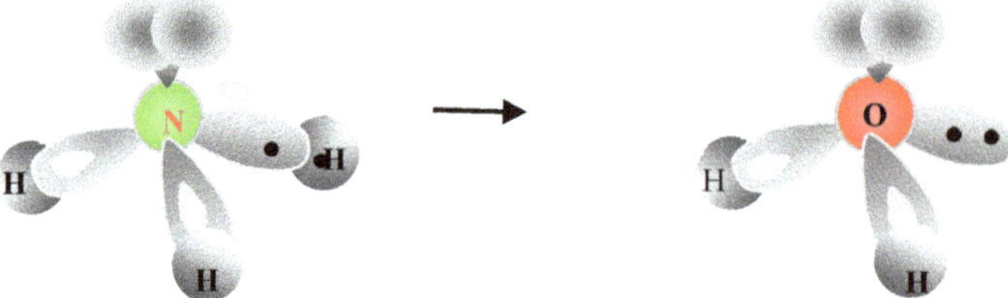

Figure 6.13. The formation of another lone pair of electrons on the central oxygen atom.

Now that one of the hydrogen atoms from this molecule is removed, the number that remains is what is present in the H_2O molecule. *The bond would disappear but the pair of electrons forming it would now become a second lone pair on the central oxygen atom.*

If the bond finally collapses, we will end up with the molecular geometry of the H_2O molecule as follows:

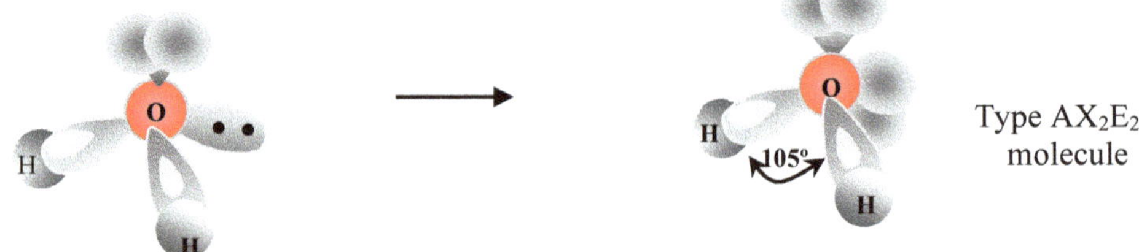

Figure 6.14. The final transformation of the bond into another lone pair.

It should be noted that the second lone pair of electrons on the central oxygen atom occupies the same position where a third bond would be, if the molecule were an AX_3E type. However, the bond angle is less than 105° because there are now two lone pairs of electrons very close to the other two O — H bonds, occupying even more space and pushing on them with an even greater force of repulsion than in the ammonia molecule. Based on the space-filling model and other empirical evidence, the molecular geometry of H_2O is found to have a **v-shape or bent** conformation.

A general ruleis that in any AX₂E₂ type of molecule, since the electron pairs (bond pairs and lone pairs) have tetrahedral arrangement around the central atom, the geometry will be V- shaped or bent.

(a)

(b)

Figure 6.15(a). The space filling model of the H_2O molecule.

Figure 6.15(b). The ball and stick model of the H_2O molecule.

The AX₂ Type of Molecules

The carbon dioxide (CO_2) is an example of this of molecule and its molecular geometry can be determined by following the following steps:

Step 1. Draw the Lewis dot diagram for the molecule.

$$\ddot{\underset{..}{O}} = C = \ddot{\underset{..}{O}}$$

The CO_2 molecule is different from all the others that have been studied so far. The following should be noted about molecules like this:

- It has multiple bonds, in this case, double bonds.
- **Multiple bonds are considered single bonds in determining molecular geometry.**

Since multiple bonds are considered single bonds, this molecule can be considered to have only two bonds. Also, since the central atom has no lone pairs, the molecule is classified as an AX_2 type. The logical geometry of this molecule would be **linear** as the bonds would repel each other as far as possible sideways with bond angles of 180°.

A general rule is that in any AX_2 type of molecule the valence electron pairs on the central atom are arranged linearly and its geometry would be linear.

Step 2. Draw the bonds around the central carbon atom.

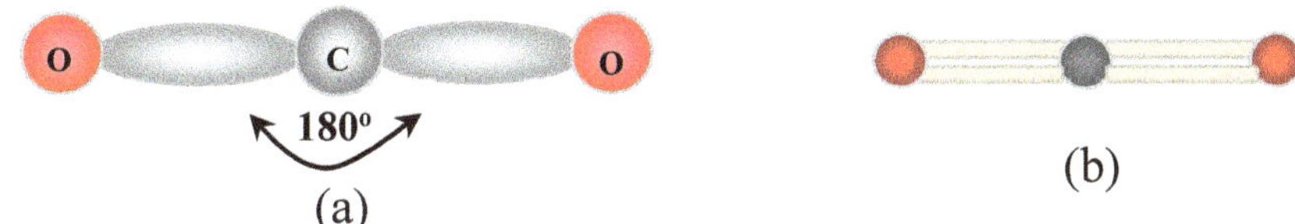

Figure 6.16a. The space filling model of the CO_2 molecule

Figure 6.16b. The ball and stick model of the CO_2 molecule

The AX₃ Type of molecules

The boron trihydride (BH_3) is an example of this of molecule and its molecular geometry can be determined by following the following steps:

Step 1. Draw the Lewis dot diagram for the molecule.

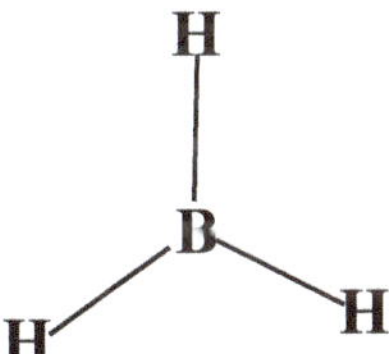

Notice that the element boron has only three electron pairs around itself in this molecule. This means it does not have the octet. The reason for this is because *boron is an electron deficient element.* This molecule is classified as an AX_3 type.

Since there only three bonds **with no lone electron pairs**, the electron pairs would have a trigonal planar arrangement around the central atom as illustrated below. Its molecular geometry would also be **trigonal planar.**

Step 2. Draw the three bonds around the central boron atom each with a bond angle of 120°.

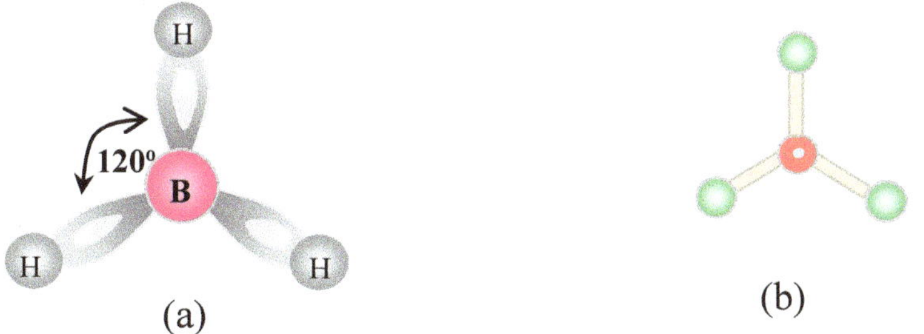

Figure 6.17(a). The space filling model of the BH_3 molecule.

Figure 6.17(b). The ball and stick model of the BH_3 molecule.

A general rule is that in any AB_3 molecule, since the electron pairs (only bond pairs) have trigonal arrangement around the central atom, the molecule would also have a trigonal planar geometry.

 Polarity of Polyatomic Molecules

Dipole moment: *This is a measure of the degree of charge separation in a covalent bond.* Diatomic molecules such as HCl, for example, are polar because of the charge separation that takes place in their bonds.

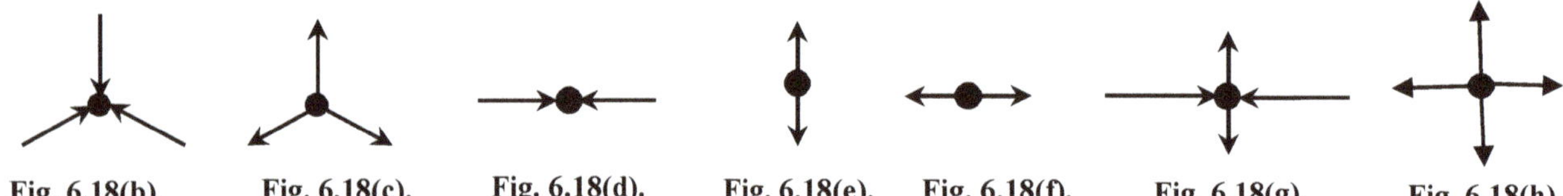

Figure 6.18(a). The bond dipole of the HCl molecule.

In the above example, the partial charges gained are represented by the, $\delta+$ and $\delta-$ symbols. The bond dipole is indicated by an **arrow with a positive sign** at the end opposite to its head. The arrow is made to point away from the partial positive charge towards the partial negative charge. Dipole moment is a vector quantity; it has both magnitude (the degree of charge separation) and direction.

How to determine if a polyatomic molecule is polar or non-polar
A bond dipole can be likened to a vector. If the peripheral atoms are more electronegative than the central atom, we can say that each peripheral atom is exerting a pulling force on the central atom. If the central atom is more electronegative than the peripheral atoms, then we can say that the central atom is exerting a pulling force on the peripheral atoms. The resolution of these forces on the central atom can be used to determine if the molecule as a whole, is polar or not. To do this, the following highlighted points must be considered:

➤ **If the resultant force on the central atom is zero, the molecule will be non-polar.** In this case, there is no resultant dipole. The following vector diagrams are used to illustrate this point. A black dot is used to represent the central atom and lines are used to indicate the pull or push force on it. The lengths of the lines represent the magnitude of the forces (degrees of charge separation).

Fig. 6.18(b).   Fig. 6.18(c).   Fig. 6.18(d).   Fig. 6.18(e).   Fig. 6.18(f).   Fig. 6.18(g).   Fig. 6.18(h).

According to each of the vector diagrams above, the resultant force on the black dot is zero. The forces are equal and their unique directions make this possible. In real molecules the bond dipoles would cancel out and the molecules would non-polar.

➤ **If the resultant force on the central atom is greater than zero, the molecule will be polar.** In this case there is a net shift of the electron clouds in one direction with a resultant dipole.

Fig. 6.18(i).   Fig. 6.18(j).   Fig. 6.18(k).   Fig. 6.18(l).

In every case above the resultant force on the dot is greater than zero. The directions of the resultant forces are indicated by red arrows. This is analogous to what happens in molecules where the central atom is surrounded by peripheral atoms with differences in electronegativity or when the molecules are non-symmetrical. In real molecules, the bond dipoles would not cancel out and the molecules would be polar.

Polarity of AX₄ Type of Molecules
Using carbon tetrachloride as an example, we have the following arrangement of the dipoles in its molecule

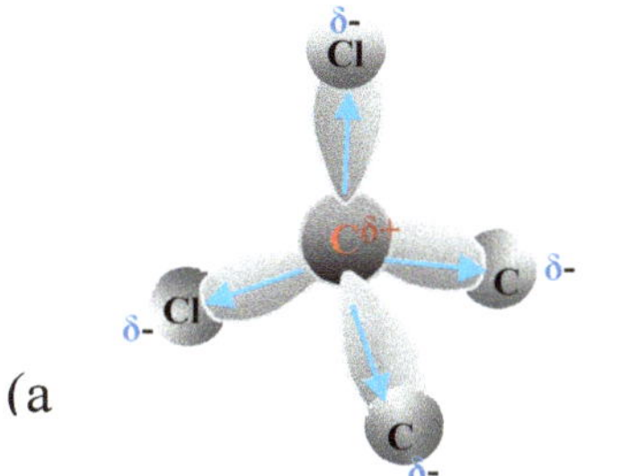
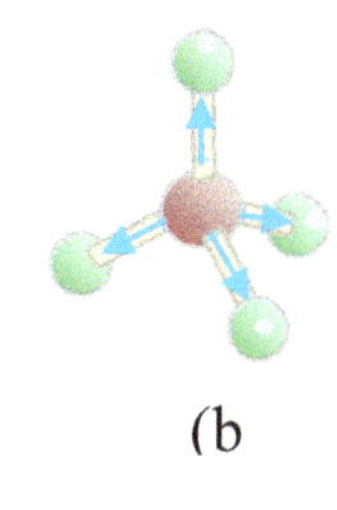

Figure 6.19(a). Arrangement of the dipoles in the CCl₄ molecule.

Figure 6.19(b). Ball and stick model showing no net dipole in the CCl₄

The bonds in the CCl_4 molecule are all polar as the electronegativity difference between carbon and chlorine is 0.5. Each chlorine atom in the C-Cl bond would have partial a positive charge (δ-) and the carbon atom would have partial negative charge (δ+). However, because of the symmetry of the molecule, where each of its four bonds is having the exact bond angle of 109.5°, the bond dipoles cancel, leaving the molecule with no net dipole. There is no resultant force on the central atom and as such there is no net electron shift; the molecule remains non-polar.

However, if one of the chlorine atoms is replaced by a more electronegative atom such as fluorine, the molecule becomes polar.

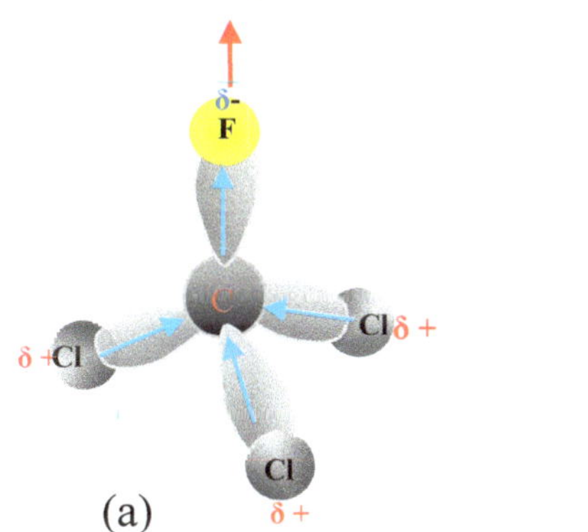

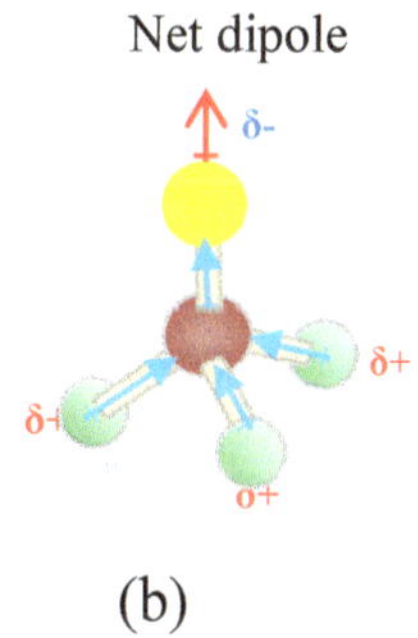

Figure 6.20(a). Arrangement of the dipoles in the CCl₃F molecule.

Figure 6.20(b). Ball and stick model showing a net dipole in CCl₃F molecule.

In CCl_3F molecule there is net dipole in the direction of the fluorine atom. This is because of the higher relative electronegativity of the fluorine atom when compared to the other atoms. Thus, there is a net shift of the electron cloud in the direction of the fluorine atom away from the others, making the molecule polar. A partial negative charge is acquired by the fluorine atom and a small positive charge is acquired by each of the four chlorine atoms.

Polarity of AX₃E Type of Molecules

Using ammonia as an example, we have the following arrangement of the dipoles in its molecule.

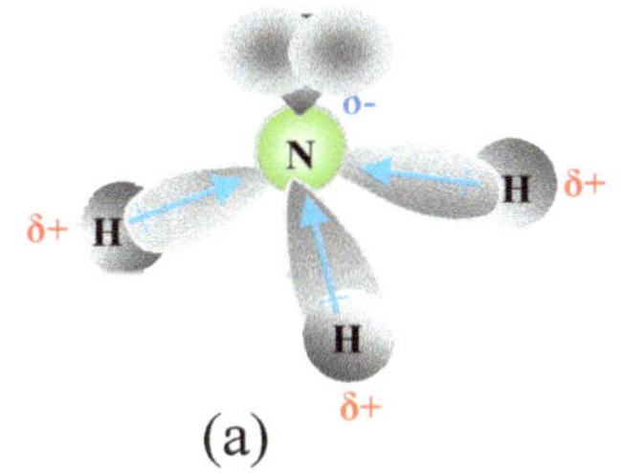

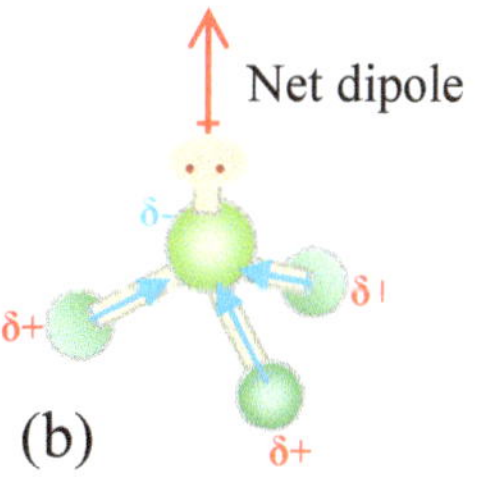

Figure 6.21(a). Arrangement of the dipoles in the NH₃ molecule.

Figure 6.21(b). Ball and stick model showing net dipole in the NH₃ molecule.

Unlike the CCl_4 molecule, where there are four bond dipoles opposing the other and nullifying their effects, it is different in the NH_3 molecule; one bond dipole is absent. In the NH_3 molecule, there are three bond dipoles in the direction of the more electronegative nitrogen atom. There is, therefore, a net dipole in the direction of the nitrogen atom; there is a net shift of the electron cloud in the direction of the nitrogen atom away from the others, making the molecule polar. A partial negative charge is acquired by the nitrogen atom and a small positive charge is acquired by each of the three hydrogen atoms.

Polarity of AX_2E_2 Type of Molecules

Using water as an example, we have the following arrangement of the dipoles in its molecule.

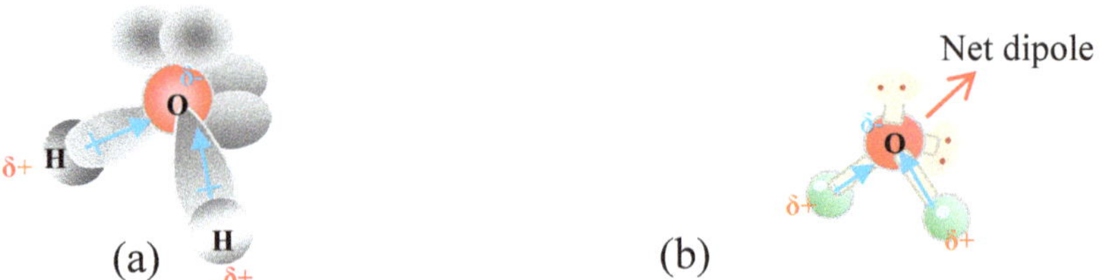

Figure 6.22(a). Arrangement of the dipoles in H_2O molecule.

Figure 6.22(b). Ball and stick model showing net dipole H_2O molecule.

Again, unlike the CCl_4 molecule, where there are four bond dipoles opposing the other and nullifying their effects, it is different in the water molecule; two bond dipoles are missing. In this molecule, the two bond dipoles present are in the direction of the oxygen atom. There is therefore a net dipole in the direction of the oxygen atom; there is a net shift of the electron cloud in the direction of the oxygen atom away from the others, making the molecule polar. A partial negative charge is acquired by the oxygen atom a small positive charge is acquired by each of the two hydrogen atoms.

Polarity of AX_2 Type of Molecules

Using carbon dioxide as an example, we have the following arrangement of the dipoles in its molecule.

Figure 6.23(a). Arrangement of the dipoles in CO_2 molecule.

Figure 6.23(b). Ball and stick model showing no net dipole in CO_2 molecule.

In the CO_2 molecule, there are two dipoles of equal magnitude pointing in opposite directions. As a result of this, they cancel each other leaving the molecule with no net dipole. Since each oxygen atom is pulling more of the electron cloud away from the central carbon atom in exactly opposite directions, their effects cancel out. No partial negative charge would remain on either oxygen atom nor would any partial positive remain on the central carbon atom. The molecule thus remains non-polar.

Polarity of AX₃ Type of Molecules

Using boron trifluoride as an example, we have the following arrangement of the dipoles in its molecul

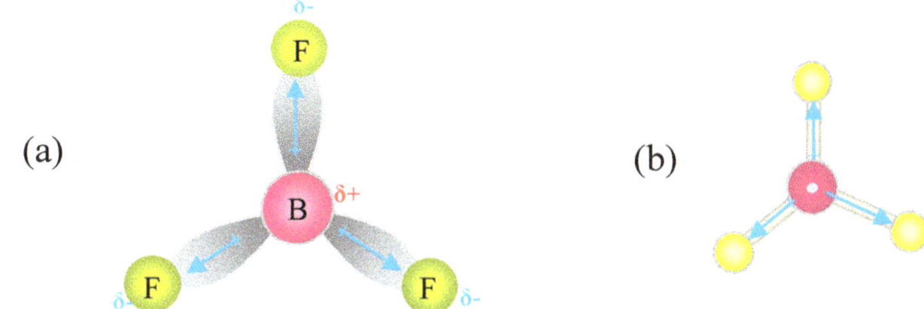

Figure 6.24(a). Arrangement of the dipoles in BF₃ molecule.

Figure 6.24(b). Ball and stick model showing no net dipole in BF₃ molecule.

In the BF₃ molecule, there are three dipoles of equal magnitude each pointing away from the central boron atom at an angle of 120°. As a result of this, they cancel each other leaving the molecule with no net dipole. Since each fluorine atom is pulling more of the electron cloud away from the central boron atom at an angle of 120°, their effects cancel out. No partial negative charge would remain on the fluorine atom, nor would any partial positive charge remain on the central boron atom; the molecule thus remains non-polar.

However, if one or two of the peripheral fluorine atoms are replaced by an atom of lower electronegativity such as hydrogen, the molecule becomes polar. The following figures depict how this happens when two fluorine atoms are replaced by two hydrogen atoms.

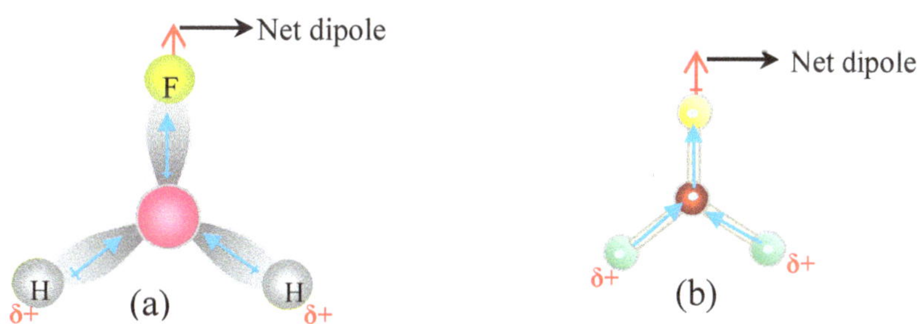

Figure 6.25(a). Arrangement of the dipoles in BH₂F molecule.

Figure 6.25(b). Ball and stick model showing net dipole in BH₂F molecule.

In the BH₂F molecule, the two B-H bonds are non-polar because there is an electronegativity difference of only 0.1. However, because only the B-F bond is polar, the **molecule has only one dipole, which is pointing away from the other atoms, towards the fluorine** atom. The result is the lone fluorine atom pulling more of the electron cloud towards itself away from the other atoms. The fluorine atom acquires a small negative charge, and each of the other two hydrogen atoms acquires a partial positive charge making the molecule polar.

Table 6.2. Summary of all concepts in the VSEPR theory.

Type of molecule	Lewis structure	Bonded atoms (number)	Lone pairs (number)	Arrangement of electron pairs	Molecular geometry	Bond angles	Examples
AX_4	CH_4	4	0	Tetrahedral	Tetrahedron	$109.5°$	CCl_4, CH_4 SiH_4, CBr_4
AX_3E	NH_3	3	1	Tetrahedral	Trigonal pyramidal	$107°$	NH_3, PH_3 PCl_3, PBr_3
AX_2E_2	H_2O	2	2	Tetrahedral	V-shaped or bent	$105°$	H_2O, H_2S OCl_2
AX_3	BH_3	3	0	Trigonal planar	Trigonal planar	$120°$	BH_3, BF_3 $AlCl_3$
AX_2	$O=C=O$	2	0	Linear	Linear	$180°$	CO_2, $BeCl_2$

Exercise 6.3

Table 6.3. Complete the following table. K/U T/I

Compounds	Lewis dot Structure	A_xX_y Type	Electron Arrangement	Molecular Geometry	Bond Angle	Polarity
SiH_4						
PCl_3						
H_2S						
$BeCl_2$						
AlF_3						
SO_2						
NO_3^-						

1. The following is a reaction that takes between BF_3 and NH_3.

$$BF_3 \quad + \; \overset{\bullet\bullet}{}NH_3 \quad \longrightarrow \quad BF_3NH_3$$

Note that boron is an electron deficient atom and nitrogen has an available lone pair of electrons.
a) Draw the structural diagrams of the three molecules above. K/U

b State the molecular geometry of the BF_3 and NH_3 molecules. K/U T/I

c) What new shapes do the bonds around the B and N atoms take in the BF_3NH_3 molecule, if any? K/U T/I

d What type of bond is formed between the N and B atoms? K/U T/I

2. H_2S is heavier than H_2O, yet it is a gas and water is a liquid at room temperature. Explain the difference in states of these two compounds using suitable diagrams, electronegativity values, polarity and intermolecular forces. K/U T/I C

3. The I-Cl and Br-Br molecules have the same number of electrons, but I-Cl has a higher boiling point than Br-Br. Explain why this is so. K/U C

4. Which of these molecules: CH_4 or CH_3Cl would have a higher boiling point? Explain your answer using suitable diagrams. K/U C

5. For the following molecules, draw their structural diagram, decide which is polar or non-polar and indicate the direction of the net diplole. K/U

a) $BCl_3 \longrightarrow$ b) $O F_2 \longrightarrow$ c) $BeI_2 \longrightarrow$ d) PBr_3

6.9   Intermolecular Force of Attraction

Intermolecular force of attraction refers to the force of attraction ***between*** molecules. This should not be confused with *intramolecular* force of attraction that refers to the force of attraction ***between the atoms that make up the molecule.***

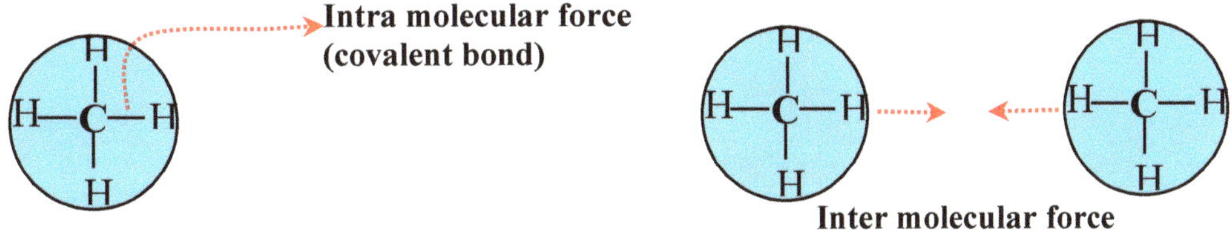

Figure 6.26. The difference between intermolecular and intramolecular forces.

Intramolecular force of attraction is strong because the covalent bonds are relatively strong. However, intermolecular force of attraction is relatively weak. The reason for this is due to the fact that the molecules are neutral, and they **do not have** whole positive and negative charges to attract each other strongly, as happens in ionic compounds.

Intermolecular forces of attraction are of two types, and together, they are called van der Waals forces, after Johannes van der Waals. In 1873, Johannes van der Waals attributed the attraction of molecules of gases to each other as a deviation from the ideal gas model. He believed that gas molecules themselves have infinitesimally small volumes, compared to the volume of space that they occupy, and they attract each other. The two types of forces of attraction between molecules are; *dipole-dipole forces* and *London dispersion forces.*

Polar molecules, as can be recalled, have *bond dipoles,* and if two of these molecules are close enough together, they will attract each other. This happens when the partial positive charge on the end of one molecule is attracted to the partial negative charge on one end of a neighboring molecule. The attraction between the dipole of one molecule and the dipole on other molecules is called *dipole-dipole force*. This concept is illustrated below between two polar molecules. It shows how the partial **partial positive charge** on molecule **1** is attracted to the **partial negative charge** on neighbouring molecule **2**.

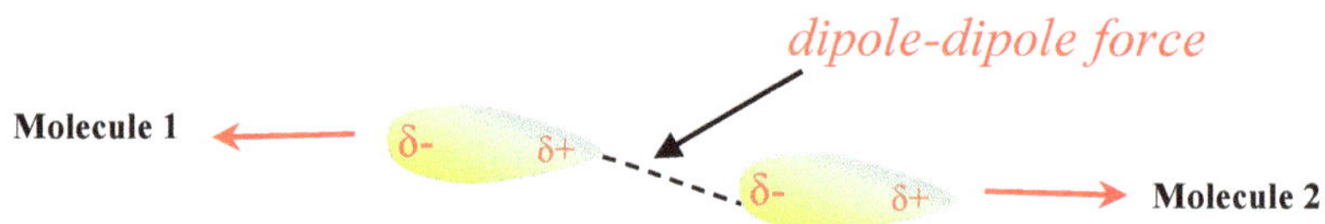

Figure 6.27. Dipole-dipole attraction between two polar molecules.

In solids, the molecules align themselves in a regular pattern so the positive end of one is much closer to the negative end of the other, than what happens in liquids.

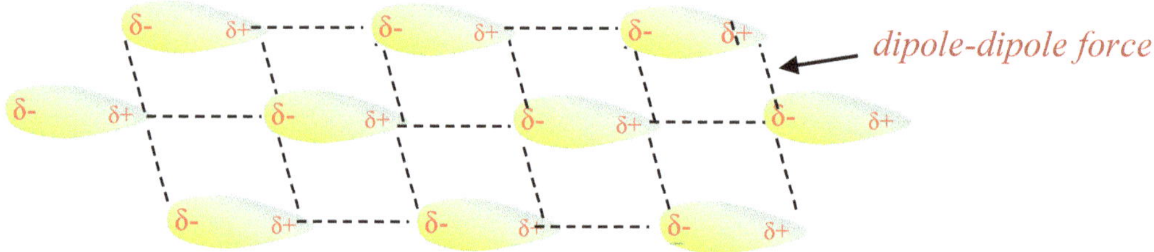

Figure 6.28. The arrangement of dipoles in a solid.

In liquids, because the molecules are in random motion, the dipoles are not aligned in any regular pattern. The dipole-dipole forces of the molecules in the liquid state are thus weaker than those in the solid state. Because of the stronger intermolecular forces of attraction in solids, the boiling and melting points of these solids would be expected to be higher than for liquids.

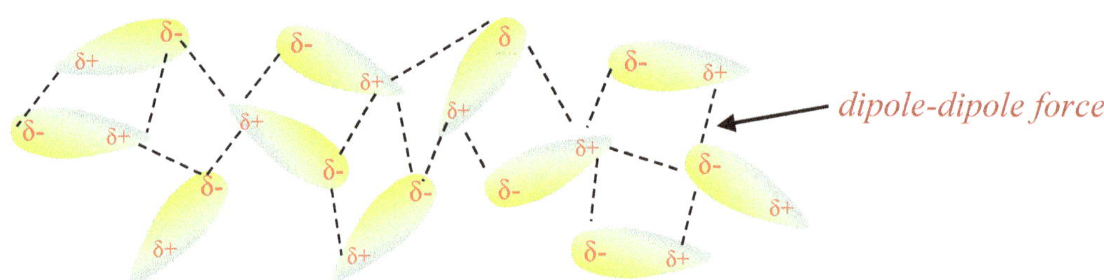

Figure 6.29. The arrangement of dipoles in a liquid.

Empirical evidence shows the non-polar molecules such as $CH_{4(g)}$ and $CO_{2(g)}$, when cooled to very low temperatures, become liquid and solid respectively. These observations indicate that there is considerable attraction between their molecules at low temperatures. Since these molecules do not have any charges on them to attract each other as happens in polar molecules, what could account for their attraction? The answer was provided by the German physicist, Fritz Wolfgang London (1900-1954). He explained that since the electrons in molecules are in constant motion, if they were to be frozen at any time, it would reveal that their distribution over the molecules would be nonsymmetrical. This indicates the electrons swarm more to one side of any molecule than the other at any given time. The tendency of electrons in a molecule to do this describes its *polarizability*. As a result of this constant non symmetrical distribution of electrons over a molecule, an *instantaneous dipole* is created on the molecule; a small negative charge is created on one end of the molecule and a small positive charge is created on the oppositive end. This is illustrated in the following figures:

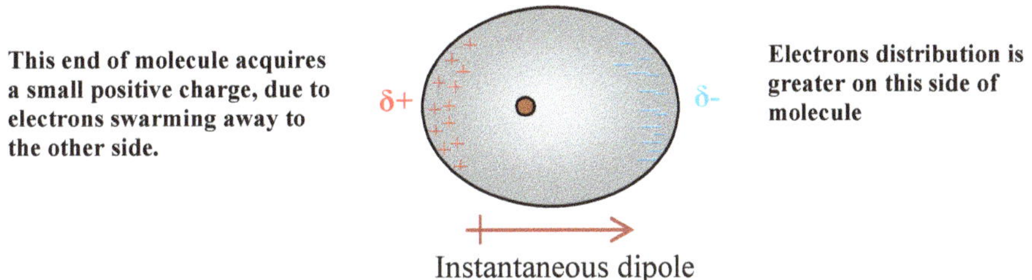

Figure 6.30.   The instantaneous dipole on a non-polar molecule.

According to Fritz London, the instantaneous dipole on one molecule can induce a dipole on a neighboring molecule making the two molecules attract each other. The figure below shows how the instantaneous dipole on molecule **1** induces a dipole on molecule **2**; the negative end of molecule 1 repells electrons on molecule 2 to it opposite end but induces a small positive charge next to it. Because the charges facing each other on the two molecules are now oppositive, the molecules attract each other.

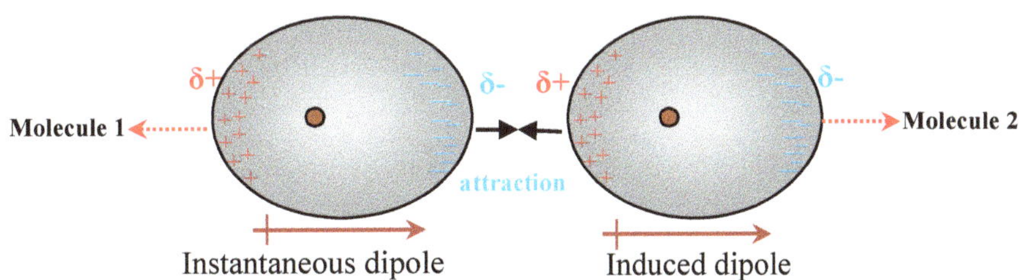

Figure 6.31. London dispersion force between two non-polar molecules.

This attraction between molecules, that results from an instantaneous dipole on one end of a molecule inducing a dipole in a neighboring molecule is called **London dispersion forces.** These are forces are very short-lived, appearing and disappearing continually with lightning speed.

Intermolecular Force of Attraction and Boiling Points

The following table shows how the boiling points of non-polar molecules vary as their molar masses change.

Table 6.3. Boiling point variations in non-polar molecules with change in molar mass.

Molecules	Molar mass g/mol	Boiling Point °C	Number of electrons in molecule
Iodine (I_2)	253.8	184.2	106
Bromine (Br_2)	149.9	58.8	70
Carbon dioxide (CO_2)	44.01	-34.6	24
Methane (CH_4)	16.5	-162.5	16
Hydrogen (H_2)	2.02	-257.8	2

The molecules in Table 6.3 are all non-polar, so their only intermolecular force of attraction is London dispersion forces. The table also shows that as the molar mass of the molecules increases so do their boiling points. This means that their intermolecular force of attraction increases accordingly also. The explanation for this is that as the molecules increases in mass so do their number of electrons. *As the number of electrons increases, a larger number of them are at a distance farther away from the nucleus. This makes it easier for them to keep shifting from side to side of a molecule, creating a greater instantaneous dipole.* When this happens, the molecule is said to be more **temporarily polarized**. This greater *polarizability* in a molecule results in a greater induced dipole on a neighboring molecule with greater intermolecular attraction.

The trends in boiling points of non-polar molecules can therefore be predicted, based on the relatively number of electrons present in the molecules and also their sizes and shapes. Linear molecules having the same number of electrons as spherical molecules would have higher boiling points. This is because the linear molecules would share greater surface with each other than spherical molecules do, and this creates stronger London dispersion forces.

London Dispersion Forces in Polar Molecules

The following table shows how the boiling points of some hydrogen halides vary with increasing molar mass.

Table 6.4. Boiling Points of Some Hydrogen Halides

Compound	Electronegativity Difference	Boiling Point (°C)	Bond Polarity	London dispersion Forces
$HI_{(g)}$	0.4	-36		
$HBr_{(g)}$	0.7	-67°C		
$HCl_{(g)}$	0.9	-85°C		

According to the information given in the table above, all of the compounds are diatomic and polar, based on electronegativity differences. Based on what we know of attraction due to dipole-dipole forces, the greater the bond polarity, the greater should the force of attraction between the molecules be. Also, the greater the force of attraction between the molecules of a compound, the higher should be its boiling point. For the compounds above, however, this is not the case. We see that boiling point *decreases* with *increase in bond polarity*. *This anomaly* can be explained in terms of **London dispersion forces.** For the compounds above, as the bond polarity decreases, there is a corresponding increase in molar mass and number of electrons in their molecules. As a result of these, there is a significant increase in intermolecular attraction due to London dispersion forces. It can therefore be concluded that *polar molecules have both London dispersion forces and dipole-dipole forces.*

6.12 **Hydrogen Bonding**

Hydrogen bonding is a special type of attractive force between the hydrogen atom covalently bonded to a highly electronegative atom, such as oxygen, nitrogen or fluorine, and a lone pair of electrons on a highly electronegative atom on the same molecule or a neighboring molecule. The following figure shows how hydrogen bonding takes place between water molecules.

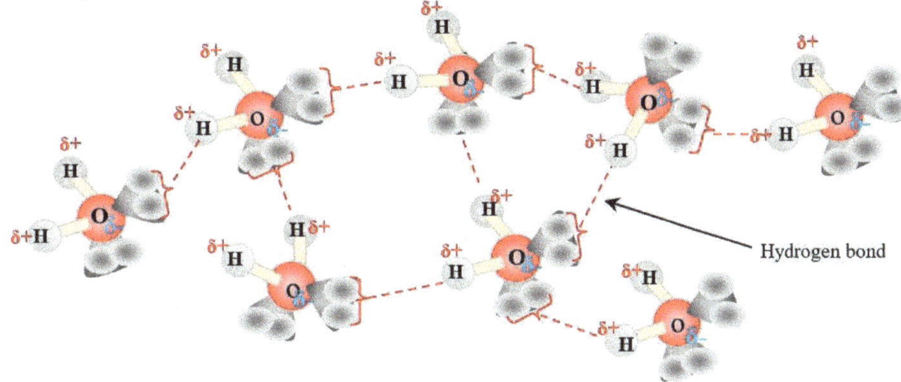

Figure 6.32. Hydrogen bonding in water molecules

Because the oxygen atom is highly electronegative compared to the hydrogen atom, the O — H bond is highly polar. The hydrogen atom which has a **small positive charge ($\delta+$),** is now attracted to **one lone pair of electrons** on the **oxygen atom** of a neighboring molecule. The same may happen between the hydrogen and other atoms having lone pairs of electrons such as **nitrogen** and **fluorine.** The following molecules depict this.

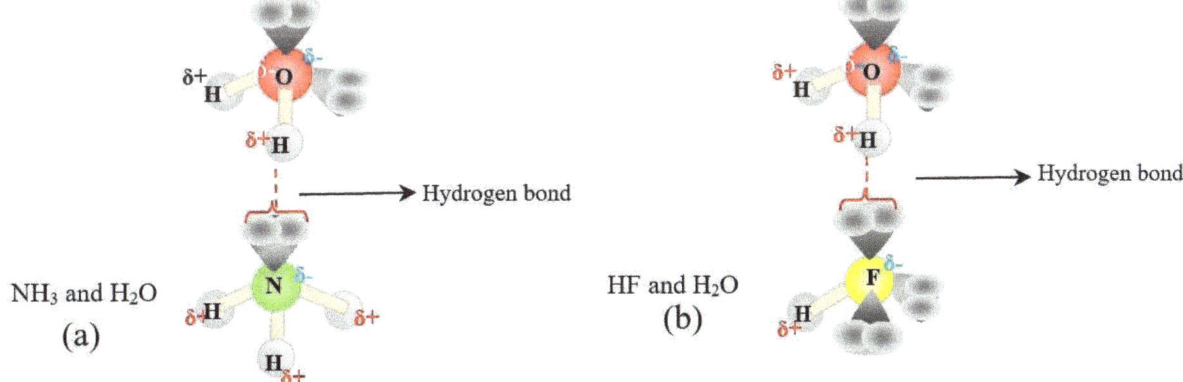

Figure 6.33. Hydrogen bonding between ammonia and hydrogen fluoride with water.

Evidence for Hydrogen Bonding

Comparison of the boiling points between the following covalent hydrides of Group IVA elements is as follows:

Table 6.5. Trends in boiling point and molar mass for some hydrides.

Compounds	Trends in Boiling Points	Trends in Molar Mass
CH_4		
SiH_4	Increase ↓	Increase ↓
GeH_4		

Here we see, as expected, the greater the molar mass of a hydride the higher is its boiling point. The difference in boiling points is attributed to the increase in London dispersion forces. *There is no hydrogen bonding in these molecules **since there are no lone pairs of electrons on the molecules.***
However, for the following hydrides, in the table below, there is a remarkably higher boiling point for the first hydride in this group of the periodic table (VIA). There is a difference of 160 °C in boiling point between H_2O and H_2S even though the molar mass of H_2S is greater than that of H_2O. This is so because there is hydrogen bonding in H_2O but not in H_2S, H_2Se nor H_2Te. The reason for this is due to the fact that the S, Se and Te atoms in H_2S, H_2Se and H_2Te molecules respectively, are **neither small** nor **sufficiently electronegative** to bring about hydrogen bonding.
*Note. Even though hydrogen bonding is a relatively strong intermolecular force of attraction; it is approximately **only 5% as strong as a single covalent bond**.*

Table 6.6. Trends in boiling point and molar mass for the first hydrides in their group.

Hydrides	Boiling Points (°C)	Molar Mass g/mol
H_2O	100	
H_2S	-60	
H_2Se	-40	Increase ↓
H_2Te	-1	

Increase ↑ (Boiling Points column)

Exercise 6.5

1. List the following molecules in order of **increasing** boiling points. K/U T/I

   (a) C_4H_{10} (b) CH_4 (c) C_3H_8 (d) C_2H_6

2. List the following molecules in order of **increasing** boiling points. K/U T/I

   (a) Br_2 (b) ICl

 Testing for Polarity

To distinguish between a liquid that is polar from that which is not polar, a simple test can be done. A stream of the liquid is placed near to charged rod and observations are made. If the stream of liquid is attracted to the charged rod, then it can be concluded that it is polar, if not, it is non-polar. The following figures illustrate the results of these tests.

Observation:

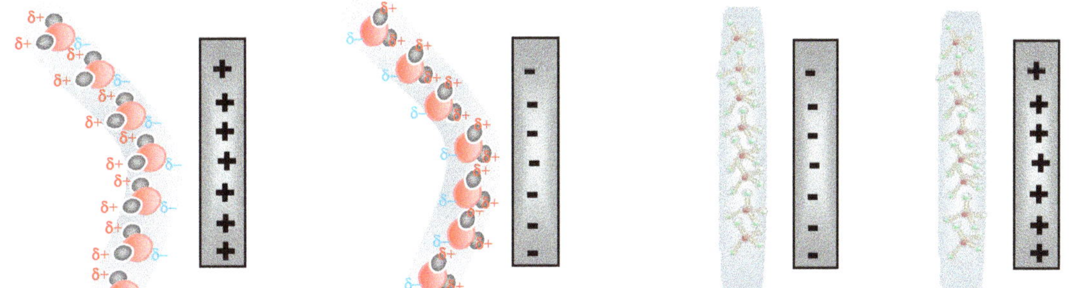

Figure 6.34(a). A stream of water being bent by both negatively and positively charged rods.

Figure 6.34(b). A stream of carbon tetrachloride not affected by either charged rods.

In figure **6.34(a),** above, the polar molecules in water are attracted to both of the charged rods. This happens because the molecules are able to change their orientations, so that they face the charged rods with their ends that have charges opposite to the rods. Because of this, the stream of water was attracted to the rods. When the rod was positive, the water molecules faced it with their negative ends; the opposite happens when the rod was negative.

This would not happen in liquids that are non-polar; no charges are present in them to be attracted to the charged rods.

Analysis:

1. How were the water molecules oriented when tested with the different rods? Use the laws of electrostatics to explain why the streams of water were attracted to both rods. K/U T/I

2. Why weren't the streams of carbon tetrachloride affected by the charged rods? K/U T/I

Exercise 6.6

1. Draw the structural diagrams for the following molecules and then decide whether or not if a stream of each would be affected if placed next to a charged rod: K/U T/I

a) $C_6H_{6(l)}$ b) $Br_{2(l)}$ c) $CH_3COOH_{(l)}$ d $CH_3OH_{(l)}$

CHAPTER REVIEW BONDING (5-6)

Matching

Match each term in the table below with the correct statements that follow

A	Octet	G	Instantaneous dipole
B	Electron cloud	H	Bonding capacity
C	Electric field	I	Polyatomic ion
D	Electronegativity	J	Coordinate covalent bond
E	Ionic bond	K	Heteronuclear
F	Polar	L	Dipole-dipole

1.	This is produced by non-polar molecules as they polarize.
2.	This is used to describe diatomic molecules with two different atoms.
3.	This results when electrons are transferred from a metal to a non-metal.
4.	The force of attraction that is present in polar molecules.
5.	The number of covalent bonds an atom forms to complete its octet.
6.	The volume of space around an atom or in a molecule where there is a probability of finding electrons.
7.	An arrangement of eight electrons in the outer shell of an atom or ion.
8.	A covalent bond where the two atoms have an electronegativity difference of more than 0.4.
9.	A measure of the ability of an atom in a molecule to draw bonding electrons to itself.
10.	A covalent bond that is formed one of the two bonded atoms contributes both electrons.
11.	A number of non-metal atoms bonded together and having a charge.
12.	This consists of electric field lines that are responsible for attracting opposite charges.

True or False

Read each of the following statements and then decide if it is *True* or *False*

1.	Atoms of elements combine with each other to achieve an inert gas electron configuration.
2.	To gain an inert gas electron configuration, non-metals lose electrons.
3.	When two or more nonmetals react they form molecular compounds.
4.	The bond formed when a metal reacts with a non-metal is ionic.
5.	Covalent bonds are formed when electrons are transferred from one atom to another.
6.	Covalent bonds are very weak.
7.	A compound dissolves in water, so it must be ionic.
8.	The smallest unit of an ionic compound is a molecule.

9.	London Dispersion Forces are found in all molecular compounds.
10.	The larger the molecule the weaker is the London Dispersion Force.
11.	Ionic compounds dissolve in water to form electrolytes.

Multiple Choice

Choose the letter that best answers questions:

1. Atoms of element attain an inert gas electron configuration by

a	sharing electrons	*d*	all of the above
b	gaining electrons	*e*	none of the above
c	losing electrons		

2. To attain an inert gas electron configuration, atoms of metallic elements

a	share electrons	*d*	all of the above
b	gain electrons	*e*	none of the above
c	lose electrons		

3. The bonding capacity of nitrogen is

a	1	*d*	4
b	2	*e*	5
c	3		

4. The valence attained by an ion depends on

a	the number of electrons an atom gains	*d*	a and b
b	the number of electrons an atom loses	*e*	none of the above
c	the number of electrons an atom shares		

5. The formula for Iron(III) chloride is $FeCl_3$. In forming this compound,

a	the iron atom loses 3 electrons	*d*	all of the above
b	each chlorine gains one electron	*e*	a and b
c	only the chlorine atoms the octet		

6. Covalent bonds are formed by reaction

a	between metals and non-metals	*d*	between non-metals and metalloids only
b	between non-metals only	*e*	none of the above
c	between metals only		

7 A covalent bonds is said to be negatively charged because

a	in its formation one atom becomes negatively charged
b	a high concentration of negative cloud builds up between the two bonded atoms
c	a negative charge appears on the more electronegative atom
d	a and c
e	none of the above

8. The number of **electron pairs** around the central nitrogen atom in NH_3 is?

a	4		*d*	1
b	3		*e*	None of the above
c	6			

9. In the H_2CO_3 molecule, the central carbon atom is surrounded by

a	2 single bonds and 1 double bond		*d*	2 double bonds
b	3 single bonds		*e*	None of the above
c	4 single bonds			

10. Which of the following compounds is **not** ionic?

a	NaI		*d*	KI
b	HI		*e*	MgI_2
c	LI			

11. In which of the following compounds is the bond most polar?

a	HI		*d*	HBr
b	HCl		*e*	H_2S
c	HF			

12. The bonds in the CCl_4 molecule are more polar than in the NH_3 molecule, yet NH_3 is polar but CCl_4 is not. The reason for this is due to the fact that

a	CCl_4 molecule has no residual dipole but the NH_3 molecule has
b	the molecular geometry of CCl_4 is that of a tetrahedron while that of NH_3 is a trigonal pyramid
c	nitrogen is more electronegative than chlorine
d	a and b only
e	none of the above

13. The arrangement of the valence shell electron pairs in the PH_3 is?

a	trigonal planar		*d*	v-shaped
b	tetrahedral		*e*	None of the above
c	trigonal pyramidal			

14. The shape of the SO_2 molecule is

a	linear	d.	Tetrahedral
b	v-shaped	e.	none of the above
c	trigonal planar		

15. Which of the following compounds is polar?

a	$CO_{2(g)}$	c.	PCl_3
b	CH_4	d.	BF_3
c	SiH_4		

16. H_2O is an example of a

a	AX_4 type of molecule	d.	AX_1E_3 type of molecule
b	AX_3E_1 type of molecule	e.	None of the above
c	AX_2E_2 type of molecule		

17. Which of the following compounds has the highest boiling point?

a	CH_4	d.	C_2H_6
b	C_3H_8	e.	C_5H_{12}
c	C_4H_{10}		

18. London Dispersion Forces (L.D.S.) are associated with

a	electrostatic force	d.	instantaneous dipoles
b	permanent dipoles	e.	both c and d
c	induced dipoles		

19. Carbon dioxide is 2.44 times as heavy as water; yet it is a gas and water is a liquid at STP. The reason for this is due to the fact that

a	$CO_{2(g)}$ molecules are attracted by only weak London dispersion forces
b	$H_2O_{(l)}$ molecules are attracted only by stronger London dispersion forces
c	$H_2O_{(l)}$ molecules are attracted by both London dispersion forces and hydrogen bonds
d	$H_2O_{(l)}$ molecules are attracted by greater intermolecular forces than $CO_{2(g)}$ molecules
e	a, c and d

20. HI is less polar than HCl, yet it has a higher boiling point than HCl. This is because

a	HI has stronger dipole-dipole forces	c.	all of the above
b	HI has very strong L.D.F.	d.	none of the above
c	HCl has no L.D.F.		

CHAPTER 7
Chemical Reactions

Chapter Content:

Every second, in living organisms and society, hundreds of chemicals reactions are taking place. Respiration and photosynthesis are two natural processes that are most important to us and they entail many reactions such as in the Krebs's and the Calvin-Benson cycles. The process of cellular respiration, which includes a series of reactions, is responsible for providing us with the energy that sustains our lives. During this process, glucose is combined with oxygen to produce water, carbon dioxide and energy. During the process of photosynthesis, molecules of carbon dioxide and water are combined in a series of reactions to make glucose and oxygen. The energy for the latter process is provided by sunlight. Other organisms, called consumers, survive on the food manufactured by plants. Without these two processes, life would come to a standstill.

However, the vast majority of other reactions are man-made. Human activities run the gamut from combustion of various types of fossil fuels to the synthesis of a myriad of compounds such as chemical fertilizers, pharmaceuticals, fabrics, paints and plastics.

While these activities and their products are beneficial to us in most ways, they are also having profound effects on our environment and the quality of our lives. For example, the combustion of fossil fuels that powers automobiles, aircrafts and factories, produces gaseous oxides such as $CO_{2(g)}$, $NO_{2(g)}$ and $SO_{2(g)}$ that combine with water to form acidic rain. Acid rain destroys vegetation, aquatic organisms, soils, metal and marble infrastructures. Carbon dioxide is one of the gases mostly is responsible for global warming. The incidence of lung cancer, emphysema, silicosis, pulmonary fibrosis and some types of cardiovascular diseases is much more prevalent in highly industrialized areas than in non-industrialized areas.

7.1   Recognizing Chemical Changes

In chemical changes, the original substance(s) is changed to one or more new substances that have different physical and chemical properties than the original one(s). The reaction below exemplifies this.

$$2\,Mg_{(s)} \quad + \quad O_{2(g)} \quad \rightarrow \quad 2\,MgO_{(s)} \quad + \quad energy$$

Shiny metal   Colourless gas   White solid

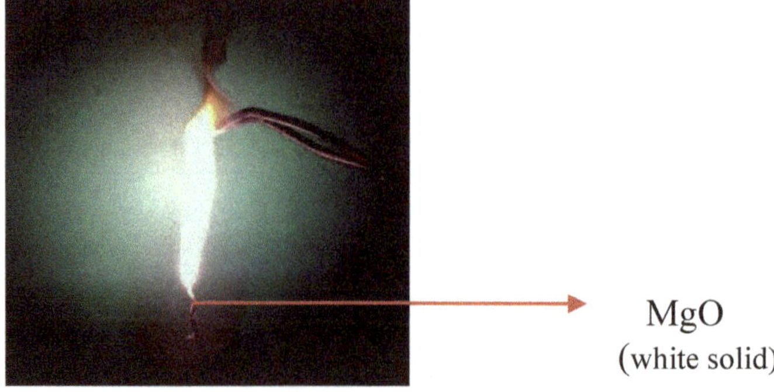

Figure 7.1. The burning of magnesium ribbon in air.

In this reaction, the element magnesium, a shiny metal burns brilliantly with oxygen in air to produce a white solid and energy. Here, the reactants have different physical properties from the products; neither magnesium nor oxygen reacts with cold water, but the magnesium oxide formed produces a basic solution if added to water.

Evidence for Chemical Reactions

To know if a chemical change takes place, one or more of a number of observations are made.

Evidence:
1. Formation of a solid (precipitate):

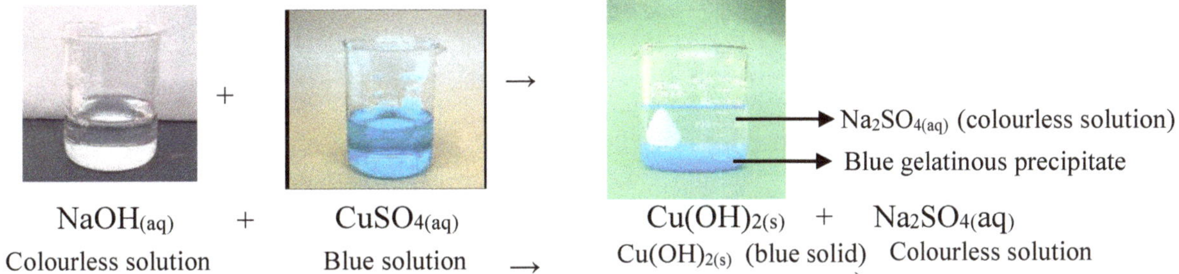

Figure 7.2. The formation of $Cu(OH)_{2(s)}$ by the reaction between $NaOH_{(aq)}$ and $CuSO_{4(aq)}$.

During precipitate formation, one of the products of the chemical reaction does not dissolve in water and thus remains a **solid**. In this case, it is the $Cu(OH)_{2(s)}$, a blue gelantinous precipitate.

2. **Evolution of a gas:**

Figure 7.3. The production of $CO_{2(g)}$ from the reaction between $Na_2CO_{3(aq)}$ and $HCl_{(aq)}$.

3. **Energy changes:**

$$CH_{4(g)} + 2\,O_{2(g)} \rightarrow 2\,CO_{2(g)} + 2\,H_2O_{(g)} + \text{heat} + \text{light}$$

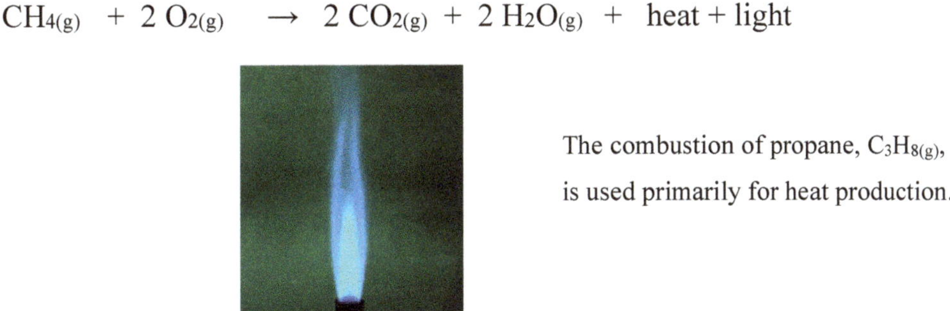

Figure 7.4. The production of heat and light from the burning of propane gas.

4. Colour changes:

$$Pb(NO_3)_{2(aq)} \quad + \quad 2\,KI_{(aq)} \quad \rightarrow \quad PbI_{2(s)} \quad + \quad 2\,KNO_{3(aq)}$$

(colourless solution) (colourless solution) (yellow precipitate) (colourless solution)

Figure 7.5. The production of yellow $PbI_{2(s)}$ from the reaction between $Pb(NO_3)_{2(aq)}$ and $KI_{(aq)}$.

5. Odour formation:

$$S_{(s)} \quad + \quad O_{2(g)} \quad \rightarrow \quad SO_{2(g)}$$

In this reaction, there is a distinct smell of sulphur dioxide gas. The acidic nature of this gas can be demonstrated by burning a match and placing a wet piece of blue litmus paper or wet pH paper above it. The paper would change its colour to red indicating acidity.

Exercise 7.1

In each of the equations below, identify the evidence for the chemical change. K/U

1. $Zn_{(s)} \quad + \quad 2\,HCl_{(aq)} \quad \rightarrow \quad ZnCl_{2(aq)} \quad + \quad H_{2(g)}$

2. $2\,AgCl_{(aq)} \quad + \quad H_2SO_{4(aq)} \quad \rightarrow \quad Ag_2SO_{4(s)} \quad + \quad HCl_{(aq)}$

3. $2\,KNO_{3(s)} \quad \rightarrow \quad 2\,KNO_{2(s)} \quad + \quad O_{2(g)}$

4. $2\,H_2O_{2(l)} \quad \rightarrow \quad 2\,H_2O_{(l)} \quad + \quad O_{2(g)}$

5. $2\,H_{2(g)} + O_{2(g)} \quad \rightarrow \quad 2\,H_2O_{(g)} + Heat + Sound + Light$

6. $NH_4Cl_{(s)} \quad \rightarrow \quad HCl_{(g)} \quad + \quad NH_{3(g)}$

7. $2\,AgNO_{3(aq)} \quad + \quad BaCl_{2(aq)} \quad \rightarrow \quad 2\,AgCl_{(s)} \quad + \quad Ba(NO_3)_{2(aq)}$

8. $Na_2CO_{3(aq)} \quad + \quad CaCl_{2(aq)} \quad \rightarrow \quad CaCO_{3(s)} \quad + \quad 2\,NaCl_{(aq)}$

9. $FeCl_{3(aq)} \quad + \quad NH_4OH_{(aq)} \quad \rightarrow \quad Fe(OH)_{3(s)} \quad + \quad NH_4Cl_{(aq)}$

10. $Al(OH)_{3(s)} \quad + \quad H_2SO_{4(aq)} \quad \rightarrow \quad Al_2(SO_4)_{3(aq)} + \quad H_2O_{(l)} + heat$

7.2   Representing Chemical Change

There are two ways of representing chemical changes. One way is to use **word equations**. **A word equation** uses the actual names of the reactants and products in a chemical change. For example, the decomposition of hydrogen peroxide is represented as follows:

$$\underbrace{\text{hydrogen peroxide}}_{\textbf{Reactant}} \quad \rightarrow \quad \underbrace{\text{water} \;+\; \text{oxygen}}_{\textbf{Products}}$$

In these equations the arrow symbol, "$\rightarrow$" means to produce and the plus sign, "+" is used when two or more different compounds are reacting or when two or more different compounds are produced. The other way of representing chemical changes is the use of **chemical equations**.

In a chemical equation, the actual formulae for compounds and symbols for elements are used. For the previous reaction, we have the following **skeleton equation**:

$$H_2O_{2(l)} \quad \rightarrow \quad H_2O_{(l)} \quad + \quad O_{2(g)}$$

7.3   The Law of Conservation of Mass

The law of conservation of mass states that, in any chemical change, atoms simply rearrange and are not created or destroyed; *the mass of the reactants must therefore be equal to the mass of the products.*

Applying the law of conservation of mass for the following skeleton equation, it can be observed that mass of the products is greater than that of the reactants; more are in the products than there is in the reactants.

$$H_2O_{2(l)} \quad \rightarrow \quad H_2O_{(l)} \quad + \quad O_{2(g)}$$

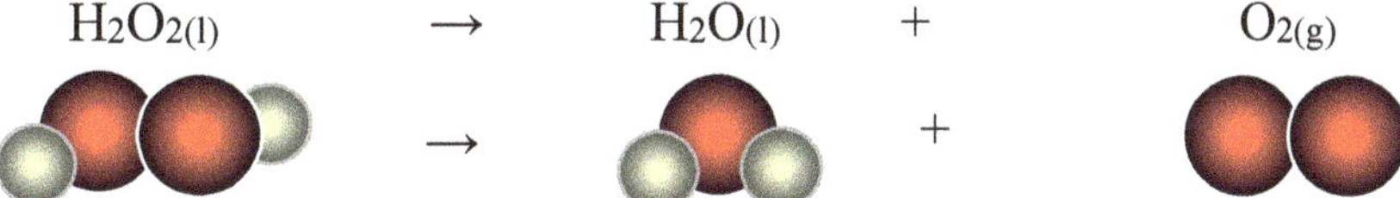

Figure 7.6. Space-filling models for the hydrogen peroxide, water and oxygen molecules.

To rectify this inaccuracy, coefficients are used in front of the compounds and elements in the chemical equations. Using appropriate coefficients, the equation above is now balanced

$$2\,H_2O_{2(l)} \quad \rightarrow \quad 2\,H_2O_{(l)} \quad + \quad O_{2(g)}$$

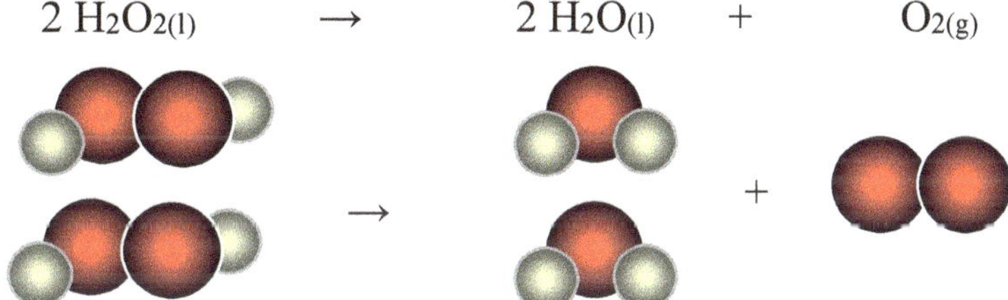

Figure 7.7. Space-filling models for the decomposition of hydrogen peroxide. According to the above figure, it can be seen that the mass of the products will be the same as the mass of the reactants; the number of atoms of each type of element is now the same on both sides of the equation.

Note that it is also important to indicate the physical states of all the reactants and products in the chemical equation.

The following represents the balanced chemical equation for the decomposition of nitrogen dioxide. Beneath it are the space filling models for the reactants and products. According to the number atoms present in the models, the equation is not balanced. Complete the models to make the equation balanced, as was done in Figure 7.7.

$$2\,NO_{2(g)} \quad \longrightarrow \quad N_2 \quad + \quad 2\,O_{2(g)}$$

7.4   Guidelines for Balancing Chemical Equations

As noted before, balancing chemical equations requires that the number of atoms of one element must be the same on both sides of the equation. *However, this must not be done by changing the formula of any substance in the equation.* The only way of achieving this is to place numbers, called **coefficients** in front of the formulas for compounds and symbols for elements; *it must not be done by changing the formulas of any substances in the equation.*

The following examples are done to illustrate how chemical equtions are balanced.

Magnesium reacts with oxygen according to the following skeleton equation:

$$Mg_{(s)} \quad + \quad O_{2(g)} \quad \rightarrow \quad MgO_{(s)}$$

In this example, there are two atoms of oxygen on the left side of the equation but only one atom of it on the right. To balance the oxygen, the coefficient, **2** is placed in front of the product, MgO.

$$Mg_{(s)} \quad + \quad O_{2(g)} \quad \rightarrow \quad 2\,MgO_{(s)}$$

However, by placing the coefficient, **2** there, the Mg atom is also multiplied by it**.** This means that the products have twice the number of Mg atoms as the reactants. To rectify this, the coefficient, **2** must now be placed in front of the Mg on the left side of the equation; all atoms on both sides are now equal.

$$2\,Mg_{(s)} \quad + \quad O_{2(g)} \quad \rightarrow \quad 2\,MgO_{(s)}$$

The following general guidelines should be used when balancing equations:

- ➢ Balance the **one** element, other than hydrogen and oxygen, that has the greatest number of atoms either in the reactants or product.
- ➢ Balance the other elements, other than hydrogen and oxygen.
- ➢ Balance oxygen or hydrogen, whichever one is present in **the combined state** (as in a compound)
- ➢ Leave, until last, to balance whichever element is present in the uncombined state.
- ➢ Check that the equation is balanced by counting the number of atoms of each element on both sides of the equation, to ensure that they are equal.

Example 1.
The following other reactions will be used to illustrate how to balance equations.

$$FeCl_{3(aq)} \quad + \quad NaOH_{(aq)} \quad \rightarrow \quad Fe(OH)_{3(s)} \quad + \quad NaCl_{(aq)}$$

- ➢ The element chlorine must be balanced by putting the coefficient, **3**, in front of $NaCl_{(aq)}$, that has only one. Doing this also triples the sodium on the right side of the equation.

$$FeCl_{3(aq)} \quad + \quad NaOH_{(aq)} \quad \rightarrow \quad Fe(OH)_{3(s)} \quad + \quad 3\,NaCl_{(aq)}$$

- ➢ To balance the OH and Na, the coefficient, **3**, must now be placed in front of $NaOH_{(aq)}$.

$$FeCl_{3(aq)} \quad + \quad 3\,NaOH_{(aq)} \quad \rightarrow \quad Fe(OH)_{3(s)} \quad + \quad 3\,NaCl_{(aq)}$$

Counting all the atoms of elements on both sides of the equation shows that they are equal; the equation is thus balanced.

Note:
When dealing with radicals, do not break them up, treat them as a unit of atoms bonded together.

Example 2.

The following reaction will be used to illustrate how another equation is balanced.

$$AgNO_{3(aq)} \quad + \quad CaCl_{2(aq)} \quad \rightarrow \quad AgCl_{(s)} \quad + \quad Ca(NO_3)_{2(aq)}$$

- ➢ The element chlorine must be balanced by putting the coefficient, **2** in front of $AgCl_{(s)}$. Doing this also doubles the silver on the right side of the equation.

$$AgNO_{3(aq)} \quad + \quad CaCl_{2(aq)} \quad \rightarrow \quad 2\,AgCl_{(s)} \quad + \quad Ca(NO_3)_{2(aq)}$$

- ➢ The NO_3^- radicals must be balanced by putting the coefficient, **2** in front of $AgNO_{3(aq)}$.

$$2\,AgNO_{3(aq)} \quad + \quad CaCl_{2(aq)} \quad \rightarrow \quad 2\,AgCl_{(s)} \quad + \quad Ca(NO_3)_{2(aq)}$$

The coefficient, **2** balances both the Ag atoms and NO_3^- radicals.
Counting all the atoms of elements on both sides of the equation shows that they are equal; the equation is thus balanced.

For the following word equations, write the chemical formulas of the reactants and products and then balance them. K/U

a) sodium hydroxide + sulphuric acid $\rightarrow$ sodium sulphate + water

b) calcium hydroxide + nitric acid $\rightarrow$ calcium nitrate + water

c) hydrochloric acid + magnesium hydroxide $\rightarrow$ magnesium chloride + water

d) sodium + water $\rightarrow$ sodium hydroxide + hydrogen (gas)

e) potassium + oxygen $\rightarrow$ potassium oxide

f) zinc + sulphuric acid $\rightarrow$ zinc sulphate + hydrogen (gas)

g) magnesium + nitric acid $\rightarrow$ magnesium nitrate + hydrogen (gas)

h) copper + oxygen $\rightarrow$ copper (II) oxide

i) sodium carbonate + hydrochloric acid $\rightarrow$ sodium chloride + water + carbon dioxide

Balance the following equations: K/U

1. $P_{(s)} + O_{2(g)} \rightarrow P_2O_{3(s)}$

2. $Zn_{(s)} + HCl_{(aq)} \rightarrow ZnCl_{2(aq)} + H_{2(g)}$

3. $C_{(s)} + O_{2(g)} \rightarrow CO_{(g)}$

4. $Na_2CO_{3(aq)} + HCl_{(aq)} \rightarrow NaCl_{(aq)} + CO_{2(g)} + H_2O_{(l)}$

5. $KNO_{3(s)} \rightarrow KNO_{2(s)} + O_{2(g)}$

6. $KClO_{3(s)} \rightarrow KCl_{(s)} + O_{2(g)}$

7. $H_2O_{2(l)} \rightarrow H_2O_{(l)} + O_{2(g)}$

8. $Al_{(s)} + CuCl_{2(aq)} \rightarrow AlCl_{3(aq)} + Cu_{(s)}$

9. $Ca(OH)_{2(aq)} + HNO_{3(aq)} \rightarrow Ca(NO_3)_{2(aq)} + H_2O_{(l)}$

10. $AgNO_{3(aq)} + BaCl_{2(aq)} \rightarrow AgCl_{(s)} + Ba(NO_3)_{2(aq)}$

12. $Na_2CO_{3(aq)} + CaCl_{2(aq)} \rightarrow CaCO_{3(s)} + NaCl_{(aq)}$

13. $CO_{(g)} + O_{2(g)} \rightarrow CO_{2(g)}$

14. $Fe_2O_{3(s)} + CO_{(g)} \rightarrow Fe_{(s)} + CO_{2(g)}$

15. $Fe_2O_{3(s)} + H_{2(g)} \rightarrow Fe_{(s)} + H_2O_{(l)}$

16. $Al(OH)_{3(s)} + H_2SO_{4(aq)} \rightarrow Al_2(SO_4)_{3(aq)} + H_2O_{(l)}$

17. $CO_{2(g)} + NH_{3(g)} \rightarrow CO(NH_2)_2 + H_2O_{(l)}$

7.5 Types of Chemical Reactions

Generally, chemical reactions fall into one of four categories. These are as follows:

Synthesis (Combination) Reactions.

In synthesis reactions, two or more substances combine to form a single more complex compound. In this type of reactions, smaller species combine to form a more complex one. Letters are used to represent the species present.

$$A + B \rightarrow AB$$

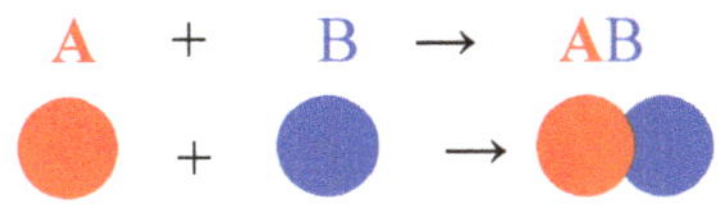

Synthesis reactions can be achieved in a number of ways as follows:

- **Element**   **+**   **Element** $\rightarrow$ **Compound**

(a)	$C_{(s)}$	$+$	$O_{2(g)}$	$\rightarrow$	$CO_{2(g)}$ *(as in combustion of coal)*
(b)	$2\,H_{2(g)}$	$+$	$O_{2(g)}$	$\rightarrow$	$2\,H_2O_{(g)}$
(c)	$S_{(s)}$	$+$	$O_{2(g)}$	$\rightarrow$	$SO_{2(g)}$ *(as in combustion of fossil fuels)*
(d)	$2\,Mg_{(s)}$	$+$	$O_{2(g)}$	$\rightarrow$	$2\,MgO_{(s)}$
(e)	$4\,Fe_{(s)}$	$+$	$3\,O_{2(g)}$	$\rightarrow$	$2\,Fe_2O_{3(s)}$ *(as in rusting of iron)*

The rusting of iron is a synthesis reaction that takes place very slowly and it may go on undetected in important structures. This reaction is important because it can cause considerable destruction to iron infrastructure such as buildings and bridges. Invariable collapse of these sometimes cause fatal accidents.

- **Compound**   **+**   **Element** $\rightarrow$ **Compound**

(f)	$2SO_{2(g)}$	$+$	$O_{2(g)}$	$\rightarrow$	$3SO_{3(g)}$
(g)	$2NO_{(g)}$	$+$	$O_{2(g)}$	$\rightarrow$	$2NO_{2(g)}$ *(atmospheric oxidation)*

- **Compound**   **+**   **Compound** $\rightarrow$ **Compound**

(h)	$NH_{3(g)}$	$+$	$HCl_{(g)}$	$\rightarrow$	$NH_4Cl_{(s)}$
(i)	$CaO_{(s)}$	$+$	$CO_{2(g)}$	$\rightarrow$	$CaCO_{3(s)}$
(j)	$SO_{3(g)}$	$+$	$H_2O_{(l)}$	$\rightarrow$	$H_2SO_{4(aq)}$ (as in *acid rain formation*)

The formation of acid rain is a synthesis reaction between the oxides of non-metal and water. The reaction between the oxides produced by the combustion of sulphur-containing fuel and water from falling rain is an example of this. Acid rain can cause adverse environmental problems if not contained.

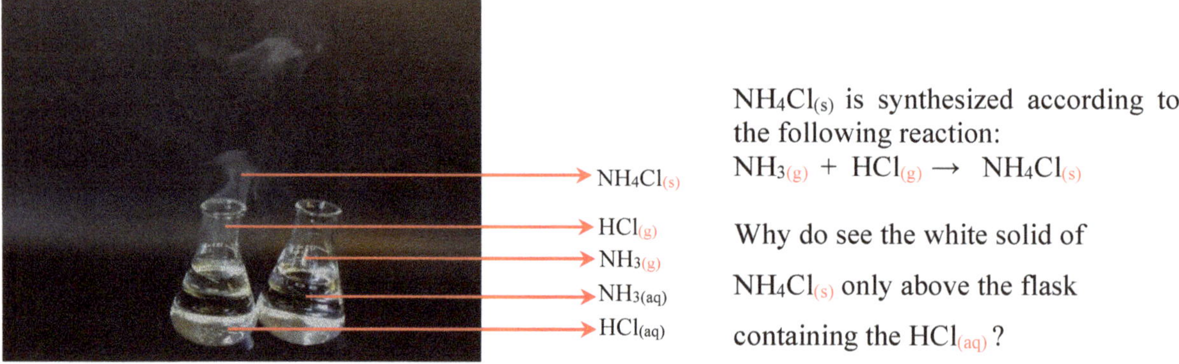

$NH_4Cl_{(s)}$ is synthesized according to the following reaction:

$$NH_{3(g)} + HCl_{(g)} \rightarrow NH_4Cl_{(s)}$$

Why do see the white solid of $NH_4Cl_{(s)}$ only above the flask containing the $HCl_{(aq)}$?

Figure 7.8. The formation of $NH_4Cl_{(s)}$ from the reaction between $NH_{3(g)}$ and $HCl_{(g)}$.

Decomposition Reactions

In decomposition reactions, a single complex compound breaks up into two or more simpler substances. These reactions can be generally represented as follows:

$$AB \quad \rightarrow \quad A \quad + \quad B$$

Decomposition reactions can be considered to be the opposite of synthesis reactions; a more complex species is broken down into less complex ones. This type of reaction can be achieved in a number of ways:

- **Compound** $\rightarrow$ **Element** + **Element**

(a) $2\,H_2O_{(l)} \xrightarrow{\text{electricity}} 2\,H_{2(g)} + O_{2(g)}$

(b) $2\,NaCl_{(l)} \xrightarrow{\text{electricity}} 2\,Na_{(s)} + Cl_{2(g)}$

(c) $2\,NO_{(g)} \xrightarrow{\text{Pt/Pd}} N_{2(g)} + O_{2(g)}$

The last reaction above is an example of a catalyzed decomposition reaction. It takes place in the catalytic converters of automobiles. The platinum/palladium catalyst breaks down nitrogen monoxide gas into harmless nitrogen and oxygen, before it escapes the exhaust and enters the atmosphere. If nitrogen monoxide mixes with air, it spontaenously forms nitrogen dioxide which is carcinogenic and also produces acid rain.

- **Compound** $\rightarrow$ **Compound** + **Element**

(c) $3\,SO_{3(g)} \xrightarrow[\Delta t]{/MnO_2} 2\,SO_{2(g)} + O_{2(g)}$

(d) $2\,KClO_{3(s)} \xrightarrow{MnO_2} 2\,KCl_{(s)} + 3\,O_{2(g)}$ *(This reaction can supply O_2 for combustion if required)*

(e) $2\,H_2O_{2(l)} \rightarrow 2\,H_2O_{(l)} + O_{2(g)}$

- **Compound** $\rightarrow$ **Compound** + **Compound**

(f) $NH_4NO_{3(s)} \xrightarrow{\Delta} N_2O_{(g)} + 2\,H_2O_{(g)}$

(g) $NH_4Cl_{(s)} \xrightarrow{\Delta} NH_{3(g)} + HCl_{(g)}$

Exercise 7.4

1. For the following reactions, determine if they are synthesis or decomposition. K/U

(a) $3\,H_{2(g)} + N_{2(g)} \rightarrow 2\,NH_{3(g)}$
(b) $2\,HgO_{(s)} \rightarrow 2\,Hg + O_{2(g)}$
(c) $H_2CO_{3(l)} \rightarrow H_2O_{(l)} + CO_{2(g)}$
(d) $CaO_{(s)} + H_2O_{(l)} \rightarrow Ca(OH)_{2(s)}$
(e) $2\,NaHCO_{3(s)} \rightarrow Na_2CO_{3(s)} + CO_{2(g)} + H_2O_{(l)}$

2. The following are synthesis reactions. Complete the equations by first predicting the product, and then balancing the equation. K/U T/I

(a) $Ba_{(s)} + O_{2(g)} \rightarrow$
(b) $H_{2(g)} + O_{2(g)} \rightarrow$
(c) $Fe_{(s)} + O_{2(g)} \rightarrow$
(d) $Na_2O_{(s)} + H_2O_{(l)} \rightarrow$
(e) $SO_{3(g)} + H_2O_{(l)} \rightarrow$
(f) $P_{(s)} + Cl_{2(g)} \rightarrow$

3. The following are decomposition reactions. Complete the equations by first predicting the product, and then balancing the equations. K/U T/I

(a) $Ag_2O_{(s)} \rightarrow$
(b) $H_2O_{2(l)} \rightarrow$
(c) $K_2S_{(s)} \rightarrow$
(d) $CaCO_{3(s)} \rightarrow$
(e) $AlBr_{3(s)} \rightarrow$
(f) $CuO_{(s)} \rightarrow$

Single Displacement Reactions.

In a single displacement reaction, one element displaces another element from a compound. These reactions can be generally represented as follows: The more reactive element, A displaces the less reactive one, B from the compound, BC.

$$A + BC \rightarrow AC + B$$

An example of a single displacement type of reaction is the reaction between magnesium metal and a solution of Copper (II) sulphate.

- **Metal** + **Salt solution**

$$Mg_{(s)} \quad + \quad CuSO_{4(aq)} \quad \rightarrow \quad MgSO_{4(aq)} \quad + \quad Cu_{(s)}$$

shiny metal   blue solution   colourless solution   brown solid

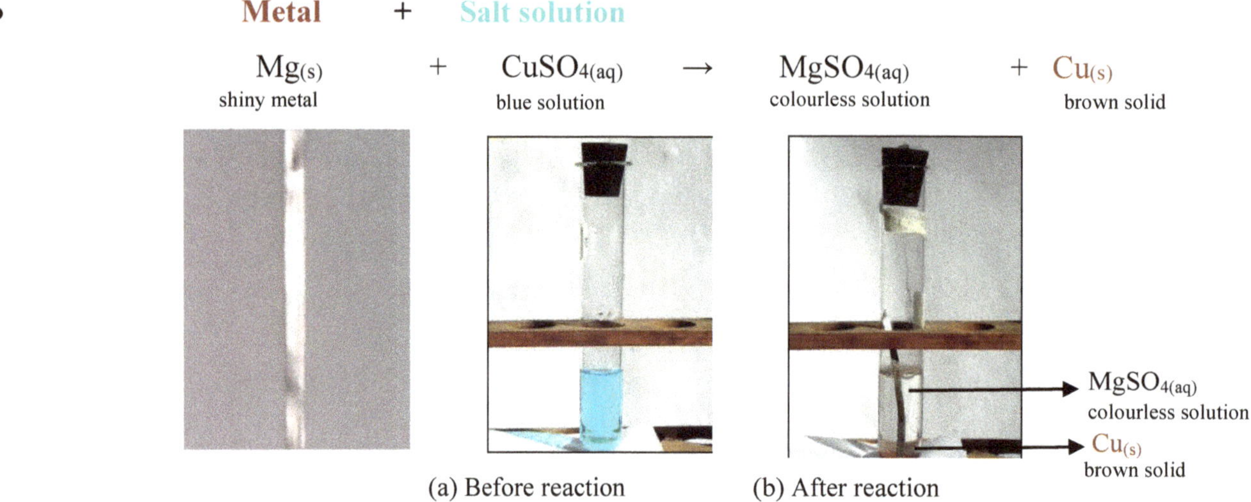

Figure 7.9. The reaction between magnesium and aqueous copper(II) sulphate.

In the example above, the element magnesium which is a reactive metal, displaces the less reactive element copper, which is also a metal. Single displacement reactions are not limited to only metals displacing fellow metals from aqueous compounds; non-metals can also displace fellow non-metals from aqueous compounds. The following is an example of a more reactive non-metal displacing a less reactive one.

- **Non-metal** + **Aqueous salt solution**

$$Cl_{2(g)} \quad + \quad 2\,NaBr_{(aq)} \quad \rightarrow \quad 2\,NaCl_{(aq)} \quad + \quad Br_{2(l)}$$

Greenish-yellow gas   colourless solution   colourless solution   reddish-brown liquid

In the example above, the more reactive non-metallic element, chlorine displaces the less reactive nonmetallic element, bromine from an aqueous solution of sodium bromide.

Ionic Equations for Reactions

In writing an ionic equation for any reaction, the following guidelines should be followed:
- Dissociate the aqueous compounds into their respective cations and anions.
- Do not separate the cations and anions of any precipitate formed, since they are insoluble and thus do not dissociate in water.
- **Leave all molecules, elements and precipitates as they are.**

Writing the ionic equation for the first example of single displacement reaction (Fig.7.9), we get:

$$Mg_{(s)} \quad + \quad Cu^{2+}_{(aq)} + SO_4^{2-}_{(aq)} \quad \rightarrow \quad Mg^{2+}_{(aq)} + SO_4^{2-}_{(aq)} + Cu_{(s)}$$

Net Ionic Equations for Reactions

As can be observed from the ionic equation, some species are not actively involved in the reaction. These are called **spectator ions** and are common in both sides of the equations. If these are removed from the equation, what is left is called the **net ionic equation** and it represents the net change that is taking place.

$$Mg_{(s)} \quad + \quad Cu^{2+}_{(aq)} + \cancel{SO_4^{2-}}_{(aq)} \quad \rightarrow \quad Mg^{2+}_{(aq)} + \cancel{SO_4^{2-}}_{(aq)} + Cu_{(s)}$$

Spectator ion   spectator ion

The net ionic equation is:

$$Mg_{(s)} \quad + \quad Cu^{2+}_{(aq)} \quad \rightarrow \quad Mg^{2+}_{(aq)} \quad + \quad Cu_{(s)}$$

According the net ionic equation, one $Mg_{(s)}$ atom is changed to a $Mg^{2+}_{(aq)}$ ion and the blue $Cu^{2+}_{(aq)}$ ion is changed to a brown **$Cu_{(s)}$** atom. This reaction can also be explained in terms of electron transfer, using the following half equations:

$$Mg_{(s)} \rightarrow Mg^{2+}_{(aq)} + 2e^-$$
$$Cu^{2+}_{(aq)} + 2e^- \rightarrow Cu_{(s)}$$

In this reaction, the $Mg_{(s)}$ atom loses two electrons and is changed to $Mg^{2+}_{(aq)}$. The **$Cu^{2+}_{(aq)}$** ion then accepts these two electrons and becomes a **$Cu_{(s)}$** atom. The insoluble $Mg_{(s)}$ now becomes soluble as $Mg^{2+}_{(aq)}$ and the soluble **$Cu^{2+}_{(aq)}$** is changed to the insoluble, **$Cu_{(s)}$**.

Other types of single displacement reaction are as follows:

- **Metal + Acid**

$$Zn_{(s)} + 2\,HCl_{(aq)} \rightarrow ZnCl_{2(aq)} + H_{2(g)}$$

The ionic equation for this reaction is as follows:

$$Zn_{(s)} + 2\,H^+_{(aq)} + 2\,Cl^-_{(aq)} \rightarrow Zn^{2+}_{(aq)} + 2\,Cl^-_{(aq)} + H_{2(g)}$$

After removing the spectator ions, we have the following net ionic equation:

$$Zn_{(s)} + 2\,H^+_{(aq)} \rightarrow Zn^{2+}_{(aq)} + H_{2(g)}$$

This reaction can also be illustrated in terms of electron transfer:

$$Zn_{(s)} \rightarrow Zn^{2+}_{(aq)} + 2e^-$$
$$2\,H^+_{(aq)} + 2e^- \rightarrow H_{2(g)}$$

- **Metal + Water**

$$2\,K_{(s)} + 2\,H_2O_{(l)} \rightarrow 2\,KOH_{(aq)} + H_{2(g)}$$

Note: Only the metals: K, Na, Ca, Ba and Li react with cold water; the rest are not so reactive to do this.

7.6 Developing a Reactivity Series for Metals

In single displacement reactions involving the displacement of one metal by another, it is observed that only if a metal is more reactive than another one, then the former can displace the latter from an aqueous solution of its salt. For example, in the following reaction, the element magnesium, being more reactive than copper, displaces it from an aqueous solution of one of its salts.

$$Mg_{(s)} + CuSO_{4(aq)} \rightarrow MgSO_{4(aq)} + Cu_{(s)}$$

To demonstrate this, a strip of magnesium is dipped into a solution of copper (II) sulphate. Resulting from this reaction, is **a coating of the displaced copper** that appears on the strip of magnesium strip. The magnesium atoms that reacted, dissolved in the solution as $Mg^{2+}_{(aq)}$ ions.

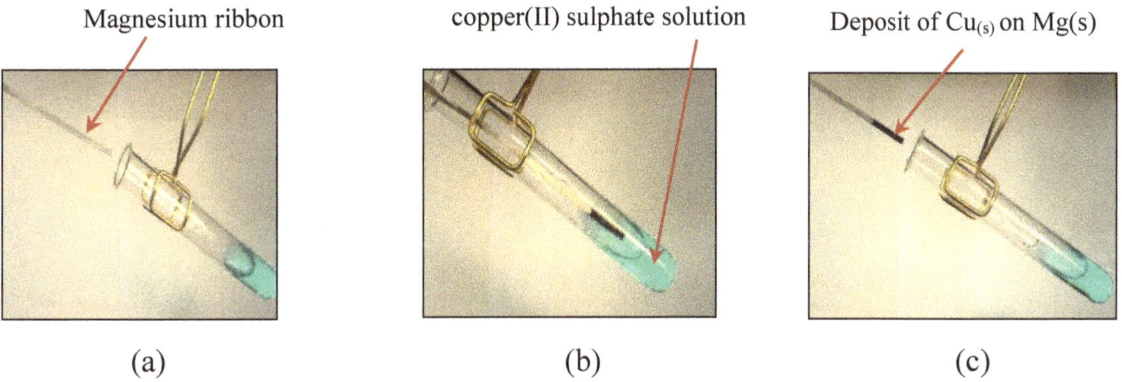

Figure 7.10. The displacement of copper by magnesium from aqueous copper (11) sulphate

The figures before show what would be observed for any single displacement of one metal by another; a coating of the less reactive metal is formed on the more reactive one (the metal strip being dipped). To know if **one metal** is more reactive than **another**, **a strip of that metal** is dipped into an **aqueous salt solution of the other metal,** and observation is made. If a coating is formed on the metal strip being dipped, it is more reactive, but if not, it is less reactive. To develop a mini activity series for a few metals, a number of single displacement reactions are made, and the metals are ranked according to their ability to displace others from their aqueous solutions. The following activity demonstrates how this is done.

Materials:

Aprons	**Aqueous solutions of the following compounds:**
Eye protection	$Mg(NO_3)_{2(aq)}$, $Al(NO_3)_{3(aq)}$,
6 250 mL beakers	$Fe(NO_3)_{3(aq)}$, $Sn(NO_3)_{2(aq)}$, $Zn(NO_3)_{2(aq)}$,
Strips of Mg, Al, Zn,	$Pb(NO_3)_{2(aq)}$, $Cu(NO_3)_{2(aq)}$
Sn, Fe, Cu, and Pb.	

Procedure:

1. Label each of the 250 mL beakers with the names of the aqueous solutions provided.
2. Pour about 50 mL of each solution from their containers into their matching beakers.
3. Dip any of the listed metal strip into a beaker with any solution for a few seconds. Observe and record the results as in the table below. Clean this strip and repeat this step with the other 5 solutions.
4. Repeat step 3, using the other 5 strips.

Results:

The following table shows the results that were obtained from these tests. Wherever a displacement reaction occurs a checkmark is made in a table.

Table 7.1. Single displacement reactions of some metals.

Metals	Aqueous Solutions of Salts						
	$Cu(NO_3)_{2(aq)}$	$Mg(NO_3)_{2(aq)}$	$Al(NO_3)_{3(aq)}$	$Fe(NO_3)_{3(aq)}$	$Sn(NO_3)_{2(aq)}$	$Zn(NO_3)_{2(aq)}$	$Pb(NO_3)_{2(aq)}$
$Mg_{(s)}$	√	N.R	√	√	√	√	√
$Al_{(s)}$	√	N.R	N.R	√	√	√	√
$Fe_{(s)}$	√	N.R	N.R	N.R	√	N.R	√
$Zn_{(s)}$	√	N.R	N.R	√	√	N.R	√
$Sn_{(s)}$	√	N.R	N.R	N.R	N.R	N.R	√
$Pb_{(s)}$	√	N.R	N.R	N.R	N.R	N.R	N.R
$Cu_{(s)}$	N.R	N.R	N.R	N.R	N.R	N.R	N.R

Analysis:

When the results from these test were analysed, the following deductions were made:

- Mg$_{(s)}$ is able to displace all the other metals from their aqueous salt solutions.
- Al$_{(s)}$ is able to displace all the other metals from their aqueous salt solutions except Mg.
- Zn$_{(s)}$ is able to displace all the other metals from their aqueous salt solutions except Mg and Al.
- Fe$_{(s)}$ is able to displace all the other metals from solution except Mg, Al and Zn.
- Pb$_{(s)}$ is able to displace only Cu from its aqueous solution.
- Cu$_{(s)}$ is unable to displace any of the metals from their aqueous salt solutions.

Conclusion:

Based on the results, the metals under investigation can be now ranked in terms of chemical reactivity. Starting with the most reactive to the least reactive, we have the following order: **Mg, Al, Zn, Fe, Sn, Pb and Cu.**
 This is what would be obtained by students using a small sample of metals and their corresponding salt solutions. However, scientists have already investigated most of the metals and have come up with a much bigger reactivity series as follows:

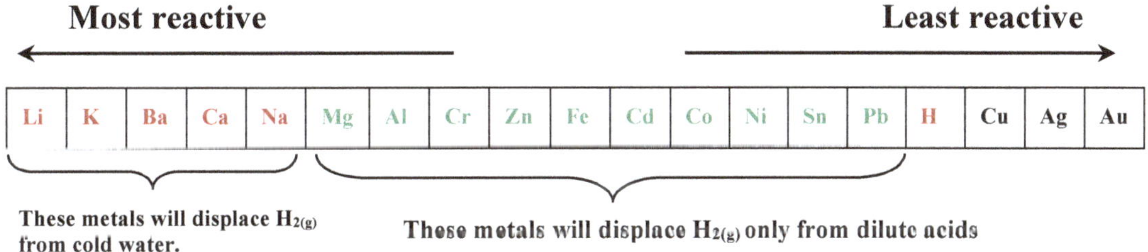

Figure 7.11. The reactivity series of metals.

7.7 Developing a Reactivity Series for the Halogens

Materials:

Aprons	**Aqueous solutions of the following compounds:**
Eye protection	NaF$_{(aq)}$, KCl$_{(aq)}$, NaBr$_{(aq)}$, and KI$_{(aq)}$ (**Halides**)
16 10-mL test tubes	Sources of F$_{2(g)}$, Cl$_{2(g)}$, Br$_{2(l)}$, and I$_{2(s)}$ (**Halogens**)

Procedure:

1. Label **4** 10 mL test tubes with the names of the halides solutions listed above.
2. Pour 2 mL samples of each of the halide solutions into their matching test tubes.
3. Pass one source of one of the gaseous halogen into each test tube and observe.
4. Record all observations.
5. Discard the contents of the test tubes safely into a waste container.
6. Clean the test tubes and repeat steps 2, 3, 4 and 5, using the other 3 halogens.

Results: The following table simulates the results that were obtained from these tests were actually done.

Table 7.2. Single displacement reaction of non-metals by non-metals.

Halogens	Aqueous solutions of Halides			
$F_{2(g)}$ A clear yellow gas	$F_{2(g)}$ NaF$_{(aq)}$ **NR**	$F_{2(g)}$ KCl$_{(aq)}$ $2KCl + F_{2(g)} \rightarrow 2KF + Cl_{2(g)}$	$F_{2(g)}$ NaBr$_{(aq)}$ $2NaBr + F_{2(g)} \rightarrow 2NaF + Br_{2(g)}$	$F_{2(g)}$ KI$_{(aq)}$ $2KI + F_{22(g)} \rightarrow 2KBr + I_{2(s)}$
$Cl_{2(g)}$ A greenish-yellow gas	$Cl_{2(g)}$ NaF$_{(aq)}$ **NR**	$Cl_{2(g)}$ KCl$_{(aq)}$ **NR**	$Cl_{2(g)}$ NaBr$_{(aq)}$ $2NaBr + Cl_{2(g)} \rightarrow 2NaCl + Br_{2(g)}$	$Cl_{2(g)}$ KI$_{(aq)}$ $2KI + Cl_{2(g)} \rightarrow 2KCl + I_{2(s)}$
$Br_{2(g)}$ A reddish-brown vapour	$Br_{2(g)}$ NaF$_{(aq)}$ **NR**	$Br_{2(g)}$ KCl$_{(aq)}$ **NR**	$Br_{2(g)}$ NaBr$_{(aq)}$ **NR**	$Br_{2(g)}$ KI$_{(aq)}$ $2KI + Br_{2(g)} \rightarrow 2KBr + I_{2(s)}$
$I_{2(g)}$ A dark-brown vapour	$I_{2(g)}$ NaF$_{(aq)}$ **NR**	$I_{2(g)}$ KCl$_{(aq)}$ **NR**	$I_{2(g)}$ NaBr$_{(aq)}$ **NR**	$I_{2(g)}$ KI$_{(aq)}$ **NR**

Analysis:

When an analysis of the results was made, the following deductions were made:

- $F_{2(g)}$ was able to displace $Cl_{2(g)}$, $Br_{2(g)}$, and $I_{2(s)}$ from solutions of their metal halides.
- $Cl_{2(g)}$ was able to displace $Br_{2(g)}$, and $I_{2(s)}$ from solutions of their metal halides but not $F_{2(g)}$.
- $Br_{2(g)}$ was able to displace only $I_{2(s)}$ from a solution of its metal halide but not $F_{2(g)}$ nor $Cl_{2(g)}$.
- $I_{2(g)}$ was unable to displace any of the other halogens from solutions of their metal halides.

Conclusion:

Based on the results, the halogens under investigation can be now ranked, in terms of chemical their reactivity. Starting with the most reactive to the least reactive, we have:

$F_{2(g)} > Cl_{2(g)} > Br_{2(g)} > I_{2(s)}$.

Figure 7.12. The single displacement of hydrogen from water by potassium.

In this reaction, the element potassium is displacing hydrogen gas from cold water according to the following equation: $K_{(s)} + 2\ H_2O_{(l)} \rightarrow 2\ KOH_{(aq)} + H_{2(g)}$.

The ionic equations for this reaction are as follows:

$$2\ H_2O_{(l)} \rightleftharpoons 2\ H^+_{(aq)} + 2\ OH^-_{(aq)}$$

$$2\ K_{(s)} \rightarrow 2\ K^+_{(aq)} + 2\ e^-$$

$$2\ H^+_{(aq)} + 2\ e^- \rightarrow H_{2(g)}$$

The heat produced by this reaction is so great it causes spontaneous combustion of the hydrogen gas produced. The flame is lilac because during combustion some potassium atoms become excited by the heat produced.

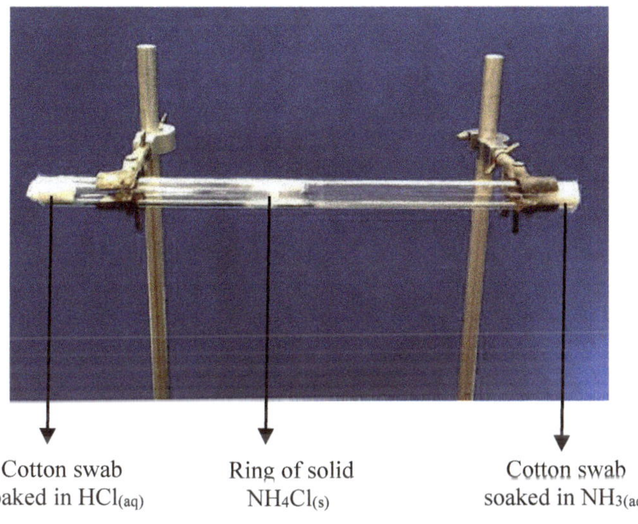

In this synthesis reaction, two cotton swabs soaked separately in $NH_{3(aq)}$ and concentrated $HCl_{(aq)}$ and are placed simultaneously at the opposite ends of a glass tube. As the colourless gases diffuse into the tube from the opposite ends, they meet **approximately** in the middle, forming a white solid of ammonium chloride.

$$NH_{3(g)} + HCl_{(g)} \rightarrow NH_4Cl_{(s)}$$
$$\text{colourless} \quad \text{colourless} \quad \text{white}$$

Figure 7.13. The synthesis of $NH_4Cl_{(s)}$ from the reaction between $NH_{3(g)}$ and $HCl_{(g)}$.

Why do think that the solid was formed more towards the end of the tube with the cotton swab soaked with $HCl_{(aq)}$? Use the following formula below to help with your answer.

$$r \propto \frac{1}{\sqrt{d}}$$

where   r = rate of diffusion of a gas
          d = density of the gas

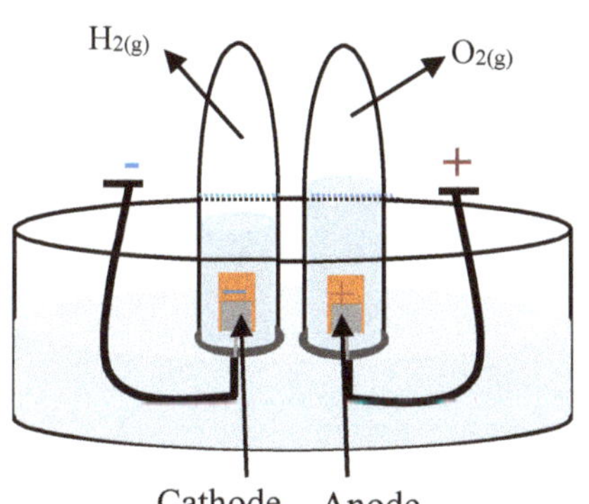

In this decomposition reaction, the following happens:

At the cathode (-)	**At the anode (+)**
$4\ H^+_{(aq)} + 4e^- \longrightarrow 2\ H_{2(g)}$	$4\ OH^-_{(aq)} \longrightarrow O_{2(g)} + 2\ H_2O_{(l)} + 4e^-$
2 volumes	1 volume

*Note that the $H^+_{(aq)}$ ions and the $OH^-_{(aq)}$ ions are from the auto-ionization of water as follows:

$$H_2O_{(l)} \rightleftharpoons H^+_{(aq)} + OH^-_{(aq)}$$
$$1.0 \times 10^{-7}\ \text{mol/L} \quad 1.0 \times 10^{-7}\ \text{mol/L}$$

Water conducts electricity poorly because the concentration of the $H^+_{(aq)}$ and $OH^-_{(aq)}$ ions produced are extremely low.

Figure 7.14. The decomposition of water by electricity to produce $H_{2(g)}$ and $O_{2(g)}$.

1. Balance the following single displacement reactions. K/U

(a) $Cu_{(s)} + AgNO_{3(aq)} \rightarrow Cu(NO_3)_{2(aq)} + Ag_{(s)}$

(b) $Cl_{2(g)} + NaBr_{(aq)} \rightarrow NaCl_{(aq)} + Br_{2(l)}$

(c) $Fe_{(s)} + SnCl_{2(aq)} \rightarrow FeCl_{3(aq)} + Sn_{(s)}$

(d) $Mg_{(s)} + HClO_{3(aq)} \rightarrow Mg(ClO_3)_{2(aq)} + H_{2(g)}$

(e) $Zn_{(s)} + HI_{(aq)} \rightarrow ZnI_{2(aq)} + H_{2(g)}$

2. The following are single displacement reactions. Complete the following equations by first predicting the products, and then balancing them. K/U T/I

(a) $Zn_{(s)} + HBrO_{3(aq)} \rightarrow$

(b) $Br_{2(g)} + NaI_{(aq)} \rightarrow$

(c) $Cu_{(s)} + AgNO_{3(aq)} \rightarrow$

(d) $Al_{(s)} + H_2SO_{4(aq)} \rightarrow$

(e) $Ca_{(s)} + H_2O_{(l)} \rightarrow$

(f) $F_2O_{3(s)} + CO_{(g)} \rightarrow$

3. Use the reactivity series to determine whether or not the following single reactions can happen. If a reaction is possible, complete the equation by first predicting the products, and then balancing the equation. If a reaction cannot happen simply write 'NR'. K/U T/I

(a) $Cu_{(s)} + SnCl_{2(aq)} \rightarrow$

(b) $Cl_{2(g)} + 2NaI_{(aq)} \rightarrow$

(c) $Ag_{(s)} + Cu(NO_3)_{2(aq)} \rightarrow$

(d) $Zn_{(s)} + H_2O_{(l)} \rightarrow$

(e) $Na_{(s)} + H_2O_{(l)} \rightarrow$

(f) $Au_{(s)} + H_2SO_{4(aq)} \rightarrow$

(g) $Br_{2(g)} + NaCl_{(aq)} \rightarrow$

(h) $Al_{(s)} + SnSO_{4(aq)} \rightarrow$

7.8 Investigating the Reaction of Metals with Dilute Acids

Hydrogen gas can be liberated from dilute acids by metals which are more reactive than hydrogen. For example, tin reacts with hydrochloric acid as follows:

$$Sn_{(s)} + 2\,HCl_{(aq)} \longrightarrow SnCl_{2(aq)} + H_{2(g)}$$

The net ionic equation for this is: $Sn_{(s)} + 2\,H^+_{(aq)} \longrightarrow Sn^{2+}_{(aq)} + H_{2(g)}$

The half reactions for these reactions for these are:

$$Sn_{(s)} \longrightarrow Sn^{2+}_{(aq)} + 2e^-$$
$$2\,H^+_{(aq)} + 2e^- \longrightarrow H_{2(g)}$$

The rate with which this reaction tales place depends on how easily the electrons are given off from the metals, so that the $2\,H^+_{(aq)}$ ions will accept them and produce $H_{2(g)}$. The ease with which the electrons come off metals is related to their reduction potentials; the more negative the reduction potential, the more reactive is the metal. The following table gives is a list of reduction potentials for a few metals and hydrogen.

Table 7.3. Standard electron potential for a few metals and hydrogen.

Metals	Mg	Cu	Zn	Fe	Al	H
Ionization energy E^0(V)	-2.37	0.34	-0.67	-0.45	-1.66	0.0

Hypothesis: Make a hypothesis about the relative reactivity of the metals in the table. What empirical evidence would you use to monitor the rates of reaction for each metal with $HCl_{(aq)}$.

Materials:

Safety goggles: 5 test tubes;1 test tube rack; 1 mol/L $HCl_{(aq)}$;
granulated forms of Mg, Cu, Zn, Fe and Al; 10 mL measuring cylinder

Procedure:

To test your hypothesis, carry out the following procedures:
1. Put on your protective equipment.
2. Obtain 5 test tubes and label them Mg, Cu, Zn, Fe and Al.
3. Place these test tubes in the rack.
4. Using the measuring cylinder, pour 4 mL portions of the acid of the same temperature into each test tube.
5. Measure out, using a balance, 0.25 g samples of each of the five metals and place each on a separate piece of paper.
6. Match each sample of metals to the labelled test tubes and simultaneously drop each sample into its matched test tube and observe.
7. Record your observation in the table provided.

Table 7.4.

Metals	Observation
Mg	
Cu	
Zn	
Fe	
Al	

Analysis and Evaluation: T/I

1. Rank the reactions from the least reactive to the most reactive.
2. Did your results agree with your hypothesis? If not, provide an explanation for any unexpected results.
3. Wherever a reaction takes place, write a balanced equation for it.
4. Identify the dependent and independent variables.
5. What were the controlled variables?

Conclusion:

Rank the metals in terms of relative reactivity from the least to the most.

Extension:

Exercise 7.6

1. For the following reactions, decide whether or not a reaction would take place base on the results that you obtained from the experiment above. If a reaction takes balance it, if not write NR. K/U T/I

a) $Mg_{(s)}$ + $CuSO_{4(aq)}$ $\longrightarrow$

b) $Cu_{(s)}$ + $ZnSO_{4(aq)}$ $\longrightarrow$

c) $Al_{(s)}$ + $Mg(NO_3)_{2(aq)}$ $\longrightarrow$

d) $Fe_{(s)}$ + $CuSO_{4(aq)}$ $\longrightarrow$

e) $Zn_{(s)}$ + $Fe(NO_3)_{3(aq)}$ $\longrightarrow$

2. Five freshly polished metal strips of the metals used in the experiment above were obtained, but they had no identifying label; the metals were labelled A, B, C, D, and E, instead. In an effort to identify these, each was reacted separately with aqueous salt solutions as shown in the table below. Wherever a reaction took place, a check mark was used and an X was used where no reaction took place. T/I

Table 7.5 Results from reactions between metals and aqueous salt solutions.

Metals	Aqueous Salts Solution				
	$Cu(NO_3)_2$	$Mg(NO_3)_2$	$Zn(NO_3)_2$	$Fe(NO_3)_3$	$Al(NO_3)_3$
A	√	X	√	√	X
B	X	X	X	X	X
C	√	X	√	√	√
D	√	X	X	√	√
E	√	X	X	X	X

    a) Rank these metals from least reactive to most reactive.
    b) Based on your ranking in the previous experiment, identify each of these metals.

Making Connection A

3. The Goldschmidt or Thermite process involves a single displacement reaction between some metal oxides such as ferric oxide and aluminum. Thermite is made by mixing ferric oxide and powdered aluminum in a 3:1 ratio by mass. When this mixture is ignited in crucible, the following explosive reaction takes place that can reach a temperature up to 4500 °F.

$$Fe_2O_{3(s)} + 2\ Al_{(s)} \longrightarrow Al_2O_{3(s)} + 2\ Fe_{(l)} + heat$$

For this reaction to start it needs an initial source of heat; in most demonstrations a strip of burning magnesium suffices. *Note, this reaction is very dangerous should be demonstrated outdoors or under strict supervision.

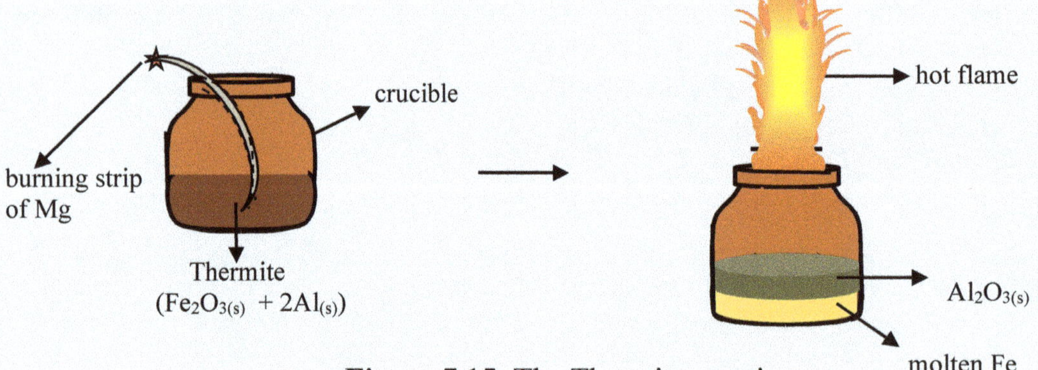

Figure 7.15. The Thermite reaction

The heat generated from this reaction and the molten iron produced can be put to various uses. Research the Thermite reaction and its various applications. Suggest some reasons for the extreme amount of heat that is produced from this reaction.

4. Iron is extracted from its ore using a huge blast furnace. The following equation summarizes the changes that take place during the extraction process:

$$2\ Fe_2O_{3(s)} + 3\ C_{(s)} \longrightarrow 4\ Fe_{(s)} + 3\ CO_{2(g)}$$

Iron ore   molten iron

Research this process and report on all the impacts that the extraction has on the environment. Suggest ways that these impacts can be minimized.

5. A simple electrochemical cell makes use of a single displacement reaction to create a difference of electric potential which is then used to generate electricity. An example of this is the displacement of copper from an aqueous copper compound by metallic zinc. This reaction takes in two separate half cells; the following equations show what happens during this reaction:

$$Zn_{(s)} + CuSO_{4(aq)} \longrightarrow ZnSO_{4(aq)} + Cu_{(s)}$$

Net ionic equation: $Zn_{(s)} + Cu^{2+}_{(aq)} \longrightarrow Zn^{2+}_{(aq)} + Cu_{(s)}$

Half reactions: $\qquad\qquad Zn_{(s)} \longrightarrow Zn^{2+}_{(aq)} + 2e^- \quad V = -0.76$

$$Cu^{2+}_{(aq)} \longrightarrow Cu_{(s)} \qquad\qquad V = +0.34$$

The difference in electrical potential between the two half-cell electrodes is the voltage of the cell and it is calculated as follows: $E^{o}_{cell} = -0.76V - (+0.34V) = -1.1 V = 1.1 V$

The voltage of any electrochemical cell depends on what electrodes are used in the two half-cells. If lithium is used instead of zinc, the half reactions would be as follows:

$$2 Li_{(s)} \longrightarrow 2 Li^{+}_{(aq)} + 2e^- \quad V = -3.03$$

$$Cu^{2+}_{(aq)} \longrightarrow Cu_{(s)} \qquad\qquad V = +0.34$$

The voltage of this new cell will be: $-3.03 V - (+0.34 V) = -3.37 V = 3.37 V$
Research how an electrochemical cell works.

What trend exists between the electrode potentials and voltage of a cell?

The half reaction for lithium produces the highest voltage when compared to any other metal. Why do you think lithium batteries are so powerful?

 Double Displacement Reactions

In a double displacement reaction, the cation of one aqueous ionic compound exchanges place with the cation of another aqueous ionic compound. Double displacement reactions are generally represented as follows, using letters:

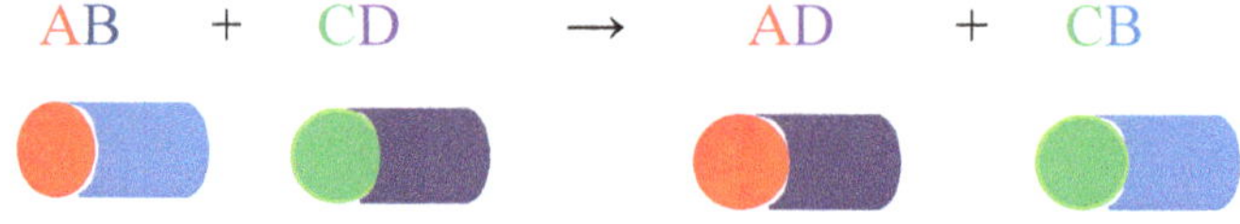

$$AB \; + \; CD \; \rightarrow \; AD \; + \; CB$$

The following reaction exemplifies a double displacement reaction between aqueous solutions of potassium iodide and silver nitrate; the cation $K^+_{(aq)}$ and $Ag^+_{(aq)}$ exchange places.

$$KI_{(aq)} \; + \; AgNO_{3(aq)} \; \rightarrow \; KNO_{3(aq)} \; + \; AgI_{(s)}$$
$$\text{(yellow precipitate)}$$

Double displacement reactions can be recognized in one of three ways:

- formation of a precipitate (a solid)
- production of a gas
- formation of water accompanied by heat production

Double Displacement Reactions that Form a Precipitate

When certain cations are allowed to team up with certain anions, they form ionic compounds that do not dissolve in water. Using the above reaction as an example, both potassium iodide, KI and silver nitrate, $AgNO_3$ are soluble in water. However, when these are mixed together they react to produce potassium nitrate, $KNO_{3(aq)}$ which is **soluble** in water, and silver iodide, $AgI_{(s)}$ which is **insoluble** in water. The $AgI_{(s)}$ formed thus comes out of solution as a **solid.** The question that can be asked is, why do some compounds dissolve in water easily, while others do not? The answer to this has to do with the magnitude of electrostatic force of attraction between the cations and anions of the ionic compounds in their crystalline form (refer to chapter 12 for a description of the dissolving of ionic compounds in water). If the force of attraction is relatively weak, the ions of that compound easily dissociate in water. If the force of attraction is relatively strong, the ions do not dissociate easily in water, so the compound remains insoluble. In the above example, it can be inferred that the electrostatic force of attraction between the silver cation, Ag^+, and the iodide anion, I^- is very strong to prevent them from dissociating in water at room temperature; AgI is thus insoluble in water. Some precipitate can be made to dissolve in water if they are heated in the presence of water. The heat provides the additional energy to separate the cations and ions from each other and become solvated. For example, at room temperature, lead chloride is insoluble in water, but if heated in water it dissolves.

Table 7.3. Solubility Chart of Ionic Compounds at 25 ^oC.

Anions 1	Cations 2 Metal ions that form **Sparingly soluble** compounds with anions in the left column	Cations 3 Metal ions that form **soluble** compounds with anions in the extreme left column
1 Cl^- Br^- I^-	$Ag^+, Pb^{2+}, Tl^+, Hg_2^{2+}$ $Ag^+_{(aq)} + Cl^-_{(aq)} \rightarrow AgCl_{(s)}$ $Pb^{2+}_{(aq)} + 2Cl^-_{(aq)} \rightarrow PbCl_{2(s)}$ $Tl^+_{(aq)} + I^-_{(aq)} \rightarrow TlI_{(s)}$	All other metal ions will form soluble compounds with the anions on the extreme left column **1**. For example, $Na^+_{(aq)}$ is not amongst the cations in column **2**; NaCl therefore is soluble in water.
2 OH^-	All cations will form insoluble hydroxides with $OH^-_{(aq)}$ except those in the right column eg. $Cu^{2+}_{(aq)} + 2\ OH^-_{(aq)} \rightarrow Cu(OH)_{2(s)}$ $Cu(OH)_{2(s)}$ is insoluble because $Cu^{2+}_{(aq)}$ is not in column **3**.	$Li^+, Na^+\ K^+, Rb^+, Cs^+,$ Ba^{2+}, Sr^{2+}, NH_4^+ For example, NaOH is soluble
3 SO_4^{2-}	$Ag^+, Pb^{2+}, Sr^{2+}, Ba^{2+}, Ca^{2+}$ $Ba^{2+}_{(aq)} + SO_4^{2-}_{(aq)} \rightarrow BaSO_{4(s)}$	All other metal ions, except from column **2**, will form soluble compounds with the anions on the extreme left column **1**. For example, $CuSO_4$ is soluble
4 SO_3^{2-} PO_4^{3-} CO_3^{2-}	Most cations will form precipitates with anions on the left except those in the right column eg. $Ba^{2+}_{(aq)} + SO_3^{2-}_{(aq)} \rightarrow BaSO_{3(s)}$ $Mg^{2+}_{(aq)} + CO_3^{2-}_{(aq)} \rightarrow MgCO_{3\ (s)}$ $3\ Mg^{2+}_{(aq)} + 2\ PO_4^{3-}_{(aq)} \rightarrow Mg_3(PO_4)_{2(s)}$	$Li^+, Na^+\ K^+\ Cs^+, Rb^+,$ NH_4^+ For example, Li_2CO_3, K_3PO_4 and Na_2SO_3 are soluble.
5 S^{2-}	All cations will form precipitates except those in the right column	**NH_4^+ and ions of groups 1 and 2 elements**
6 $C_2H_3O_2^-$	Ag^+	All others metal ions will form soluble acetates. For example, $LiC_2H_3O_2$ is soluble.
7 NO_3^-	**None**	All metal ions will form soluble nitrate For example, $Ba(NO_3)_2$ is soluble.

 ## Solubility of Ionic Compounds

To predict whether or not a precipitate will form when aqueous solutions of two ionic compounds are mixed, it is important to understand the rules that govern the solubility of ionic compounds in water. The previous table provides useful information about the relative solubility of various ionic compounds in water.

Predicting Precipitate Formation

The following example is done to explains how the table is used to predict whether or not a precipitate forms when two aqueous compounds are mixed.

Aqueous solutions of the following pairs of ionic compounds are mixed, predict whether or not a precipitate will form:

$$MgCl_{2(aq)} + Na_2CO_{3(aq)}$$

Solution: To do this, the following steps are carried out.

- Predict the products and write the skeleton equation for the reaction.

$$MgCl_{2(aq)} + Na_2CO_{3(aq)} \rightarrow NaCl + MgCO_3$$

- Find the ions of the products: $NaCl \longrightarrow Na^+ + Cl^-$, $MgCO_3 \longrightarrow Mg^{2+} + CO_3^{2-}$

- Check to see if the products are soluble or not, using the solubility table.

To determine if NaCl is soluble or not, inspect row **1** of column **1** where the $Cl^-_{(aq)}$ anion is located, and then inspect column **2** of row **1** to see if the cation, $Na^+_{(aq)}$ is present there. Since $Na^+_{(aq)}$ is absent there, it means that it must belong to the soluble cations of column **3** of row **1**; **NaCl is thus soluble in water.**

For MgCO₃, inspect row **4** of column **1** where the $CO_3^{2-}_{(aq)}$ anion is located, and then inspect column **3** of row **4** to see if the cation, $Ca^{2+}_{(aq)}$ is present there. Since $Ca^{2+}_{(aq)}$ is absent there, where soluble cations are found, it means that it must belong to the sparingly soluble cations of column **2** of row **4**. **Magnesium carbonate is thus insoluble and it will form a precipitate.**

The process of elimination is used to determine which of the two columns a cation belongs. Check the column where a set of cations are given to see if the one in question is present there or not. **If it is present in that set,** *its solubility is stated at the top of the column.* **If not,** *it belongs to the other column and its solubility is stated at the top of that column.* The balanced equation for this reaction can now be rewritten as follows:

$$MgCl_{2(aq)} + Na_2CO_{3(aq)} \rightarrow 2\,NaCl_{(aq)} + MgCO_{3(s)}$$

The ionic equation for this reaction is:

$$Mg^{2+}_{(aq)} + 2Cl^-_{(aq)} + 2Na^+_{(aq)} + CO_3^{2-}_{(aq)} \rightarrow 2Na^+_{(aq)} + 2Cl^-_{(aq)} + MgCO_{3(s)}$$

The net ionic equation is: $Mg^{2+}_{(aq)} + CO_3^{2-}_{(aq)} \rightarrow MgCO_{3(s)}$

Double Displacement Reactions that Form Water

Another type of double displacement reaction involves the reaction between a base and an acid and it called a **neutralization reaction.** During a neutralization reaction, one of the products is water and the other is a salt that dissolves in water. For example, the reaction between hydrochloric acid and sodium hydroxide produces aqueous sodium chloride and water. This reaction is represented as follows:

$$NaOH_{(aq)} + HCl_{(aq)} \rightarrow NaCl_{(aq)} + H_2O_{(l)} + heat$$

Unlike double displacement reactions where precipitates are visibly outstanding, there is no visible change during neutralization reactions, since the salts produced are soluble in water. Therefore, the only way to know if a reaction takes place is to check if heat is produced or not. Also, a pH meter placed in the mixture will show a change in pH, indicating a chemical change.

Double Displacement Reactions that Produce a Gas

In some double displacement reactions where no precipitate or energy is produced, one of the products is very unstable and decomposes spontaneously, producing a gas. The following are examples of these types of reactions:

> $2HCl_{(aq)}$ + $Na_2CO_{3(aq)}$ → $2NaCl_{(aq)}$ + $H_2CO_{3(aq)}$
>
>   carbonic acid
>
>   Acids + Carbonates → salts + $CO_{2(g)}$ ⇄ H_2O

In this reaction, the carbonic acid formed decomposes spontaneously to produce water and carbon dioxide gas. This is evident by a fizzing effect. This entire reaction can now be written as follows:

$$2\,HCl_{(aq)} + Na_2CO_{3(aq)} \rightarrow 2\,NaCl_{(aq)} + CO_{2(g)} + H_2O_{(l)}$$

This reaction is similar to what takes place in our stomach when we drink soda. The burping that takes place is due to the evolution of $CO_{2(g)}$ from the reaction between stomach acid and the carbonated solution in soda.

>   Salts of Ammonia + Strong Bases $NH_{3(g)}$ ⇄ $H_2O_{(l)}$

$$NH_4Cl_{(aq)} + KOH_{(aq)} \rightarrow KCl_{(aq)} + NH_4OH_{(aq)}$$

                                            ammonium hydroxide

In this reaction, the ammonium hydroxide formed spontaneously decomposes to produce water and ammonia gas. This reaction is evident by the pungent smell of ammonia gas. This entire reaction can now be rewritten as follows:

$$NH_4Cl_{(aq)} + KOH_{(aq)} \rightarrow KCl_{(aq)} + NH_{3(g)} + H_2O_{(l)}$$

Exercise 7.7   **Table 7.3.**   Predicting double displacement reactions. T/I

Reagents	$HCl_{(aq)}$ ⌐ ⌐ ions	$K_2SO_{4(aq)}$ K^+ SO_4^{2-} ions	$NaOH_{(aq)}$ Na^+ OH^- ions	$ZnI_{2(aq)}$ ions	$Li_3PO_{4(aq)}$ ions	$K_2S_{(aq)}$ ions
$CuCl_{2(aq)}$ ions						
$Ba(NO_3)_{2(aq)}$ Ba^{2+} NO_3^- ions		$BaSO_{4(s)} + KNO_{3(aq)}$ √				
$AgNO_{3(aq)}$ ions						
$HNO_{3(aq)}$ H^+ NO_3^- ions			$H_2O_{(l)} + NaNO_{3(aq)}$ + Energy √			
$Al(NO_3)_{3(aq)}$ ions						
$Na_2CO_{3(aq)}$ Na^+ CO_3^{2-} ions			$NaOH_{(aq)} + Na_2CO_{3(aq)}$ NR			

This activity involves reacting pairs of aqueous solutions and predicting if a double reaction will happen.

To do this exercise, react each reagent in left column with each of those in the upper row of the table. Start by dissociating each compound into their respective ions and then do the double displacement of their cations. Use the solubility table to predict precipitate formation, and the other evidence to decide if a reaction happens. If a reaction occurs, place a check mark below the equation, if not write NR. As a guide, a few examples have been done in the table 7.4.

Exercise 7.8

For the following pairs of aqueous ionic compounds, complete the skeleton equations. If a double displacement reaction occurs, balance the equation. If no reaction occurs, write NR. K/U/T/I

(a) $NaOH_{(aq)}$ + $H_2SO_{4(aq)}$ $\rightarrow$

(b) $KI_{(aq)}$ + $Pb(NO_3)_{2(aq)}$ $\rightarrow$

(c) $MgCl_{2(aq)}$ + $NaNO_{3(aq)}$ $\rightarrow$

(d) $NH_4Cl_{(aq)}$ + $KOH_{(aq)}$ $\rightarrow$

(e) $H_2SO_{4(aq)}$ + $CsOH_{(aq)}$ $\rightarrow$

(f) $CaS_{(aq)}$ + $HCl_{(aq)}$ $\rightarrow$

(g) $ZnBr_{2(aq)}$ + $HNO_{3(aq)}$ $\rightarrow$

(h) $K_2CO_{3(aq)}$ + $HCl_{(aq)}$ $\rightarrow$

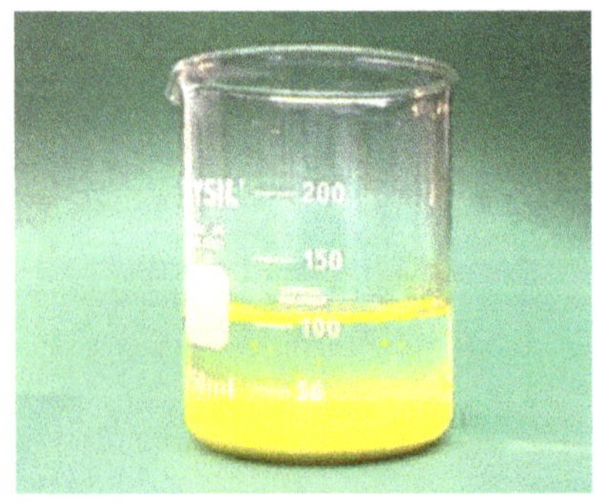

The figure to the left shows hows a yellow precipitate of $PbI_{2(s)}$ is formed from the double displacement reaction between $Pb(NO_3)_{2(aq)}$ and $KI_{(aq)}$.

$$Pb(NO_3)_{2(aq)} + 2\ KI_{(aq)} \rightarrow PbI_{2(s)} + 2\ KNO_{3(aq)}$$

Its ionic equation is as follows:

$$Pb^{2+}_{(aq)} + 2\ NO_3^-{}_{(aq)} + 2\ K^+_{(aq)} + 2\ I^-_{(aq)} \rightarrow PbI_{2(s)} + 2\ NO_3^-{}_{(aq)} + 2\ K^+_{(aq)}$$

Its net ionic equation is as follows:

$$Pb^{2+}_{(aq)} + 2\ I^-_{(aq)} \rightarrow PbI_{2(s)}$$

Figure 7.15. Yellow precipitate of lead (II) iodide.

7.12 Combustion

Combustion reactions are very important as they produce energy that is essential for a host of activities such as the running of automobiles and aeroplanes, heating buildings and powering factories. During complete combustion, some type of fuel is burnt in oxygen to produce oxides and energy, in the form of light and heat. The type of oxides produced depends on the nature of the fuel that is burnt. The following reactions illustrate a few types of combustion reactions:

- $S_{(s)}$ + $O_{2(g)}$ $\rightarrow$ $SO_{2(g)}$ + energy

- $C_{(s)}$ + $O_{2(g)}$ $\rightarrow$ $CO_{2(g)}$ + energy

- $CH_{4(g)} + 2\ O_{2(g)}$ $\rightarrow$ $CO_{2(g)} + 2\ H_2O_{(g)}$ + energy
  methane

Hydrocarbons belong to a class of compounds extracted from crude petroleum. These compounds contain only the elements hydrogen and carbon. The combustion of fuels. including those from fossils, has serious implications for the environment. The combustion of fossil fuels as exemplified by the burning of methane above, produces carbon dioxide and water. Carbon dioxide is one of the gases responsible for global warming. Also, fossil fuels have small amounts of impurities in the form of sulphur and nitrogen. As such, when fossil fuels are burnt, carbon dioxide and oxides of sulphur and nitrogen are also produced, along with some hydrogen sulphide and thiols. Since the oxides of non- metals are acidic, they combine with water in rain, snow etc. to form acid precipitations. The following equations exemplify how acid rain is formed:

Rain water, hail or snow

- $CO_{2(g)}$ + $H_2O_{(l)}$ $\longrightarrow$ $H_2CO_{3(aq)}$ (carbonic acid)

- $4\,NO_{2(g)}$ + $2\,H_2O_{(l)}$ + $O_{2(g)}$ $\longrightarrow$ $4\,HNO_{3(aq)}$ (nitric acid)

- $2SO_{2(aq)}$ + $2\,H_2O_{(l)}$ + $O_{2(g)}$ $\longrightarrow$ $2\,H_2SO_{4(aq)}$ (sulphurous acid)

Acid rain

Incomplete Combustion of Fuels

Carbon burns completely according to the following equation:

$$C_{(s)} + O_{2(g)} \longrightarrow CO_{2(g)} + \textbf{393.5 kJ/mol}$$

However, in the absence of an adequate amount of oxygen, carbon may burn incompletely according to the following equation:

$$C_{(s)} + \frac{1}{2}O_{2(g)} \longrightarrow CO_{(g)} + \textbf{110.5 kJ/mol}$$

Some Problems associated with Incomplete Combustion

➤ The incomplete combustion of carbon to carbon monoxide produces less energy than complete combustion. This represents inefficient burning of carbon. Carbon monoxide can further be burnt according to the following equation to release energy:

$$CO_{(g)} + \frac{1}{2}O_{2(g)} \longrightarrow CO_{2(g)} + \textbf{283.5 kJ/mol}$$

Whenever carbon monoxide is released from the combustion of fuels and escapes into the air, ***it thus represents*** *a loss of 283.5 kJ* ***of energy per mol of CO.***

➤ The $CO_{(g)}$ produced is a deadly poisonous gas. This gas combines with hemoglobin molecules of the red blood cells obstructing the take up of molecular oxygen from the lungs; oxygen is necessary for cellular respiration. Prolong exposure to this gas could lead to death, since it has no colour, odour or smell for personal detection. It is essential that all furnaces be regularly inspected to ensure that they burn their fuels efficiently and to eliminate the risk of $CO_{(g)}$ poisoning.

➤ For some fuels, such as hydrocarbons, incomplete combustion may even lead to the production of elemental carbon (**soot),** in addition to carbon monoxide. The incomplete combustion of propane exemplifies this concept.

$$2\,C_3H_{8(g)} + 7\,O_{2(g)} \longrightarrow 8\,H_2O_{(g)} + 2\,CO_{2(g)} + 2\,CO_{(g)} + 2\,C_{(s)}$$

propane   + oxygen $\longrightarrow$ water   + carbon dioxide + carbon monoxide + carbon

The soot produced, affects air quality. When inhaled, it can cause a series of cardiovascular diseases. Soot can also destroy the surfaces of leaves if it precipitates on them, reducing their ability to carry out photosynthesis effectively. In large quantities it may even affect weather patterns.

7.14 Activity on Combustion of Acetylene

Purpose:

The following activity was carried out to investigate how efficiently a fuel would burn, using different ratios of fuel and air combinations, by volume. Acetylene gas that is used in welding is used as the fuel. It is produced by the reaction of calcium carbide with water according to the equation:

$$CaC_{2(s)} + 2\,H_2O_{(l)} \rightarrow C_2H_{2(g)} + Ca(OH)_{2(s)}$$

Materials:

*safety goggles
*calcium carbide, $CaC_{2(s)}$
*Four 10 mL test tubes with stoppers
*one test tube rack
*one 1L beaker
*matches and splints
*1L of distilled water
*one pair of tweezer
*latex gloves

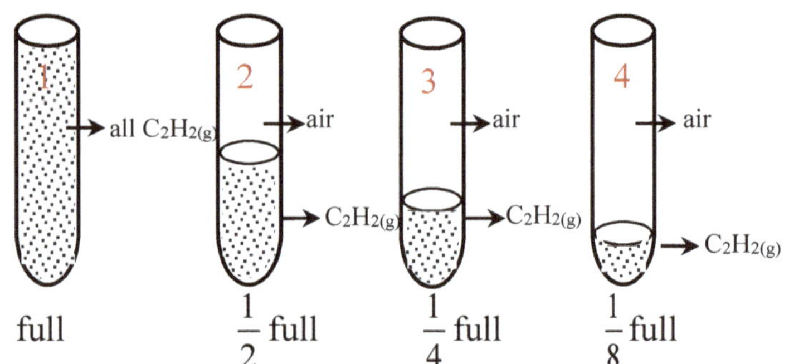

Figure 7.16. Test tubes indicating amounts of acetylene gas and volume of air to be ignited.

Procedure:

1. Wear protective clothing, goggles and gloves.
2. Using an indelible marker, label the four test tubes: 1, 2, 3, and 4; then make marks on these test tubes: $\frac{1}{2}$ full, $\frac{1}{4}$ full, and $\frac{1}{8}$ full and full, as indicated by the figure above, to indicate the relative volumes of acetylene gas and air that would be inserted into them.
3. Fill the test tubes completely with water and place them in the test tube rack.
4. Fill the 1L beaker with cold water up about three quarters full.
5. Using the tweezer, take a 2-gram sample of calcium carbide and place it in the beaker of water. Fizzing will take place spontaneously with the evolution of acetylene gas. *** Never hold calcium carbide with bare fingers.**
6. Place your thumb over the mouth of test **tube 1** and invert it in the beaker of water until the mouth of the test tube is directly above the lump of reacting calcium carbide. As you remove your finger, the acetylene gas that is produced will rise into the test tube to displace the water downwards. Allow **all** the water to be displaced from this test tube. Keeping the mouth of the test tube facing down remove it from the water and quickly stopper it. Return this test tube to the rack.
7. Repeat procedure 6 with test tube 2, but allow only half of the water to be displaced by the gas.
8. Repeat step 6 with test tube 3, but allow only one quarter of the water to be displaced by the gas.
9. Repeat step 6 with test tube 4, but allow only one eighth of the water to be displaced by the gas.
10. Invert the stoppered test tubes several times to allow the gas and air to mix well.
    - Note that when the test tubes were removed from the beaker, with their mouth pointed downwards, as the warter in them fell out, air moved to occupy the same volume as the water that fell out.

 Make hypotheses as to which test tubes there will be complete and incomplete combustion.

11. Ignite the gas in each test tube successively, by placing a burning splint quickly into the mouth of each test after the stopper is removed and record your results.

Results:

The following observations were made when this activity was carried out.

Test tube 1.

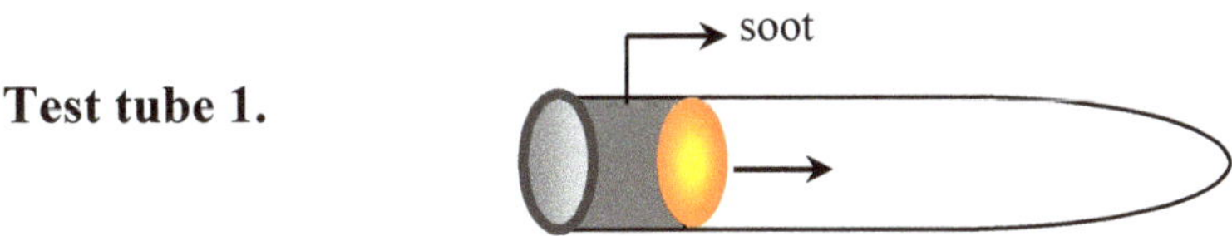

Figure 7.16(a). Burning of a full test tube of acetylene

The gas ignited at the mouth of the test tube and burnt with a small yellow flame that moved gradually down the test tube but became extinguished about less than a quarter the way down. The inside of the upper part of test tube was partially layered with soot. A very small amount of heat was produced.

Test tube 2.

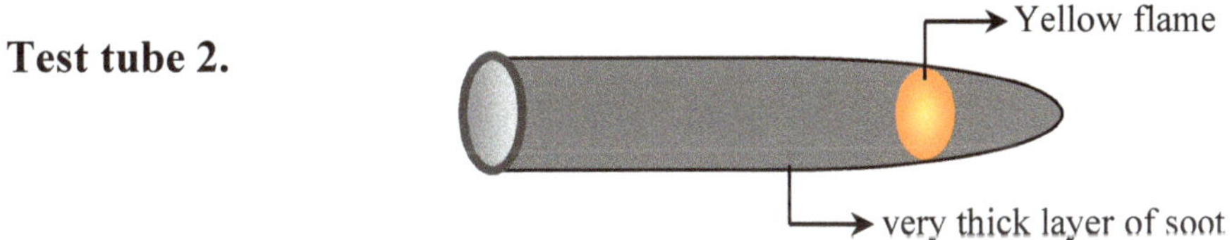

Figure 7.16(b). Burning of a half-full test tube of acetylene

The gas ignited at the mouth of the test tube, with a yellow flame which quickly moved down to the bottom of the tube and went out. The walls of the test tube were thickly layered with soot. More heat was felt than in test tube 1.

Test tube 3.

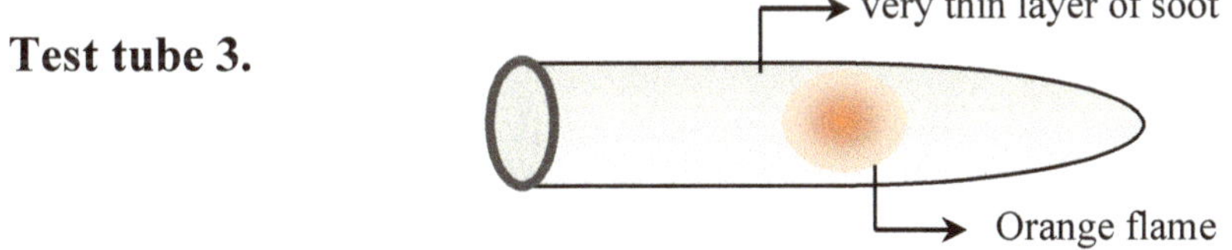

Figure 7.16c. Burning of a one quarter-full test tube of acetylene gas.

The gas ignited at the mouth of the test tube and burnt throughout the test tube with an orange flame. The walls of the test tube were thinly layered with soot. The test tube felt very hot; more energy was produced than in test tube 2.

Test tube 4.

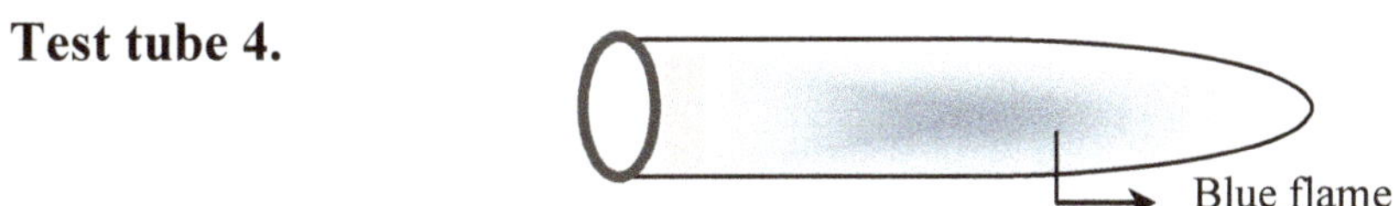

Figure 7.16(d). Burning of a one eight-full test tube of acetylene

The gas ignited at the mouth of the test tube and burnt throughout the test tube with a blue flame, producing a huge explosion. There was no evidence soot in the test tube. The test tube felt extremely hot.

1. In which test tube was the least amount of energy produced?
2. In which test tube was the most amount of energy produced?
3. In which test tube(s) was there complete combustion?
4. In which test tube(s) was there incomplete combustion?
5. How could you tell the answer to questions 3 and 4?
6. What relationships exist between the amount of oxygen present and the degree of combustion of the fuel?
7. Explain your observations of what happened in the four test tubes. Include in your answers the possible products of combustion that might be present in each teat tube.
8. What relationship exists between colour of the flame produced and the degree of combustion?
9. In which test tube was there the most efficient burning of the fuel?

Making Connection A

10. Why do you think it is important to tune up motor vehicles regularly? This process involves changing clogged up air filters, and old spark plugs etc.
11. People should not do certain types of burnings in enclosed dwelling areas. Explain the wisdom of this statement.
12. Home owner are advised to have their furnaces regularly inspected and serviced and also to have carbon monoxide detectors installed in their homes. Discuss the wisdom of having these things done.

7.15 Ontario's Drive Clean

This is an automobile emission control programme that had been instituted by the Government of Ontario in April, 1999, to reduce smog and other types of air pollution. It was recognised that exhaust emissions from aging vehicles, for various reasons, were one of the major sources of smog causing pollutants in Ontario. In an endeavour to reduce pollution, which is responsible for thousands of deaths, and the billions of dollars in yearly medical expenses incurred by the government, it was legislated that all vehicles, depending on their age, must undergo an annual or biennial emission test before their license plate sticker can be renewed. An emission test in done by a special licensed mechanic workshop during which time the vehicle is doing an assimilated drive but is not moving. Its tail pipe is connected to a hose that feeds the emission into a system that does a computer analysis of it. After this test, a printout is made showing all the components of the emission and their concentrations. These values are then compared to the acceptable limits set by the government. The test fails if any component of the emission is higher than the set limit. The owner of the vehicle is then advised to consult a qualified auto mechanic to remedy any fault that is affecting the emission. Poor emission could be to a number of factors such as:

- Faulty catalytic converter
- Imbalance of fuel to air mixture
- Leakage in exhaust system
- Unclean air filters
- Dysfunctional spark plugs

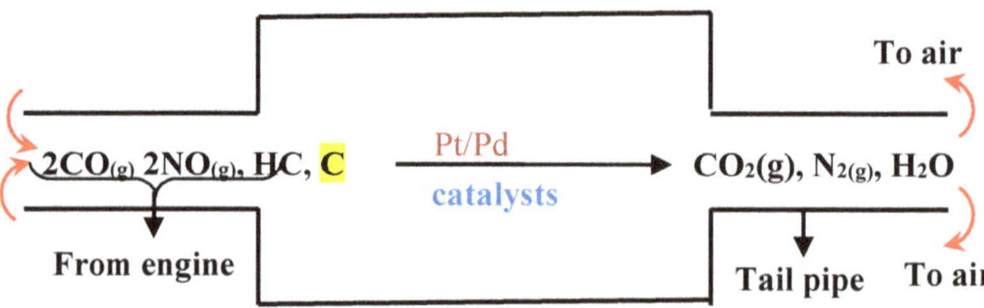

Figure 17.17. A catalytic converter changing soot, C unburnt fuel (HC), CO, and oxides of nitrogen into safer $CO_{2(g)}$, $N_{2(g)}$ and $H_2O_{(g)}$.

Fig.17.17 illustrates how a catalytic converter works. Any carbon momoxide, oxides of nitrogen, unburnt hydrocarbons, and solid carbon existing the combustion engine, are converted to safer levels in the form of carbon dioxide, nitrogen, and water vapour. Catalysts in the form of platinum and palladium are used to effect these changes. The converter is situated somewhere in the exhaust systems of vehicles before the emissions are allowed to enter the air.

*Note. *Since April 1, 2019,* Ontario drivers are not required to get Drive Clean emission tests for their **light vehicles,** however, emission standards still apply to all vehicles

Consider the following tables and then answer the questions below:

Table 7.4 (a). Limits set by government.

Components	Percent /volume
CO_2	0.5
CO	0.010
Sulphur oxides	0.0010
Formaldehyde	0.0005
Nitrogen oxides	0.0005

Table 7.4 (b). Emission from vehicle A.

Components	Percent /volume
CO_2	0.47
CO	0.009
Sulphur oxides	0.0009
Formaldehyde	0.004
Nitrogen oxides	0.0004

Table 7.4 (c). Emission from vehicle B.

Components	Percent /volume
CO_2	0.55
CO	0.021
Sulphur oxides	0.002
Formaldehyde	0.0005
Nitrogen oxides	0.0008

13. a) Which vehicle failed and which passed the test?

   b) What type of diagnosis could you make for the vehicle that failed if you were an auto mechanic?

   c) For the vehicle that failed, what risks to human health would vehicles like these

   cause, if they are allowed to keep driving?

14. If you were the minister of the environment, what other laws would you legislate with regards to the emissions from other sources of combustions such industries?

15. Not all carbon monoxide is converted to carbon dioxide. Why it would be dangerous for anyone to warm up their vehicles in a closed garage for any extended period of time?

As can be recalled, ionic compounds are composed of metal ions (*cations*) and nonmetal ions (*anions*). The process by which these ions are identified and separated in ionic compounds is called ***qualitative analysis***.

Cations of transition metals (except zinc) form complexes with colours specific for each cation. When their cations form complexes with **ligands,** (some ions or moelcules) these interfere with some ***d orbitals*** because of their close proximity to them than others. This causes splitting of the d orbitals with some (**dz²** **and dx²-y²**) having higher energy than the others (*dxy, dxz and dyz*), if the ligands form octahedral complexes **Fig 7.18.** To put an electron into the split *dz² and dx²-y²* orbitals would require more energy than if the orbitals were degenerate. For this to happen, a quantum of energy from the visible spectrum corresponding to a certain colour is absorbed by the electron(s). The light that is emitted from the compound is complimentary to the colour absorbed. For example, $CuSO_{4(aq)}$ is blue because it **absorbs red light** from the visible spectrum **leaving blue (cyan)** which is complementary to red. The colours emitted may depend on the nature of ligands that are attached to the central cation. For example, the light blue colour of $CuSO_{4(aq)}$ is changed from cyan to a **deeper blue** if excess aqueous ammonia is poured into it. This happens because some of the water molecules are replaced by ammonia molecules that cause greater splitting of the d orbitals; the electron clouds of ammonia ligands repel the dz² and dx²-y² orbitals more than water ligands do. This requires more energy (**yellow** photons) for the promoting of the electron(s) resulting with complementary **deep blue** colour being seen, figures 7.19 (a) and 7.19 (b).

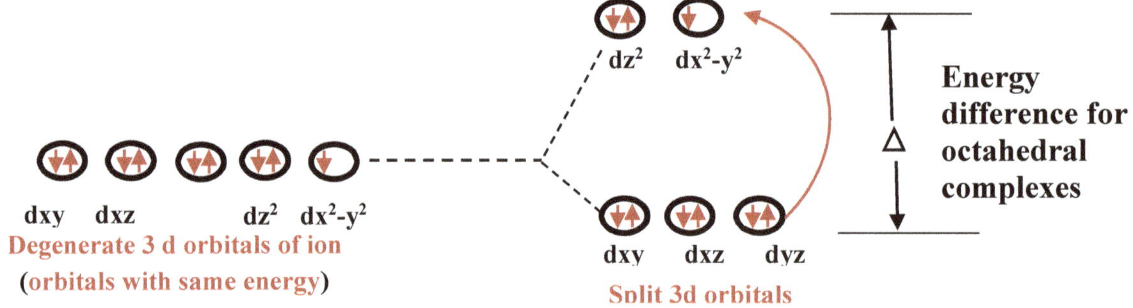

Figure 7.18.   Showing the splitting of the d orbitals of Cu^{2+} in $CuSO_{4(aq)}$.

The above explanation is very simplistic of how these complexes are formed by transition metals. A full explanation of this concept is beyond the scope of this book. Students are not required to know this. This was done to facilitate the teacher.

Table: 7.5. Colours of some aqueous transition metal cations.

Cations	Colour in aqueous solution
$Cu^{+}_{(aq)}$	colourless
$Cu^{2+}_{(aq)}$	Blue
$Fe^{2+}_{(aq)}$	light green
$Fe^{3+}_{(aq)}$	brownish-yellow
$Mn^{2+}_{(aq)}$	light pink
$Mn^{7+}_{(aq)}$	Magenta
$Cr^{3+}_{(aq)}$	Green
$Cr^{6+}_{(aq)}$	Orange
$V^{3+}_{(aq)}$	green
$V^{4+}_{(aq)}$	Blue
$V^{5+}_{(aq)}$	Orange
$Ni^{2+}_{(aq)}$	Green
$Co^{2+}_{(aq)}$	Pink
$Ti^{2+}_{(aq)}$	Purple

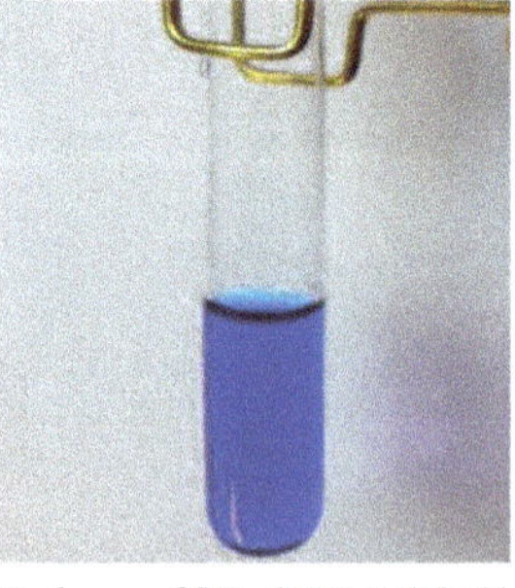
Figure 7.19(b). Colour of $[Cu(NH_3)_4(H_2O)_2]^{2+}$ complex

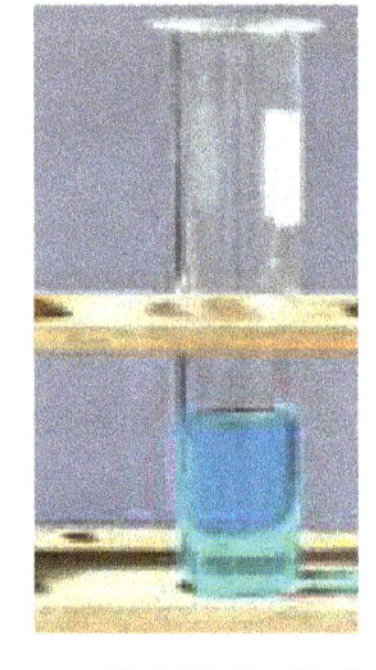
Figure 7.19(a). Colour of $[Cu(H_2O)_6]^{2+}$ complex

Figure 7.20. Colours of a few transition metal

Analysis of Cations

Cations of ionic in aqueous solutions can be identified in a number of ways:

- By inspection of the colours of the solutions, as in the case of the transition metal ions (table 7.5.)
- By the colours produced when their solutions are heated in a Bunsen flame pg.25. This is more applicable to metal ions that do not form coloured solutions. The following table gives the colours produced by the flame test method.

Table 7.6. Flame test for a few cations

Element	Ions	Colour in Flame
Potassium	K^+	Lilac
Sodium	Na^+	yellow orange
Lithium	Li^+	carmine red
Cesium	Sc^+	violet
Calcium	Ca^{2+}	yellow red
Strontium	Sr^{2+}	crimson red
Barium	Ba^{2+}	pale green
Rubidium	Rb^+	grey purple
Zinc	Zn^{2+}	bluish green
Lead	Pb^{2+}	Bluish white

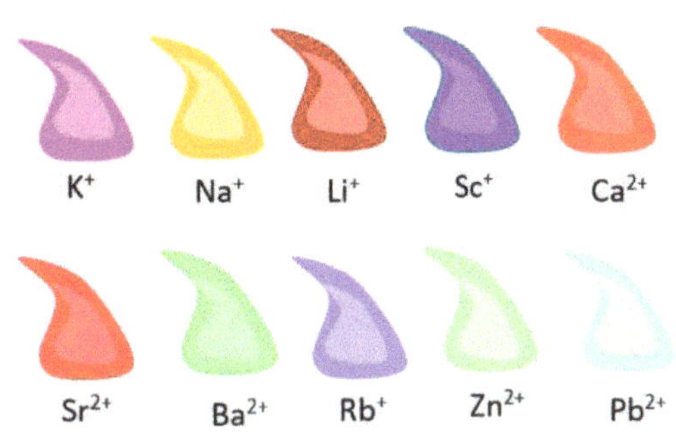

Figure 7.21. Assimilated flame tests for a few cations

- By chemical analysis. The following table shows how a few metal ions can be identified by chemical means.

Table: 7.7.   Chemical analysis for a few cations

Ions	Test	Observation
$Fe^{2+}_{(aq)}$	Add $NaOH_{(aq)}$ to a sample of solution	Formation of a green gelatinous precipitate (**ppt**) $Fe^{2+}_{(aq)} + 2\,NaOH_{(aq)} \longrightarrow Fe(OH)_{2(s)} + 2\,H_2O_{(l)}$
$Fe^{3+}_{(aq)}$	Add $NaOH_{(aq)}$	Formation of a green reddish precipitate $Fe^{3+}_{(aq)} + 3\,NaOH_{(aq)} \longrightarrow Fe(OH)_{3(s)} + 3\,H_2O_{(l)}$
$Cu^{2+}_{(aq)}$	Add $NaOH_{(aq)}$ or $NH_{3(aq)}$ to a sample solution	Formation of a blue gelatinous precipitate with $NaOH_{(aq)}$. With $NH_{3(aq)}$ the blue precipitate re-dissolves to form a deep blue solution $Cu^{2+}_{(aq)} + 2\,NaOH_{(aq)} \longrightarrow Cu(OH)_{2(s)} + 2\,H_2O_{(l)}$
$Zn^{2+}_{(aq)}$	Add $NaOH_{(aq)}$ or $NH_{3(aq)}$ to a sample of solution	Formation of white precipitate that re-dissolves in excess $NaOH_{(aq)}$ or $NH_{3(aq)}$. $Zn^{2+}_{(aq)} + 2\,NaOH_{(aq)} \longrightarrow Zn(OH)_{2(s)} + 2\,H_2O_{(l)}$
$NH_4^+{}_{(aq)}$	Add $NaOH_{(aq)}$ to a sample solution	Evolution of a pungent smelling gas ($NH_{3(aq)}$) that turns damp pink litmus paper blue. $NH_4^+{}_{(aq)} + OH^-{}_{(aq)} \longrightarrow NH_4OH_{(aq)} \longrightarrow NH_{3(g)} + H_2O_{(l)}$

Analysis of Anions

Table: 7.8.   Chemical analysis for a few anions.

Anions	Test	Observation
$Cl^-_{(aq)}$	Add $AgNO_{3(aq)}$ to a sample solution	Formation of white precipitate that re-dissolves in dilute $NH_{3(aq)}$. $Ag^+_{(aq)} + Cl^-_{(aq)} \longrightarrow AgCl_{(s)}$ (white precipitate)
$Br^-_{(aq)}$	Add $AgNO_{3(aq)}$ to a sample solution followed by dilute $NH_{3(aq)}$.	Formation of cream precipitate that is sparingly soluble in dilute $NH_{3(aq)}$. It re-dissolves only in concentrated $NH_{3(aq)}$. $Ag^+_{(aq)} + Br^-_{(aq)} \longrightarrow AgBr_{(s)}$ (cream precipitate)
$I^-_{(aq)}$	Add $AgNO_{3(aq)}$ to a sample solution	Formation of yellow precipitate that is insoluble in $NH_{3(aq)}$. $Ag^+_{(aq)} + I^-_{(aq)} \longrightarrow AgI_{(s)}$ (yellow precipitate)

Ion	Test	Observation
$CO_3^{2-}{}_{(aq)}$	Add dilute $HCl_{(aq)}$ to a sample solution	Effervescence takes place with the evolution of a colourless, odourless gas ($CO_{2(g)}$) that turns lime water milky. $CO_3^{2-}{}_{(aq)} + 2\,H^+{}_{(aq)} \longrightarrow H_2CO_{3(aq)} \longrightarrow CO_{2(g)} + H_2O_{(l)}$
$SO_4^{2-}{}_{(aq)}$	Add $BaCl_{2(aq)}$ to a sample solution	Formation of white precipitate that does not re-dissolves in dilute $HCl_{(aq)}$. $Ba^{2+}{}_{(aq)} + SO_4^{2-}{}_{(aq)} \longrightarrow BaSO_{4(s)}$ (white precipitate)
$SO_3^{2-}{}_{(aq)}$	Add $BaCl_{2(aq)}$ to a sample solution followed by $HCl_{(aq)}$.	Formation of white precipitate that re-dissolves in dilute $HCl_{(aq)}$ producing a gas ($SO_{2(g)}$) that turns pink acidified $KMnO_{4(aq)}$, colourless. $Ba^{2+}{}_{(aq)} + SO_3^{2-}{}_{(aq)} \longrightarrow BaSO_{3(s)} \longrightarrow BaSO_{3(s)} + 2\,HCl_{(aq)}$ $BaCl_{2(aq)} + H_2SO_{3(aq)} \longrightarrow H_2SO_{3(aq)} \longrightarrow H_2O_{(l)} + SO_{2(g)}$
$NO_3^-{}_{(aq)}$	Add $NaOH_{(aq)}$ to sample of solution. Add pieces of $Al_{(s)}$ to the mixture and warm.	Evolution of a pungent smelling gas, $NH_{3(aq)}$ that turns damp pink litmus paper blue. $3\,NO_3^-{}_{(aq)} + 8\,Al_{(s)} + 5\,H_2O_{(l)} \longrightarrow 3NH_{3(g)} + 8\,[Al(OH)_4]^-{}_{(aq)}$

Exercise 7.9

1. What would be the colours of the following aqueous solutions? K/U

   a) $NaBr_{(aq)}$ b) $FeCl_{3(aq)}$ c) $AlBr_{3(aq)}$ d) $K_2Cr_2O_{7\,(aq)}$ f) $CoCl_{2(aq)}$

   g) $CuSO_{4(aq)}$ h) $Fe(NO_3)_{2(aq)}$ i) $Mg(NO_3)_{2(aq)}$ j) $NiI_{2(aq)}$ h) $CaBr_{2(aq)}$

Analysis of Ionic Compounds

Table: 7.9.   Analysis of the ions of a few ionic compound

Compounds	Test	Observation	Inference	Formula
potassium chloride	-Do flame test	- Lilac flame produced.	$K^+{}_{(aq)}$ present	
	- Add $AgNO_{3(aq)}$ followed by $NH_{3(aq)}$.	-A white ppt formed that re-dissolved in $NH_{3(aq)}$.	$Cl^-{}_{(aq)}$ present	KCl
copper (II) sulphate	- Add $NaOH_{(aq)}$ to sample of solution.	- A blue gelatinous ppt is produced.	$Cu^{2+}{}_{(aq)}$ present	
	- Add $BaCl_{2(aq)}$ to a sample solution followed $HCl_{(aq)}$.	- formation of a white ppt that does not re-dissolve in dilute $HCl_{(aq)}$.	$SO_4^{2-}{}_{(aq)}$ present	$CuSO_4$
zinc bromide	- Add excess $NaOH_{(aq)}$ to sample of solution. - Do flame test	- A white ppt formed that re-dissolved in excess $NaOH_{(aq)}$. - Bluish green flame formed.	$Zn^{2+}{}_{(aq)}$ present	
	- Add $AgNO_{3(aq)}$ to a sample of solution followed by conc. $NH_{3(aq)}$.	- A cream ppt formed that re-dissolved in conc. $NH_{3(aq)}$.	$Br^-{}_{(aq)}$ present	$ZnBr_2$
ammonium carbonate	- Add $NaOH_{(aq)}$ to a sample of solution.	- A pungent smelling gas that turns damp pink litmus paper blue was produced.	$NH_4^+{}_{(aq)}$ present	
	- Add dilute $HCl_{(aq)}$ to a sample solution. - Pass any gas produced into lime water.	The evolution of a colourless gas that turned lime water milky.	$CO_3^{2-}{}_{(aq)}$ present	$(NH_4)_2CO_3$
lithium sulphite	- do flame test	- A carmine red flame was produced.	$Li^+{}_{(aq)}$ present	
	- Add $BaCl_{2(aq)}$ to a sample solution followed by $HCl_{(aq)}$. Pass any gas produced into acidified $KMnO_{4(aq)}$.	- Formation of white precipitate that re-dissolved in dilute $HCl_{(aq)}$ producing a gas ($SO_{2(g)}$ that turns pink acidified $KMnO_{4(aq)}$.	$SO_3^{2-}{}_{(aq)}$ present	Li_2SO_3

7.17   Activity: Analysis of unknowm Compounds

A bottle containing an aqueous solution of an ionic compound has lost its label. The compound is believed to be either *barium chloride* **or** *lithium carbonate*. Describe how this compound can be identified using the various tests described above.

The following procedure outlines the steps that are carried out analyse and determine solutions of ionic compounds.

Procedure:
1) Pour samples of the solution into four test tubes and label these A, B, C, and D.
2) To each of these samples carry out the following tests as outlined in the following table.

Table: 8.0. Analyses of ionic compounds

Sample	Tests	Observation	Inference
A	-Add $Na_2SO_{4(aq)}$ followed by dilute $HCl_{(aq)}$.	-If a white ppt. forms that does not re-dissolve in $HCl_{(aq)}$. $$BaCl_{2(aq)} + Na_2SO_{4(aq)} \longrightarrow BaSO_{4(s)} + 2\,NaCl_{(aq)}$$	-$Ba^{2+}_{(aq)}$ ions are present.
	- Do the flame test	-If a pale green flame is produced then	-$Ba^{2+}_{(aq)}$ ions are confirmed present.
		-If neither of these are produced then	- $Ba^{2+}_{(aq)}$ ions are absent.
B	- Do the flame test	-If a carmine red flame is produced then.	- $Li^{+}_{(aq)}$ ions are confirmed present.
		- If not then	- $Li^{+}_{(aq)}$ ions are absent.
C	- Add $AgNO_{3(aq)}$ followed by $NH_{3(aq)}$.	- If a white ppt forms that re-dissolves in $NH_{3(aq)}$ then $$AgNO_{3(aq)} + BaCl_{2(aq)} \longrightarrow Ba(NO_3)_{2(aq)} + 2AgCl_{(s)}$$	-$Cl^-_{(aq)}$ ions are confirmed present.
		- If these do not take place then	-$Cl^-_{(aq)}$ ions are absent.
D	-Add dilute $HCl_{(aq)}$	- If effervescence takes place with the evolution of a gas that turns lime water milky then $$Li_2CO_{3(aq)} + HCl_{(aq)} \longrightarrow 2LiCl_{(aq)} + H_2O_{(l)} + CO_{2(g)}$$	-$CO_3^{2-}_{(aq)}$ ions are confirmed present.
		- If these do not take place then	- $CO_3^{2-}_{(aq)}$ ions are absent.

Conclusion:

- If samples A and C tested positive, the compound is **BaCl₂**.
- If samples B and D tested positive, the compound is **Li₂CO₃**

Exercise 7.10

1. Describe how you would do the same, as previously done, for a bottle of a reagent that has lost its label that could be either potassium iodide or Iron (III) sulphite. T/I C

2. Barium sulphate is given to patients in a form of a drink to create an image of some parts of the gastrointestinal tract, during X-ray imaging. It helps in this process because wherever its particles are, they able to absorb the X-rays thereby creating a visible image of the outline of the organ; without it no image would be seen. Why do you think that is barium sulphate used even though it is toxic? AK/U

3. A mixture consists of iron filings, calcium carbonate and potassium bromide. Describe how would first separate each component of the mixture and test the *compounds* to confirm their identities. CT/I

4. During the breathalyzer test for alcohol, a subject is required to breathe into a testing device that has potassium dichromate solution. A positive test results in a change in the dichromate solution from orange to varying shades of green depending on the concentration of alcohol in the breath sample. Research this test and determine what property of the testing reagents allows for a change in colour of the solution. T/I A

5. The results of forensic testing of various substances at a crime scene are significant in various ways: they can be used to indict suspects; convict or exonerate those charged with crimes. Research why the crime scene should be cordoned off as soon as possible and all protocols must be stringently adhered to when collecting, handling, storing and testing samples. A T/I

6. Chlorine is one of the agents that municipalities use to kill pathogens when purifying water for domestic use. High levels of chlorine in drinking water is associated with increased risks for bowel cancer. It has been established that people living in two different municipalities, A and B, consume similar types of food and beverages, but drink water from two different municipal sources. However, the people living in locality A experience a much higher incidence of bowel cancer than people living in locality B. The presence of chlorine in water is indicated by the addition of aqueous silver nitrate solution to it when a white precipitate of silver chloride is formed. Depending on the amount of chlorine in water, the higher the intensity of whiteness produced. Describe how you would use the scientific method to investigate this problem? T/I A

   *Note for the teacher.
   Use tap water, but put a bit of sodium chloride in one sample and make it water from source A. Leave the other one as is for sample B. Provide these samples and get students to do the activity with.

CHAPTER 8
Writing Names and Formulas for Chemical Compounds

Chapter Content:

Since each chemical compound has its own unique physical and chemical properties, it is essential that it be known internationally by a name that is unique to it. Up to the eighteenth century, there was no system of naming compounds; some compounds were even named after places and people. It was only in 1787 that the French chemist, Guyton de Morveau devised his **classical system** of naming compounds. Morveau used words from both Greek and Latin to names compounds. In his system, he named the metal of a compound first and ending it with a suffix such as **-ous** and **-ic**. For example, the element iron in a compound can be named ferr**ous** or ferr**ic**. For some compounds, the suffix can be **-ite and -ate** and the prefix can be **hypo- and per-.** However, with the discovery of more complex compounds the classical system became limited in its usefulness. For example, some metals have ions that have more than two valences, as such it fell short in its capacity to name compounds with these large number of valences.

In the year 1919, the Prussian chemist Alfred Stock devised an alternative system of naming compound which became known after him as the **Stock system.** This system uses Roman numerals in parentheses to represent the specific charges on the metal ions rather than prefixes which have some degree of ambiguity. For example, the name cup**rous** is replaced by copper (I) and fer**rous** is replaced by iron (II). In these cases, (I) and (II) represent the 1+ and 2+ valences of the copper and iron ions respectively.

The International Union of Pure and Applied Chemistry (IUPAC) which was formed in1920 has overseen the standardization of rules for naming compounds, and its most recent revision was published in 1990, with regards to naming compounds. One of its main objectives was to ensure that there is always international agreement to naming compounds; scientists must have precise knowledge of the different elements and the number of atoms of each element that a compound is composed of. The recommendations provided allow nations of the world the flexibilities of using the classical or Stock systems in naming compounds**.**

Names and Formulas of Chemical Compounds

8.1 Binary Ionic Compounds

A **binary ionic compound** is made up of only two kinds of elements; a metal atom(s) that has lost an electron(s) and a non-metal atom(s) that has gained an electron(s). The two resulting types of ions are called cations and anions respectively. The formula for any binary ionic compound can be arrived at in many different ways. The formula of the compound, lithium oxide for example, can be arrived at as follows:

Bohr-Rutherford diagrams

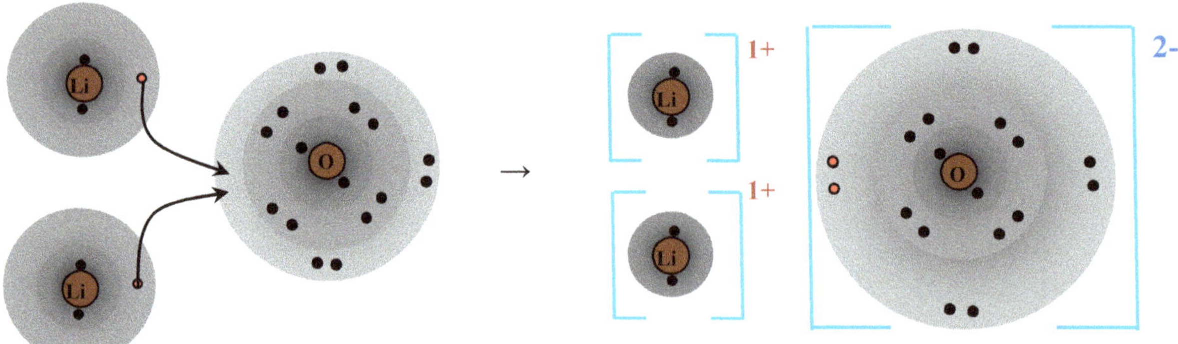

Figure 8.1. The formation of Li_2O using Bohr-Rutherford diagrams of Li and O.

The above figures show how two lithium ions combine with one oxygen ion to form the ionic compound, lithium oxide. **Subscripts** are used to indicate how many ions are present for each element in the compound.

➢ **Lewis dot diagrams**

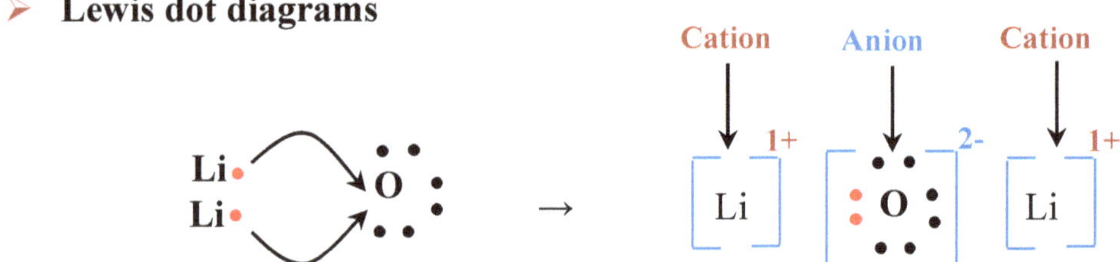

Figure 8.2. The formation of Li_2O using Lewis dot diagram.

In Figure 8.2, it can once again be seen how two lithium ions combine with one oxygen ion to form the ionic compound, lithium oxide. The charges on the ions are referred to as their **valences**. It should be noted that for any ionic compound, *the sum of the positive valences must be equal to the sum of the negative valences.* An ionic compound must therefore be neutral, since the sum of these charges results in zero charge.

➢ **The Crisscross Method**

The crisscross method is base on the assumption that the sum of the positive valences and negative valences of the ions in an ionic compound is zero. This concept can be exemplified, using the compound, lithium oxide.

$$Li^{1+}$$
$$O^{2-}$$
$$Li^{1+}$$

Sum of positive valences for 2 Li^{1+} ions = 2 (1+) = **2+**

Sum of negative valences for 1 O^{2-} ion = 1 (2-) = 2-

Total charge = 0

The following steps are carried out, using the crisscross method to derive the formula of any ionic compound, using lithium oxide as an example.

Step 1. Write the symbol for each element of the compound, with the metal first followed by the non-metal.

$$Li \qquad O$$

Step 2. Write the respective valence above each element.

$$\overset{1+}{Li} \qquad \overset{2-}{O}$$

Step 3. Crisscross the valences, making them subscripts for the elements. *Do not include the charges.*

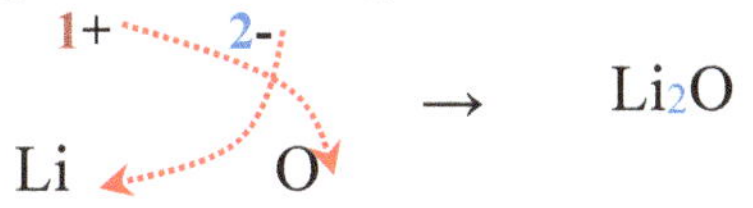

$$\overset{1+}{\text{Li}} \quad \overset{2-}{\text{O}} \quad \rightarrow \quad Li_2O$$

Step 4. If a subscript of **1** is present, remove it. *Any element without a subscript is assumed to have a subscript of one.* The formula of lithium oxide is thus **Li_2O.**

Using calcium oxide as other example, we have:

$$\overset{2+}{\text{Ca}} \quad \overset{2-}{\text{O}} \quad \rightarrow \quad Ca_2O_2$$

Reducing the subscripts to the lowest values, we get the formula, CaO.

8.2 Metal Ions with Standard Valences versus Multiple Valences

Some metals have only one valence which we shall call **standard valence.** But, there are some metals, such as the transition metals, that can have more than one valence. These are called **multivalent** elements. The common non-metal atoms also have standard valences when they form binary compounds. The following section of the periodic table shows the valences of some elements.

Table 8.1. Part of the periodic table showing the valences for a few common elements.

1																	18
H^{1+}	2											13	14	15	16	17	He^0
Li^{1+}	Be^{2+}											B^{3+}		N^{3-}	O^{2-}	F^{1-}	Ne^0
Na^{1+}	Mg^{2+}	3	4	5	6	7	8	9	10	11	12	Al^{3+}		P^{3-}	S^{2-}	Cl^{1-}	Ar^0
K^{1+}	Ca^{2+}	Sc^{3+}	Ti^{3+} Ti^{4+}	Vd^{3+} Vd^{4+} Vd^{5+}	Cr^{2+} Cr^{3+} Cr^{6+}	Mn^{2+} Mn^{7+}	Fe^{2+} Fe^{3+}	Co^{2+} Co^{3+}	Ni^{2+} Ni^{4+}	Cu^{1+} Cu^{2+}	Zn^{2+}					Br^{1-}	Kr^0
Rb^{1+}	Sr^{2+}									Ag^{1+}	Cd^{2+}		Sn^{2+} Sn^{4+}			I^{1-}	Xe^0
Cs^{1+}	Ba^{2+}								Pt^{2+} Pt^{4+}	Au^{1+}	Hg^{1+} Hg^{2+}		Pb^{2+} Pb^{4+}				

Metal ions in Chemical Compounds

If it were required to write the formula for the compound, calcium oxide, the only formula that would come to mind is **CaO.** This is so because the valences of the calcium and oxygen are fixed. However, writing the formula for the compound, copper oxide, would be an ambiguous task since copper is a multivalent element with valence of either **1+ or 2+.** The oxides of copper could thus be written as either **CuO or Cu_2O.**

To distinguish between the various valences for metal ions having multiple valences, scientists have derived a few methods of naming such metal ions in compounds.

➤ **The Classical System**

This method of naming multivalent ions in a compound uses the Latin name for the element, followed by a suffix. To indicate the metal ion with the lower valence, the suffix, *ous* is used and for the metal ion with the higher valence the suffix, *ic* is used. The disadvantage of this method is that it is a limited to naming metal ions with only two valences. For example, in the compound, Cu_2O, the copper ion is named *cuprous,* and in CuO, it is named *cupric.*

The following figure illustrates the classical names for the ions formed from a few elements that form only two different valences.

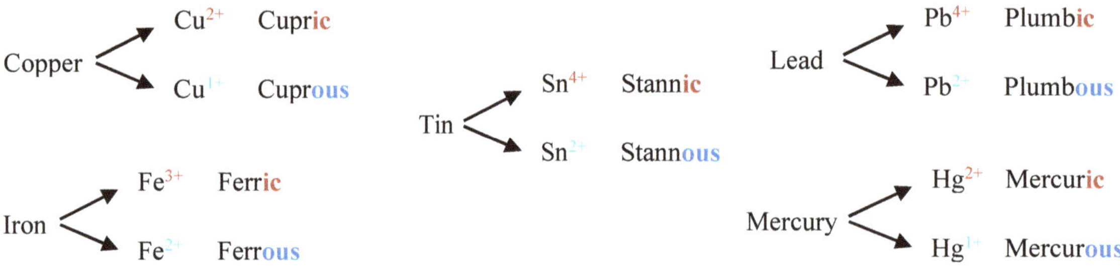

Figure 8.2. The classical way of naming metal ions with two valences.

> **The Stock System**

Alfred Stock, the German Chemist, used Roman numerals in parenthesis to indicate the valence on the metal ion. For example, in the compound, Cu_2O, the copper ion is named Copper (I) and in CuO, it is named Copper (II).

Table 8.2. The Roman equivalents of Arabic numeral.

Arabic Numerals	1	2	3	4	5	6	7
Roman Numerals	I	II	III	IV	V	VI	VII

Table 8.3. Stock System for naming metal ions with multiple valences.

Metal	Ion of Lower Valence	Ion of Higher Valence
Copper	Copper (**I**)	Copper (**II**)
Iron	Iron (**II**)	Iron (**III**)
Tin	Tin (**II**)	Tin (**IV**)
Lead	Lead (**II**)	Lead (**IV**)
Mercury	Mercury (**I**)	Mercury (**II**)

Note:
When writing the formula for a binary ionic compound containing a metal ion of standard valence, do not use Roman numerals to indicate the charge on the ion. This is redundancy since the charge on the ion is already known and it does not change.

8.3   Naming the Non-metal in Binary Ionic Compounds

To name the non-metal in binary ionic compounds, the last syllable in the name of non-metal is dropped and is replaced by the syllable **–ide**. For example, oxygen becomes **oxide.** The following table shows how this is done for the rest of the non-metal elements.

Table 8.4. Naming the non-metal in binary ionic compounds.

Oxygen	*Sulphur*	*Nitrogen*	*Phosphorus*	*Fluorine*	*Chlorine*	*Bromine*	*Iodine*	*Hydrogen*
Oxide	Sulphide	Nitride	Phosphide	Fluoride	Chloride	Bromide	Iodide	Hydride

Write the formula for the following compounds: K/U

Compound	Formula	Compound	Formula
(a) magnesium fluoride		(k) aluminum nitride	
(b) lead (II) oxide		(l) tin (IV) bromide	
(c) potassium sulphide		(m) iron (II) chloride	
(d) zinc iodide		(n) strontium nitride	
(e) iron (III) sulphide		(o) sodium posphide	
(f) calcium bromide		(p) aluminum oxide	
(g) barium chloride		(q) cesium bromide	
(h) copper (II) oxide		(r) rubidium nitride	
(i) silver iodide		(s) gold(I) chloride	
(j) lead (IV) oxide		(t) mercury (II) oxide	

8.4   Naming Binary Ionic Compounds with Multiple Valences

When writing the name of a binary ionic compound, it is important to ask the following big question:

Is the valence of metal ion in the compound multivalent or standard?

If the metal ion in the compound is of standard valence, just state its name and continue naming the compound. For example, in the formula ZnO the zinc ion is of standard valence, so the name of the compound is zinc oxide. However, if the metal ion in the compound is multivalent the following must be done.

➤ First, deduce its valence.
Using the formula, PbO_2, for example, where the element lead is multivalent, its valence is found as follows:
Since the compound is ionic, the sum of the charges on its negative ion(s) and that for its positive ions must be zero. Since there are two oxide ions with standard valence of 2-, then their sum is: $2O^{2-} = 4-$
Since there is only one lead ion, its valence must be $4+$ for the compound to neutral.

$$\begin{aligned} 2\,O^{2-} &= 4- \\ 1\,Pb &= 4+ \\ &= 0 \end{aligned}$$

➤ Second, place the specific Roman numeral in parenthesis to indicate the valence of the metal and then proceed with the naming process. The name of the compound is thus lead (IV) oxide, using the IUPAC system, or plumbic oxide using the classical system.
For the compound Fe_2O_3, if the same steps are applied, we have:

$$\begin{aligned} 3\,O^{2-} &= 6 \\ 2\,Fe &= 6+ \\ &= 0 \end{aligned}$$

Since two Fe ions have a total of $6+$, then each must have a valence of $3+$; the ion is written as Fe^{3+}. The name of the compound, Fe_2O_3 is Iron (III) oxide using the IUPAC system or ferric oxide using the Classical system.

Student Difficulty

Students sometimes experience difficulty in distinguishing between the multivalent metal ions from those with standard valence. To assist with this, the following table is provided with the various common metal ions and their valences, both standard and multiple.

Table 8.5. **List of metal with standard and multiple valences.**

Group IA	Li^{1+} Na^{1+} K^{1+} Rb^{1+} Cs^{1+}	Along with	Metal ions with standard valence
Group IIA	Be^{2+} Mg^{2+} Ca^{2+} Sr^{2+} Ba^{2+}	Ag^{1+}	
Group IIIA	B^{3+} Al^{3+}	Zn^{2+}	
Transition metals	Fe^{2+} Fe^{3+} Cu^{1+} Cu^{2+} Sn^{2+} Sn^{4+} Pb^{2+} Pb^{4+} Au^{2+} Au^{1+} Co^{3+} Co^{2+} Hg^{2+} Hg^{1+} ......etc.		Metal ions with multiple valences

If the metal ions with the standard valences; Group IA, Group IIA, Group IIIA, along with Ag^{1+} and Zn^{2+} are committed to memory, then by the process of elimination the multivalent one can easily be identified. If an ion is not standard it must be multi-valent.

Exercise 8.2

Write the names of the following binary ionic compounds using the IUPAC system of nomenclature. K/U

(a)	**Li_2O**	**(i)**	**HgO**
(b)	**CaO**	**(j)**	**Ag_2S**
(c)	**$AlCl_3$**	**(k)**	**$PbCl_4$**
(d)	**MgS**	**(l)**	**BaI_2**
(e)	**Na_2S**	**(m)**	**Cs_2O**
(f)	**FeO**	**(n)**	**$SnBr_4$**
(g)	**$CuCl_2$**	**(o)**	**KF**
(h)	**Na_3N**	**(p)**	**$CoCl_2$**

8.5 Writing Formulas for and Naming Binary Molecular Compounds.

A binary molecular compound is one that is composed of two non-metallic elements. To name these types of compounds, prefixes are used to indicate the number of each type of atom in the compound. The following table gives a list of the prefixes that are used.

Table 8.6. Corresponding prefixes for the Arabic numerals.

Number	1	2	3	4	5	6	7	8	9	10
Prefix	mono-	di-	tri-	tetra-	penta-	hexa-	hepta-	octa-	nona-	deca-

If there is only one atom of each element in the compound, the prefix *mono-* is written *only for the second element*. For example, NO is called nitrogen monoxide. For poetic reasons, the "o" is dropped from the prefix, "mono-". The following are the names of a few binary molecular compounds.

Table 8.7. Names and formulae for a few binary molecular compounds. K/U

Formula	Name
N_2O_4	dinitrogen tetroxide
SCl_2	sulphur dichloride
N_2O	dinitrogen monoxide

Exercise 8.3

Write the formulas for the following binary molecular compounds. K/U

(a)	carbon tetrachloride	(e)	nitrogen dioxide
(b)	phosphorus trichloride	(f)	diphosphorus trioxide
(c)	dinitrogen pentoxide	(g)	sulphur trioxide
(d)	sulphur hexachloride	(h)	ammonia

Exercise 8.4

Write the IUPAC name for each of the following binary molecular compounds. K/U

(a)	SBr_6	(e)	CF_4
(b)	CS_2	(f)	P_2O_5
(c)	SiO_2	(g)	HCl
(d)	CO	(h)	N_2O_4

Student Difficulty

So far, it was relatively easy to write the names and formulas for binary compounds. The reason being that each type of compound was categorized; they were either only binary ionic or binary molecular. However, when the different types of binary compounds (ionic and molecular) are mixed up, and students are required to name them, some degree of difficulty arises. Take the following compounds for example: PCl_5, $FeCl_2$ and $CaCl_2$. To minimize difficulty in naming a set of binary compounds like the ones above, it is useful to do the following:

> **Decide if the compound is molecular or ionic.**

To do this, inspect the compound to see if both elements are non-metal. If both are non-metal, then the compound is molecular. If not, it must contain a metal and a non-metal. If this is the case, the compound is ionic.

In our list above, it can be deduced that PCl_5 is molecular and both $FeCl_2$ and $CaCl_2$ are ionic. The compound, PCl_5 being molecular is named **phosphorus pentachloride**.

If the compound is ionic, then the next step would be:

> **Decide if the metal ion in the ionic compound is multi-valent or standard.**

In the case of $CaCl_2$, the calcium ion is of standard valence **as it belongs to group IIA** of the periodic table. It is thus named **calcium chloride**.

In the case of $FeCl_2$, since the iron ion **is not from the list of ions with standard valence**, it must be multivalent. To name this compound it is important to first find the valence of iron in the compound. Since the charge on the anion is known, the valence of iron is found as follows:

$$2\ Cl^- = 2-$$
$$1\ Fe = 2+$$
$$0$$

The calculated valence of iron is 2+ so the ion is written as Fe^{2+}. To name this compound, the appropriate Roman numeral is placed in parenthesis after the metal followed by the name of the anion. **$FeCl_2$ is thus named, iron (II) chloride.**

Write the IUPAC name for each of the following binary compounds. K/U

(a)	$AuCl$		(f)	SnO_2
(b)	BaO		(g)	NCl_3
(c)	PbI_2		(h)	K_3N
(d)	ZnO		(i)	CuO
(e)	HBr		(j)	Ag_2S
(K)	Zn_3N_2		(p)	Hg_2O
(l)	$AuCl_3$		(q)	N_2O
(m)	Pb_3N_2		(r)	HI
(n)	O_3		(s)	OCl_2
(o)	Ag_3N		(t)	ZnI_2

Matching

Match each term in the table below with the correct statements that follow

A	$CH_{4(g)} + 2O_{2(g)} \rightarrow CO_{2(g)} + 2H_2O_{(l)} + energy$	G	Decomposition
B	Catalyst	H	$Cu_{(s)} + 2HCl_{(aq)} \rightarrow CuCl_{2(aq)} + H_{2(g)}$
C	Li, K Na, Ba and Ca	I	$SO_{3(g)} + H_2O_{(l)} \rightarrow H_2SO_{4(aq)}$
D	$C_{(s)} + \dfrac{1}{2}O_{2(g)} \rightarrow CO_{(g)} + energy$	J	$Mg_{(s)} + Cu^{2+}_{(aq)} \rightarrow Mg^{2+}_{(aq)} + Cu_{(s)}$
E	$Zn_{(s)} + 2\,HCl_{(aq)} \rightarrow ZnCl_{2(aq)} + H_{2(g)}$	K	$4\,Fe_{(s)} + 3O_{2(g)} \rightarrow 2\,Fe_2O_{3(s)}$
F	$HCl_{(aq)} + NaOH_{(aq)} \rightarrow NaCl_{(aq)} + H_2O_{(l)} + E$	L	A change in temperature

1. A reaction in which two or more substances combine to form a more complex compound.

2. A reaction in which a single complex compound breaks up into two or more simpler substances.

3. An example of a single displacement reaction.

4. An example of a double displacement reaction.

5. The net ionic equation of a single displacement reaction

6. One of the evidences of a chemical change.

7. An example of a combustion reaction.

8. An example of an incomplete combustion reaction.

9. An example of acid rain formation reaction.

10. This speeds up the rate of a chemical reaction without being used up.

11. Only these elements will displace hydrogen from water.

12. This reaction is not possible.

True or False

Read each of the following statements and then decide if it is *True* or *False*

1. When aqueous solutions of two ionic compounds are mixed a reaction must happen.

2. Vinegar, when added to baking powder produced fizzing. A reaction thus took place.

3. In a single displacement reaction, a single element reacts with an aqueous solution of a compound.

4. For a single displacement reaction to occur, the added element must be more reactive than one in the compound to which it added.

5. Any element above hydrogen in the reactivity series will displace hydrogen from dilute acids.

6. In decomposition reactions less complex compounds are changed into two or more complex compounds.

7. In a double displacement reaction one of the products may be a precipitate.

8.	Fluorine is the most reactive non-metal.
9.	If aqueous solutions of silver nitrate and potassium iodide are mixed a precipitate of silver iodide is formed.
10.	The reduction of iron (III) oxide by carbon dioxide is a double displacement reaction.
11.	Carbon monoxide is produced as a result of incomplete combustion of carbon and hydrocarbons.
12.	Neutralization reactions are examples of double displacement reactions in which there is no visible change.
13.	Rusting of iron is a synthesis reaction.
14.	Rusting of iron can be slowed down by coating it with a more reactive metal.
15.	Acid rain is produced when non-metal oxides react with water in rain.

Multiple Choice

Choose the letter that best answers the question:

1. The indicator for a chemical change is

a	energy change	*d*	gas evolution
b	precipitate formation	*e*	all the above
c	colour change		

2. $Mg_{(s)} + Zn(NO_3)_{2(aq)} \longrightarrow Zn_{(s)} + Mg(NO_3)_{2(aq)}$. The preceding equation is an example of a

a	decomposition reaction	d.	combustion reaction
b	synthesis reaction	e.	single displacement reaction
c	double displacement reaction		

3. In the presence of the catalyst manganese dioxide, hydrogen peroxide fizzes. This is an example of a

a	synthesis reaction	d.	single displacement reaction
b	combustion reaction	e.	double displacement reaction
c	decomposition reaction		

4. The following the following equation represents a single displacement reaction:

$$Cu_{(s)} + 2\,AgNO_{3(aq)} \rightarrow Cu(NO_3)_{2(aq)} + 2\,Ag_{(s)}$$

The observations made of this reaction over a time will be

a	the formation of solid silver
b	the disappearance of the solid copper
c	the solution changes from colourless to light blue
d	the formation of a gas
e	a, b and c

5. Chlorine gas (greenish-yellow) is passed through a clear aqueous salt solution which then turned dark grey. The solution is most likely to be

a	$NaCl_{(aq)}$	d	$KF_{(aq)}$
b	$NaBr_{(aq)}$	e	none of the above
c	$KI_{(aq)}$		

6. A white solid compound dissolves in water, has a low melting point but does not conduct electricity. This compound is most likely to be

a	polar and molecular	d	glucose
b	ionic	e	a and d
c	only molecular		

7. Which of the following combinations will produce a single displacement reaction?

a	aqueous nitric acid and solid copper
b	aqueous sodium nitrate and solid aluminum
c	aqueous zinc iodide and solid lead
d	aqueous copper(II) chloride and solid iron
e	aqueous potassium nitrate and solid zinc

8.
$$Fe_2O_{3(s)} + 2\,Al_{(s)} \longrightarrow Al_2O_{3(s)} + 2\,Fe_{(l)} + heat.$$

The above reaction can be described as

a	double displacement	d	b and c
b	single displacement	e	decomposition
c	the Thermite reaction		

9.
Methane burns in oxygen according to the following equation:
$$CH_{4(g)} + 2\,O_{2(g)} \longrightarrow CO_{2(g)} + 2\,H_2O_{(l)} + 890 \text{ kJ/mol}$$
In the absence of adequate amount of oxygen, the above reaction may

a	produce less energy	d	all of the above
b	also produce some carbon dioxide	e	none of the above
c	also produce some unburnt carbon		

10.
The following is an example of a double displacement reaction:
$$Na_2CO_{3(aq)} + 2\,HCl_{(aq)} \longrightarrow 2\,NaCl_{(aq)} + H_2CO_{3(aq)} \text{ (unstable)}$$
The evidence that this is a double displacement reaction will be

a	the formation of a precipitate	d.	a change in colour
b	the production of heat	e.	none of the above
c	the evolution of carbon dioxide gas		

11.

The following reaction can be classified as a

$$Ca(OH)_{2(aq)} + 2\ HNO_{3(aq)} \rightarrow Ca(NO_3)_{2(aq)} + 2\ H_2O_{(l)}$$

a	neutralization reaction	d.	double displacement reaction
b	single displacement reaction	e.	a and d
c	decomposition reaction		

12. Which of the following mixtures of aqueous solutions will produce a double displacement reaction?

a	nitric acid and potassium hydroxide solutions
b	sodium nitrate and potassium chloride solutions
c	zinc iodide and magnesium sulphate
d	potassium sulfate and iron(II) chloride solutions
e	sodium chloride and ammonium sulphate solutions

13. The following the following equation represents a double displacement reaction:

$$2\ KI_{(aq)} + Pb(NO_3)_{2(aq)} \rightarrow PbI_{2(s)} + 2\ KNO_{3(aq)}$$

The observation made of this reaction will be

a	the formation of two aqueous solutions
b	the formation of yellow precipitate
c	the formation of a gas
d	the production of heat
e	none of the above

14. Which of the following reactions represents acid rain formation?

a	$CO_{2(g)} + H_2O_{(l)} \rightarrow H_2CO_{3(aq)}$	d	all of the above
b	$SO_{2(aq)} + H_2O_{(l)} \rightarrow H_2SO_{3(aq}$	e	only a and b
c	$4\ NO_{2(g)} + 2\ H_2O_{(l)} \rightarrow 4\ HNO_{2(aq)}$		

15. Which of the following may enter the atmosphere due to incomplete combustion of fuels or faulty catalytic converters in automobiles?

a	$CO_{(g)}$	d	$C_{(s)}$
b	$NO_{(g)}$	e	all of the above
c	hydrocarbons		

16.

The addition of $AgNO_{3(aq)}$ to a sample of an aqueous salt solution produces a white precipitate that readily dissolves in dilute $NH_{3(aq)}$. The solution most likely contains

a	$SO_4^{2-}{}_{(aq)}$		$Br^-{}_{(aq)}$
b	$CO_3^{2-}{}_{(aq)}$		$I^-{}_{(aq)}$
c	$Cl^-{}_{(aq)}$		

17. Zinc reacts with dilute hydrochloric acid according to the following equation:
$Zn_{(s)} + 2\ HCl_{(aq)} \rightarrow ZnCl_{2(aq)} + H_{2(g)}$. The net ionic equation for this reaction is as follows:
$Zn_{(s)} + 2H^{+}_{(aq)} \rightarrow Zn^{2+} + H_{2(g)}$. The half reactions for this are as follows:
* $Zn_{(s)} + energy \rightarrow Zn^{2+}_{(aq)} + 2e^{-}$
* $2\ H^{+}_{(aq)} + 2\ e^{-} \rightarrow H_{2(g)}$

If the ionization energy of four metals: A, B, C and D are ranked: A > B > C > D > E.
Which of these will liberate hydrogen gas from dilute $HCl_{(aq)}$ at the fastest rate
if they are all above hydrogen in the reactivity series?

a	A		D
b	B		E
c	C		

18. Carbonated sodas produce burping in the stomach when ingested. The reactions that take place in the stomach to produce this sensation can be described as:

a	double displacement	d.	single displacement reaction
b	decomposition	e.	both a and b
c	synthesis		

19. The following reaction: $2\ KClO_3 \rightarrow 2\ KCl + 3\ O_{2(g)}$, can be classified as

a	synthesis	d.	single displacement
b	combustion	e.	double displacement
c	decomposition		

20. A solid compound produces a lilac flame when heated in Bunsen burner. When a sample of its aqueous solution reacts $BaCl_{2(aq)}$, a white precipitate is formed that re-dissolves in $HCl_{(aq)}$. The compound is most likely to be

a	Na_2CO_3	d	K_2SO_3
b	$CuSO_4$	e	K_2SO_4
c	KI		

21. A sample of a blue aqueous salt solution reacts with $NH_{3(aq)}$ to produce a blue gelatinous precipitate that re-dissolves in excess $NH_{3(aq)}$ to form a deep blue solution. Another sample of it reacts with $BaCl_{2(aq)}$, to form a white precipitate that does no re-dissolve in excess $HCl_{(aq)}$. The salt solution is most likely to be

a	Na_2CO_3	d	$CuSO_3$
b	KI	e	K_2SO_4
c	$CuSO_4$		

22.		The deep blue colour of the $[Cu(NH_3)_4(H_2O)_2]^{2+}$ complex is seen because the complex
	a	absorbs yellow and reflects deep blue
	b	absorbs red and reflects deep blue
	c	absorbs green and reflects deep blue
	d	absorbs orange and reflects deep blue
	e	none of the above

23.		Ammonium chloride solution is mixed with a slotion of o sodium hydroxide. If a colourless gas with a pungent odour that turns damp pink litmus paper bule, is produced, then the reaction can be described as		
	a	double displacement	d.	single displacement reaction
	b	decomposition	e.	none of the above
	c	synthesis		

24.		The element iron form the following two compounds: $FeCl_2$ and $FeCl_3$. From this information, iron can be described as being
	a	multivalent
	b	a transition element
	c	univalent
	d	a non-metal
	e	a and b

25.		In ionic compounds, the sum of the charges on the positive charges on the cations is equal to the sum of the negative charges on the anions. For the ion, X^{3+}, which of the following negative ion combination will produce a neutal compound with it.
	a	one Cl^- ion with one X^{3+} ion
	b	three Br^- ions with one X^{3+} ion
	c	three O^{2-} ion with two X^{3+} ions
	d	One N^{3-} ion X^{3+} ion
	e	b,c and d

UNIT 2
Quantities in Chemical Reactions

Unit 2 Content

CHAPTER 9
The Mole Concept

Chapter Content:

It would be inconceivable for someone doing masonry to order a quantity of sand in terms of number of grains. In fact, sand is commonly supplied in convenient quantities such as 20 kg sacks, tons or truck loads. In chemistry a similar thing happens; a quantity is sought that avoids the measuring of masses of chemicals in terms of number of atoms, ions or molecules. This quantity is the **mole**. A mole of a substance contains 6.022×10^{23} atoms, ions or molecules, as the case may be. This number, 6.022×10^{23} is also called Avogadro's number. The mass of **any one** of these entities, atoms, ions or molecules is incredibly small and only a phenomenally large cluster of these will have a mass that is easily measurable and convenient to work with. The element carbon, for example, has an average atomic mass of 12.01 u; expressing this mass into grams give the following: $\dfrac{12.01\,g}{6.02 \times 10^{23}}$. As can be seen, this value is indeed very small. However, **one mole** of carbon translates into 12.01 g. As in the case of carbon, quantities in moles of all the other elements and compounds translate into masses that are convenient to measure and use.

Figure 9.1. One mole of carbon weighing 12.01 g.

    Atomic Number, Mass Number and Atomic Structure

The following atomic notation is for an atom of the element, lithium:

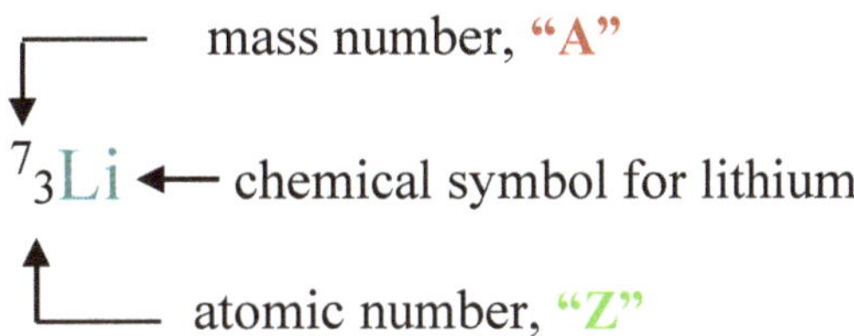

The atomic number, Z *indicates the number of protons that is present in the nucleus of an atom.* According to the model above, there are three protons present in a lithium atom. Since each proton carries a positive charge, each lithium atom will have a total of 3 positive charges. Since every atom is electrically neutral, its number of protons must be equal to its number of electrons. In this case, there will be 3 electrons present since each electron carries a negative charge.

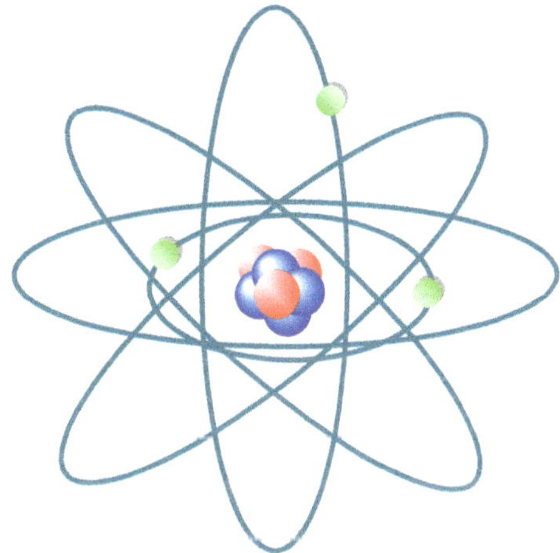

Figure 9.2. Planetary model of the lithium atom.

The following table shows some properties of the three basic sub-atomic particles.

Table 9.1. Properties of protons, neutrons and electrons.

Sub-atomic particle	Mass in atomic mass unit (u)	Relative charge	Where found?
Proton	1	1+	In the nucleus
Neutron	1	0	In the nucleus
Electron	$\dfrac{1}{1839}$	1-	Outside the nucleus

For the purpose of this book, since the electron has such a small mass compared to that of either a neutron or a proton, its mass will be neglected and considered to be **zero**. Also, it should be noted the mass of the neutron is heavier than the mass of the proton by the mass of an electron. This can be illustrated as follows:

1 Neutron ⟶ 1 Proton + 1 electron

Since this difference in mass between a neutron and a proton is so small, it is assumed that their masses are the same; each proton and neutron has a unit of mass called *atomic mass unit*, symbolized **u.**

Based on the information given before, it can be inferred that *the mass of any atom is due primarily to the mass of the nucleus*. **The mass of any atom is thus found by adding the number of protons and neutrons and expressing it in atomic mass unit, "u".** This is also referred as its mass number, **A.**

In all cases, the atomic notation does not give the number of neutrons that an atom has; this has to be determined. It number is found by subtracting its atomic number, Z from its mass number, **A.**

The following table gives a summary of the number and type of sub-atomic particles for a few elements and their atomic masses.

Table 9.2. Sub-atomic particles found in the atoms of a few elements.

Elements	Hydrogen	Carbon	Nitrogen
Standard atomic notation	$_1^1H$	$_6^{12}C$	$_7^{14}N$
# protons	1	6	7
# neutrons	0	6	7
# electrons	1	6	7
mass number	1	12	14
atomic mass	1 u	12 u	14 u

1. Indicate if the following statements are *true* or *false*. K/U
   (a) The mass of one proton is equal to the mass of a neutron.
   (b) The proton is a neutral particle.
   (c) The electron has a mass of 1u.
   (d) The mass number is the sum of the number of protons and neutrons.
   (e) The neutron is negatively charged particle.
   (f) The number of neutrons is found by subtracting the number atomic number from the mass number.

2. Consider the following atoms which are isotopes of the element carbon. K/U

   $$_6^{12}C \qquad _6^{13}C \qquad _6^{14}C$$

   (a) What is common about these atoms?
   (b) How many protons does each atom have?
   (c) How many electrons does each atom have?
   (d) How many neutrons does each atom have?
   (e) What makes these atoms different?

Use a periodic table of elements to complete the following table: K/U

Atomic Notation	Number of Protons	Number of Neutrons	Number of Electrons	Mass Number
$^{7}_{3}Li$				
		12		23
	20	20		
		14	13	
K			19	39
	15			

9.2 Isotopes and Relative Abundance

Isotopes are atoms of the same element having the same atomic number, Z, but different mass number, A. For example, the element carbon has three isotopes as indicated in the following table:

Table 9.3. Relative number of protons and neutrons present in the isotopes of carbon.

Isotopes	$^{12}_{6}C$	$^{13}_{6}C$	$^{14}_{6}C$
Number of protons	6	6	6
Number of neutrons	6	7	8

As can be seen in the table above, the difference between these three isotopes of the element carbon is due to the difference in their number of neutrons since they have the same number of protons. The $^{13}_{6}C$ and $^{14}_{6}C$ isotopes of the element carbon are heavier than the $^{12}_{6}C$ isotope by a mass of one and two neutrons respectively. Therefore, the atomic mass of the element, carbon, cannot be **12 u, 13 u or 14 u**, but an average of the atomic masses of the isotopes, $^{12}_{6}C$, $^{13}_{6}C$ and $^{14}_{6}C$. One other factor that must be taken into consideration, when finding their average mass, is their relative abundance. To learn how to calculate the average mass of elements having isotopes, it is useful to follow the analogy below:

Example:
Suppose that there are ten marbles and nine of them have a mass of 10.0 g each and one has a mass of 12.0 g. Then the average mass of the marbles would be:

$$9 \, marbles \times \frac{10.0 \, g}{marble} = 90.0 \, g$$

$$1 \, marble \times \frac{12.0 \, g}{marble} = 12.0 \, g$$

$$Total \, mass \quad = \, 102.0 \, g$$

$$Average \, mass \quad = \frac{102.0 \, g}{10 \, marbles}$$

$$= \, 10.20 \, g \, / \, marble$$

*In this problem, the masses and relative abundance of the marbles were provided so their average mass could be found. In a similar way, if the **relative abundance** (*the percentage of one isotope in a natural sample of a mixture of isotopes of an element*) and masses of the isotopes for the atoms for any element are provided, then its average atomic mass can be calculated. For example, the three isotopes of carbon have the following natural abundance:

$$^{12}_{6}C \; = \; \mathbf{98.90\%}$$

$$^{13}_{6}C \; = \; \mathbf{1.10\%}$$

$$^{14}_{6}C \; = \; \mathbf{<\,0.001\%} \; \leftarrow \quad \text{(Since this value is so small and insignificant, it is}$$

thus not used in calculating the average atomic mass of carbon.)

The average atomic mass of the element carbon would be calculated as follows:

$$\text{Average mass} \;\; = (12.00 \text{ u} \times 98.90\% + 13.00 \text{ u} \times 1.10\%) \div 100\%$$

$$= (1186.8\% \text{ u} + 14.3\% \text{ u}) \div 100\%$$

$$= (1201.1\% \text{ u}) \div 100\%$$

$$= \mathbf{12.01 \text{ u}}$$

This means that any natural random sample of carbon having $^{12}_{6}C$, $^{13}_{6}C$, and $^{14}_{6}C$ isotopes with 98.90%, 1.10% and less than 0.001% abundance, respectively, will always have an average atomic mass of 12.01 u. *** Recall that atomic mass unit, u, is the smallest unit of mass and it is that of either a proton or neutron,** since the mass of the electron is negligible.

Table 9.4. Atomic masses of the elements in the second row of the periodic table.

6.94	9.01	10.81	12.01	14.01	16.00	19.00	20.18
Li	**Be**	**B**	**C**	**N**	**O**	**F**	**Ne**
lithium	beryllium	boron	carbon	nitrogen	oxygen	fluorine	neon

Student Misconception

Sometimes students wonder why there are decimals in the atomic mass for some elements. The common misconception is that there are fractions of protons and neutrons. Students should note that the number above each element in Table 9.4 and in the periodic table, indicates the average atomic mass for all the isotopes of that particular element, and it is not obtained from any fractions of protons or neutrons.

- Note. In most periodic tables the average masses of the atoms of elements are indicated below their symbols while their atomic numbers are above.

The Atomic Nuclei of Some Isotopes:

The following table shows the standard atomic notations for **one of the isotopes** of each of the elements: hydrogen, carbon and magnesium. It also shows the respective number of protons and neutrons in their nuclei. Electrons are excluded, since they do not contribute mass and are thus not relevant in calculating atomic mass.

Table 9.5. Standard atomic notations and the number of sub-atomic particles in the atoms: $_1^1H$, $_6^{12}C$ and $_{12}^{24}Mg$ isotopes.

Standard Atomic Notation	$_1^1H$	$_6^{12}C$	$_{12}^{24}Mg$
Number of protons	1	6	12
Number of neutrons	0	6	12
Atomic mass	1 u	12 u	24 u

Table 9.6. The relative mass and size of the nuclei for the atoms: $_1^1H$, $_6^{12}C$ and $_{12}^{24}Mg$.

Atomic notation	$_1^1H$	$_6^{12}C$	$_{12}^{24}Mg$
Atomic nuclei			
Atomic mass	1 u	12 u	24 u

Practice:

1. Based on the information above, how many atoms of hydrogen isotope, $_1^1H$ are required to weigh the same as one atom of carbon, $_6^{12}C$ isotope?

9.3 Avogadro's Number and Molar Mass for Elements

In this section, the atoms, $_1^1H$, $_6^{12}C$ and $_{12}^{24}Mg$ will be used to illustrate how the molar masses of the various elements can be found. As can be recalled, one mole of anything contains Avogadro's number of its entities. Table 9.6 displays the relative sizes and masses of the nuclei of the atoms: $_1^1H$, $_6^{12}C$ and $_{12}^{24}Mg$.

Table 9.7(a). Atomic mass and size of the nuclei of the atoms, $_1^1H$, $_6^{12}C$ and $_{12}^{24}Mg$

Elements	Atomic notation	Atomic nuclei	Atomic mass in atomic mass units (u) and grams	Molar mass (mass of 6.022x10²³ atoms) or 1 mole of atoms
Hydrogen	$_1^1H$	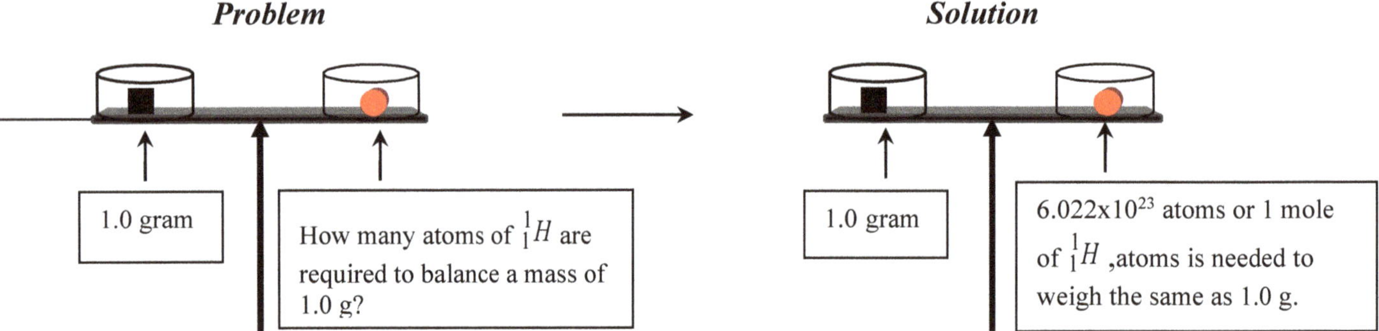	1 u or mass in g?___	?
Carbon	$_6^{12}C$		12 u or mass in g ?___	?
Magnesium	$_{12}^{24}Mg$		24 u or mass in g?___	?

From the tables before, it can be seen that the $_6^{12}C$ and $_{12}^{24}Mg$ atoms are 12 and 24 times as heavy as the $_1^1H$ atom, respectively. Based on these relationships in mass among the three elements, a number of illustrations are created below, to facilitate our understanding of how to find the molar mass of any element.

In the following illustrations, the atomic nuclei of the elements will be used to represent their atoms, *since the mass of the atoms are represented primarily by their nuclei.*

Problem   Solution

Figure 9.3. Balancing illustrations showing how 1 mole of $_1^1H$ atoms weighs 1.0 g.

From the analysis of the illustrations before, the following points can be deduced:

1. One mole or Avogadro's number (6.022×10^{23}) of $_1^1H$ atoms **weighs 1.0 g.**

2. Since such a large number of $_1^1H$ atoms weigh only 1.0 g, then the mass of one $_1^1H$ atom must be extremely small.

3. Since 6.022×10^{23} atoms of $_1^1H$ weigh 1.0 g, **the mass of one its atom** can be expressed as:

$$\frac{1.0 \; g/mol}{6.022 \times 10^{23} \; atoms/mol} = \frac{1.0 \; g}{6.022 \times 10^{23} \; atoms} \quad \text{or} \; 1u$$

Based on these finding, the first row of table 9.7(b) can now be completed.

Table 9.7(b). **Atomic** mass versus **molar** mass for the isotope $_1^1H$.

Elements	Atomic notation	Atomic nuclei	Atomic mass in atomic mass units (u) and grams	Molar mass (mass of 6.022 ×10²³ atoms) or 1 mole of atoms
Hydrogen	$_1^1H$		$1\,u$ **or** $\dfrac{1.0\,g}{6.022\times10^{23}\,atoms}$	$\dfrac{1.0\,g}{6.022\times10^{23}\,atoms}\times\dfrac{6.022\times10^{23}\,atoms}{mol}$ $= \; 1.0\,g/mol$
Carbon	$_6^{12}C$		$12\,u$ or mass in g?___	
Magnesium	$_{12}^{24}Mg$		$24\,u$ or mass in g?___	

Suppose that balancing is done with $_1^1H$ and $_6^{12}C$ atoms instead of weights, what new relationships can be established between these two atoms?

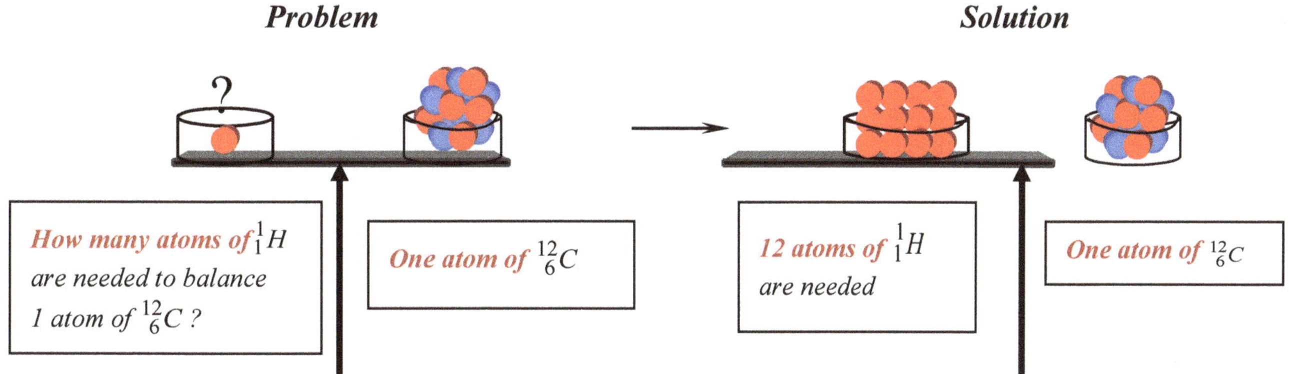

Figure 9.4(a). Illustrations showing how 12 $_1^1H$ atoms weigh the same as 1 $_6^{12}C$ atom.

Suppose moles of $_1^1H$ and $_6^{12}C$ atoms were used in balancing instead of atoms, what new relationships could be established between these two elements, in terms of moles?

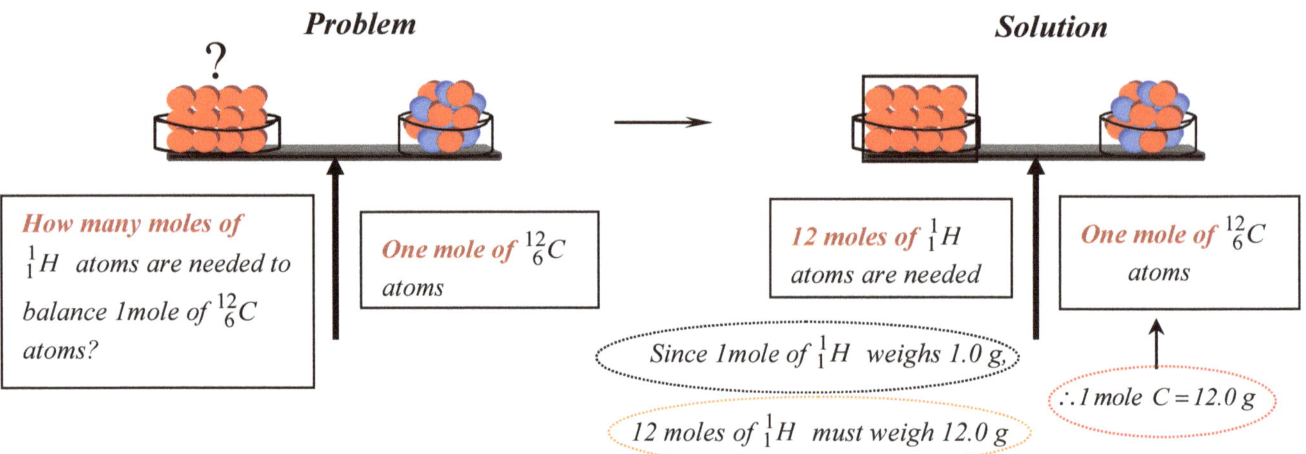

Figure 9.4(b). Illustrations show how **12 moles $_1^1H$ atoms** weigh the **same as 1 mole $_6^{12}C$** atoms.

From the analysis of the illustrations above, the following points can be deduced.

1. **12 moles of** $_1^1H$ atoms must weigh **12.0 g,** since the molar mass of $_1^1H$ is 1.0 g/mol.

2. Since 12 moles of $_1^1H$ atoms weigh 12.0 g, and these are used to balance 1 mole of $_6^{12}C$ atoms, **it can therefore be concluded that the molar mass of $^{12}_6C$ atoms must be 12 g/mol.**

3. Since 6.022×10^{23} atoms of $_6^{12}C$ isotopes weigh 12.0 g, the **mass of one atom** can be *expressed as:*

$$\frac{12.0 \text{ g / mol}}{6.022 \times 10^{23} \text{ atoms / mol}}$$

$$= \frac{12.0 \text{ g}}{6.022 \times 10^{23} \text{ atoms}} \quad \text{or } 12 \text{ U}$$

Using the above information, Table 9.7 (c) can now be continued.

Table 9.7(c). Atomic mass versus molar mass for the isotopes, $_1^1H$ and $_6^{12}C$.

Elements	Atomic notation	Atomic nuclei	Atomic mass in atomic mass units (u) and grams	Molar mass (mass of 6.022×10^{23} atoms, or 1 mole of atoms
Hydrogen	$_1^1H$	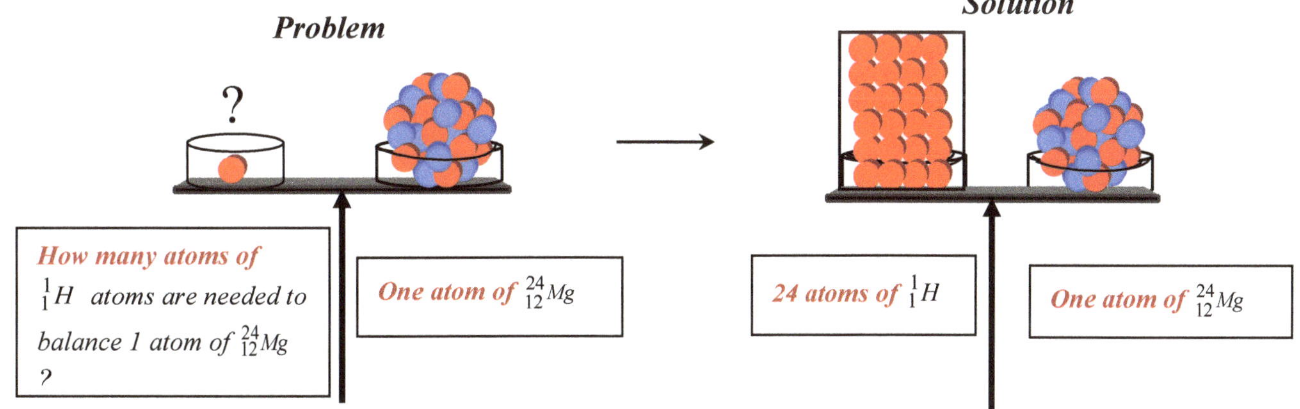	**1 u** or $\dfrac{1.0\ g}{6.022 \times 10^{23}\ atoms}$	$\dfrac{1.0\ g}{6.022 \times 10^{23}\ atoms} \times \dfrac{6.022 \times 10^{23}\ atoms}{mol}$ **1.0 g/mol**
Carbon	$_6^{12}C$		**12 u** or $\dfrac{12.0\ g}{6.022 \times 10^{23}\ atoms}$	$\dfrac{12.0\ g}{6.022 \times 10^{23}\ atoms} \times \dfrac{6.022 \times 10^{23}\ atoms}{mol}$ **12.0 g/mol**
Magnesium	$_{12}^{24}Mg$		**24 u** or mass in g?___	

Suppose that balancing is done with $_1^1H$ and $_{12}^{24}Mg$ **atoms** instead of $_6^{12}C$ **atoms**, what relationship could be established between these two atoms?

Figure 9.5(a). Illustrations showing how **24** $_1^1H$ **atoms** weigh the **same as 1** $_{12}^{24}Mg$ **atom.**

Suppose that **moles** of $_1^1H$ and $_{12}^{24}Mg$ atoms were used in balancing **instead of only atoms,** what new relationships could be established between these two elements in terms *of moles?*

Problem   Solution

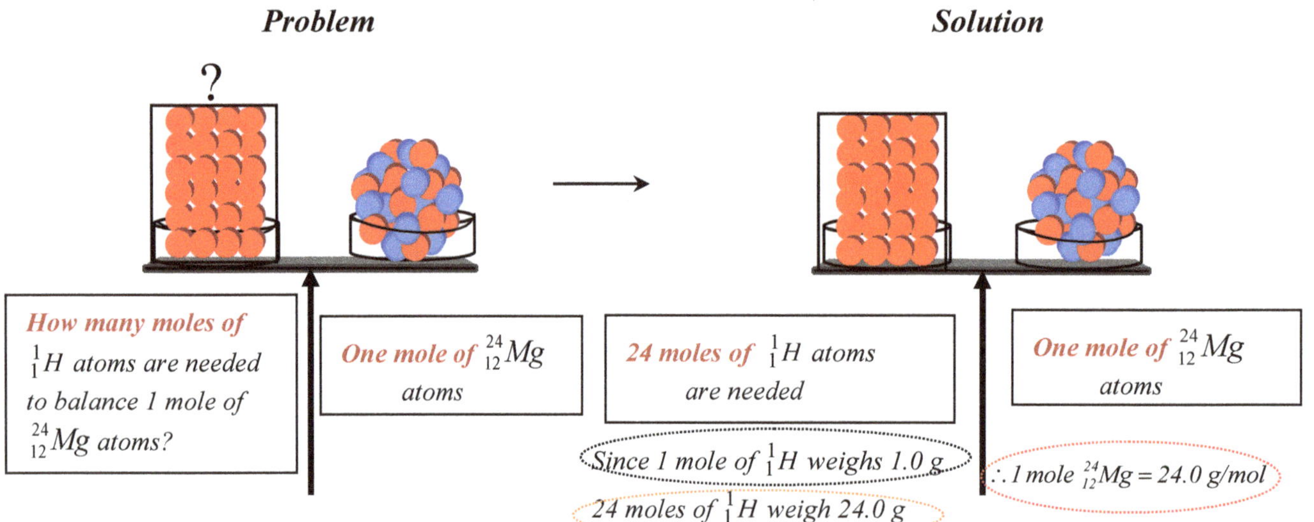

Figure 9.5(b). Illustrations showing how **24 moles of** $_1^1H$ **atoms** weigh the **same as 1 mole of** $_{12}^{24}Mg$ **atom.**

From the analysis of the illustrations before the following can be concluded:

1. 24 moles of $_1^1H$ atoms must weigh 24.0 g, since its molar mass is 1.0 g/mol.

1 mole of $_1^1H$ = 1.0 g   ∴ 24 moles of $_1^1H$ = 24.0 g

2. Since 24 moles of $_1^1H$ atoms weigh 24.0 g, and these are used to balance 1 mole of $_{12}^{24}Mg$ atoms, **then it can be concluded that the molar mass of** $_{12}^{24}Mg$ **atoms must be 24.0 g/mol.**

24 moles of $_1^1H$ atoms = 24.0 g   ∴ 1 mole of $_{12}^{24}Mg$ atoms = 24.0 g

3. Since 6.022×10^{23} atoms of $_{12}^{24}Mg$ weigh 24.0 g, **the mass of one atom** can be expressed as:

$$\frac{24.0 \text{ g / mol}}{6.022 \times 10^{23} \text{ atoms / mol}}$$

$$= \frac{24.0 \text{ g}}{6.022 \times 10^{23} \text{ atoms}} \quad \text{or } 24 \text{ U}$$

Based on the preceding analysis the rest of our Table 9.7(d) can now be completed.

9.4 **Table 9.7(d).** Atomic mass versus molar mass for the isotopes, $_1^1H$ and $_6^{12}C$ and $_{12}^{24}Mg$.

Elements	Atomic Notation	Atomic Nuclei	Atomic mass in atomic mass units (U) and grams	Molar mass mass of 6.022×10²³ atoms) or 1 mole of atoms
Hydrogen	$_1^1H$		**1 u** **or** $\dfrac{1.0\ g}{6.022\times10^{23}\ atoms}$	$\dfrac{1.0\ g}{6.022\times10^{23}\ atoms}\times\dfrac{6.022\times10^{23}\ atoms}{mol}$ $= 1.0\ g/mol$
Carbon	$_6^{12}C$		**12 u** **or** $\dfrac{12.0\ g}{6.022\times10^{23}\ atoms}$	$\dfrac{12.0\ g}{6.022\times10^{23}\ atoms}\times\dfrac{6.022\times10^{23}\ atoms}{mol}$ $= 12.0\ g/mol$
Magnesium	$_{12}^{24}Mg$		**24 u** **or** $\dfrac{24.0\ g}{6.022\times10^{23}\ atoms}$	$\dfrac{24.0\ g}{6.022\times10^{23}\ atoms}\times\dfrac{6.022\times10^{23}\ atoms}{mol}$ $= 24.0\ g/mol$

Analysis:

From the above activities, it can be seen that the following relationships exist:

Elements:	$_1^1H$	$_6^{12}C$	$_{12}^{24}Mg$
	↓	↓	↓
Molar mass:	1.0 g/mol	12.0 g/mol	24.0 g/mol
	↓	↓	↓
# of atoms:	6.022×10^{23} atoms/mol	6.022×10^{23} atoms/mol	6.022×10^{23} atoms/mol

Note: In any natural random sample of anyone of these elements, there is **a mixture of all the various types of isotopes**; carbon, for example, exists in the form of $_6^{12}C$, $_6^{13}C$ and $_6^{14}C$. Calculated separately, the molar mass of each of these isotopes would be; 12.0 g/mol, 13.0 g/mol and 14.0 g/mol respectively. Since there is mixture of these isotopes in an sample of carbon, it would not be accurate to say that the molar mass of carbon is either, 12.0 g/mol, 13.0 g/mol 14.0 g/mol, but it would be the average of these. The illustrations below in Figure 9.5 help to clarify this concept.

Molar Mass for Elements with Isotopes

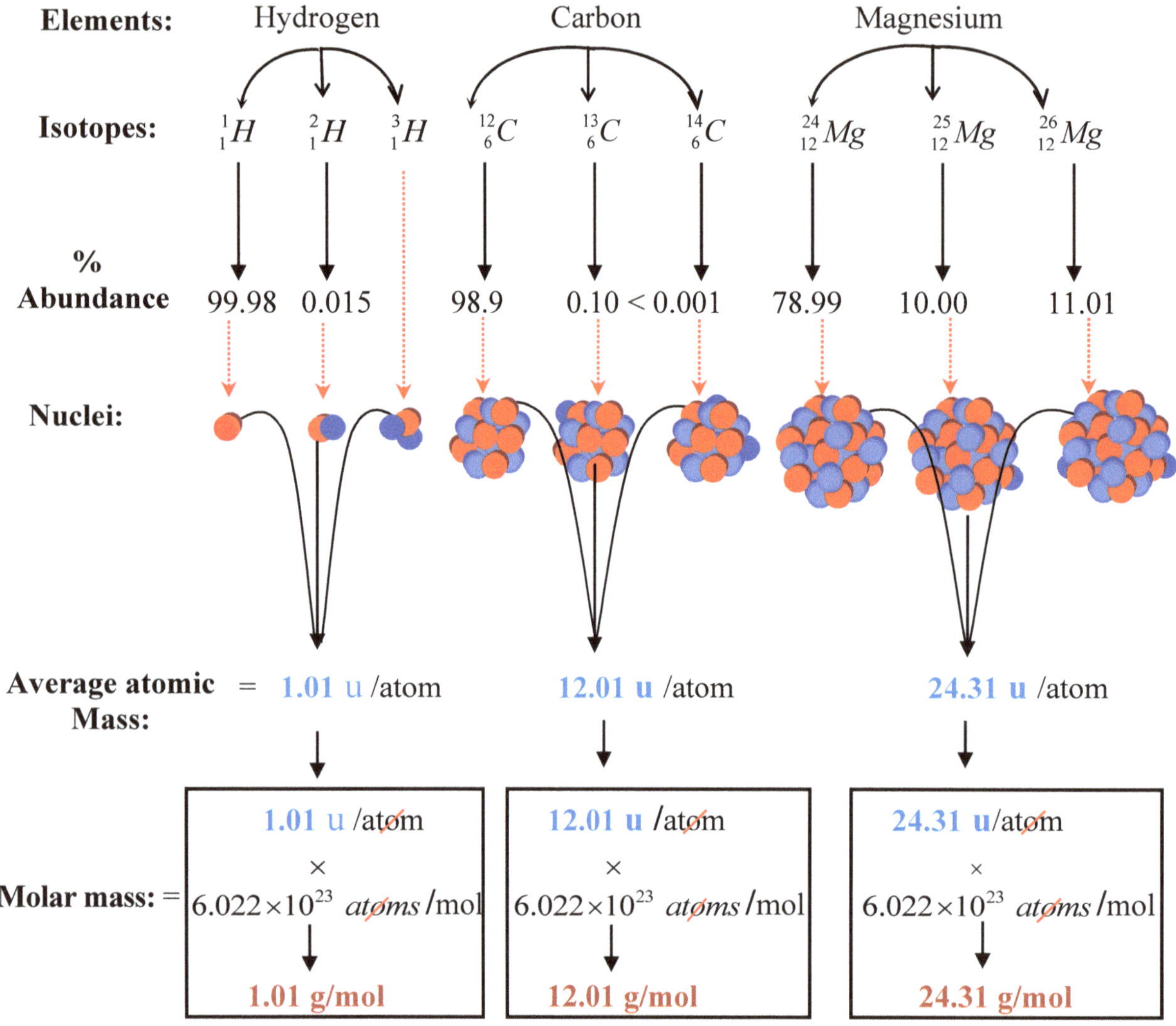

Figure 9.6. Illustrations of how the average atomic and molar masses of the elements, hydrogen, carbon and magnesium are calculated.

In these illustrations, the molar mass of the element carbon, for example, is *12.01 g/mol* and *not 12.0 g/mol.* However, if we are dealing only with the isotope, $^{12}_6$C, then it would be absolutely perfect to say that its molar mass **is 12.0 g/mol**. In fact, according to the **I.U.P.A.C**, *one mole of a substance is defined as the amount of it that has the same number of entities (**atoms, molecules, ions, or formula units**) as there are atoms in 12 g of $^{12}_6C$ atoms.*

However, in nature there is a homogeneous mixture of all the isotopes for each element. So, in order to determine the molar mass of a given element, it is important to take the average atomic mass of that element's isotopes, taking into account the relative abundance of each isotope. The values given in the periodic table below for each element is the average mass for all the isotopes for each element.

To find the molar mass, symbolised M, for any element take the atomic mass as indicated in the periodic table and express it in grams.

Unit and Symbol: The SI unit and symbol are **g/mol** and **M**, respectively. We shall now revisit a part of the periodic table.

Table 9.8. The atomic masses of the elements in the second row of the periodic table.

^3Li	^4Be	^5B	^6C	^7N	^8O	^9F	^{10}Ne
6.94	9.01	10.81	12.01	14.01	16.00	19.00	20.18

Using the above elements to illustrate this, we would have the following calculations to get their molar masses.

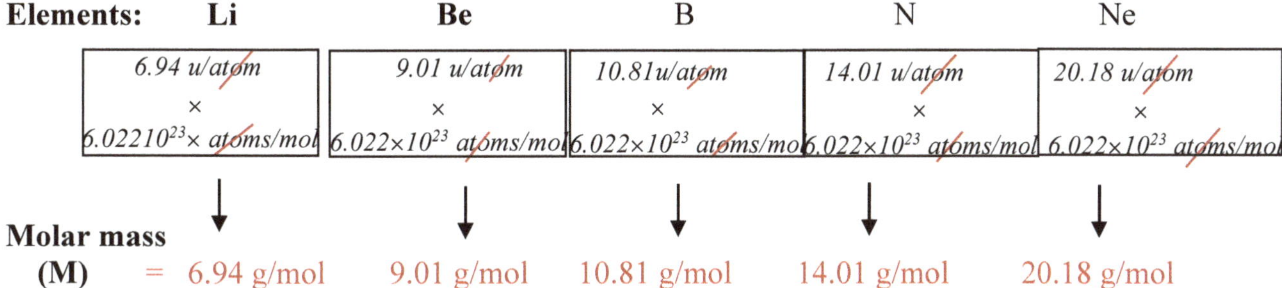

Exercise 9.3

1. Find the molar mass of the following elements as they appear in nature: K/U

(a) Na (b) Ca   (c) Cl   (d) Zn

2. How many atoms will be in each of the above masses?

3. Rank these elements in the above queation, based on mass, from heaviest to lightest.

4. The following are spherical objects having the same density. Which one would have the greatest molar

mass?   A   B   C

5. Form the following two isotopes of carbon: C-12 and C-14, which would have the greater molar mass?

The Mole Concept and Calculations

9.6 Converting Amount in Moles to Amount in Mass

In these types of calculation, the following symbols are used:

Mass is symbolized by **(m)**.

Moles is symbolized by **(n)**

Molar mass is symbolized by **(M)**

Sample problem 1:

Calculate the mass of 4.000 moles of carbon.

Solution:

To do this, it is useful to follow the steps below:

Step 1. Look up the molar mass of the element carbon, (M_C) from the periodic table.

$$M_C = 12.01 \ g/mol$$

Step 2. Since 1 mole of C weighs 12.01 g, then the mass of 4.00 moles is obtained by multiplying that number (12.01 g) by 4.000.

mass = moles x molar mass

$$m = n \times M \longrightarrow \boxed{\textbf{(This is the general equation that must be used)}}$$

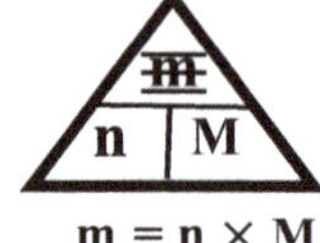

$$m_C = 4.000 \ mol \times \frac{12.01 \ g}{mol}$$

$$= 48.04 \ g$$

Exercise 9.4

1. Find the mass of the following: K/U

   (a) 5.00 mol of calcium

   (b) 2.50 mol of potassium

   (c) 3.00 mol of hydrogen **atoms**

   (d) 4.00 mol of chlorine **atoms**

Sample problem 2:

Calculate the number of moles of atoms in 48.04 g of carbon.

Solution:

To do this, it is useful to follow the steps below:
Step 1. Look up the molar mass of the element carbon from the periodic table.
$$M_C = 12.01\,g/mol$$

Step 2. Divide the given mass by the molar mass of that element. This will give you the number of moles of the element. The equation below is simply a rearrangement of the equation introduced previously ($m = n \times M$).

$$Moles = \frac{mass}{molar\ mass}$$

$$n = \frac{m}{M} \longrightarrow \boxed{\textbf{(This is the general equation that must be used)}}$$

$$n_C = \frac{48.04\,g}{12.01\,g/mol}$$
$$= 4.000\,mol$$

The following triangle is also useful in determining the relationships between moles, molar mass and mass.

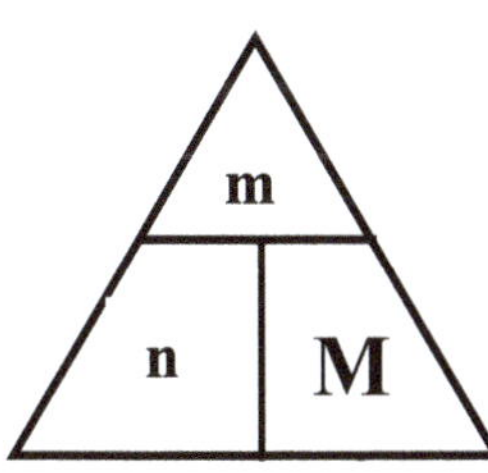

Figure 9.7. The mole triangle.

From this triangle the following relationships can be made:

$$m = n \times M$$
$$n = \frac{m}{M}$$
$$M = \frac{m}{n}$$

1. Find the number of **moles** of each element in the following masses: K/U

(a) 18.02 g Be

(b) 72.93 g Mg

(c) 60.02 g Ca

(d) 68.97 g Na

 Calculating the Number of Entities (N)

In chemistry, **entities** refer to the number of atoms, molecules, ions or formula units in a sample of a substance. However, in this section calculations will be restricted to atoms only, since molecules and ions have not been examined yet.

Entities are symbolized as **(N)**
Avogadro's number is symbolized as **(N_A)**

Sample problem 3:

Calculate the number of atoms in 4.00 moles of carbon.

Solution:

Since it is known that one mole of any element contains Avogadro's number of atoms, (6.022×10^{23}), to solve this problem, all that is needed is to multiply the number of moles by the number, 6.02×10^{23} atoms/mol.

$$N = n \times N_A$$

Entities = number of moles × Avogadro's number

$$N_C = 4.00 \; \cancel{mol} \times \frac{6.022 \times 10^{23} \; atoms}{\cancel{mol}}$$

$$= 2.41 \times 10^{24} \; atoms$$

Exercise 9.6

Find the number of **atoms** (entities) in the following: K/U

(a) 6.00 moles of nitrogen *atoms*　　　(d) 3.00 moles of potassium *atoms*

(b) 2.50 moles of calcium *atoms*　　　(e) 5.00 moles of helium *atoms*

(c) 4.00 moles of hydrogen *atoms*

9.9　**Converting Amount in Mass to Number of Entities**

In the previous section, it was easy to convert moles to number of entities. However, it becomes a bit more challenging to convert mass of an element to number of atoms. The following problem will illustrate how this is done.

Sample problem 4:

Calculate the number of atoms in 20.04 g of calcium.

The number of entities is found using the formula, $N = n \times N_A$

We know that N_A is 6.022×10^{23} atoms/mol, but the number of moles, n is unknown.

To solve this problem, the two steps below are followed:

Step 1. First find the number of moles of the element in the mass given.

This is found using the formula, $n = \dfrac{m}{M}$

$$n_{Ca} = \dfrac{20.04\ g}{40.08\ g/mol}$$

$$= 0.5000\ mol$$

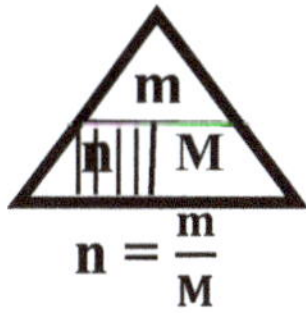

$$n = \dfrac{m}{M}$$

Step 2. Find the number of atoms present using the formula,

$$N = n \times N_A.$$

Substituting *0.5000 mol* for *n,* in the equation, $N = n \times N_A$, we get:

$$N_{Ca} = 0.5000\ mol \times \dfrac{6.022 \times 10^{23}\ atoms}{mol}$$

$$= 3.011 \times 10^{23}\ atoms$$

Find the number of **atoms** in each of the following masses: K/U

(a) 4.00 g of oxygen   (c) 92.9 g of phosphorus

(b) 13.49 g of aluminum   (d) 126.78 g of copper

9.10 Converting Number of Entities to Amount in Moles

How many moles of atoms are there in 1.204×10^{24} atoms of hydrogen?

Step 1. Divide the number of atoms by Avogadro's number.

Using the equation, $N = n \times N_A$ and making *n* the unknown in the equation, we get:

$$n = \dfrac{N}{N_A}$$

$$n_H = \dfrac{1.204 \times 10^{24}\ atoms}{6.022 \times 10^{23}\ atoms\ /\ mol}$$

$$= 2.000\ mol\ of\ hydrogen\ atoms$$

1. Convert the following number of **atoms to amounts in moles** in the following: K/U

a) 3.01×10^{23} atoms of Zn

b) 1.81×10^{24} atoms of Cu

c) 1.51×10^{23} atoms of K

d) 6.02×10^{22} atoms of Li

9.11 Converting Number of Entities to Amount in Mass

Sample problem 6:

What mass of copper would have 3.011×10^{23} atoms?

Solution:

To solve these types of problems, it is useful to follow the steps below:

Step 1. Convert the number of entities to amount in moles.

Using the formula, $n = \dfrac{N}{N_A}$

We get: $\quad n_{Cu} = \dfrac{3.011 \times 10^{23} \ atoms}{6.022 \times 10^{23} \ atoms / mol}$

$$= 0.5000 \ mol$$

Step 2. Convert the amount in moles to amount in mass.

Using the formula, $m = n \times M$ and substituting for n and M, we get:

$$m_{Cu} = 0.5000 \ mol \times \dfrac{63.55 \ g}{mol}$$

$$= 31.78 \ g$$

Exercise 9.9

1. Convert the following **entities to amount in mass:** K/U

(a) 1.204×10^{24} atoms of silver

(b) 6.02×10^{22} atoms of zinc

(c) 3.01×10^{23} atoms of lead

(d) 2.408×10^{22} atoms of carbon

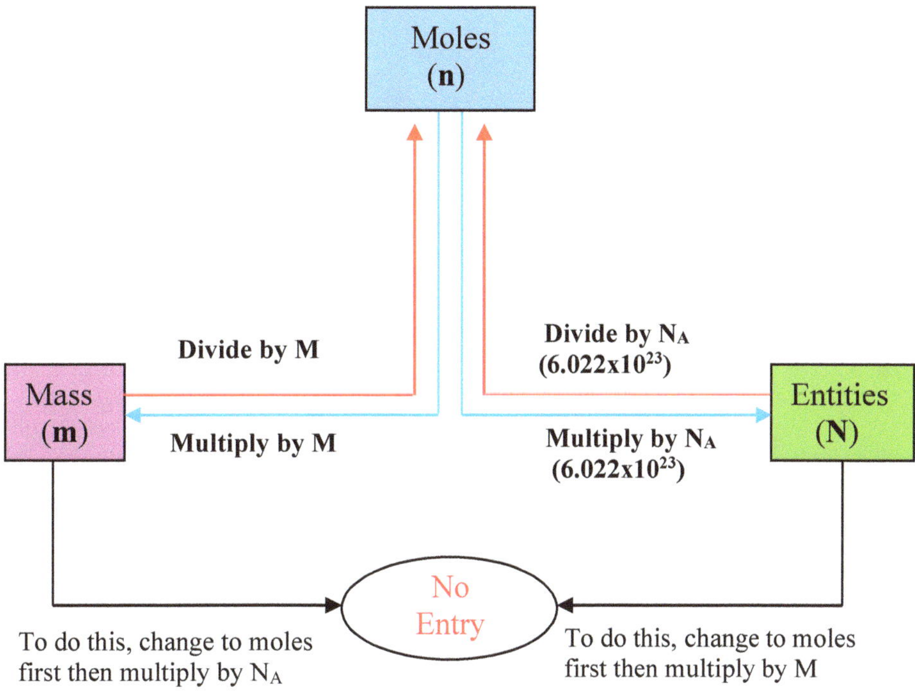

Figure 9.8. The Mole hill summarizing the relationships between moles, mass and molar mass.

9.12   SUMMARY

Note: As evident from Figure 9.7 before, two steps are needed when converting from mass to entities or vice versa. These are as follows:

- To change mass to number of entities, it first required to divide the given mass of the substance by its molar mass to find the number of moles of the substance. Then the number of mole is multiplied by Avogadro's number.
- To change number of entities to mass, it first required to divide the number of entities of the substance by Avogadro's number to find the number of moles of the substance. Then the number of moles is multiplied by its molar mass.

The following example uses a mass of 24.02 g of carbon to summarize this concept.

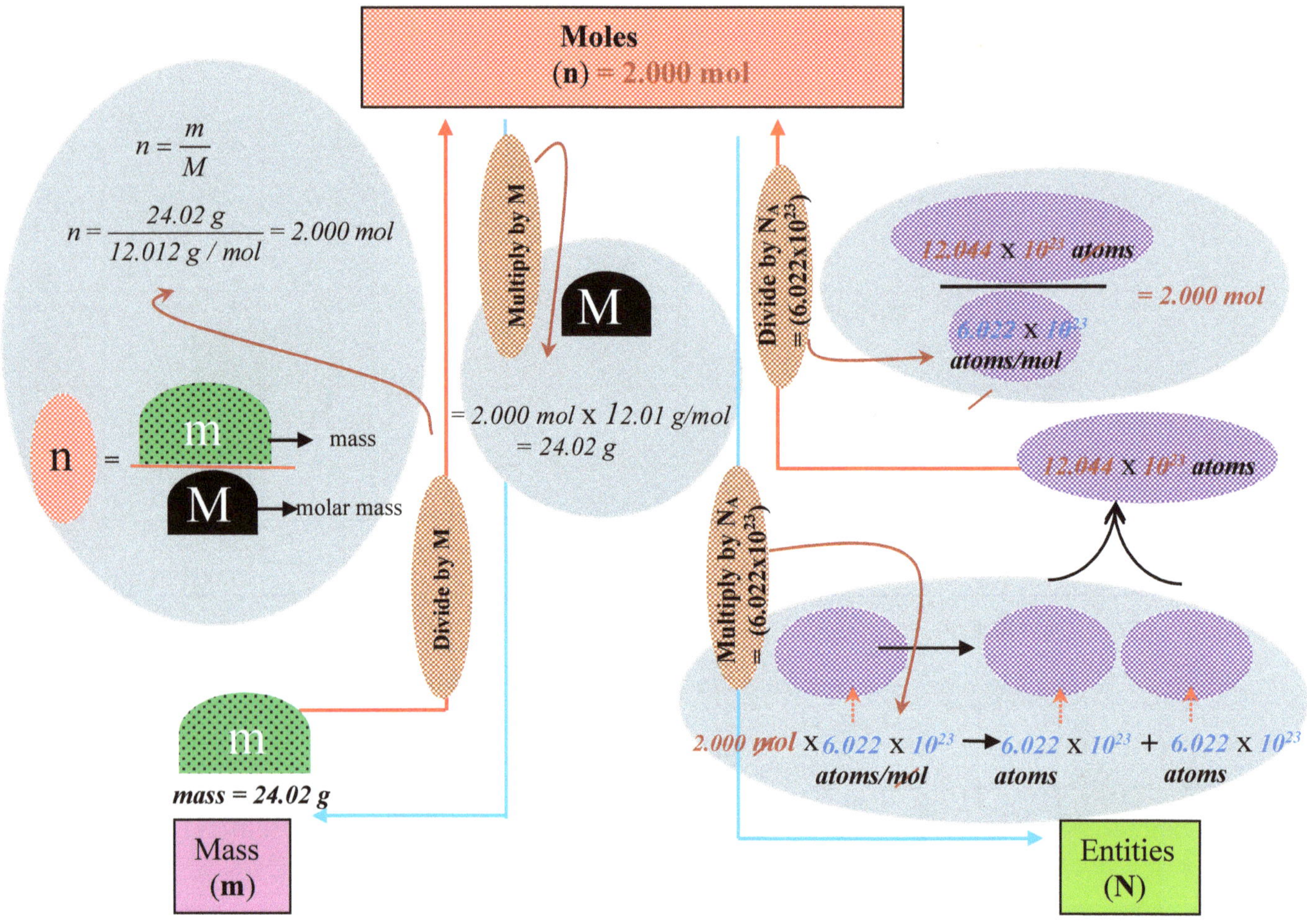

Figure 9.9. Using a mass of (carbon) to illustrate the relationships between moles, mass and molar mass.

Finding Molar Mass for Compounds

Molecules

A molecular compound is formed when two or more non-metal elements combine during a chemical change. It is useful to note that it is actually the **atoms** of the elements that are involved in chemical changes. The following examples will illustrate this concept:

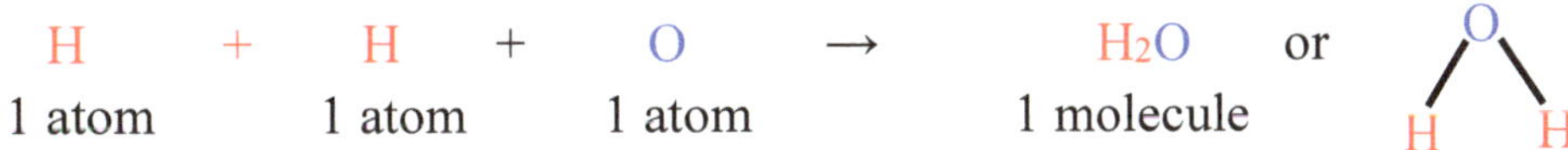

$$H \quad + \quad H \quad + \quad O \quad \rightarrow \quad H_2O \quad \text{or}$$

$$\text{1 atom} \qquad \text{1 atom} \qquad \text{1 atom} \qquad \text{1 molecule}$$

In this case, two **atoms** of hydrogen combine with one **atom** of oxygen to form one *molecule* of water. Note that H_2O is a molecular compound since it is formed by non-metals only. In this compound, the repeating unit would be H_2O molecules represented as:

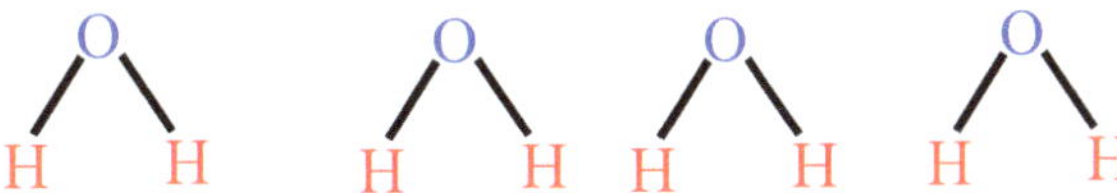

This other example illustrates how one molecule of carbon dioxide is formed.

$$C \quad + \quad O \quad + \quad O \quad \rightarrow \quad CO_2 \quad \text{or} \quad O=C=O$$

$$\text{1 atom} \qquad \text{1 atom} \qquad \text{1 atom} \qquad \text{1 molecule}$$

Formula Units

$$\textbf{Na} \quad + \quad \textbf{Cl} \quad \rightarrow \quad \textbf{[NaCl]}$$

$$\text{1 atom} \qquad \text{1 atom} \qquad \text{1 formula unit}$$

In this case, one *atom* of sodium combines with one *atom* of chlorine to form **one** *formula unit* of sodium chloride. Note that sodium chloride is an ***ionic compound*** since it is formed by a metal (sodium) and a non-metal (chlorine). In ionic compounds, **formula units** are **analogous** to **molecules** in molecular compounds; they are the smallest repeating units in ionic compounds. In the compound sodium chloride for example, the repeating formula unit is **[NaCl]**. The following other example shows how one formula unit of calcium bromide is formed:

$$\text{Ca} \quad + \quad 2\,\text{Br} \quad \rightarrow \quad [CaBr_2]$$

$$\text{1 atom} \qquad \text{2 atoms} \qquad \text{1 formula unit}$$

Fill in the blanks in the table below.

Exercise 9.10 Table 9.9. Finding the types of particles in a few compounds. K/U

Compounds	Type of particles [molecules or formula units] present?
$CaCl_2$	
CO_2	
$Ca(NO_3)_2$	
PCl_3	

 Finding Molar Mass of Compounds

Calculate the molar mass of carbon dioxide (CO_2).

Solution:

The following illustrations show how this is done:
Creating a molecule of this compound from its atoms, we have:

$$C \quad + \quad O \quad + \quad O \quad \longrightarrow \quad CO_2$$

1 atom   *1 atom*   *1 atom*   *1 molecule*

Atomic mass: 12.01 u   +   16.0 u   +   16.0 u   =   44.01 u

Molar mass:

$$12.01\,u \times 6.022 \times 10^{23} \quad 16.0\,u \times 6.022 \times 10^{23} \quad 16.0\,u \times 6.022 \times 10^{23} = 44.01\,u \times 6.022 \times 10^{23}$$

$$= 12.01\,g/mol \quad + \quad = 16.0\,g/mol \quad + \quad = 16.0\,g/mol \quad = \quad 44.01\,g/mol$$

of   *of*   *of*   *of*

atoms   *atoms*   *atoms*   ***molecules***

The above illustrations show that in one mole of the compound CO_2, there are 2 moles of oxygen atoms and 1 mole of carbon atoms. Adding the molar masses of all of these together gives the molar mass of the compound.

Summarizing the above illustrations, we can find the molar mass of CO_2 as follows:

$$2\,mol\,Oxygen \times 16.0\,g/mol = 32.0\,g$$
$$1\,mol\,Carbon \times 12.01\,g/mol = 12.01\,g$$
$$Molar\,mass = 44.0\,g/mol$$

Conclusion: To find the molar mass for any compound, add up the individual masses for all the elements in the compound and then express it as amount in grams per mole.

Finding Molar Mass for Compounds with Radicals

Finding the molar mass of the radical-containing compound, $Ca(NO_3)_2$.

Solution:

The following illustrations show how this is done:

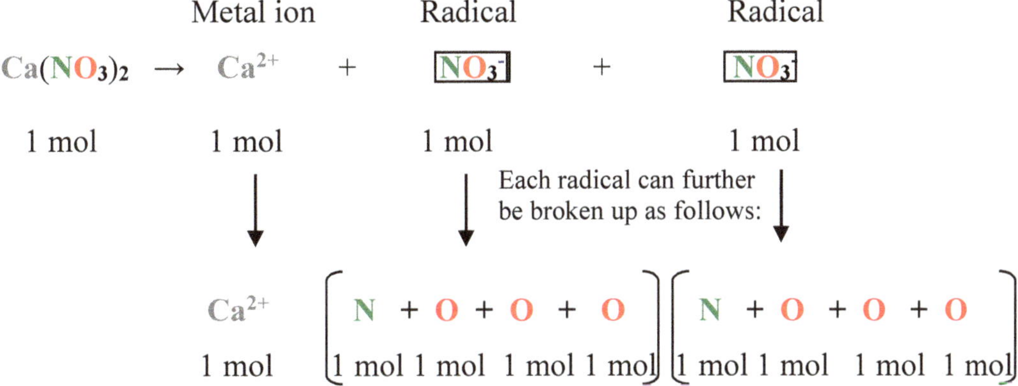

Total number of atoms $= 1$ mol Ca^{2+} $+ 2$ mol $N + 6$ mol O

The above illustrations show how one mole of Ca $(NO_3)_2$ dissociates to produce one mole of calcium ions and two moles of NO_3^- radicals; note how each mole of NO_3^- radical is shown to have one mole of nitrogen atoms and three moles of oxygen atoms. The two moles of NO_3^- radicals would therefore have a total of 2 moles of nitrogen and 6 moles of oxygen. Using the illustrations above, the molar mass of Ca $(NO_3)_2$ is calculated as follows:

$$1\ mol\ Ca \times 40.08\ g\,/\,mol = 40.08\ g$$
$$2\ mol\ N \times 14.01\ g\,/\,mol = 28.02\ g$$
$$6\ mol\ O \times 16.00\ g\,/\,mol = 96.00\ g$$
$$Molar\ mass = 164.1\ g\,/\,mol$$

To further explain this concept, we shall practice finding the number of moles of atoms in one mole of another radical containing compound, $(NH_4)_2SO_4$.

Sample problem 8:

Finding the molar mass of $(NH_4)_2SO_4$.

Solution:

The following illustrations show how this is done:
This compound dissociates as follows:

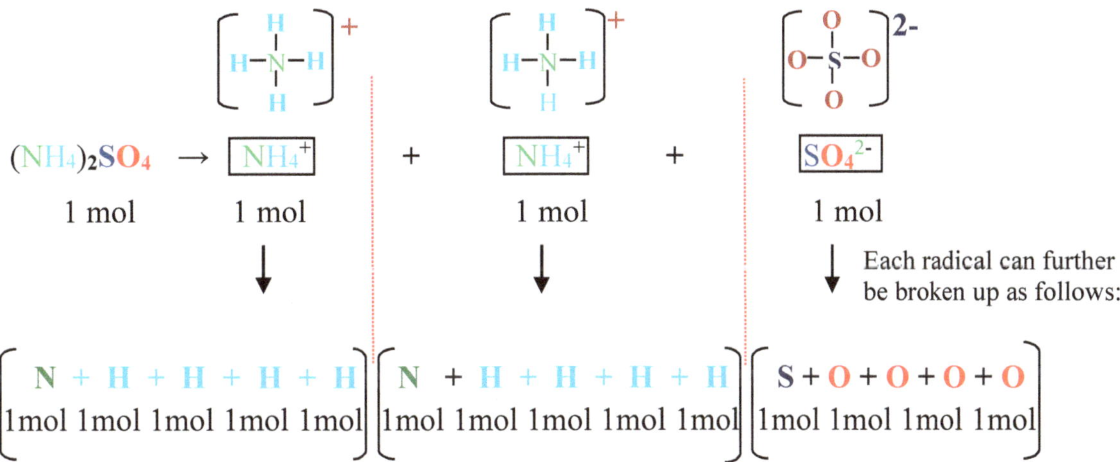

The above illustrations show how **one mole** of $(NH_4)_2SO_4$ dissociates to produce **two moles** of NH_4^+ radicals and **one mole** of SO_4^{2-} radical. Each mole of SO_4^{2-} radicals is shown to have **one mole** of sulphur atoms and **four moles** of oxygen atoms. Also, **each mole** of NH_4^+ radicals is shown to have **four moles** of hydrogen atoms and **one mole** of nitrogen atoms.

Counting up all the elements, it can be seen that there are **two moles** of nitrogen, **eight moles** of hydrogen, **one mole** of sulphur and **four moles** of oxygen atoms respectively **in one mole** of $(NH_4)_2SO_4$.

As can be deduced from the examples above, to find the exact number of moles of elements in the radicals, **multiply the number of moles of each of the elements in the parentheses by the subscript outside of it**. For example, in one mole of $(NH_4)_2SO_4$, there are two moles of nitrogen and eight moles of hydrogen because the number of moles of nitrogen and hydrogen in the parentheses are multiplied by the subscript **2**, outside $[(NH_4)_2]$.

The molar mass is thus found as follows:

$$
\begin{aligned}
2 \ mol \ N \times 14.01 g/mol &= 28.02 \ g \\
8 \ mol \ H \times 1.01 \ g/mol &= 8.08 \ g \\
1 \ mol \ S \times 32.06 \ g/mol &= 32.06 \ g \\
4 \ mol \ O \times 16.00 \ g/mol &= 64.00 \ g \\
Molar \ mass &= 132.16 \ g/mol
\end{aligned}
$$

- For these calculations, ignore the charges on the ions since they are due to a deficiency or an excess of electrons, which have negligible mass.

Molar Mass for Diatomic Molecular Elements

Many elements exist in nature as diatomic molecules instead of atoms. The following are examples of these:

Table 9.10. Example of common diatomic molecules.

Element	Hydrogen	Oxygen	Nitrogen	Fluorine	Bromine	Chlorine	Iodine
Molecule	H_2	O_2	N_2	F_2	B_2	Cl_2	I_2

Note that whenever these elements are produced or used in chemical reactions or are in their natural states, **they will always be in their molecular forms**.

Chemical equations must therefore be balanced in order to show them in their molecular forms. The following equations will illustrate this concept:

$$2HCl_{(aq)} \ + \ Mg_{(s)} \ \rightarrow \ MgCl_{2(aq)} + \textbf{2H}_{(g)} \quad \textbf{(Incorrect)}$$

The previous equation is not acceptable since hydrogen is in its **atomic form**. The following equation is the correct one:

$$2HCl_{(aq)} \ + \ Mg_{(s)} \ \rightarrow \ MgCl_{2(aq)} + \textbf{H}_{2(g)} \quad \textbf{(Correct)}$$

Question:
Why do these elements prefer to be as molecules rather than atoms?

Answer:
As can be recalled from your unit in bonding, the elements such as ***hydrogen, oxygen, nitrogen, chlorine, bromine, fluorine and iodine*** form **diatomic molecules** to attain a noble gas electron configuration by

covalent bond formation. In their molecular forms they have a greater degree of stability than in their atomic form.

$$H\bullet \ + \ \bullet H \qquad \longrightarrow \qquad H-H$$
$$(H_2)$$

Calculate the molar mass for molecular oxygen.

To form one mole of molecular oxygen, two moles of atomic oxygen are required.

Atoms	+	Atoms		Molecules
O	+	O	$\longrightarrow$	$O_{2(g)}$
O	+	O	$\longrightarrow$	$O=O$
1 mol	+	1 mol		1 mol
16.0 g/mol	+	*16.0 g/mol*	=	*32.0 g/mol*

$$M_{O_2} = \ M_O \ + \ M_O$$
$$= \ 16.0 \text{ g/mol} \ + \ 16.0 \text{ g/mol}$$
$$= 32.0 \text{ g/mol}$$

Finding Molar Mass for Hydrates

Hydrates are compounds that incorporate molecules of water into their crystal lattice. Each mole of these compounds needs to have a specific number of moles of water for proper formation of their crystals. If any of these compounds are analysed, it would be found that each is accompanied by a specific number of moles of water per mole of the compound. The following are examples of some hydrates:
$CaCl_2 \bullet 2H_2O$, $Na_2SO_4 \bullet 10H_2O$, $MgSO_4 \bullet 7H_2O$, $Na_2S_2O_3 \bullet 5H_2O$ and $Fe_2O_3 \bullet 3H_2O$.

To calculate the molar mass of any specific hydrate, it is important to first find the molar mass of the compound, and then add to it the mass of the specific number of moles of water present in the compound.

Calculate the molar mass for $MgSO_4 \bullet 7H_2O$

Magnesium sulphate heptahydrate, $MgSO_4 \bullet 7H_2O$ can be separated as follows:

$$MgSO_4 \bullet 7H_2O \quad \longrightarrow \quad MgSO_4 \quad + \quad 7 \ H\!-\!O\!-\!H$$

$$\text{1 mol} \qquad\qquad \text{1 mol} \quad + \quad \text{7 mol}$$

Step 1: Find the molar mass of the **anhydrous** compound, $MgSO_4$. *The anhydrous compound is the part of the compound that does not have water of crystallization.*

Molar mass $MgSO_4$ =

$$1 \ mol \ Mg \times 24.31 \ g / mol = 24.31 \ g$$
$$1 \ mol \ S \times 32.03 \ g / mol = 32.03 \ g$$
$$4 \ mol \ O \times 16.00 \ g / mol = 64.00 \ g$$
$$M_{MgSO_4} = 120.34 \ g/mol$$

Step 2: Find the molar mass of water

Molar mass for H_2O =

$$2 \ mol \ H \times 1.01 \ g/mol = 2.02 \ g$$
$$1 \ mol \ O \times 16.00 \ g/mol = 16.00 \ g$$
$$M_{H_2O} = 18.02 \ g/mol$$

Step 3: Find the mass for **7 moles of water**

Mass for 7 moles of H_2O =

$$7 \ mol \ x \ 18.02 \ g/mol = 126.14 \ g$$

Step 4: Find the molar mass of the compound, $MgSO_4 \bullet 7H_2O$ by adding the molar mass of $MgSO_4$ to the mass of 7 moles of H_2O.

Molar mass for $MgSO_4 \bullet 7H_2O$ = 1 mol $MgSO_4$ + 7 mol H_2O
$$120.04 \ g \ + \ 126.14 \ g$$
$$= 246.18 \ g/mol$$

Calculating Mass for Mole Fractions of Hydrates

What is the mass for 0.5000 moles of $MgSO_4 \bullet 7H_2O$?

To solve this problem, it is useful to follow the steps above.

Step 1.
In this case, the calculated molar mass for $MgSO_4 \bullet 7H_2O$ is 246.18 g/mol. Therefore, the mass for 0.5000 moles of $MgSO_4 \bullet 7H_2O$ will be:

$$m = n \times M$$
$$m_{MgSO_4 . 7H_2O} = \frac{246.18 \ g}{mol} \times 0.5000 \ mol$$
$$= 123.1 \ g$$

However, the mole ratio of the magnesium sulphate and water in 0.5000 moles of $MgSO_4 \bullet 7H_2O$ can be illustrated as follows:

$$MgSO_4 \bullet 7H_2O \quad \rightarrow \quad MgSO_4 \quad + \quad 7H_2O$$

1 mol	*1 mol* +	*7 mol*
0.5000 mol	*0.5000 mol* +	*3.500 mol*

Or generally: *X mol*   *X mol*   *7X mol*

Finding the respective masses of $MgSO_4$ and H_2O in the sample can be done as follows:

$$m_{MgSO_4 \cdot 7H_2O} = 0.5000 \; mol \; MgSO_4 \times \frac{120.04 \; g}{mol \; MgSO_4} \quad + \quad 3.500 \; mol \; H_2O \times \frac{18.01 \; g}{mol \; H_2O}$$

$$= 60.02 \; g \; MgSO_4 \quad + \quad 63.07 \; g \; H_2O$$

$$= \textbf{123.1 g} \; in \; 0.5000 \; mol \; MgSO_4.7H_2O$$

Exercise 9.11

1. Find the molar mass for each of the following compounds: K/U

(a)	$Ca(OH)_2$	(j)	$Mg(NO_3)_2$
(b)	KOH	(k)	$Ca(IO_3)_2$
(c)	Fe_2O_3	(l)	$KMnO_4$
(d)	$K_2Cr_2O_7$	(m)	$CuSO_4 \bullet 5H_2O$
(e)	$C_6H_{12}O_6$	(n)	$Ca(HCO_3)_2$
(f)	$Ba(HCO_3)_2$	(o)	H_2O_2
(g)	$Al_2(SO_4)_3$	(p)	$NaClO_3$
(h)	N_2	(q)	$Ba(NO_3)_2$
(i)	CO_2	(r)	$AgIO_3$

CHAPTER 10
Composition of Compounds

Chapter Content:
10.1 Finding the Percentage Composition of Compounds
10.2 The Empirical Formulas of Compounds
10.3 Experimental Determination of Emperical Formulas
10.4 Finding Molecular Formulas of Compounds

Before the usefulness of any compound can be evaluated, its molecular composition must first be ascertained. This chapter briefly discusses how the empirical and molecular formulae of simple compounds are determined, based on the elements present, and their respective percentage compositions by mass, in the compounds.

The percentage composition of compounds, by mass, not only leads to the determination of their molecular formulae, but it is also useful in making comparisons of the percentages of the same or different elements in different compounds. For example, nitrogenous fertilizers are made and sold in different forms, such as KNO_3 and $(NH_4)_2SO_4$. Knowing the percentage composition of each element in these compounds, by mass, would allow someone to choose which of the two would be more suitable to replenish a soil that is depleted of nitrogen. In the case of treating anemia, it important to know the percentage composition of iron in a supplemental compound, such that the correct amount of it can be safely administered to patients.

By knowing the molecular formula of a compound, its molecular geometry can be achieved, which in turn, can be useful in deducing its physical and chemical properties. Knowing these properties can provide useful information about a compound's ability to change into more useful ones.

More importantly, molecular formulas of compounds lead to the calculations of their molar masses which is essential in stoichiometry.

The compound, ethene (C_2H_4), for example, has the following features; knowledge of these are important in its combustion and other chemical reactions.

Table 10.1. Some features of the compound, ethene.

Percentage Composition by mass		Empirical formula	Molecular formula	Molar mass:
% C	% H	CH_2	C_2H_4	28.06 g/mol
85.60	14.40			

10.1 Finding the Percentage Composition of Compounds

The law of definite proportions

Scientists discovered that compounds contain elements in fixed proportions by mass. Regardless of how many different samples of the same compound are analyzed, results show that the elements contained in each different sample are in fixed ratios, by mass. The French chemist, Joseph Louis Proust (174-1794), having analyzed various samples of copper (II) carbonate, found that the elements copper, carbon and oxygen were always in the same proportion by mass. He formulated his **Law of Definite Proportions,** which states that: *In any specific compound, the elements are always present in definite proportions by mass.*

Percentage Composition

Suppose that a box has the following composition of marbles:

20 blue, 30 red, 10 white and 40 black marbles

The total number of marbles would be: $20 + 30 + 10 + 40 = 100$

Then the percentage composition of marbles would be:

$$\% \text{ blue marbles} = \frac{20\ marbles}{100\ marbles} \times 100\% = 20\%$$

$$\% \text{ red marbles} = \frac{30\ marbles}{100\ marbles} \times 100\% = 30\%$$

$$\% \text{ white marbles} = \frac{10\ marbles}{100\ marbles} \times 100\% = 10\%$$

$$\% \text{ black marbles} = \frac{40\ marbles}{100\ marbles} \times 100\% = 40\%$$

$$Total\ = 100\%$$

Percentage Composition of Compounds

Each different chemical compound has its own characteristic molar mass and percentage composition for its elements. The compounds, sodium hydroxide and calcium hydroxide will be used to illustrate how the percentage compositions of compounds are calculated.

<table>
<tr><th>Molar mass (M) for NaOH</th><th>Molar mass (M) for Ca(OH)$_2$</th></tr>
<tr><td>$Na = 22.99\ g\ /\ mol$</td><td>$Ca = 40.08 g/mol$</td></tr>
<tr><td>$O = 16.0\ g\ /\ mol$</td><td>$2O = 16 g/mol \times 2 mol = 32g$</td></tr>
<tr><td>$H = 1.01\ g\ /\ mol$</td><td>$2H = 1.01 g/mol \times 2 mol = 2.01g$</td></tr>
<tr><td>$Molar\ mass = 22.99\ g + 16.0\ g + 1.01\ g$</td><td>$Molar\ mass = 40.08g + 32g + 2.02g$</td></tr>
<tr><td>$= 40.0\ g\ /\ mol$</td><td>$= 74.10 g/mol$</td></tr>
</table>

The percentage composition by mass of the elements in one mole of ***sodium hydroxide*** would be:

$$\% \text{ Na} = \frac{22.99\ g}{40.00\ g} \times 100\% = 57.48\%$$

$$\% \text{ O} = \frac{16.00\ g}{40.00\ g} \times 100\% = 40.00\%$$

$$\% \text{ H} = \frac{1.01\ g}{40.00\ g} \times 100\% = 2.52\%$$

$$Total\ = 100\%$$

The percentage composition by mass of the elements in one mole of ***calcium hydroxide*** would be:

$$\% \text{ Ca} = \frac{40.08 \text{ g}}{74.10 \text{ g}} \times 100\% = 54.09\%$$

$$\% \text{ O} = \frac{32.00 \text{ g}}{74.10 \text{ g}} \times 100\% = 43.18\%$$

$$\% \text{ H} = \frac{2.02 \text{ g}}{74.10 \text{ g}} \times 100\% = 2.73\%$$

$$Total = 100\%$$

Exercise 10.1

Find the percentage composition by mass of the elements in the following compounds: K/U

a) KNO_3

b) NH_4NO_3

c) $(NH_4)_2 SO_4$

d) As can be seen, NH_4NO_3, $(NH_4)_2SO_4$ and KNO_3 are compounds containing the element, nitrogen. Nitrogen is an essential element for plant growth and is sometimes supplied to the soil in the form of nitrogenous fertilizers. Which of these three compounds would provide the greatest amount of nitrogen per gram of fertilizer?

e) Use your knowledge of percentage composition to find the mass of oxygen in the following compounds:

   i) 10.0 g of $NaOH_{(s)}$

   ii) 49.0 g of $H_2SO_{4(aq)}$.

10.2 The Empirical Formula of Compounds

The empirical formula gives the smallest whole number ratio of the elements in a compound. This is what would be found if any compound is analyzed experimentally. *The molecular formula* represents the actual number of atoms of each element that make up a molecule or a formula unit of a compound. The following table will help distinguish between the empirical and molecular formulas for a few compounds.

Table 10.1. Empirical and Molecular formulas for some compounds.

Name of compound	Molecular formula	Empirical formula	Smallest ratio of elements
dinitrogen tetroxide	N_2O_4	NO_2	**1:2**
glucose	$C_6H_{12}O_6$	CH_2O	**1:2:1**
ethene	C_2H_4	CH_2	**1:2**
water	H_2O	H_2O	**2:1**
sodium carbonate	Na_2CO_3	Na_2CO_3	**2:1:3**
benzene	C_6H_6	CH	**1:1**
nitrogen dioxide	NO_2	NO_2	**1:2**

Finding the Empirical Formulae of compounds from Percentage Composition

Calculate the empirical formula of a compound that has the following percentage composition by mass: **2.02%** hydrogen, **32.65%** sulphur and **65.31%** oxygen.

As can be recalled, the percentage composition of all the elements in any compound must add up to 100%. From this, we can infer that if there was a 100 g of the compound, then the elements present would have the same ratio by mass as their percentage composition. In this case, there would be:

  2.02 g of H   **32.65 g** of S   **65.31 g** of O **in 100.0 g** of the compound.

To find the empirical formula of this compound, it is useful to follow the next steps:

Step 1.
Assuming 100 g of the compound is present, convert percentage composition to amount in mass in grams.

Elements:	**H**	**S**	**O**
Mass (g)	2.02 g	32.65 g	65.31 g

Step 2.
Find the number of moles of atoms for each of the elements present. This is done by dividing the mass of each element by their respective molar masses.

$$\frac{2.02\ g}{1.01\ g/mol} = 2.00\ mol \qquad \frac{32.65\ g}{32.06\ g/mol} = 1.020\ mol \qquad \frac{65.31\ g}{16.00\ g/mol} = 4.080\ mol$$

Do not round off the decimals at this point

Step 3.
To get rid of the decimals, choose the smallest number and divide all the others by this value. In this case, the smallest number is 1.02. Dividing all the others by this number we get:

$$\frac{2.00\ mol}{1.02} \qquad\qquad \frac{1.020\ mol}{1.02} \qquad\qquad \frac{4.080\ mol}{1.02}$$
$$1.96\ mol \qquad\qquad\quad 1.00\ mol \qquad\qquad\quad 4.00\ mol$$

 If decimals still persist after this step and they are extremely close to a whole number, then round them off to the nearest whole number. Rounding off the decimals to whole numbers in this problem indicates that there are two moles of hydrogen, one mole of sulfur and four moles of oxygen per mole of the compound. Making each of these numbers the subscript of the element from which it was obtained gives us the empirical formula, **H_2SO_4.**

 In other cases, where the decimals are not very close to whole numbers, rounding off of the decimals would lead to erroneous results. In these cases, an additional step is necessary to get rid of the decimals. The following sample problem illustrates how this is done.

A compound contains 16.30% carbon, 32.10% chlorine and 51.60% fluorine by mass. What is its empirical formula?

Solution:

Step 1.
Assuming 100 g of the compound present, convert percentage composition to amount in grams.

C	Cl	F
16.3 g	*32.1 g*	*51.6 g*

Step 2.
Find the number of moles of atoms for each of the elements present. This is done by dividing the mass of each element by their respective molar mass.

$$\frac{16.3\ g}{12.01\ g/mol} \qquad \frac{32.1\ g}{35.45\ g/mol} \qquad \frac{51.6\ g}{19.00\ g/mol}$$

$$= 1.36\ mol \qquad\qquad 0.910\ mol \qquad\qquad 2.72\ mol$$

Step 3. To get rid of the decimals, choose the smallest number and divide each answer by this number.

$$\frac{1.3\ mol}{0.910} \qquad\qquad \frac{0.910\ mol}{0.910} \qquad\qquad \frac{2.72\ mol}{0.910}$$

$$=1.50\ mol \qquad\qquad = 1.00\ mol \qquad\qquad =3.00\ mol$$

In this case, to get rid of the decimal, **0.5** in 1.5, it is necessary to multiply all the other values by 2. In doing this, we get:

C	Cl	F
1.50 mol × 2	1.00 mol × 2	3.00 mo × 2
3 mol	2 mol	6 mol

The empirical formula for this compound is $C_3Cl_2F_6$.
In this case, it was easy to get rid of the decimal of **0.5.** Other decimals may be more challenging.

For example, suppose there were the following numbers with their respective decimals:

1.2	2.25	1.33	2.4	1.5	2.5	1.6	1.67	2.27

Rounding any one of these values to a whole number would lead to an erroneous empirical formula. From a mathematical standpoint, multiplying numbers with such decimals by specific numbers would change them to whole numbers. For example, to change 1.2 to a whole number, the following mathematical operation is carried out: 1.2 x 5 = 7.

Here, multiplying the decimal number, 1.2 by 5 gives the whole number of 7. This example is done in column 2 in the table below. Other factors for eliminating decimals are also provided.

Table 10.2 Common factors used to eliminate decimals.

Decimals	1.2	X.25	X.33	X.4	X.5	X.6	X.67	X.75	X.8
Factors to multiply by	1.2x 5 = 7	4	3	5	2	5	3	4	5

In the table above, (**X**) represents any whole number.

In the previous sample problem, the mole ratios of C, Cl, and F were 1.5, 1.0 3.0 respectively. To get rid of the decimal (0.5) in the element carbon, the number of moles *for each element* was multiplied by the whole number **2,** producing the formula, $C_3Cl_2F_6$.

10.3 Experimental Determination of Empirical Formula

There are a number of ways in which the empirical formula of compound can be determined. Two of these are examined as follows.

Method 1: Synthesis of a Compound

In this method, a known mass of a given element (**A**) is allowed to react completely with another element (**B**) of unknown mass to form a compound (**C**). The mass of the compound formed is then measured. The mass of the element **B** is found by subtracting the mass of element **A,** from the mass of compound **C**. Now that the masses of both elements **A** and **B** are known, the empirical formula of the compound **C** can be found.

Experimental Design:

The determination of the empirical formula of the compound magnesium oxide utilizes this method. In this example, a known mass of magnesium strip is placed in a crucible and is burnt in air, in a controlled way to produce magnesium oxide. It is important that it be done in a controlled way to ensure that all the magnesium is converted to its oxide. If not some magnesium can be converted to other compounds such as Mg_3N_2, which can escape as a gas. This could lead to erroneous results. The following are actual results that were obtained from an experiment for this reaction:

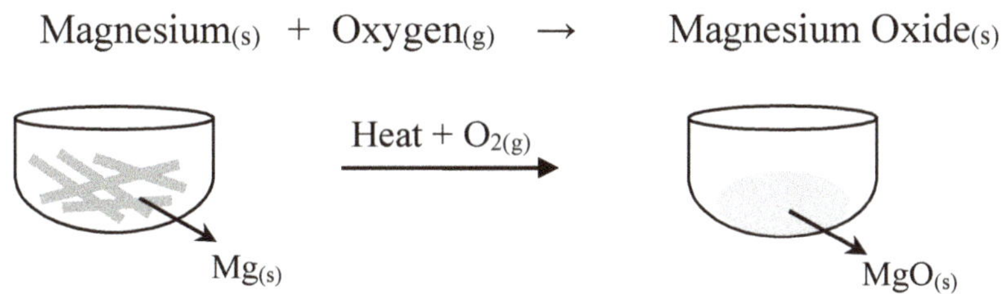

Figure 10.1. Controlled burning of magnesium in a crucible to produce only MgO(s).

Mass of empty crucible + lid   = *254.000 g*
Mass crucible + lid + Mg$_{(s)}$ (before heating)   = *24.100 g*
Mass crucible + lid + Mg$_{(s)}$ (after heating)   = *254.164 g*

Analysis:

Mass of magnesium used = *254.100 g - 254.000 g = 0.100 g*
Mass of oxygen used   = *254.164 g - 254.100 g = 0.064 g*

Solution:

Step 1. *Find the number of moles of each element present in the compound.*

Magnesium

$$n_{Mg} = \frac{0.100\ g}{24.31\ g\ /\ mol}$$

0.00410 mol

Oxygen

$$n_{O} = \frac{0.0640\ g}{16.0\ g\ /\ mol}$$

0.00400 mol

*Note that **16.0 g/mol** is used and not **32.0 g/mol** for oxygen, since **atoms** are used **instead of molecules.**

Step 2. *Divide each of the answers by the smallest number to get rid of the decimals.*

$$\frac{0.004100\ mol}{0.00400}$$

1.02 mol

$$\frac{0.00400\ mol}{0.00400}$$

1.00 mol

Rounding off the decimal to the nearest whole number, the following mole ratio is obtained:

Magnesium   *Oxygen*
1 mol   :   *1 mol*

The formula of magnesium oxide is therefore ***MgO.***

Method 2: Decomposition of a Compound

In this method, a known mass of a compound (**C**) is obtained and is then decomposed into its component elements (**A** and **B**). The masses of elements **A** and **B** are then measured and used to calculate the empirical formula of the compound. For example, Copper has two oxides; copper (I) oxide and copper (II) oxide. If a known mass of either of these solids is provided, and it is required to determine which of the two oxides of copper it is, the decomposition method can be used to do this.

Both hot copper (I) oxide and copper (II) oxide react with gaseous hydrogen according to the following word equation:

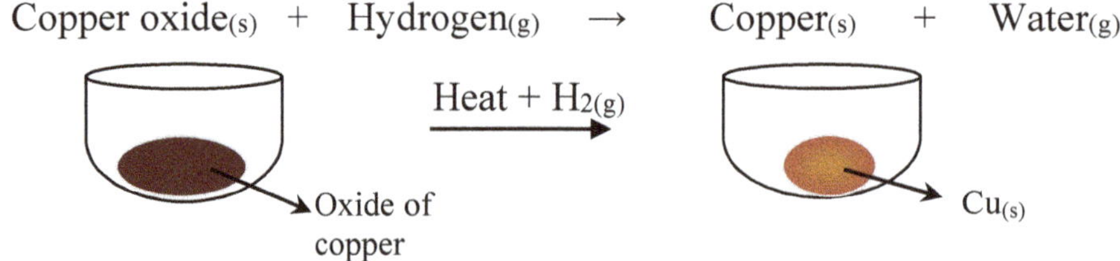

Figure 10.2. Decomposition of copper oxide by heating it in a crucible with hydrogen.

A known mass of the solid oxide is placed in a crucible which is then heated thoroughly in a stream of hydrogen gas. After the reaction is complete, the difference between the original and final mass of the compound would be the mass of oxygen originally present in the oxide. This is so because only the oxygen was removed from the reaction between the oxide and hydrogen to form water vapour that escaped the crucible. The mass of substance left in the crucible would be that of the copper present originally in the oxide. The following were the results actually obtained from an experiment to determine the correct formula for one of the oxides of copper.

Results:

Mass of empty crucible	$= 310.00\ g$
Mass of crucible + oxide	$= 312.00\ g$
Mass of crucible after reaction with hydrogen	$= 311.60\ g$

Analysis:

Mass of oxygen present $= 312.00\ g - 311.60\ g = 0.40\ g$

Mass of copper present $= 311.60\ g - 310.00\ g = 1.60\ g$

Solution:

Step 1.

Find the number of moles of each element present in the compound.

Copper	**Oxygen**	
$n_{Cu} = \dfrac{1.60\ g}{63.55\ g/mol}$	$n_{O} = \dfrac{0.400\ g}{16.0\ g/mol}$	*Note again that **16.0 g/mol** is used and not **32.0g/mol** for oxygen as **atoms** are used instead of molecules.
$0.0250\ mol$	$0.0250\ mol$	

Step 2.

Divide each of the answers by the smallest number to get rid of the decimals.

$$\frac{0.0250\ mol}{0.0250} \qquad \frac{0.0250\ mol}{0.0250}$$

$$1.00\ mol \qquad 1.00\ mol$$

Since there are no decimals to round off and the mole ratio of copper to oxygen is **1:1**, it can be concluded that the formula of the oxide is **CuO**.

If the mole ratio had worked out to be **2:1**, then the formula of the oxide would have been **Cu$_2$O**.

10.4 Finding Molecular Formulas of Compounds

As defined before, the empirical formula gives the smallest whole number ratio of the elements in a compound. As an example, the hydrocarbon ethene has the chemical formula, **C$_2$H$_4$** and a molar mass of **28.06 g/mol.** If the percentage composition of this compound is calculated, we get the following:

$$\% C = \frac{24.02\ g}{28.06\ g} \times 100\% \qquad : \% H = \frac{4.04\ g}{28.06\ g} \times 100\%$$

$$= 85.60\% \qquad\qquad 14.40\%$$

The empirical formula of the compound ethene, C_2H_4 is CH_2. However, there is an entire series of hydrocarbons that have the same percentage composition and empirical formula as the compound, C_2H_4. The following table shows some of these:

Table 10.3 Percentage composition of some alkenes

Hydrocarbons (alkenes)	Empirical formula	% C	% H
C_2H_4	CH_2	85.60	14.40
C_3H_6	CH_2	85.60	14.40
C_4H_8	CH_2	85.60	14.40
C_5H_{10}	CH_2	85.60	14.40
C_6H_{12}	CH_2	85.60	14.40

Suppose that the problem is reversed, and it was required to find the empirical formula of a compound that has 85.60 % carbon and 14.4 % hydrogen by mass, what would the result be?

Following the previous guidelines, the following is obtained:

Step 1.
Convert percentage composition to amount in mass (g).

$$\textbf{Element:} \quad C \quad : \quad H$$
$$\textbf{Mass:} \quad 85.60\ g \quad\quad 14.40\ g$$

Step 2. Find the number of moles of each element present in the compound.

$$\frac{85.60\ g}{12.01\ g\,/\,mol} \quad : \quad \frac{14.40\ g}{1.01\ g\,/\,mol}$$
$$= 7.127\ mol \quad\quad = 14.26\ mol$$

Step 3. To get rid of the decimals, choose the smallest and divide all the other values by it. The smallest number in this case is **7.127.**

$$\frac{7.127\ mol}{7.127} \quad : \quad \frac{14.26\ mol}{7.127}$$
$$= 1.000\ mol \quad\quad = 2.000\ mol$$

The empirical formula of the compound is CH_2 *as expected.*

In this example, the empirical formula obtained, as can be seen, would be the same for all the compounds in Table 10.3 on the previous page. To know which one of those compounds is under investigation, some *additional information* is needed, the molar mass of the compound.

Additional information

If **the molar mass** of the compound **is 56.12 g/mol,** what is its molecular formula? To find the molecular formula of the compound, it useful to follow the additional steps below:

Step 4. Find the molar mass of the compound's empirical formula, CH_2.

$$1 \text{ mol of } C = 1 \text{ mol} \times \frac{12.01 \text{ g}}{\text{mol}} = 12.01 \text{ g}$$

$$2 \text{ mol of } H = 2 \text{ mol} \times \frac{1.01 \text{ g}}{\text{mol}} = 2.02 \text{ g}$$

$$\textbf{Molar mass of } CH_2 = 12.01 \text{ g} + 2.02 \text{ g} = \textbf{14.03 g/mol}$$

Whatever the molar mass of the compound is, *it must be a whole number multiple of the molar mass of its empirical formula.* To find out what this **multiple** is, carry out the step below.

Step 5. Divide the molar mass of the compound by the molar mass of its empirical formula.

$$\frac{Molar\ mass}{Empirical\ formula\,mass} = \frac{56.12 \text{ g/mol}}{14.03 \text{ g/mol}} = 4\ (\textbf{Multiple})$$

The molar mass of the compound is thus *4 times* as heavy as its empirical formula mass. To find its molecular formula, *multiply the subscript of the elements in the empirical formula by the multiple, 4.*

Step 6. Multiply the subscripts **of each element of the empirical formula** by the multiple obtained from **step 2.**

$$= C_{1\times4}H_{2\times4}$$

The molecular formula of the compound is therefore **C_4H_8.**

Step 7. Check our solution to see if the molar mass of **C_4H_8** is **56.12 g/mol.**

$$\left(4.000 \text{ mol } C \times 12.01 \text{ g/mol}\right) + \left(8.000 \text{ mol } H \times 1.01 \text{ g/mol}\right) = 56.12 \text{ g/mol}$$

Exercise 10.2

1. Glucose has the following percentage composition by mass: 39.99% carbon, 6.73% hydrogen, and 53.28% oxygen. If the molar mass of glucose is 180.18 g/mol, what is its molecular formula? How many moles of glucose are required if each of 200 athletes is to be given a glucose tablet of having a mass 10.0 g? K/U T/I

2. The hydrocarbon, butane, has the following percentage composition by mass: 82.63% carbon, and 17.37% hydrogen. If the molar mass of butane is 58.14 g/mol, what is its molecular formula? K/U T/I

3. Caffeine (trimethylxanthine), the stimulant found in tea and coffee, has the following percentage composition by mass: 49.47% carbon, 5.20% hydrogen, 28.85% nitrogen, and 16.48% oxygen. What is its empirical formula? K/U T/I

4. Vitamin C, a deficiency of which causes the disease scurvy, has the following percentage composition by mass: 40.91% carbon, 4.59% hydrogen, and 54.50% oxygen. If the molar mass of vitamin C is 176.14 g/mol, what is its molecular formula? If you take a 60.0 mg dose of this vitamin daily, how long will it take for you to consume one mole? K/U T/I

5. If the molecular formula for vitamin D which is responsible for calcium metabolism is $C_{56}H_{88}O_2$, find its percentage composition by mass of the elements present. K/U T/I

6. Folic acid, (one of the 8 B vitamins), is important for good health and for the reduction of spina bifida in unborn children. If it has the molecular formula, $C_{19}H_{19}N_7O_6$, find its percentage composition by mass of the elements present. If the daily requirement for the male adult is 400 micrograms, how many vitamin B tablets can be manufactured from 10 moles of folic acid? T/I A

7. Lactic acid is the substance that causes muscle fatigue in athletes. Its percentage composition by mass is: 40.03% carbon, 6.72% hydrogen, and 53.25% oxygen. If its molar mass is 90.15 g/mol, determine its molecular formula. T/I

8. If the artificial sweetener, aspartame, has the following percentage composition by mass: 57.12% carbon, 6.18% hydrogen, 9.52% nitrogen, and 27% oxygen determine its empirical formula. How many 0.5 g packages can be made from 10 moles of aspartame? T/I

9. Morphine is one of the most widely used drugs to kill pain. Its percentage composition by mass is: 71.56% C, 6.71% H, 4.91% N, and 16.82 % O. If its molar mass is 285g/mol, find its molecular formula. T/I

10. Viagara is one of a series of drugs that is used to treat men suffering from erectyle dysfunction. Its percentage composition by mass is: 55.67% C, 6.38% H, 17.71% N, 13.50 % O and 6.76% S. If its molar mass is 474.58 g/mol, find its molecular formula. T/I

11. Atorvastatin (lipitor) is a member of a class of drugs used for lowering blood cholestrol. Its percentage composition by mass is: 70.95% C, 6.33% H, 3.40% F, 5.01% N, 14.29 % O. If its molar mass is 558.64 g/mol, find its molecular formula. T/I

12. Caffeine, present in tea and coffee stimulates the central nervous system alleviating the systoms fatigue and drowsiness; in premature babies it is used as a treatment to increase their lower than normal heart rates. Its percentage composition by mass is: 49.47% C, 5.20% H, 28.86% N,16.47 % O . If its molar mass is 194.19 g/mol, find its molecular formula. T/I

13. Combustion of a sample of a compound containing only carbon and hydrogen produced 4.81g of carbon dioxide and 1.94 g of water. If the molar mass ofthe compound is 70.0 g/mol, calculate its molecular formula. T/I

Chapter Content:

Once, in a certain country, during tough economic times, there was a chronic shortage of basic household commodities, including soap. A poor woman was told that if she boiled the fat from her slaughtered pig with sodium hydroxide, it would make her some soap. She did this and obtained some soap. However, when she bathed her children with the soap, they suffered severe skin burns and had to be hospitalized.

The above scenario exemplifies what happens when reagents **are not** mixed in correct quantities during chemical reactions. In this example, pig fat reacted with a caustic solution of sodium hydroxide neutralizing some of it and forming some soap and glycerol in the process. But, because the quantities of fat and sodium hydroxide were not accurately measured and mixed, after reaction some sodium hydroxide was left in excess, and it caused the skin burns. In chemical reactions, the reactants have chemical and physical properties which are different from the products they form. In this case, sodium hydroxide is extremely caustic and dangerous, but the fat is greasy and harmless. However, the soap formed is safe to use for cleaning and bathing and the glycerol (a byproduct) is a safe alcohol. But if one of the reactants is in excess, it remains with the product(s) and retains its original properties. This is what happened in this case with the sodium hydroxide. Had the fat been the excess reagent, it would have neutralized all sodium hydroxide, but some of it would be left over; the soap would have been greasy.

Companies that manufacture medicines, drugs and other type of treatments that people drink or use on their bodies, must ensure that no excess reagents are left in their products; this may cause harm to users.

This method of using correct quantities using balanced chemical equations to ensure that there are no leftover reagents, encompasses **stoichiometry.** This concept is further developed as this chapter progresses.

Quantities in Chemical Reactions

Balanced Chemical Equations

To show the molar relationships between reactants and products in any chemical reaction, a balanced equation must first be written for it. The following is a balanced equation representing the reaction between hydrochloric acid and sodium carbonate solution.

$$2\ HCl_{(aq)} \quad + \quad Na_2CO_{3(aq)} \quad \rightarrow \quad 2\ NaCl_{(aq)} \quad + \quad H_2O_{(l)} \quad + \quad CO_{2(g)}$$

In this balanced chemical equation, we see that **two** moles of $HCl_{(aq)}$ react with **one** mole of $Na_2CO_{3(aq)}$ to produce **two** moles of $NaCl_{(aq)}$, **one** mole of $CO_{2(g)}$ and one mole of $H_2O_{(l)}$.

The study of the relationships between the amounts of reactants used and products formed in chemical reactions is called *stoichiometry.* This concept will further be examined.

Analysis of the above reaction could lead to the following inferences:

➢ Every **2 moles** of HCl(aq) need exactly **1 mole** of Na2CO3(aq) for complete reaction and simultaneously produces exactly **2 moles** of NaCl(aq), **1 mole** of H2O(l) and **1 mole** of CO2(g).

➢ Every **1 mole** of Na2CO3(aq) needs exactly **2 moles** of HCl(aq) for complete reaction and produces simultaneously exactly **2 moles** of NaCl(aq),**1 mole** of H2O(l) and **1mole** of CO2(g).

➢ **2 moles** of NaCl(aq) are produced simultaneously with **1 mole** of H2O(l) and **1 mole** of CO2(g) when exactly **2 moles** of HCl(aq) and **1 mole** of Na2CO3(aq) react.

From the above information, it can further be deduced that, depending on the nature of the chemical reaction, the reactants react with each other in specific mole ratios and produce products in specific mole ratios. These mole ratios also make it possible for **mass ratios** to be established. This is exemplified as follows for the above reaction:

$$2\,HCl_{(aq)} \;+\; Na_2CO_{3(aq)} \;\rightarrow\; 2\,NaCl_{(aq)} \;+\; H_2O_{(l)} \;+\; CO_{2(g)}$$

Mole ratios: $2\,mol$: $1\,mol$ $2\,mol$: $1\,mol$: $1\,mol$

Masses: $72.9\,g$: $105.99\,g$ $116.68\,g$: $18.02\,g$: $44.01\,g$

The inferences that can be made on the basis of mass ratios for this reaction are as follows:

• Every **72.9 g** of HCl(aq) needs exactly **105.99 g** of Na2CO3(aq) for complete reaction and simultaneously produces exactly **116.68 g** of NaCl(aq), **18.02 g** of H2O(l) and **44.01 g** of CO2(g).

• **105.99 g** of Na2CO3(aq) needs exactly **72.9 g** of HCl(aq) for complete reaction and simultaneously produces exactly **116.68 g** NaCl, **18.02 g** of H2O(l) and **44.01 g** of CO2(g).

• **116.68 g** of NaCl are produced simultaneously with **18.02 g** of H2O(l) and **44.01 g** of CO2(g) when exactly **72.9 g** of HCl(aq) and **105.99 g** of Na2CO3(aq) react.

Exercise 11.1

For the following balanced chemical equations, fill in the missing mole ratios by inspection: K/UT/I

(a) $2\,HCl_{(aq)} \;+\; Na_2CO_{3(aq)} \;\rightarrow\; 2\,NaCl_{(aq)} \;+\; H_2O_{(l)} \;+\; CO_{2(g)}$

4 mol   __?____   __?____   __?____   __?____

(b) $2\,HCl_{(aq)} \;+\; Na_2CO_{3(aq)} \;\rightarrow\; 2\,NaCl_{(aq)} \;+\; H_2O_{(l)} \;+\; CO_{2(g)}$

__?___   __?____   __?____   __?____   **1/2 mol**

(c) $H_2SO_{4(aq)} \;+\; 2\,NaOH_{(aq)} \;\rightarrow\; 2NaCl_{(aq)} \;+\; 2H_2O_{(l)}$

1/2 mol   __?____   __?____   _?______

(d) $H_2SO_{4(aq)} \;+\; 2\,NaOH_{(aq)} \;\rightarrow\; 2\,NaCl_{(aq)} \;+\; 2H_2O_{(l)}$

___?____   ___?______   **1/2 mol**   ___?____

(e) $Ba(OH)_{2(aq)} \;+\; 2\,HNO_{3(aq)} \;\rightarrow\; Ba(NO_3)_{2(aq)} \;+\; 2H_2O_{(l)}$

__?____   __?____   **2 mol**   _?_____

(f) $Ba(OH)_{2(aq)} \;+\; 2\,HNO_{3(aq)} \;\rightarrow\; Ba(NO_3)_{2(aq)} \;+\; 2\,H_2O_{(l)}$

1/3 mol   ___?___   __?____   __?____

(g) $2\,BrO_3^-{}_{(aq)} \;+\; 5\,HSO_3^-{}_{(aq)} \;\rightarrow\; Br_{2(aq)} \;+\; 5\,SO_4^{2-}{}_{(aq)} \;+\; H_2O_{(l)} \;+\; 3\,H^+{}_{(aq)}$

___?______   __?______   **2/3 mol**   __?____   __?____   __?____

(h) $\quad 2\,BrO_3^-{}_{(aq)} \quad +\ 5\,HSO_3^-{}_{(aq)} \rightarrow\ Br_{2(aq)} +\ 5\,SO_4^{2-}{}_{(aq)} +\ H_2O_{(l)} \quad +\,3\,H^+{}_{(aq)}$

$\quad$ ___?___ $\qquad$ **2 mol** $\qquad$ __?__ $\qquad$ __?__ $\qquad$ __?__ $\qquad$ __?__

i) $\quad 3\,NO_{2(g)} \qquad +\qquad H_2O_{(l)} \rightarrow\quad 2\,HNO_{3(aq)} \qquad + \qquad NO_{(g)}$

$\quad$ ___?___ $\qquad\qquad$ **3 mol** $\qquad\quad$ __?__ $\qquad\qquad$ __?__

(j) $\quad 3\,NO_{2(g)} \qquad +\qquad H_2O_{(l)} \rightarrow\quad 2\,HNO_{3(aq)} \qquad + \qquad NO_{(g)}$

$\quad$ __?__ $\qquad\qquad$ ___?___ $\qquad\qquad$ **3 mol** $\qquad\qquad$ __?__

11.2 Calculating Mass of Reactants Required to React with Each Other *Gravimetric Stoichiometry*

In the previous exercises, where the number of moles of one reagent is given, it was relatively easy to determine the number of moles of the others in the equation, as quantities in moles were used. Calculations become more challenging when quantities in mass, instead of moles, are used in chemical reactions. The following sample problems will show how calculations are made using amounts in mass.

Sample Problem 1:

$$2\,HCl_{(aq)} +\ Na_2CO_{3(aq)} \quad \rightarrow \quad 2\,NaCl_{(aq)} +\ H_2O_{(l)} +\ CO_{2(g)}$$

The above balanced equation represents the reaction between hydrochloric acid and sodium carbonate. Use it to find what mass of $HCl_{(aq)}$ is required to neutralize 26.50 g of $Na_2CO_{3(aq)}$.

Solution:

Provided quantity: 26.50 g of $Na_2CO_{3(aq)}$
Required quantity: mass of $HCl_{(aq)}$

Step 1: Since relationships in the balanced equation are expressed as mole ratios, the first step is to convert amount in mass to amount in moles.

Using the formula, $n = \dfrac{m}{M}$ **we get:**

$$n_{Na_2CO_{3(aq)}} = \frac{26.50\ g}{105.99\ g/mol}$$

$$= 0.2500\ mol\ Na_2CO_{3(aq)}$$

Step 2: Use the chemical equation to establish the appropriate mole ratios between the two reactants. Let the letter X to represent the **required** quantity, ($HCl_{(aq)}$) in moles.

$\quad$ *(Required)* $\qquad\qquad$ *(Given)*

$$2\,HCl_{(aq)} \quad + \quad Na_2CO_{3(aq)} \quad \rightarrow \quad 2\,NaCl_{(aq)} +\ H_2O_{(l)} +\ CO_{2(g)}$$

$\qquad$ **2 mol** $\qquad$: $\qquad$ **1 mol**

$\qquad$ **X mol** $\qquad$: $\qquad$ **0.2500 mol** $\longrightarrow$ *This quantity becomes (**given**) since its its mass was provided that was subsequently converted to moles.

Rearranging this mole ratio, we get:

$$\frac{Given}{Re\,quired} = \frac{1\,mol}{2\,mol} = \frac{0.2500\,mol\,Na_2CO_{3(aq)}}{X\,mol\,HCl_{(aq)}}$$

$$X = 0.5000\,mol\,HCl_{(aq)}$$

Step 3: Convert the amount of $HCl_{(aq)}$ from moles to amount in mass, (g).

$$m_{HCl_{(aq)}} = nM$$

$$= 0.5000\,mol\,HCl_{(aq)} \times 36.45\,g/mol$$

$$= 18.23\,g\,HCl_{(aq)}$$

$$m = n \times M$$

This method of using mole ratios to calculate the mass of one reagent, which is then used to calculate others in a reaction, is called ***gravimetric stoichiometry.***

For the same reaction in ***sample problem 1***, calculate what mass of $Na_2CO_{3(aq)}$ is required to neutralize 9.110 g of $HCl_{(aq)}$.

Provided quantity: 9.110 g of $HCl_{(aq)}$
Required quantity: mass of $Na_2CO_{3(aq)}$

Following the steps outlined above we have:
Step 1: Convert the amount of $HCl_{(aq)}$ from mass to amount in moles.

$$n = \frac{m}{M}$$

$$n_{HCl_{(aq)}} = \frac{9.110\,g}{36.45\,g\,/\,mol}$$

$$= 0.2500\,mol\,HCl_{(aq)}$$

Mass
Molar mass

$$n = \frac{m}{M}$$

Step 2: Use the chemical equation to establish the appropriate mole ratio between the reactants. Let the letter X represent the **required** quantity, ($Na_2CO_{3(aq)}$), in moles.

(Given) *(Required)*

$$2HCl_{(aq)} + Na_2CO_{3(aq)} \rightarrow 2NaCl_{(aq)} + H_2O_{(l)} + CO_{2(g)}$$

2 mol   :   1 mol

0.2500 mol   :   X mol

Rearranging this mole ratio, we get:

$$\frac{Given}{Required} = \frac{2\,mol}{1\,mol} = \frac{0.2500\,mol\,HCl_{(aq)}}{X\,mol\,Na_2CO_{3(aq)}}$$

$$2X = 0.2500\,mol\,Na_2CO_{3(aq)}$$

$$X = 0.1250\,mol\,Na_2CO_{3(aq)}$$

Step 3: Convert amount of $Na_2CO_{3(aq)}$ from moles to amount in mass, (g).

$$m_{Na_2CO_{3(aq)}} = nM$$

$$= 0.1250 \ mol \ Na_2CO_{3(aq)} \times \frac{105.99 \ g}{mol}$$

$$= 13.25 \ g \ Na_2CO_{3(aq)}$$

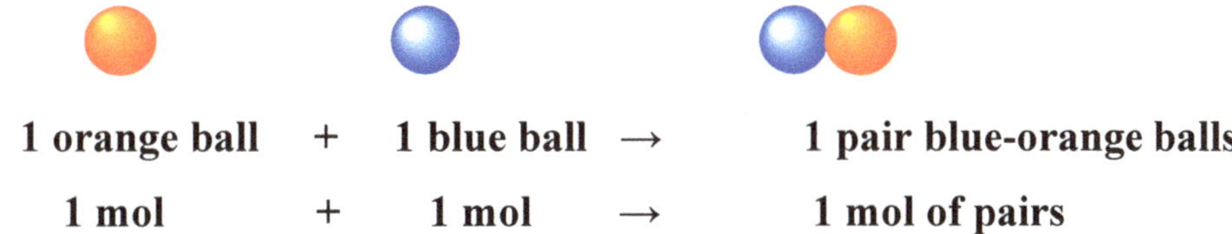

11.3 *Excess* **and** *Limiting reagents.*

In the following situation, orange and blue balls are combining in a 1:1 ratio.

1 orange ball + 1 blue ball →   1 pair blue-orange balls

1 mol + 1 mol →   1 mol of pairs

The above illustrations show how **one** mole of orange balls combine with **one** mole of blue balls to form **one** mole of blue-orange ball pairs. Suppose that the following new situation exists, how many moles of blue-orange pairs would be formed?

Problem: How many blue-orange pairs can be made from the following balls?

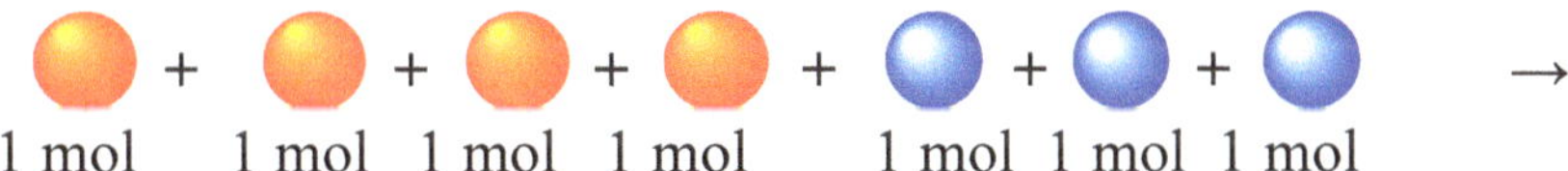

1 mol   1 mol 1 mol 1 mol   1 mol 1 mol 1 mol

Solution:

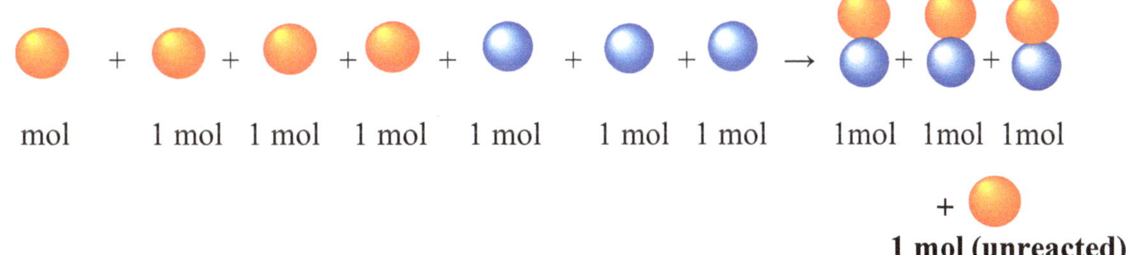

mol   1 mol 1 mol 1 mol   1 mol   1 mol 1 mol   1mol 1mol 1mol

+

1 mol (unreacted)

In the above illustrations, it can be seen how **three** moles of blue balls pair up with exactly **three** moles of orange balls to form **three** moles of blue-orange ball pairs. It also shows that **one** mole of orange balls is left over.

If the orange balls and blue balls were real reagents in a chemical reaction, since orange balls were left over, they would be called the **excess reagent**. On the other hand, since the blue balls were completely used up, they would be called the **limiting reagent**. In this case, it did not matter how many more than **three** moles of orange balls were present, they would be left over and only **three** moles of blue-orange ball pairs will be formed. There will never be any increase of blue-orange pairs unless the number of blue balls increases. Based on these observations, it can be concluded that **the limiting reagent determines two things:**

   1. The amount of the other reagent(s) that is used up.
   2. The amount of product that is formed.

***Note** that **if in any chemical reaction it is told that one reagent is in excess**, then the **other one** automatically becomes **the limiting reagent,** and it **is completely used up.**
The following sample problems are used to clarify this concept.

$$2HCl_{(aq)} + Na_2CO_{3(aq)} \rightarrow 2NaCl_{(aq)} + H_2O_{(l)} + CO_{2(g)}$$

The above balanced equation represents the reaction between hydrochloric acid and sodium carbonate. Use it to calculate the mass of $CO_{2(g)}$ produced when 26.50 g $Na_2CO_{3(aq)}$ reacts with excess $HCl_{(aq)}$.

Solution:

Step 1. Ensure that there is a balanced equation for the reaction.

$$\underset{1\,mol}{Na_2CO_{3(aq)}}^{(Given)} \qquad \underset{1\,mol}{CO_{2(g)}}^{(Required)}$$

$$2HCl_{(aq)} + Na_2CO_{3(aq)} \rightarrow 2NaCl_{(aq)} + H_2O_{(l)} + CO_{2(g)}$$

Step 2. Determine the limiting reagent.

In this case, since it is known that the $HCl_{(aq)}$ is in excess, then the $Na_2CO_{3(aq)}$ automatically becomes the limiting reagent, and it is called the ***given quantity.***

Step 3.

Find the number of moles of the limiting reagent if its amount is given in mass. **If its amount is already given in moles, then leave it as it is.**

$$n_{Na_2CO_{3(aq)}} = \frac{m}{M}$$

$$= \frac{26.50\ g\ Na_2CO_{3(aq)}}{105.99\ g/mol\ Na_2CO_{3(aq)}}$$

$$= 0.2500\ mol\ Na_2CO_{3(aq)}$$

$$n = \frac{m}{M}$$

Step 4.

Use the chemical equation to establish the appropriate mole ratio between $Na_2CO_{3(aq)}$ used and $CO_{2(g)}$ produced, **the required quantity.** This will help to determine the quantity of CO_2 produced in moles, from which its amount in mass can be found.

Provided quantity: *0.2500 mol $Na_2CO_{3(aq)}$*
Required quantity: *mass of $CO_{2(g)}$*

$$\frac{Given}{Required} = \frac{1\ mol}{1\ mol} = \frac{0.2500\ mol\ Na_2CO_{3(aq)}}{X\ mol\ CO_{2(aq)}}$$

$$X = 0.2500\ mol\ CO_2$$

Step 5.

Change the amount of $CO_{2(g)}$ produced from moles to amount in mass.

$$m = n \times M$$

$$m_{CO_2} = 0.2500\ mol \times \frac{44.01\ g}{mol}$$

$$= 11.00\ g\ CO_2$$

$$m = n \times M$$

If it were also required to find the corresponding masses of $H_2O_{(l)}$ and $NaCl_{(aq)}$ that were produced simultaneously with CO_2, then the following steps show how this can be done:

Finding the mass of NaCl produced

Provided quantity: *0.2500 mol $Na_2CO_{3(aq)}$*
Required quantity: *mass of $NaCl_{(aq)}$*

Step 6.

Since the quantity of $Na_2CO_{3(aq)}$ was already changed from amount in mass to amount in moles, the balanced chemical equation is now used to establish the appropriate mole ratios that will help to determine ***the required quantity***, *$NaCl_{(aq)}$, produced in moles.*

Since 1 mole of $Na_2CO_{3(aq)}$ produces 2 moles of $NaCl_{(aq)}$, the following ratios exist:

$$\frac{Given}{Required} = \frac{0.2500\ mol\ Na_2CO_3}{X\ mol\ NaCl} = \frac{1\ mol}{2\ mol}$$

$$X = 0.5000\ mol\ NaCl_{(aq)}$$

Step 7. *Change the amount of $NaCl_{(aq)}$ produced from moles to amount in mass, (g).*

$$m_{NaCL(aq)} = n \times M$$

$$= 0.5000\ mol\ NaCl \times \frac{58.44\ g}{mol}$$

$$= 29.22\ g\ NaCl_{(aq)}$$

Finding the mass of H_2O produced

Provided quantity: *0.2500 mol $Na_2CO_{3(aq)}$*
Required quantity: *mass of $H_2O_{(l)}$*

Step 8.

Since the $Na_2CO_{3(aq)}$ has already been changed from amount in mass to amount in moles, the balanced chemical equation is now used to establish the appropriate mole ratios. This will help to determine ***the required quantity***, *$H_2O_{(l)}$, produced in moles.*

Since 1 mole of $Na_2CO_{3(aq)}$ produces 1 mole of $H_2O_{(l)}$, the following ratios exist:

$$\frac{Given}{Required} = \frac{1\ mol}{1\ mol} = \frac{0.2500\ mol\ Na_2CO_3}{X\ mol\ H_2O}$$

$$X = 0.2500\ mol\ H_2O_{(l)}$$

Step 9. Change the quantity of $H_2O_{(l)}$ from amount in moles to mass(g).

$$m_{H_2O} = n \times M$$

$$= 0.2500\ mol\ H_2O \times \frac{18.02\ g}{mol}$$

$$= 4.510\ g\ H_2O_{(l)}$$

Exercise 11.2   **Calculating the mass of reactants, using mole ratios.**

For each of the following reactions, the mass of one of the reactant is given. Use the mole ratios of the balanced equations to calculate the mass of the other reactant indicated by the *(?)* symbol. K/U T/I

(a) $2\ AgNO_{3(aq)}$ + $BaCl_{2(aq)}$ → $2\ AgCl_{(s)}$ + $Ba(NO_3)_{2(aq)}$

     84.94 g   *?*

(b) $CuSO_{4(aq)}$ + $2\ NaOH_{(aq)}$ → $Cu(OH)_{2(aq)}$ + $Na_2SO_{4(aq)}$

     ?   8.0 g

(c) $Mg_{(s)}$ + $CuSO_{4(aq)}$ → $MgSO_{4(aq)}$ + $Cu_{(s)}$

     ?   15.96 g

(d) $Al(OH)_{3(aq)}$ + $3HCl_{(aq)}$ → $3AlCl_{3(aq)}$ + $3H_2O_{(l)}$

     ?   18.23 g

(e) $Zn_{(s)}$ + $H_2SO_{4(aq)}$ → $ZnSO_4$ + $H_{2(g)}$

     16.35 g   *?*

Exercise 11.3   **Calculating mass of products, given known mass of reactants, using mole ratios.**

In each of the following balanced chemical equations, one reactant is in excess. Use that information to calculate the mass of each product formed. K/U T/I

(a) $H_2SO_{4(aq)}$ + $2\ NaOH_{(aq)}$ → $2\ NaCl_{(aq)}$ + $2\ H_2O_{(l)}$

     20.00 g   excess   *?*   *?*

(b) $Ba(OH)_{2(aq)}$ + $2\ HNO_{3(aq)}$ → $Ba(NO_3)_{2(aq)}$ + $2\ H_2O_{(l)}$

     excess   15.75 g   *?*   *?*

(c) $3\ NO_{2(g)}$ + $H_2O_{(l)}$ → $2\ HNO_{3(aq)}$ + $NO_{(g)}$

     23.00 g   excess   *?*   *?*

(d) $CaCO_{3(s)}$ + $2\ HCl_{(aq)}$ → $CaCl_{2(aq)}$ + $H_2O_{(l)}$ + $CO_{2(g)}$

     excess   9.12 g   *?*   *?*   *?*

 Calculations Involving Excess and Limiting Reagents

We shall return to our illustrations with blue and orange balls **to explain how limiting and excess reagents are calculated in chemical reactions. For the sake of clarity, the illustrations are repeated below.**

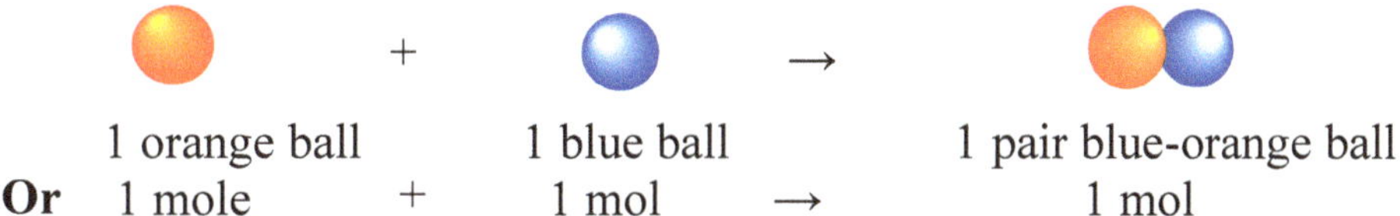

Here, we see one mole of orange balls reacting with one mole of blue balls to form one mole of blue-orange ball pairs. If we have the following new situation, how many moles of blue-orange pairs would be formed when four moles of orange balls react with three moles of orange balls?

Problem:

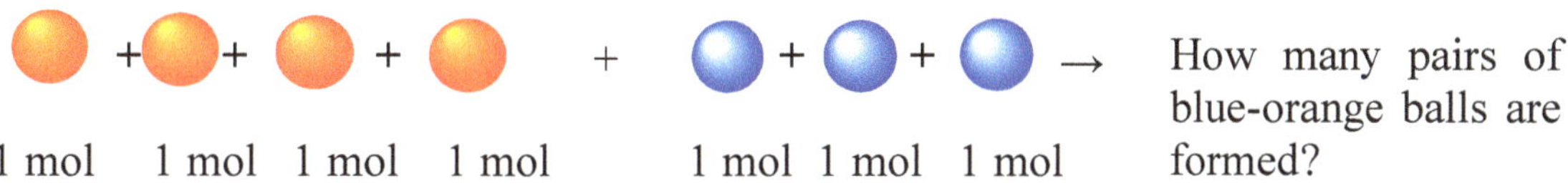

Solution:

If these two types of balls are paired up, the following situation arises.

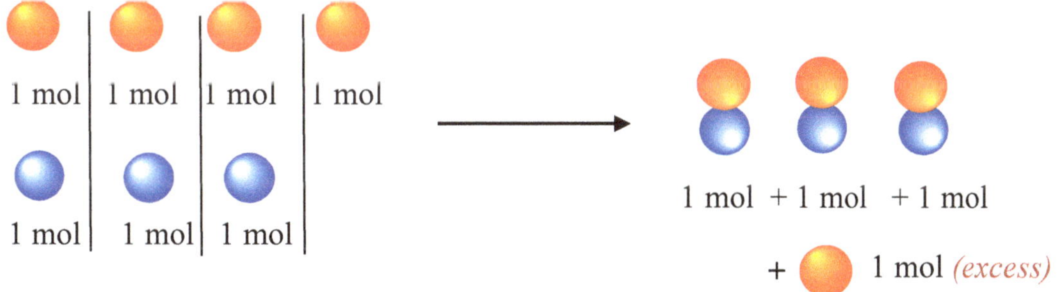

The result shows that three pairs of blue-orange balls are formed with one mole of orange balls left over.

If orange and blue balls were real reagents in a chemical reaction, then because orange balls were left over, they would be called the **excess reagent**. Since the blue balls were completely used up, they would be called the **limiting reagent.** In this case, it did not matter how many more than three moles of orange balls were present, they would be left over and only three moles of blue-orange ball pairs will be formed. There will never be any increase of blue-orange pairs unless the number of blue balls increases. If the number of moles of blue balls increases by one, then the number of moles of blue-orange ball pairs produced will be four in total. In this case, there would be complete pairing up of blue and orange balls and neither would remain in excess. **Conversely,** if the number of blue balls is increased above four moles, blue balls now become the **excess reagent,** and orange would become the **limiting reagent.**

In the previous example, it was easy to determine what reagent was limiting and what was in excess because we were dealing with balls which could be easily observed and counted. However, in real chemical reactions, it becomes more challenging to distinguish the limiting reagent from the excess reagent as the following illustrations will show.

The following is an example of a chemical reaction in which the reactants are represented as amounts in mass:

$$2\,NaCl_{(aq)} \quad + \quad Pb(NO_3)_{2(aq)} \quad \rightarrow \quad 2\,NaNO_{3(aq)} \quad + \quad PbCl_{2(s)}$$

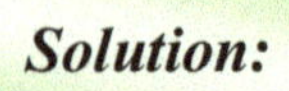

2.92 g   +   16.56 g

Unlike the previous illustrations, where marbles were used, it was easy to determine which ones were in excess and which were limiting, it it not so in this; quantities are given in mass and not in moles. In these types of problems, quantities in mass have to be converted into moles and then the mole ratios of the reactants are used to determine the limiting and excess reagents. This problem is solved below and it outlines all the relevant steps that must be taken when solving problems like this one.

Sample problem 4:

What mass of $PbCl_{2(s)}$ is produced when 2.92 g $NaCl_{(aq)}$ is reacted with 16.56 g of $Pb(NO)_{2(aq)}$?
It should be noted that 2.92 g $NaCl_{(aq)}$ and 16.56 g $Pb(NO)_{2(aq)}$ *are the actual masses of solutes dissolved in water to make the aqueous solutions and they do not include the mass of water.* *This understanding must be kept in mind for all aqueous solutions.*

Solution:

Step 1. Convert amounts of both reactants from mass to amounts in moles.

$$n_{NaCl(aq)} = \frac{2.92\ g}{58.44\ g/mol} = 0.0500\ mol \quad \textcolor{blue}{\textbf{Correct method}}$$

***Common Student mistake:**
Because the equation has the integer 2, in front of the reactant, sodium chloride in the balanced equation, students sometimes divide by twice the molar mass (**116.88 g**), *instead of* **58.44 g**. **Do not do this; it is incorrect.**

$$n_{NaCl(aq)} = \frac{2.92\ g}{116.44\ g/mol} = 0.0250\ mol \quad \textcolor{red}{\textbf{Incorrect method}}$$

$$n_{Pb(NO_3)_2(aq)} = \frac{16.56\ g}{331.22\ g/mol} = 0.05000\ mol \quad \textcolor{blue}{\textbf{Correct method}}$$

Step 2.

Find the Limiting Reagent
The limiting reagent can be found by taking two pathways. We shall return to our orange and blue ball illustrations to explain how this is done.

1 orange ball + 1 blue ball   →   1 pair blue-orange ball

Or 1 mol   +   1 mol   →   1 mol of pairs

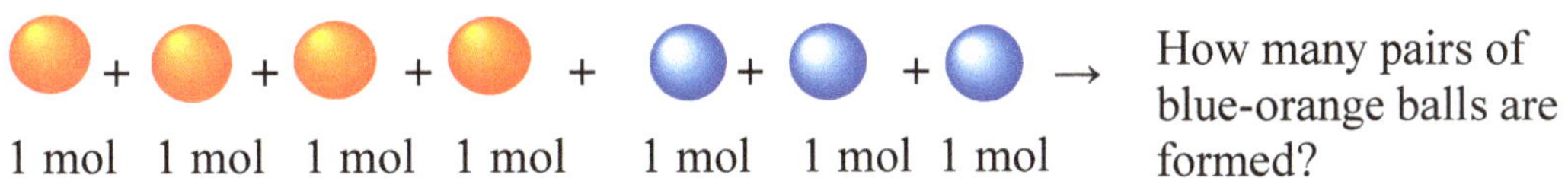

In this illustration, **four moles of orange balls** and **three moles of blue balls** are reacting. We will call these amounts *experimental values,* as they represent the amounts of reagents that are *actually used* in an experiment.

Pathway A

Suppose that a student, A **used** the number of moles of **orange balls** to try to find the corresponding number of moles of **blue balls** that is required to react completely with them, what would the result be?

Solution:

Based on our 1:1 mol ratio we get:

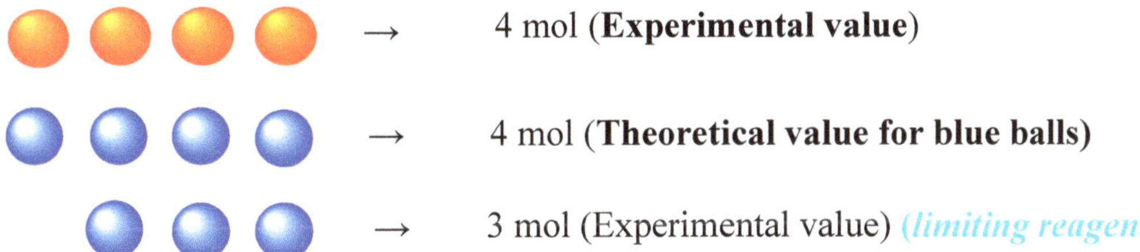

From the illustrations above it can be seen that **four moles of blue balls are theoretically** required to completely react with **four moles of orange balls.** When compared with the **experimental value, (three moles) of blue balls**, it can be deduced that the number of moles of blue balls used in the experiments was not enough to react completely with all the orange balls and blue balls were thus completely used up. We therefore conclude that **blue ball is the limiting reagent and the orange ball is the excess reagent.**

Pathway B

Suppose that another student, B **used** the number of moles **blue balls** to try **to find** the corresponding number of moles of **orange balls** that is required to react completely with them, what would the result be?

Solution:
 Based on our 1:1 mol ratio we get:

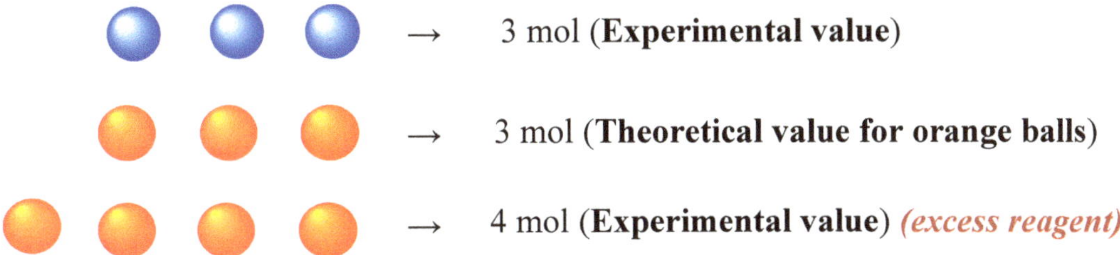

From the illustrations above it can be seen that **three moles of orange balls are theoretically** required to completely react with **three moles of blue balls.** When compared with the **experimental value (four moles) of orange balls**, it can be deduced that the number of moles of **orange balls** used in the experiments

was more than enough to react with all the blue balls and orange balls were thus left over. It can therefore be concluded **again that orange ball is the excess reagent and blue ball is the limiting reagent.**

We shall now return to finding the limiting reagent for our chemical reaction using both pathways. The following are the number of moles of reactants calculated in **step 1.**

$$n_{Pb(NO_3)_2(aq)} = 0.0500\,mol$$

$$n_{NaCl(aq)} = 0.0500\,mol$$

Pathway A to finding the Limiting Reagent

Student A, may choose the number of moles of lead (II) nitrate, $n_{Pb(NO_3)_2(aq)}$ to find the number of moles of sodium chloride, $n_{NaCl(aq)}$ that is **theoretically** required to react with it. In this case $n_{Pb(NO_3)_2(aq)}$ is the **given quantity** since it is chosen and $n_{NaCl(aq)}$ is the **required quantity**. Using the balanced chemical equation, we get the following mole ratios:

$$\frac{Given}{Required} = \frac{1\ mol}{2\ mol} = \frac{0.0500\ mol\ Pb(NO_3)_{2(aq)}}{X\ mol\ NaCl_{(aq)}}$$

$X = 0.100\ mol\ NaCl_{(aq)}$ **(Theoretical value)**

$0.0500\ mol\ NaCl_{(aq)}$ **(Experimental value)**

* The experimental value is placed next to the theoretical value for purpose of comparison. This enables one to know which reagent is limiting or is in excess.

Comparison of the experimental and theoretical values shows that the **amount of NaCl(aq) used** in the experiment (*0.0500 mol*) **is less than what is required** (*0.100 mol*) to react completely with the Pb(NO3)2(aq). It can therefore be concluded that the **NaCl(aq)** is the limiting reagent.

** If the experimental value of a reagent is less than its required theoretical value it becomes the limiting reagent.*

Pathway B to finding the Limiting Reagent

Another student, B may choose $n_{NaCl(aq)}$ to find $n_{Pb(NO_3)2(aq)}$ that is theoretically required to react completely with it. In this case, $n_{NaCl(aq)}$ becomes **the given quantity** since it is chosen and $n_{Pb(NO_3)2(aq)}$ becomes **the required** quantity. Using the balanced chemical equation, we get the following mole ratios:

$$\frac{Given}{Required} = \frac{2\ mol}{1\ mol} = \frac{0.0500\ mol\ NaCl_{(aq)}}{X\ mol\ Pb(NO_3)_{2(aq)}}$$

$$2X = 0.0500\ mol\ Pb(NO_3)_{2(aq)}$$

$X = 0.0250\ mol\ Pb(NO_3)2(aq)$ **(Theotetical value)**

$0.0500\ mol\ Pb(NO_3)2(aq)$ **(Experimental value)**

Comparison of theoretical and experimental values of **Pb(NO3)2(aq)** shows its **amount that was used** in the experiment **is more than what was required** (*0.0500 mol*) to react completely with the NaCl(aq). It can therefore be concluded that the **Pb(NO3)2(aq) is the excess reagent and the NaCl(aq) is the limiting reagent.**

If the experimental value of a reagent is more than its required theoretical value it becomes the *excess reagent.*

As can be seen, if a student used either of these pathways, he/she will end up with the same conclusion for the excess and limiting reagent; *one pathway suffices.*

Continuing to solve this problem, we have:

Step 3.
Use the Limiting Reagent to find the amount of Product(s).

Using the blue and orange ball model, it will be shown how this is done.

Since blue ball was found to be the limiting reagent, it is used to determine how many of the other reagent (orange balls) are used up and how much product(s) is formed.

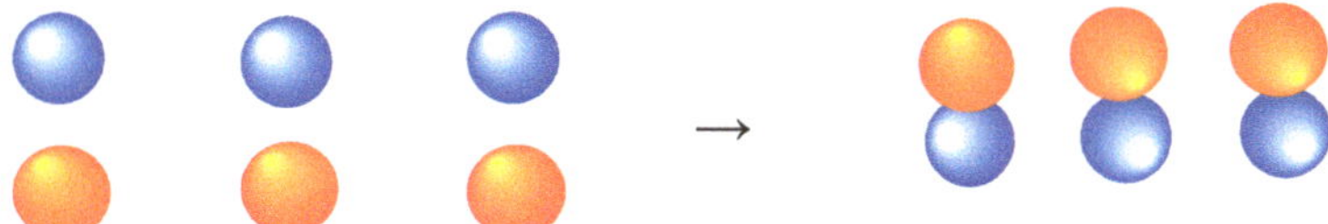

| 3 mol of each type of balls | 3 mol of blue-orange pairs are formed |

This model shows that three moles of blue balls react with exactly three moles of orange balls to produce three moles of blue-orange ball pairs.

In our real problem, $NaCl_{(aq)}$ is the *limiting reagent*. Since its amount is known, it thus becomes the *given quantity*, and *it is used to find the amount of product(s) formed.* **The products,** $PbCl_{2(s)}$ and $NaNO_{3(aq)}$) **are the *required quantities*.** This step to finding the amount of product formed has two parts since there are two different products.

Part 1: Finding the amount of $NaNO_{3(aq)}$ produced.

Using the balanced equation, we have the following mole ratios:

$$\frac{Given}{Required} = \frac{2\ mol}{2\ mol} \frac{0.0500\ mol\ NaCl}{X\ mol\ NaNO_3}$$

Cross multiply numerators with denominators

$$2X = 0.1000\ mol\ NaNO_3$$

$$X = 0.0500\ mol\ NaNO_3$$

$$m_{NaNO_{3(aq)}} = n \times M$$

$$= 0.0500\ mol \times 85.0\ g/mol \quad \textbf{Correct method}$$

$$= 4.25\ g$$

$$m = n \times M$$

*Common mistakes to watch for:

Because the balanced equation has the coefficient **2** in front of sodium nitrate, to find mass of it formed, some students would use twice its molar mass, 170 g/mol, instead of 85 g/mol. This is incorrect, as **the molar of NaNO₃ is 85 g/mol** and **not 170 g/mol.** *Note that the coefficients are used only to establish the mole ratios of reactants and products.*

$$m_{NaNO_3(aq)} = nM$$

$$= 0.0500 \; mol \times \frac{170.0 \; g}{mol} \qquad \textbf{\textcolor{red}{Incorrect method}}$$

$$= 8.500 \; g$$

Part 2: Finding the amount of $PbCl_{2(s)}$ produced.

From the balanced equation, we have the following mole ratio:

$$\frac{Given}{Required} = \frac{2 \; mol}{1 \; mol} \; \frac{0.0500 \; mol \; NaCl}{X \; mol \; PbCl_{2(s)}}$$

$$2X = 0.0500 \; mol \; PbCl_{2(s)}$$

$$X = 0.0250 \; mol \; PbCl_{2(s)}$$

$$m_{PbCl_{2(aq)}} = n \times M$$

$$= 0.0250 \; mol \times \frac{278.1 \; g}{mol}$$

$$= 6.95 \; g$$

$$m = n \times M$$

For the sake of further clarity, another problem involving limiting and excess reagent is solved as follows:

Sample problem 5:

What mass of hydrogen gas is produced between the reaction of 1.20 g of $Mg_{(s)}$ and 3.65 g of $HCl_{(aq)}$.

Solution:

Step 1. Write a balanced equation for the reaction as follows:

$$Mg_{(s)} \; + \; 2HCl_{(aq)} \; \rightarrow \; MgCl_{2(aq)} \; + \; H_{2(g)}$$

Step 2. Find the number of mol of each reactant.

$$n_{Mg} = \frac{m}{M}$$

$$= \frac{1.20 \; g}{24.3 \; g \, / \, mol}$$

$$n = \frac{m}{M}$$

$$= 0.0500 \; \textbf{\textit{mol}} \quad (\textbf{\textit{Experimental value}})$$

$$n_{HCl(aq)} = \frac{m}{M}$$

$$= \frac{3.65 \; g}{36.45 \; g \, / \, mol}$$

$$= 0.0100 \; mol \quad (\textbf{\textit{Experimental value}})$$

Step 3. Find the limiting reagent.

Pathway A

Choose the number of moles of hydrochloric acid, $n_{HCl_{(aq)}}$, to find the number of moles of magnesium, n_{Mg}, that is theoretically required to react completely with it. In this case, since $n_{HCl_{(aq)}}$ is what was chosen and its quantity is also known, it thus becomes the **given quantity** and n_{Mg} becomes the **required quantity**.
From the balanced equation we the following mole ratio:

$$\frac{Given}{Required} = \frac{2\ mol}{1\ mol} = \frac{0.0100\ mol\ HCl_{(aq)}}{X\ mol\ Mg_{(s)}}$$

$$2X = 0.0100\ mol\ Mg_{(s)}$$

$$X = 0.00500\ mol\ Mg\ \textbf{(Theoretical value)}$$
$$0.0500\ mol\ Mg\ \textbf{(Experimental value)}$$

Comparing these two values obtained, we see that the *amount of $Mg_{(s)}$ used* (**0.0500 mol**) in the *experiment **is more*** than what is required to react with **0.010 mole $HCl_{(aq)}$**; only (*0.00500 mol*) **was required.** In this case, *all the $HCl_{(aq)}$ will be used up*, **and it thus becomes** *the limiting reagent.*

Pathway B

Choose the number of moles of magnesium, n_{Mg}, to find the *number of mole of hydrochloric acid, $n_{HCl_{(aq)}}$* that is theoretically required to react completely with it.
In this case, since n_{Mg} is what was chosen and its quantity is also known, it thus becomes the *given quantity* and $n_{HCl_{(aq)}}$ becomes the *required quantity*.

From the balanced equation we have the following mole ratio:

$$\frac{Given}{Required} = \frac{1\ mol}{2\ mol} = \frac{0.0500\ mol\ Mg_{(s)}}{X\ mol\ HCl_{(aq)}}$$

$$X = 0.100\ mol\ HCl_{(aq)}\ \textbf{(Theoretical value)}$$

$$0.0100\ mol\ HCl_{(aq)}\ \textbf{(Experimental value)}$$

Here again, when we compare these two values obtained, we see that *the amount of $HCl_{(aq)}$ used* in the experiment (**0.0100 mol**) *is less than what is actually required* (**0.100 mol**) to react with **0.0500** mole $Mg_{(s)}$. We again conclude that the **$HCl_{(aq)}$** *is the limiting reagent.*

Step 4. Use the limiting reagent to find the amount of products.

In this step, **the limiting reagent becomes the *given quantity*** and it is thus used to find the amounts of product(s) formed.

***Note.** *Use the experimental value* **and not** *the theoretical value to find the amounts of other reagents involved, since it is the experimental value is what is actually used.*

$$\frac{Given}{Required} = \frac{2\ mol}{1\ mol} = \frac{0.0100\ mol\ HCl_{(aq)}}{X\ mol\ H_{2(g)}}$$

$$2X = 0.0100\ mol\ H_{2(g)}$$

$$X = 0.00500\ mol\ H_{2(g)}$$

Step 5. Convert the amount of $H_{2(g)}$ produced in moles to amount in mass.

$$m_{H_{(2)}} = n \times M$$
$$= 0.00500 \; mol \times 2.02 \; g / mol$$
$$= 0.0100 \; g$$

$$m = n \times M$$

11.5 Percentage Yield

In chemistry, the term *yield* refers to the amount of a certain product that is expected to form or what is actually formed in a reaction. Referring to sample *problem 5* above, *0.0100 g $H_{2(g)}$* is what theoretically expected to be produced, and is thus called the **expected yield.** This amount is arrived at by doing calculations. However, if the experiment is actually performed, the amount of hydrogen collected may not be the same as was expected due to various logistical problems. The amount of a product that is actually obtained in an experiment is called the ***actual yield.*** Assuming that the actual yield of hydrogen produced was 0.0080 g, the percentage yield would be calculated as follows:

$$Percentage \; yield = \frac{Actual \; yield}{Expected \; yield} \times 100\%$$

$$\% \; Yield = \frac{0.0080 \; g}{0.0100 \; g} \times 100\% = 80\%$$

In sample problem 4 above, the expected mass of $PbCl_{2(s)}$ was *6.950 g.* Supposing that the experiment was carried out and the mass of $PbCl_{2(s)}$ recovered *was 6.510 g*, the percentage yield would be:

$$\% yield = \frac{6.510 g}{6.950 g} \times 100\% = 94.00\%$$

11.6 SUMMARY OF STOICHIOMETRY

- If any one of the quantities in a chemical reaction is provided, **it becomes the given quantity** and it can be used to determine any of the other quantities. Using the following problems as examples, we have:

Mass-Mass Relationships:

Sample problem: 6

What mass of $NaOH_{(s)}$ is required to neutralize 3.65 g of $HCl_{(aq)}$?

Solution:

(Required)   (Given)

Step 1. *Balance the equation*

$$NaOH_{(s)} \; + \; HCl_{(aq)} \longrightarrow NaCl_{(aq)} \; + \; H_2O_{(l)}$$

Step 2. *Find the number of moles of $HCl_{(aq)}$*

$$n_{HCl} = \frac{3.65 \; g}{36.5 \; g/mol} = 0.100 \; mol$$

Mass

Molar mass

$$n = \frac{m}{M}$$

Step 3. *Find the number of moles of NaOH*

$$\frac{Given}{Required} = \frac{1}{1} = \frac{0.100\ mol\ HCl}{X\ mol\ NaOH}$$

$$X = 0.100\ mol\ NaOH$$

Step 4. *Find the mass of NaOH*

$$m_{NaOH} = 0.100\ mol \times 40.0\ g/mol$$

$$= 4.00\ g$$

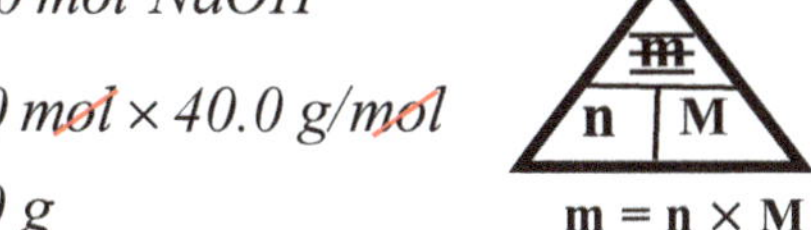

- Using HCl as the known reagent, the masses of $NaCl_{(aq)}$ and $H_2O_{(l)}$ produced can also be found.

Calculation involving aqueous Solutions (This concept will be dealt later with, in unit 3**).**

Sample Problem: 7

What volume of 0.100 mol/L $H_2SO_{4(aq)}$ is required to neutralize 50.0 mL of 0.100 mol/L $NaOH_{(aq)}$?

Solution:

Step 1. Balance the equation

$$\overset{(Required)}{H_2SO_{4(aq)}} + \overset{(Given)}{2\ NaOH_{(aq)}} \longrightarrow Na_2SO_{4(aq)} + 2H_2O_{(l)}$$

Step 2. Find the number of moles of NaOH

$$n_{NaOH(aq)} = \frac{0.100\ mol}{L} \times 0.050\ L$$

$$= 0.0050\ mol$$

Step 3. Find the number of moles of $H_2SO_{4(aq)}$

$$\frac{Known}{Unknown} = \frac{2}{1} = \frac{0.0050\ mol\ NaOH_{(aq)}}{X\ mol\ H_2SO_{4(aq)}}$$

$$2X = 0.0.0050\ mol\ H_2SO_{4(aq)}$$

$$X = 0.0025\ mol\ H_2SO_{4(aq)}$$

Step 4. Find the volume of $H_2SO_{4(aq)}$

$$V_{H_2SO_4} = \frac{0.0025\ mol\ H_2SO_{4(aq)}}{0.100\ mol\ /\ L}$$

$$= 0.025\ L = 25\ mL$$

- If any one of the quantities in a chemical reaction is provided and the rest are in excess, it becomes the given quantity and it can be used to determine any of the other quantities. To do this, follow the examples above.
- If both quantities of reactants in a chemical reaction are provided, it is first required to find which of the two is the limiting reagent, and then use it to find the quantity of any product. This involves the following steps:
  * Write a balanced equation for the reaction
  * Find the number of moles for each of the reactants
  * For pure substances, use the formula, $n = \dfrac{m}{M}$
  * For aqueous solutions, use the formula, $n = C \times V$
  * Determine which of the two reactants is **limiting.**
  * Use the limiting reagent to find the number of moles of any product(s)
  * Change quantity from moles to mass if required.

1. Find the mass of $H_{2(g)}$ produced by the reaction between 6.34 g $Zn_{(s)}$ and has 4.90 g $H_2SO_{4(aq)}$. T/I

2. Calculate the mass of $Cu(OH)_{2(s)}$ produced from a reaction between
15.96 g $CuSO_{4(aq)}$ and 4.00 g NaOH(aq). T/I

3. Calculate the masses of $CO_{2(g)}$ and $H_2O_{(l)}$ produced when 4.00 g $CH_{4(g)}$ is burnt in 8.00 g O_2. T/I

4. Calculate the mass of $AgCl_{(s)}$ produced from a reaction between 16.98 g $AgNO_{3(aq)}$ and 10.00 g $NaCl_{(aq)}$. If the actual mass of $AgCl_{(s)}$ produced is 12.10 g, find the percentage yield. T/I

5. Calculate the mass of $BaSO_{4(s)}$ produced from a reaction between
2.61 g $Ba(NO_3)_{2(aq)}$ and 2.00 g $Na_2SO_{4(aq)}$. If the actual mass of $BaSO_{4(s)}$ produced is
2.10 g, find the percentage yield. T/I

6. What mass of $PbI_{2(s)}$ is produced from the reaction between 4.00 g $NaI_{(aq)}$ and 15.00 g $Pb(NO_3)_{2(aq)}$? If the actual mass of $PbI_{2(s)}$ produced is 5.21 g, find the percentage yield. T/I

7. Aspirin is made from the reaction between salicylic acid and acetic anhydride according to the following equation: T/I

$$C_7H_6O_3 \quad + \quad C_4H_6O_3 \quad \longrightarrow \quad C_9H_8O_4 \quad + \quad C_2H_4O_2$$
salicylic acid   acetic anhydride   aspirin   acetic acid

a) What mass of aspirin can be produced if 10.0 g portions of each reactant were reacted in an experiment?

c) What mass of the excess reagent is left over?

8. One form of soap is made by the reaction between stearic acid and sodium hydroxide according to the following chemical equation: T/I

$$NaOH_{(s)} \quad + \quad C_{17}H_{35}COOH \quad \longrightarrow \quad C_{17}H_{35}COONa_{(s)} \quad + \quad H_2O_{(l)}$$

a) If 4.00 g of $NaOH_{(s)}$ is reacted with 3.00 g of $C_{17}H_{35}COOH$, calculate what mass of soap would be formed.
b) What reagent is in excess?
c) What may people who use this soap experience on their skin?

9. In an attempt to test the gravimetric stoichiometric method, a student mixed 150.0 mL of 0.15 mol/L Na_2CO_3 solution with 200.0 mL of 0.12 mol/L $BaCl_2$ solution to a white precipitate of $BaCO_{3(s)}$. The precipitate was recovered from the mixture using the filtration method. The precipitate, after being dried, had a mass of 4.42 g. Find the percent yield. Is the gravimetric stoichiometric valid? T/I C

CHAPTER REVIEW QUANTITIES IN CHEMICAL REACTIONS (9-11)

Matching

Match each term in the table below with the correct statements that follow

A	Isotopes		G	Relative atomic mass
B	Molar mass		H	Molecular formula
C	One mole		I	Avogadro's number
D	Empirical formula		J	Excess reagent
E	Formula unit		K	Stoichiometry
F	Limiting reagent		L	Hydrates

1. The study of the relationship between the quantities of reactants used and products formed in chemical reactions.

2. This represents the smallest whole number ratio of the atoms of elements present in a compound.

3. Atoms of the same element having the same atomic number but different mass number

4. This reagent is left over after a chemical reaction.

5. These compounds incorporate molecules of water in their crystals.

6. The reactant that is completely consumed in a chemical reaction.

7. This represents the actual number of atoms of each element present in one molecule of a substance.

8. This is the average mass of all the naturally occurring isotopes in an element.

9. The amount of a substance that contains the same number of entities as there are atoms in 12 g of carbon-12.

10. The smallest repeating unit in any ionic compound.

11. This number represents 6.02×10^{23} atoms, ions or molecules or other entities.

12. The mass of one mole of atoms, molecules or formula units.

True or False

Read each of the following statements and then decide if it is *True* or *False*.

1.	The mass of one proton is approximately equal to the mass of one neutron.
2.	The empirical formula of a compound represents the smallest whole number ratio of the atoms of elements present in a compound.
3.	When calculating the molar mass of a hydrated compound, the mass of water is not included.
4.	CH_2 is the empirical formula for benzene (C_6H_6).
5.	The mass of one formula unit of zinc chloride is the sum of the masses of two zinc ions and one chloride ion.
6.	The molar mass for ions is the same as that for mole of atoms for the same element.
8.	Two different compounds cannot have identical percentage compositions.
9.	The amount of the limiting reagent determines the amount of product(s) formed in a chemical reaction.
10.	If 1.20 g of magnesium oxide was produced in a reaction for which the theoretical yield was 1.40 g, then the percentage yield is 85.71%.
11.	It is of no consequences if excess reagents are left in medicines and pharma products.

Multiple Choice

Choose the letter that best answers the questions.

1. The atomic notation for carbon is $^{12}_{6}C$. The nucleus for this atom will have

a	12 neutrons and 6 protons	d	12 neutrons and 12 protons
b	6 neutrons and 6 protons	e	none of the above
c	12 protons and 6 neutrons		

2. Magnesium has three naturally occurring isotopes, as follows: 78.99% ^{24}Mg, 10.00% ^{25}Mg and 11.01% ^{26}Mg. The average atomic mass of Magnesium is

a	28.1 u	d	25.0 u
b	243.1 u	e	24.0 u

3. The mass of one hydrogen atom is approximately equal to

a	1.0 g	d	6.022×10^{23} u
b	2.0 g	e	1.0 u
c	2.0 u		

4. The mass of one $^{12}_{6}C$ atom approximately equals the mass of

a	1 $^{1}_{1}H$ atom	d	1 $^{13}_{6}C$ atom
b	12 $^{1}_{1}H$ atoms	e	6 $^{1}_{1}H$ atoms
c	1 $^{14}_{7}N$ atom		

5. The molar mass of magnesium sulphate, $MgSO_4 \cdot 7H_2O$ (Epson salt) is

a	246.5 g/mol	d	159.5 g/mol
b	120.3 g/mol	e	none of the above
c	126.1 g/mol		

6. The ratio of carbon to oxygen, by mass, in carbon dioxide is

a	2:1	d	1:3
b	3:8	e	12:16
c	12:1		

7. The number of **moles** of molecules in 8.0 g of oxygen, (molar mass = 32.0 g/mol) is

a	0.50 mol	d	32 mol
b	0.25 mol	e	1.0 mol
c	80 mol		

8. The mass of 0.250 moles of copper(II) sulphate pentahydrate, $CuSO_4.5H_2O$, is

a	62.4 g	d	39.8 g
b	124.7 g	e	22.5 g
c	249.4 g		

9. The percentage composition of nitrogen, by mass, in the compound, NH_4OH

a	47.0%	d	20.0%
b	34.2%	e	18.5%
c	39.96%		

10. The number of **molecules** in 16.0 g of oxygen is

a	6.02×10^{23}	d	3.01×10^{23}
b	1.50×10^{23}	e	1.50×10^{22}
c	1.20×10^{24}		

11.

The fertilizer, ammonium nitrate has the formula, $(NH_4)_3PO_4$. The percentage of phosphorus, by mass, in this compound is

a	8.1%	d	42.9%
b	25.5%	e	35.5%
c	20.7%		

12.

A 4.03 g sample of a compound contains 2.43 g of magnesium and 1.60 g oxygen. The empirical formula of the compound is

a	Mg_2O	d	Mg_2O_2
b	MgO_2	e	MgO_2
c	MgO		

13.

A compound is found to consist of 39.99% carbon, 6.73% hydrogen, and 53.28% oxygen, by mass. If the molar mass of the compound is 180.18 g/mol, it molecular formula is

a	CH_2O	d	$C_6H_{12}O_6$
b	$C_3H_6O_3$	e	CHO
c	C_2HO		

14.

Oxygen gas can be produced by the decomposition of potassium chlorate, as shown by the following equation: $2\ KClO_{3(s)} \longrightarrow 2\ KCl + 3\ O_2$. If 4 moles of $KClO_{3(s)}$ are decomposed, the mass of oxygen produced will be produced

a	96 g	d	16 g
b	192 g	e	64 g
c	48 g		

15.

Sulphur trioxide is produced from a reaction as shown in the equation:
$2SO_{2(g)} + O_{2(g)} \longrightarrow 2SO_{3(g)}$ The number of moles of oxygen required to produce 8 moles of sulphur trioxide is

a	16	d	8
b	4	e	12
c	32		

16.

In an experiment, substances X and Y react according the following equation:

$$X + Y \rightarrow Z.$$
$$10\ g \quad\quad 10\ g$$

The mass of Z formed is

a	20 g	d	15 g
b	10 g	e	not possible to be calculated from the information given.
c	5 g		

17. Which of the following statements is **not true** about the **limiting reagent** in a chemical reaction?

a	It is the reactant that is not consumed completely
b	It is the reactant that is consumed completely
c	It is the reactant that determines how much of the other reactant(s) is consumed
d	It is the reactant that determines how much product is formed
e	b, c and d

18. Ammonia is produced from a reaction as shown in the equation:

$$3\ H_{2(g)} + N_{2(g)} \longrightarrow 2\ NH_{3(g)}$$ The number of **moles** of hydrogen required to produce 8.0 moles of ammonia if the percentage yield is 25% is

a	12	*d*	8
b	48	*e*	4
c	32		

19. Hydrogen peroxide decomposes according to the following chemical equation:

$$2\ H_2O_{2(l)} \longrightarrow 2\ H_2O_{(l)} + O_{2(g)}$$

What mass of $H_2O_{2(l)}$ must be decomposed to produce 16.0 g of $O_{2(g)}$?

a	68.04 g	*d*	17.01 g
b	50.0 g	*e*	34.02 g
c	32.0 g		

20. Magnesium reacts with hydrochloric according to the following chemical equation:

$$Mg_{(s)} + 2\ HCl_{(aq)} \longrightarrow MgCl_{2(aq)} + H_{2(g)}$$

What mass of $Mg_{(s)}$ is required to react completely with 100 mL of 1.00 mol/L $HCl_{(aq)}$?

a	2.43 g	*d*	243 g
b	24.3 g	*e*	0.12 g
c	1.22 g		

21. What mass of calcium hydroxide is required to react completely with 200 mL of 0.100 mol/L $HCl_{(aq)}$ according to the following equation:

$$Ca(OH)_{2(s)} + 2\ HCl_{(aq)} \longrightarrow CaCl_{2(aq)} + 2\ H_2O_{(l)}$$

a	2.43 g	*d*	7.41 g
b	24.3 g	*e*	0.12 g
c	1.22 g		

22. . Determine the mass of copper that is produced when 2.43 g of magnesium reacts with excess copper (11) sulphate solution as shown in the following reaction:

$$Mg_{(s)} + CuSO_{4(aq)} \longrightarrow MgSO_{4(aq)} + Cu_{(s)}$$

a	6.36 g	*d*	3.18 g
b	63.55 g	*e*	2.43 g
c	0.63 g		

23.

Hydrochloric acid and sodium hydroxide react according to the following equation:

$$HCl_{(aq)} + NaOH_{(aq)} \longrightarrow NaCl_{(aq)} + H_2O_{(l)}$$

If 200.0 mL of 0.20 mol/L $HCl_{(aq)}$ mixes with 150.0 mL of 0.30 mol/L $NaOH_{(aq)}$.
The limiting reagent is

a	$NaOH_{(aq)}$	d	$H_2O_{(l)}$
b	$NaCl_{(aq)}$	e	None of the above
c	$HCl_{(aq)}$		

24.

Nitric acid reacts with sodium carbonate according to the following equation:

$$2\ HNO_{3(aq)} + Na_2CO_{3(s)} \longrightarrow 2\ NaNO_{3(aq)} + H_2O_{(l)} + CO_{2(g)}$$

The mass of $CO_{2(g)}$ formed when 200.0 mL of 0.10.0 mol/L $HNO_{(aq)}$ reacts with 2.0 g of $Na_2CO_{3(s)}$ is

a	2.0 g	d	0.83 g
b	8.3 g	e	0.44 g
c	4.4 g		

25.

The mass of copper produced is 12.01 g when a mass of 4.86 g of magnesium reacts with excess copper (11) sulphate solution as shown in the following reaction:

$$Mg_{(s)} + CuSO_{4(aq)} \longrightarrow MgSO_{4(aq)} + Cu_{(s)}$$

The percentage yield of copper is

a	4.9 %	d	71 %
b	94.4 %	e	58.3 %
c	60.2 %		

26.

Magnesium hydroxide, $Mg(OH)_2$, reacts with hydrochloric acid, $HCl_{(aq)}$ according to the equation: $Mg(OH)_{2(s)} + 2\ HCl_{(aq)} \longrightarrow MgCl_{2(aq)} + 2\ H_2O_{(l)}$. If an antacid tablet has $Mg(OH)_{2(s)}$ as its main ingredient, what mass of it must be present to neutralize 200.0 mL of stomach fluid that has $HCl_{(aq)}$ of concentration 0.10 mol/L?

a	0.58 g	d	71 g
b	1.16 g	e	58.3 g
c	5.83 g		

27.

Aqueous sodium carbonate reacts with aqueous barium chloride according to the following equation: $Na_2CO_{3(aq)} + BaCl_{2(aq)} \longrightarrow BaCO_{3(s)} + 2\ NaCl_{(aq)}$.
When 5.3 g of $Na_2CO_{3(aq)}$ is mixed with 15.4 g of $BaCl_{2(aq)}$ in an experiment, 9.1 g of $BaCO_{3(s)}$ was obtained after the mixture was filtered and the precipitate was dried. The percent yield of $BaCO_{3(s)}$ was?

a	95.0 %	d	75.5 %
b	100 %	e	80 %
c	92.2 %		

28.

Ethanol burns with oxygen according to the following equation:

$$C_2H_5OH_{(l)} + 3\,O_{2(g)} \longrightarrow 2\,CO_{2(g)} + 3\,H_2O_{(g)}$$

When 3.45 g of $C_2H_5OH_{(l)}$ was burnt in 7.80 g of $O_{2(g)}$, the mass of $CO_{2(g)}$ produced was found to be 6.32 g. What was the percentage yield of $CO_{2(g)}$?

a	96.9 %	*d*	90 %
b	98 %	*e*	85 %
c	92 %		

29.

Aqueous potassium nitrate reacts with aqueous silver nitrate according to the following equation: $KI_{(aq)} + AgNO_{3(aq)} \longrightarrow KNO_{3(aq)} + AgI_{(s)}$

If 2.16 g of $AgI_{(s)}$ was obtained and this represented 92 % yield, what possible mass combination of $KI_{(aq)}$ and $AgNO_{3(aq)}$ can produce this amount of precipitate?

a	1.06 g $KI_{(aq)}$: 2.49 $AgNO_{3(aq}$	*d*	all of the above
b	1.71 g $KI_{(aq)}$: 2.49 $AgNO_{3(aq}$	*e*	none of the above
c	1.06 g $KI_{(aq)}$: 2.60 $AgNO_{3(aq}$		

UNIT 3
Solutions and Solubility

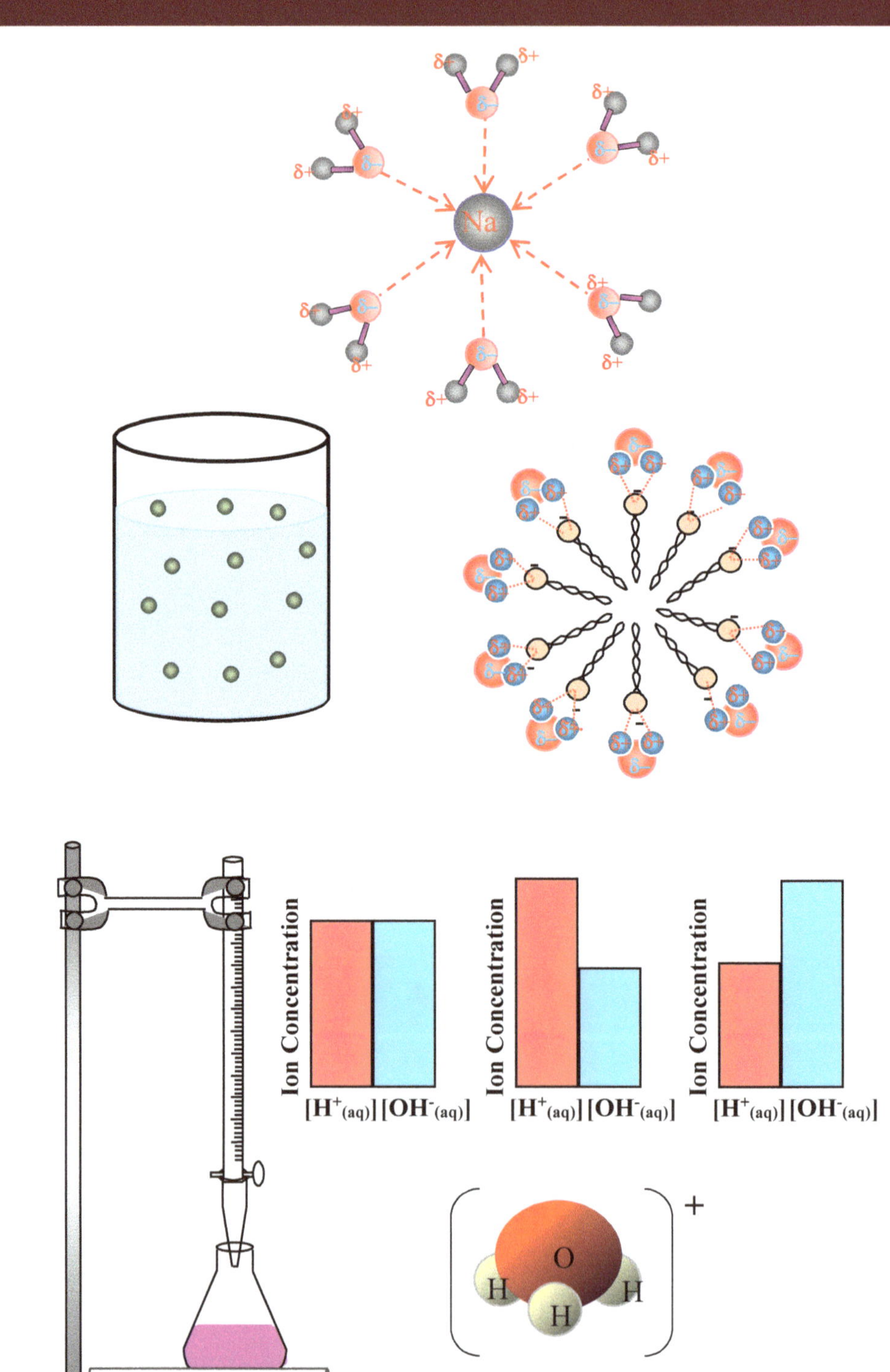

There are many covalent compounds that do not dissolve in water. T

Chapter Content:

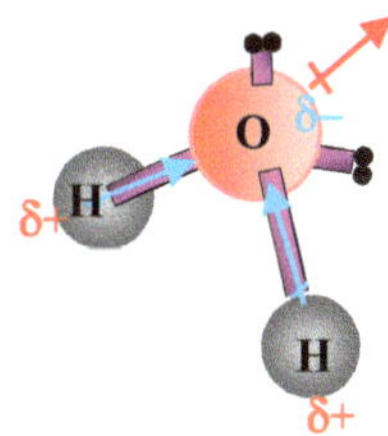

Sparkling sodas, alcoholic beverages, teas and the myriads of aqueous solutions that we consume or otherwise use, are solutions of various types of substances dissolved in water. These solutions are created because of the incredible ability of water to dissolve a wide array of substances. Because of this remarkable property, water is sometimes referred to as a 'universal solvent'. While most aqueous solutions are made for their usefulness, others can be made naturally that may be counter productive to human health and the environment. Since water does not discriminate between what substances to dissolve or not, inevitably harmful substances that are harmful to human health and other living organisms can enter its sources; water falling down through the air, as snow or rain, dissolves gases such as NO_2, SO_2 and CO_2 which eventually falls as acid precipitation. Acid precipitation can have devastating consequences on ecosystems and infrastructures made of marble or metals. Water running over the surface of the Earth dissolve organic compounds, mineral ions and other poisons in its path.

People living in poor countries of the world, with no water purification systems, have no choice but to use and drink it in these impure forms. In more affluent countries where there is scarcity of water, they recycle water from sewage. Their municipalities have the arduous task of purifying water, by removing most of these dissolved harmful substances.

This chapter describes the water molecule and explains how water dissolves various polar compounds and the energy changes accompanying the solution process. It also explains how non-polar compounds dissolve fellow non-polar compounds.

Nature and Properties of Solutions

12.1 The Water Molecule

Before any attempt is made to study aqueous solutions, it is important to have a thorough understanding of the properties of the water molecule. This section thus starts with a review of the water molecule. The following table shows some of its features.

Table 12.1. Some characteristics of the water molecule.

Formula	Lewis dot structure	AX Type?	Electron arrangement	Molecular geometry	Bond angle	Polarity
H_2O	Shown below	AX_2E_2	Tetrahedral	Bent	105.5°	Polar

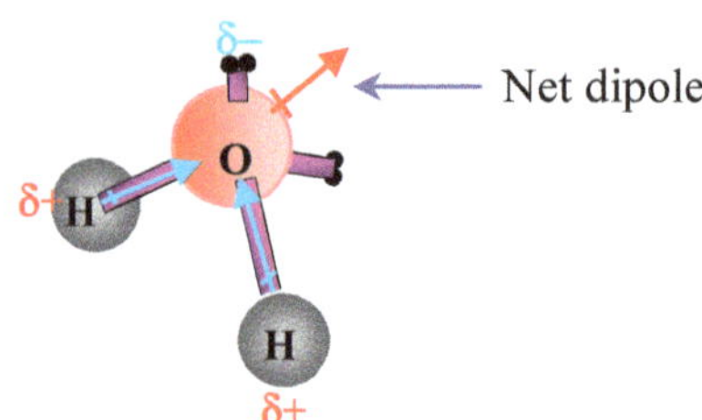

Figure 12.1. The ball and stick model of the water molecule showing different features.

The Polar O−H bond.

As can be recalled from the unit in bonding, a covalent bond is polar only if there is difference in electronegativity of at least 0.4 or more between the two atoms forming the covalent bond. The table below shows the electronegativity values assigned to the elements, hydrogen and oxygen.

Table 12.2. Electronegativity values for oxygen and hydrogen

Element	Electronegativity
Oxygen	3.5
Hydrogen	2.1

Since there is a difference in electronegativity value of 1.4 between hydrogen and oxygen, the O−H bond formed between these two atoms will be polar.

Polar Molecule

Also, as can be recalled from the unit in bonding, the bent shape of the water molecule, coupled with its two polar O−H bonds, allows it to be a polar molecule.

12.2 Hydrogen Bonding

Intermolecular forces of attraction have profound effects on the physical properties of a compound. If we compare the molar masses of the compounds water to carbon dioxide, for example, we can see that $M_{H_2O} = 18.02\ g/mol$ and $M_{CO_2} = 44.01\ g/mol$. Even though carbon dioxide molecules are 2.44 times as heavy as water molecules, water is a liquid and carbon dioxide exists as a gas at room temperature. The reason for this, as can be recalled from the unit in bonding, is due to the fact that $CO_{2(g)}$ molecules are attracted to each other only by relatively weak London dispersion forces. These forces are so weak that only the gaseous state is permissible at room temperature. Water molecules, on the other hand, are held by both London dispersion forces and relatively stronger **hydrogen bonding**. These two combined forces are strong enough to allow water molecules to be in the liquid state at room temperature. The way water molecules are held together is shown below. In ice, the water molecules assume a regular hexagonal arrangement that results in the molecules having more open space between them than in the liquid state. For this reason, ice floats on water; it is less dense than water.

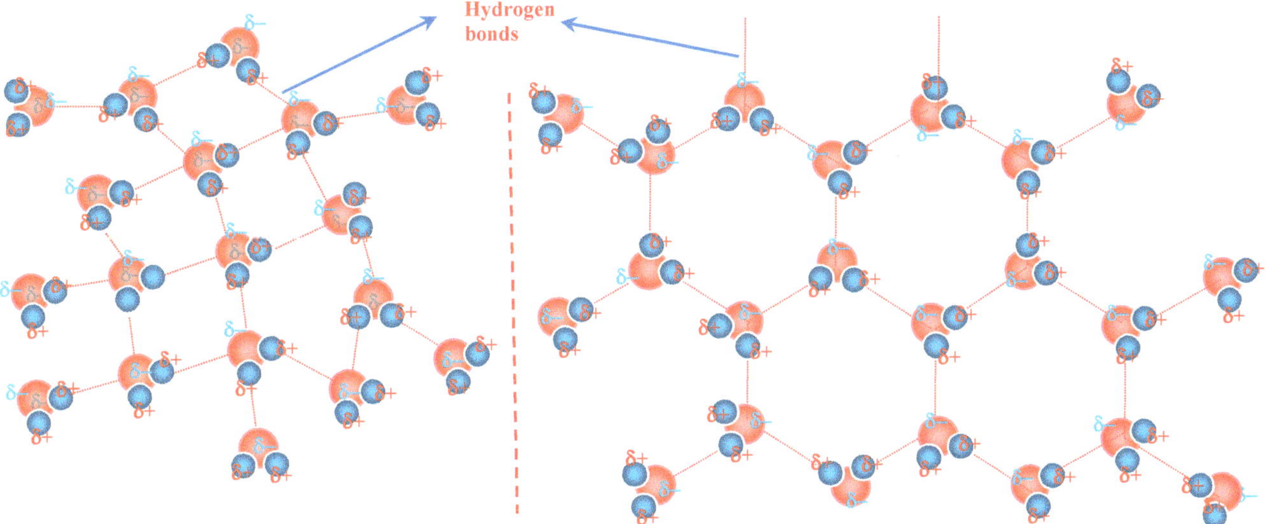

Figure 12.2. (a) Liquid water molecule being held by hydrogen bonding.

Figure 12.2. (a) Water molecules having hexagonal arrangement in ice.

The positive polar end of a water molecule is attracted to the lone pair of electrons on the oxygen atom of another water molecule, **by an electrostatic force of attraction**. A container of water can thus be imagined to be a continuous network of water molecules linked together by hydrogen bonds. However, as some water molecules gain enough energy to break these intermolecular forces of attraction, they escape and become a vapor. On the other hand, water may become solid if cooled below 0 °C.

12.3   Aqueous Solutions

A solution is formed when a solute is dissolved by a solvent.

$$\text{Solute} \; + \; \text{Solvent} \; \rightarrow \; \text{Solution}$$

In any **aqueous solution**, the **solvent is** always **water**. Since water is a polar solvent, it dissolves ionic compounds as well as polar covalent compounds. For example, alcohols are polar covalent compounds and they therefore dissolve in water. The following illustration shows how this happens between water and methyl alcohol.

12.4   Solubility of Polar Covalent Compounds in Water

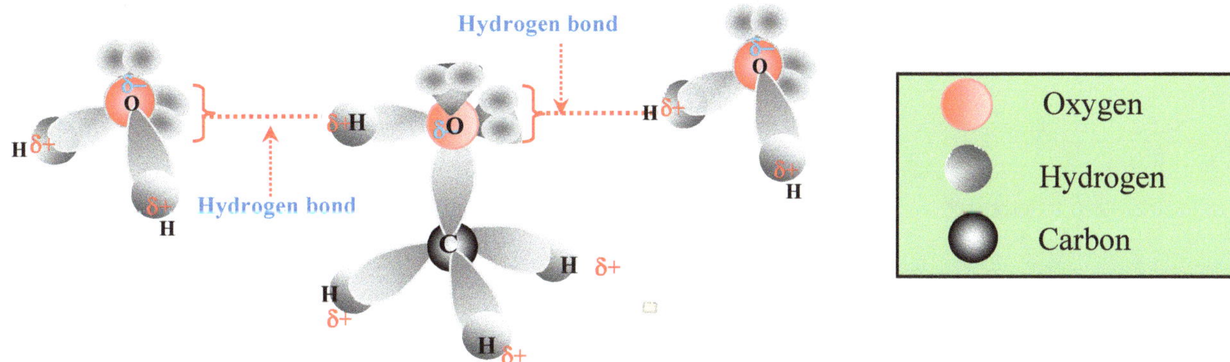

Figure 12.3.   Hydrogen bonding between water molecules and methyl alcohol.

The figure before shows how a lone pair of electrons on the oxygen atom of one water molecule forms a hydrogen bond with the positive polar hydrogen end of the alcohol molecule, and how the positive polar hydrogen end of another water molecule forms another hydrogen bond with the lone pair of electrons on the oxygen atom of the same alcohol molecule. This process repeats itself throughout the mixture allowing alcohol molecules to mix freely with water molecules. The same process happens when other polar molecular compounds such as sugars dissolve in water.

Solubility of Some Covalent Compounds in Water

he reason for this is due to the fact that these compounds do not have polar bonds to support dipole-dipole attractions or hydrogen bonding with water molecules. When these molecules are added to water, they are still held together by London dispersion forces while the water molecules are still held by hydrogen bonds; no mixing takes place. Examples of these covalent compounds are those found in grease and oils.

Figure 12.4(a).

Water with blue

food coloring.

Figure 12.4(b)

Vegetable oil.

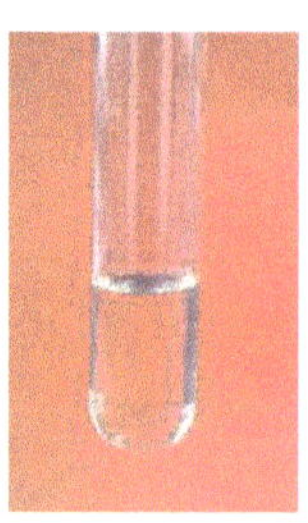

Figure 12.(4)c.

Oil floating on

water.

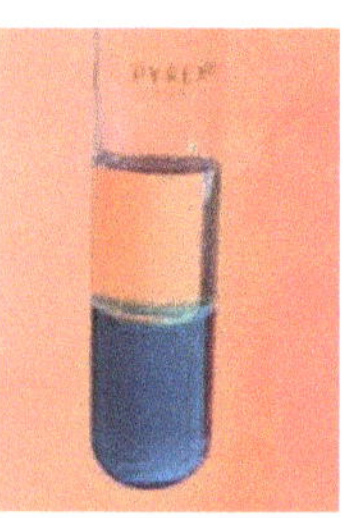

Why Do Covalent Compounds Dissolve Fellow Covalent Compounds?

Most covalent compounds, as explained before, are non-polar and thus do not dissolve in water. However, these compounds dissolve in fellow non-polar compounds. It is believed that London dispersion forces are responsible for the attraction between neighboring molecules of the compounds in the mixture, resulting in mixing. This is illustrated in the figure on the right.

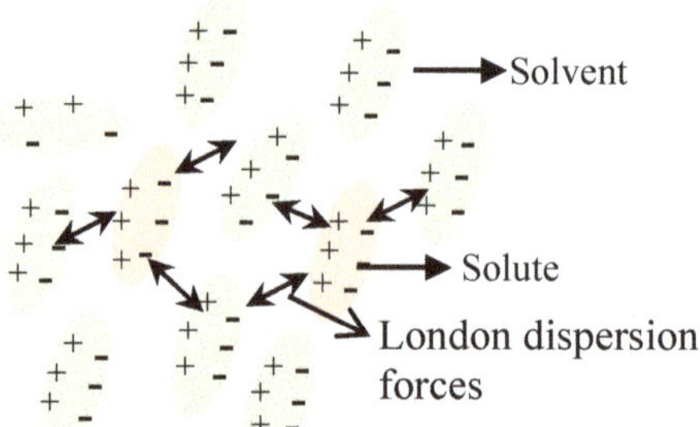

Figure 12.5. London dispersion forces explain mixing of non-polar solutes and solvents.

12.5 Dissolving of Ionic Compounds in Water

Ionic compounds have a 3-dimensional crystal lattice in which the positive and negative ions are packed in definite arrangements, depending on the types of ions present. These ions are held together by strong ionic bonds. For an ionic compound to dissolve in water, a number of processes must happen. These are outlined in the steps that follow.

Step 1. *Breaking of the hydrogen bonds in water.*

The hydrogen bonds of water must first be broken. This step is important to free up some water molecules from each other to dissociate the ions in a compound. This process **requires energy** and it is thus **endothermic** in nature.

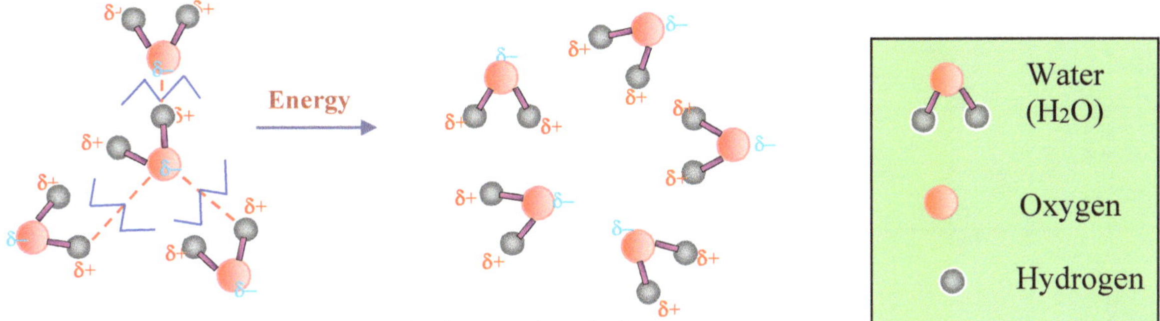

Figure 12.6.   Breaking of the hydrogen bonds in water.

Step 2. *Dissociation of the ions in the ionic crystal.*

The ionic bonds of the compound must then be broken. This is achieved by some of the free water molecules adhering to the ionic crystal with their negative oxygen ends pointed towards the positive ions of the crystal. When this happens, ***the cumulative electrostatic forces of attraction between a large number of water molecules and the positive ion, overcome the electrostatic forces of attraction between the positive ion and neighboring negative ions in the ionic crystal***. The positive ion (cation) is eventually pulled away from the crystal lattice. This process is called *dissociation*. The same thing happens to the negative ion (anion). The only difference is that ***the water molecules will now have their positive hydrogen ends directed towards the negative ions of the ionic crystal eventually pulling them away.*** These two processes go on until all the ions of the crystal separate.

Like the first step, dissociation of the ions also requires energy and is thus **endothermic** in nature.

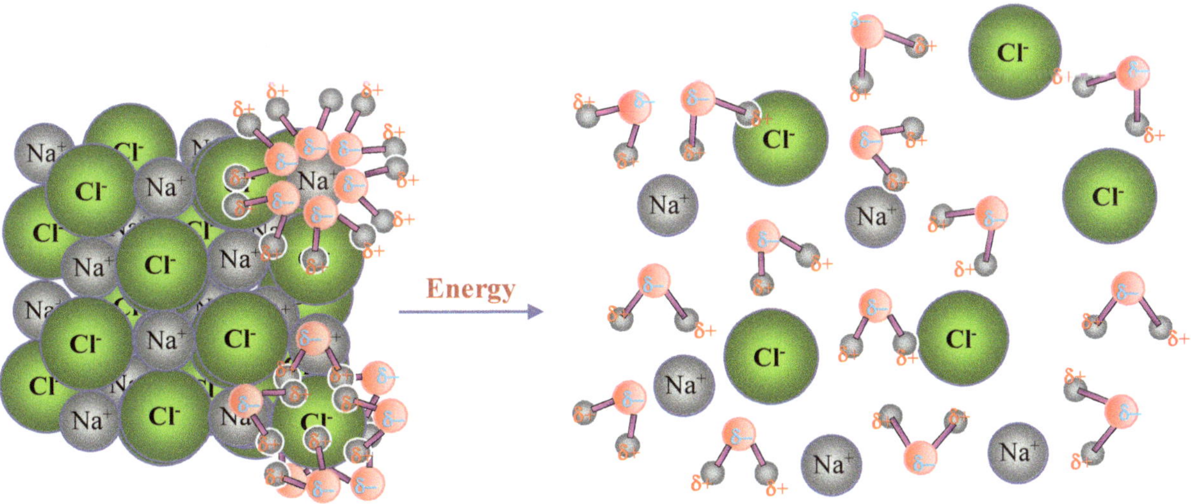

Figure 12.7. Dissociation of the ions of sodium chloride.

Step 3. *Solvation of the ions.*

As the ions dissociate from the crystal, they are immediately surrounded by water molecules due to electrostatic force of attraction between them and the polar water molecules; the negative ends of the water molecules are attracted to the positive sodium ions, (cations) and the positive ends of the water molecules are attracted to the negative chloride ions (anions).These process **releases energy** and are thus **exothermic** in nature.

The solvation of the ions is analogous to a number of lifeguards holding and suspending someone who cannot swim, in swimming pool.

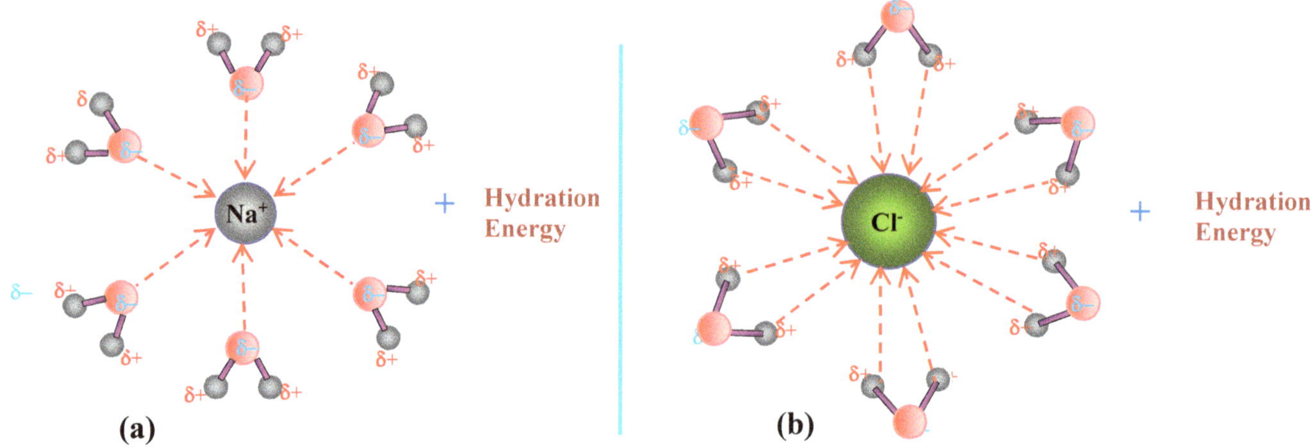

Figure 12.8(a). Hydration of the sodium ions by water molecules.

Figure 12.8(b). Hydration of the chloride ions by water molecules.

The sovation of ions by water molecules helps to explain why the copper that gets displaced in the reaction between $Mg_{(s)}$ and $CuSO_{4(aq)}$ falls to the bottom of the test tube, and the solid $Mg_{(s)}$ gets dissolved.

$$Mg_{(s)} \; + \; CuSO_{4(aq)} \longrightarrow MgSO_{4(aq)} \; + \; Cu_{(s)}$$

Using the the two half equations we have:
$$Mg_{(s)} \longrightarrow Mg^{2+}_{(aq)} \; + \; 2e^-$$
$$Cu^{2+}_{(aq)} \; + \; 2e^- \longrightarrow Cu$$

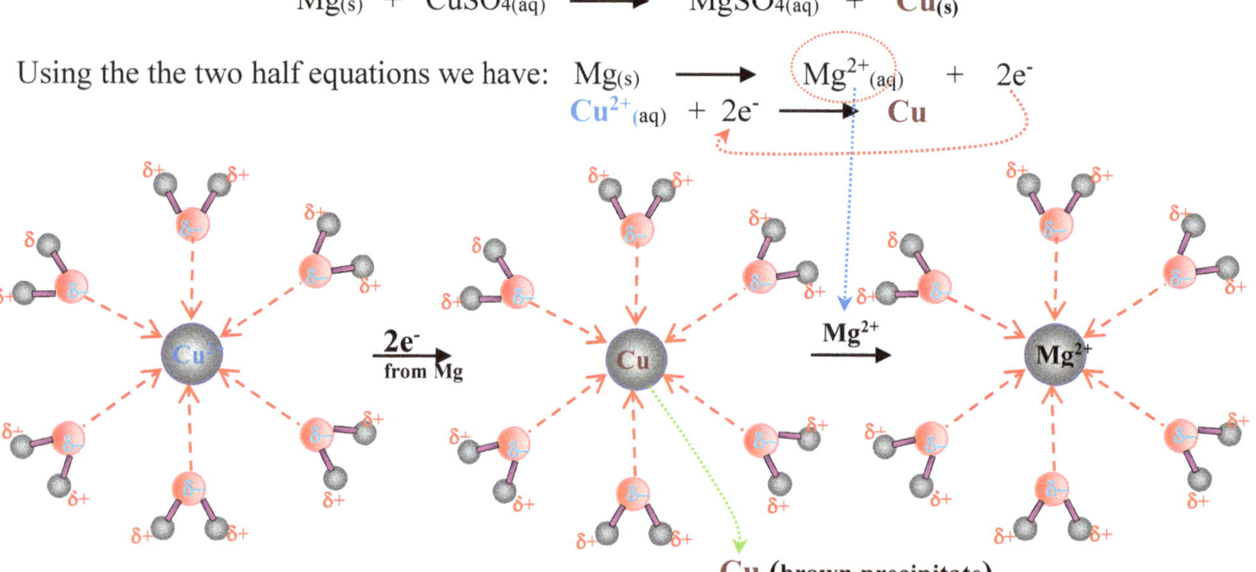

These illustrations show how the solvated $Cu^{2+}_{(aq)}$ picks up 2 electrons that are released by the Mg atom, and get discharged as **Cu** atom. The **Cu** atom that is formed, being without any charge, is not attracted anymore to the water molecules, so it falls to the bottom as a precipitate. The $Mg_{(s)}$ atom, that was insoluble, due to its lack of any charge to be attracted and be solvated by water molecules, loses two electrons and becomes positively charged. Because it is now charged, it becomes attracted the water molecules and becomes solvated. This explanation suffices for any single displacement of a less reactive metal by a more reactive one, for single displacement reactions.

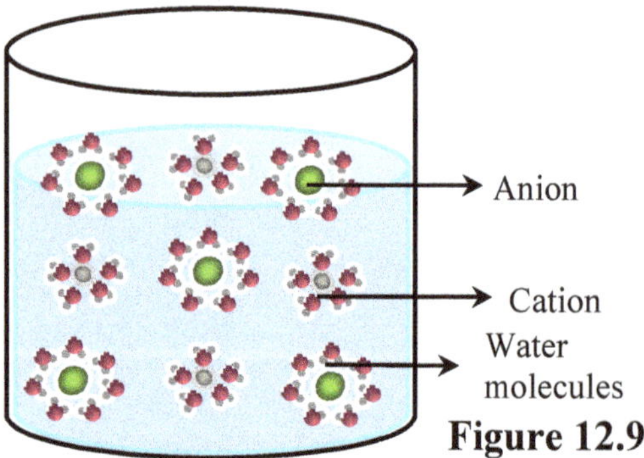

A solution of an ionic compound can thus be envisioned as a mixture where an innumerable number of cations and anions are surrounded by water molecules. This is illustrated by the figure on the left.

Figure 12.9. Solvated ions of an ionic compound when dissolved in water.

As illustrated above when an aqueous solution is formed, the first two steps absorb energy, while the third step releases energy. Depending on the amount of energy released compared with that absorbed, the solution process may be either **exothermic** or **endothermic**. *If the amount of energy absorbed from the first two steps is less than that released from the third step, the solution process is* **exothermic**. For example, when sodium hydroxide is dissolved in water, energy is released.

$$NaOH_{(s)} \xrightarrow[\text{to produce}]{\text{dissolves in water}} NaOH_{(aq)} + \textbf{Energy}$$

If the amount of energy absorbed from the first two steps is more than that released from the third step, the solution process is **endothermic**. For example, when ammonium nitrate dissolves in water, energy is absorbed.

$$NH_4NO_{3(s)} + H_2O_{(l)} + \textbf{Energy} \rightarrow NH_4NO_{3(aq)} + H_2O_{(l)}$$

Substances like ammonium nitrate that absorb heat when they dissolve in water are used for making cold packs. In the making of a cold pack, some solid ammonium nitrate is placed in a sealed delicate plastic bag, which is then inserted into a thicker water-containing plastic bag, which is then sealed. When the cold pack is needed, a presssure is applied to break the inner sealed delicate plastic bag. When this happens, the solid ammonium nitrate is allowed to dissolve in the water and the solution becomes cold.

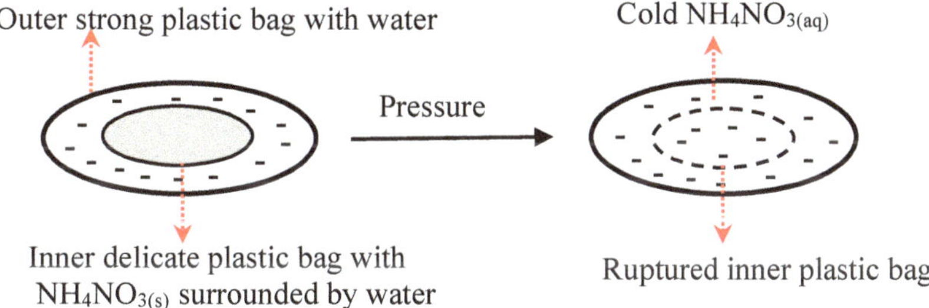

Figure 12.10.   Cold pack formation using ammonium nitrate.

Hand warmers work on the sample principle as cold packs. However, in this case, solids that produce heat as they dissolve in water are used instead. Calcium chloride is one example of such a substance.

Dissociation versus Ionization

As an ionic compound dissolves in water, it can be seen that the ions that make up the compound separate from each other. This process is called **dissociation,** as explained before. Dissociation must not be confused with the term **ionization**. In dissociation, the compound is already in the form of ions, which merely separate as the compound dissolves in water. In ionization, a **molecular** compound actually ionizes as it dissolves in water. For example, when hydrogen chloride dissolves in water, it ionizes into hydrogen and chloride ions as follows:

$$HCl_{(g)} \xrightarrow[\text{to produce}]{\text{dissolves in water}} H^+_{(aq)} + Cl^-_{(aq)}$$
hydrogen chloride   hydrogen ion   chloride ion

As stated before, a solution is made up of at least one **solute** and one **solvent**. The substance that does the dissolving is the solvent and the substance that dissolves is called the solute. In aqueous solutions, for example, the solvent is always water and the compounds such as salts, sugars, acids and bases are called solutes.

Not all solutions are aqueous; there are other solvents for substances other than water. The following table gives examples of other types of solutions.

Table 12.3. Examples of non-aqueous solutions.

Solute	Solvent	Example of solution
solid	solid	silver in gold as in gold jewelry / carbon in steel gallium arsenide and gallium phosphide in semiconductors
liquid	solid	mercury in gold (amalgams) as in gold extractions hexane in paraffin wax
liquid	gas	water in air as in atmospheric humidity (fogs)
gas	liquid	carbon dioxide in water as in carbonated drinks oxygen in water (for respiration in aquatic organisms)
gas	gas	oxygen in nitrogen as in air oxygen in helium as in heliox for scuba diving
liquid	liquid	acetic acid in water as in vinegar ethanol in water as in alcoholic beverages ethylene glycol in water as in anti-freeze

Solutions are examples of **homogeneous mixtures**. In these, the particles of the solute and solvent blend with each other so well that only one phase is observed and the solution is very clear. Every part of such mixtures is uniform in terms of their solute-solvent constituents. Examples of these are aqueous solutions of salts, acids and bases. Unlike these, in some mixtures, there are particles that remain undissolved. These large undissolved particles give the mixture a non-uniform appearance. Such mixtures are described as **heterogeneous**. Milk, blood and salad dressings are examples of these.

The following equations show how a few ionic compounds dissociate in water:

a) $CuSO_{4(s)} \xrightarrow[\text{to produce}]{\text{dissolves in water}} Cu^{2+}_{(aq)} + SO_4^{2-}_{(aq)}$

b) $Na_2SO_{4(s)} \xrightarrow[\text{to produce}]{\text{dissolves in water}} 2Na^+_{(aq)} + SO_4^{2-}_{(aq)}$

c) $Ca(NO_3)_{2(s)} \xrightarrow[\text{to produce}]{\text{dissolves in water}} Ca^{2+}_{(aq)} + 2\,NO_3^-_{(aq)}$

Not every substance that forms a solution in water dissociates into ions. Examples of these are polar covalent compounds, such as sugars and alcohols.

Complete the **dissociation equations** for the following ionic compounds in water. K/U

1. $ZnCl_{2(s)}$ $\xrightarrow{\text{dissolves in water}}$

2. $AlCl_{3(s)}$ $\xrightarrow{\text{dissolves in water}}$

3. $Al_2(SO_4)_{3(s)}$ $\xrightarrow{\text{dissolves in water}}$

4. $Ca(HCO_3)_{2(s)}$ $\xrightarrow{\text{dissolves in water}}$

5. $Mg(BrO_3)_{2(s)}$ $\xrightarrow{\text{dissolves in water}}$

6. $K_2CO_{3(s)}$ $\xrightarrow{\text{dissolves in water}}$

7. $FeBr_{3(s)}$ $\xrightarrow{\text{dissolves in water}}$

Exercise 12.2

Classify the following mixtures as either homogeneous or heterogeneous. K/U

(a) a bronze spear   (g) muddy water

(b) alcoholic beverage   (h) salad dressing

(c) smog   (i) coca cola

(d) tomato juice   (j) pizza

(e) a 12K gold ring   (k) jello

(f) distilled water   (l) air

CHAPTER 13
Concentration of Solutions

Chapter Content:

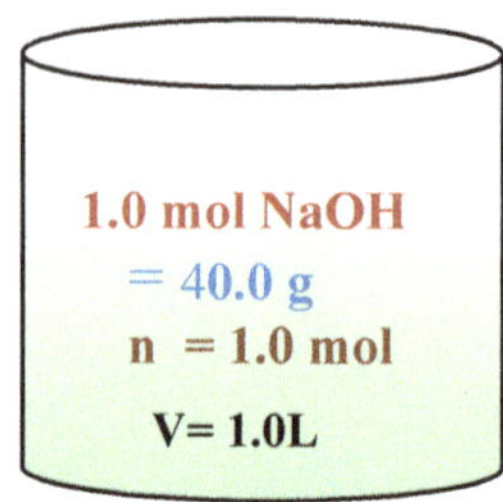

$$C = \frac{n}{v} = \frac{1.0 \text{ mol}}{1.0 \text{ L}} = 1.0 \text{ mol/L}$$

In the medical field, knowledge of the concentrations of dissolved substances in body fluids is essential; any drastic deviation in concentrations of these dissolved solutes could be detrimental to patients. Diagnosis of various diseases can be made by studying concentrations in changes of various substances in blood and urine samples; elevated blood glucose level, for example, is indicative of diabetes mellitus. Every day, millions of people who are too sick to eat, are fed intravenously by saline solutions, and reduce the risk of complications and death, technicians must ensure that the concentrations of all dissolved nutrients in saline are compatible to blood plasma; too high a concentration of sodium in the blood, for example, can cause the heart to stop beating. Other substances given to patients must also be subjected to the same type of scrutiny; doses of medicines administered to patients must be well calculated depending on the weight of patients so that their concentrations in the blood remain within acceptable safe limits; an excess of any blood thinner, for example, can cause severe hemorrhage.

Solution concentrations have other wide-ranging applications, such as in analytical chemistry, nutritional studies, cooking, soil agronomy, pharmacology and environmental science, among others. This chapter teaches how concentrations of solutions are calculated, and how solutions of required concentrations are prepared. Calculations of concentration of ions from the dissociation of ionic compounds are also taught.

Concentration of solutions

13.1 Effect of Volume of Solvent on Concentration

In the following figures, different amounts of the same solute are dissolved in equal volumes of water. Analyze these two figures to find what are same and what are different about them. How do the concentrations of the solutions produced, compare with each other?

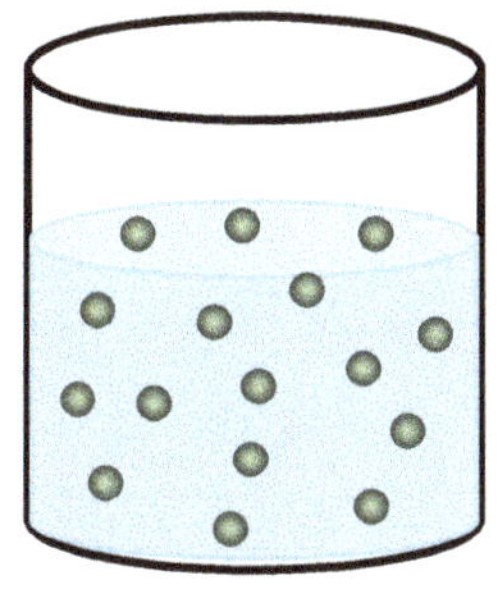

Solution A
Figure 13.1(a). sixteen
particles in volume **V.**

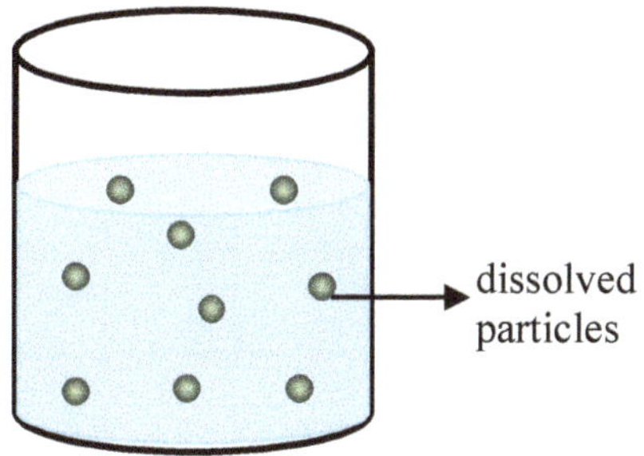

Solution B
Figure 13.1(b). eight
particles in volume **V.**

13.2 Effect of Number of Particles of on Concentration

Since in solution **A** there is twice the number of particles in the same volume of solution as in solution **B**, it can be concluded that solution **A** *is the more concentrated* one. If the solute was sugar, then solution **A** would **taste sweeter** than solution **B** because of the greater number of sugar particles dissolved in it.

The case above illustrates how concentration of a solution is directly proportional to the number of particles dissolved in it.

In the following case, the same amount of a common solute is dissolved in two different volumes of solutions:

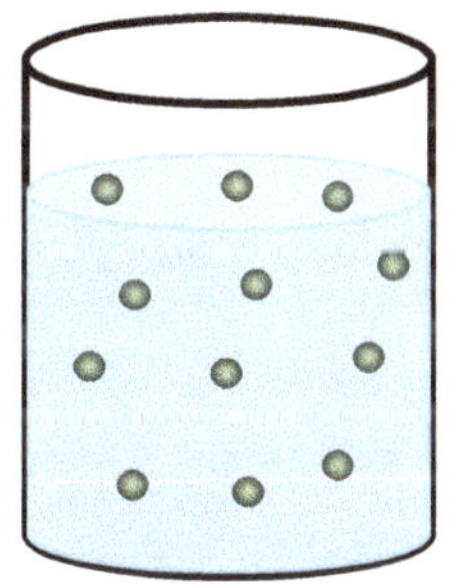

Solution A
Figure 13.2(a). twelve
particles in volume **2V.**

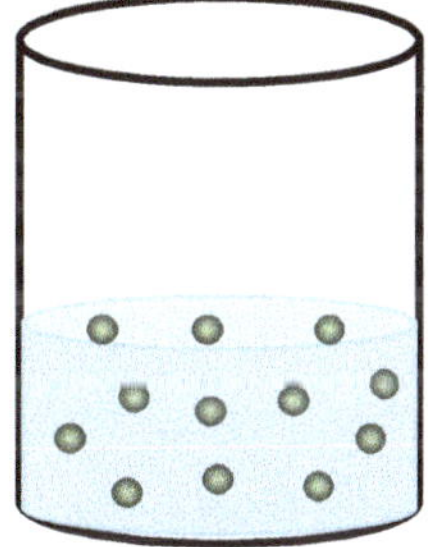

Solution B
Figure 13.2(b) twelve
particles in volume **V.**

Since in solution **A** there is the same number of particles in twice the volume as in solution **B**, it can thus be concluded that solution *B is the more concentrated* one. If the dissolved solute was sugar, then solution **B** would **taste sweeter** than solution **A**.

Solution **A** could also be considered to be a diluted version of solution **B**. Also, solution **B** could be considered to a concentrated version of solution A, which was formed by evaporating water out of it.

This case illustrates how the concentration of a solution is inversely proportional to the volume of the solution.

13.3 Molar Concentration of Solutions

Based on what was described so far, a relationship between the two factors that determine the concentration of any aqueous solution can be established as follows:

$$Molar\ concentration = \frac{amount\ solute\ (in\ moles)}{volume\ of\ solution\ (in\ litres)}$$

$$C = \frac{n}{V}$$

Where: C = *concentration*
n = *amount in moles*
v = *volume in litres*

For the following solutions below, it will be shown how the factors of amount of solutes in moles and volume in litres are used in determining their molar concentrations, using the formula, $C = \dfrac{n}{V}$

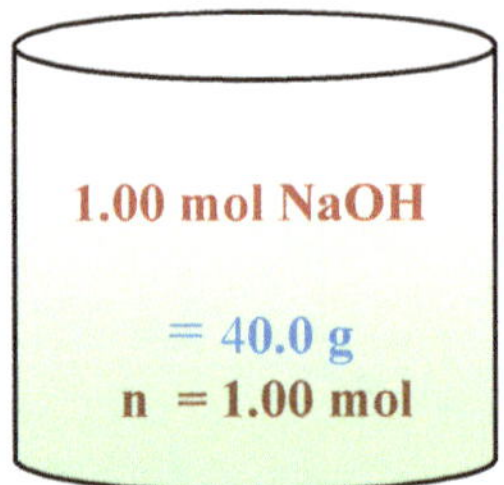

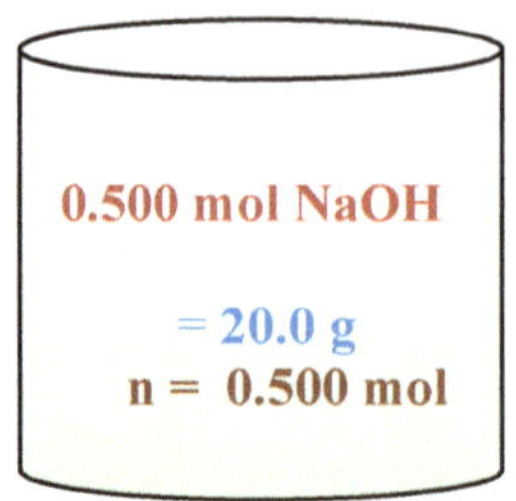

Figure 13.3(a). 1.00 litre of solution with **1.00 mol NaOH.**

Figure 13.3(b). 1.00 litre of solution with **0.500 mol NaOH.**

Figure 13.3(c). 0.500 litre of solution with **0.500 mol NaOH.**

Solution A

V = 1.00 L (1000 mL)

Concentration of A

$$C = \frac{n}{V} = \frac{1.00 \text{ mol}}{1.00 \text{ L}}$$
$$= 1.00 \text{ mol/L}$$

Solution B

V = 1.00 L (1000 mL)

Concentration of B

$$C = \frac{n}{V} = \frac{0.500 \text{ mol}}{1.00 \text{ L}}$$
$$= 0.500 \text{ mol/L}$$

Solution C

V = 0.500 L (500 mL)

Concentration of C

$$C = \frac{0.500 \text{ mol}}{0.500 \text{ L}}$$
$$= 1.00 \text{ mol/L}$$

Common Student Misconception:

Students have the common misconception that if a small sample of a stock solution is poured off, then the sample would be of different concentration as that of the stock solution. In the following calculations, it will be shown how each of the following samples has the **same** concentration **as** that of the **stock solution** from which it was poured. *The solutions from the previous illustrations **are used as the stock solutions from which the respective** samples below were poured.* In the containers below, we have 100 mL samples from solutions, A, B and C respectively.

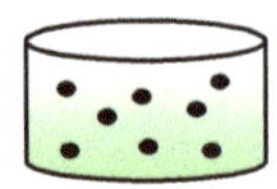

Sample A

Sample B

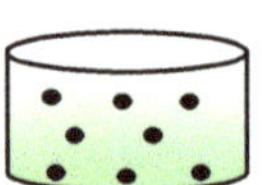

Sample C

Figure 13.4(a). 100 mL of stock solution A.

Figure 13.4(b). 100 mL of stock solution B.

Figure 13.4(c). 100 mL of stock solution C.

Since each sample has 100 mL from their respective stock solutions, the mass of dissolved solute in each is calculated as follows: *Note that mass of solute in each sample would be one tenth of its stock solution.

Sample A

$$m_{NaOH} = \frac{40.0 \text{ g}}{1000 \text{ mL}} \times 100 \text{ mL}$$
$$= 4.00 \text{ g}$$

$$n_{NaOH} = \frac{m}{M} = \frac{4.00 \text{ g}}{40.0 \text{ g / mol}}$$
$$= 0.100 \text{ mol NaOH}$$

Sample B

$$\frac{20.0 \text{ g}}{1000 \text{ mL}} \times 100 \text{ mL}$$
$$= 2.00 \text{ g NaOH}$$

$$= \frac{2.00 \text{ g}}{40.0 \text{ g / mol}}$$
$$= 0.0500 \text{ mol NaOH}$$

Sample C

$$\frac{20.0 \text{ g}}{500 \text{ mL}} \times 100 \text{ mL}$$
$$= 4.00 \text{ g NaOH}$$

$$= \frac{4.00 \text{ g}}{40.0 \text{ g / mol}}$$
$$= 0.100 \text{ mol NaOH}$$

$$C_{NaOH(aq)} = \frac{n}{V}$$

Volume of sample:	$V = 100.\ mL$ $= 0.100L$	:	$V = 100.\ mL$ $= 0.100L$	:	$V = 100.\ mL$ $= 0.100L$
$C_{NaOH(aq)} = \dfrac{n}{V} =$	$\dfrac{0.100\ mol}{0.100\ L}$	$: =$	$\dfrac{0.0500\ mol}{0.100\ L}$	$: =$	$\dfrac{0.100\ mol}{0.100\ L}$
	$= 1.00\ mol/L$	$=$	$0.500\ mol/L$		$= 1.00\ mol/L$

These results show that each sample has the same concentration as that as the original stock solutions from which they were obtained.

Based on these findings, it can be concluded that *whenever a small sample of solution is taken from a stock solution, it will have the same concentration as the stock solution, but it is the amount of solute in it that is different.* This concept is summarized by the figures below.

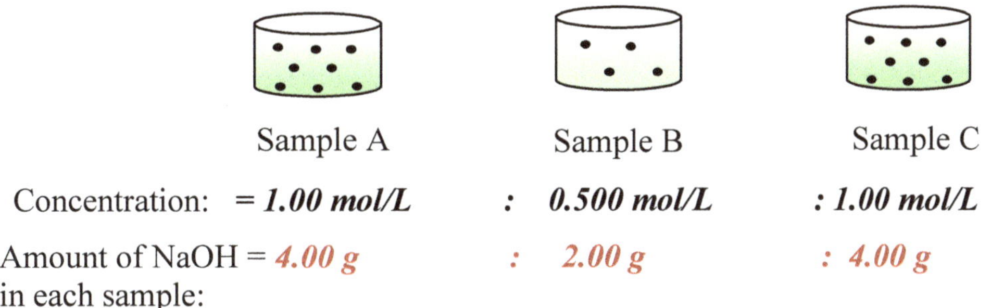

	Sample A	Sample B	Sample C
Concentration:	$= 1.00\ mol/L$	$:\ 0.500\ mol/L$	$:\ 1.00\ mol/L$
Amount of NaOH in each sample:	$= 4.00\ g$	$:\ 2.00\ g$	$:\ 4.00\ g$

Figure 13.5. Samples of solutions taken from stock solutions **having thesame original concentrations** but different masses of dissolved $NaOH_{(s)}$.

Exercise 13.1

Find the molar concentrations of the following solutions: T/I

(a) 0.50 mol of HCl in 500.0 mL of solution

(b) 0.250 mol of H_2SO_4 in 200.0 mL of solution

(c) 2.00 mol of NaCl in 1.50 L of solution

(d) 0.10 mol of NaOH in 250.0 mL of solution

(e) 0.20 mol of $Ca(NO_3)_2$ in 1.0L of solution

(f) 0.0010 mol of $KMnO_4$ in 100.0 mL of solution

13.4 Calculating Molar Concentration of Solutions

Sample Problem 1:

Calculate the molar concentration of a solution that has 24.02 g $H_2SO_{4(aq)}$ dissolved in 100.0 mL of solution.

Solution:

To solve this problem, it is useful to follow the steps below:

Step 1. Change amount of $H_2SO_{4(aq)}$ from mass to amounts in moles.

$$Recall\ that\ n = \frac{m}{M}$$

$$\therefore n_{H_2SO_{4(aq)}} = \frac{24.02\ g}{98.06\ g/mol}$$

$$= 0.2500\ mol$$

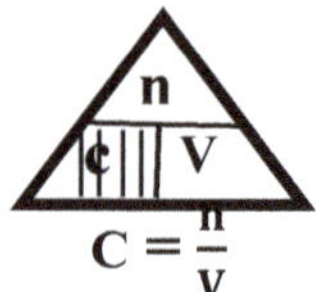

Step 2. Change volume from milliliters to volume in litres.

$$Since\ 1000\ mL = 1.00\ L$$

$$Then\ 100\ mL = \frac{100\ mL}{1000\ mL/L}$$

$$= 0.100\ L$$

Step 3. Use the formula, $C = \frac{n}{V}$ to find the molar concentration.

$$C_{H_2SO4(aq)} = \frac{0.2500\ mol}{0.100\ L}$$

$$= 2.50\ mol\ /\ L$$

Exercise 13.2

Find the molar concentrations of the following solutions that have: **T/I**

    (a) 10.0 g of NaOH dissolved in 500.0 mL of solution

    (b) 12.01 g of H_2SO_4 dissolved in 800.0 mL of solution

    (c) 6.30 g of HNO_3 dissolved in 50.0 mL of solution

    (d) 9.10 g of HCl dissolved in 200.0 mL of solution

    (e) 2.61 g of $Ba(NO_3)_2$ dissolved in 2.0 L of solution

    (f) 1.70 g of $AgNO_3$ dissolved in 500.0 mL of solution

Calculating the Amounts of Solute in Mass that is required to prepare Solutions of Known Concentrations

Sample Problem 2:

What mass of sodium hydroxide is required to prepare a 500 mL of 0.100 mol/L solution?

Solution:

Provided quantities :
    Volume(V) = 0.500 L
    Concentration(C) = 0.100 mol/L
Required quantity:
    Mass of $NaOH_{(s)}$

Step 1. Find the number of moles of $NaOH_{(aq)}$

From the formula, $C = \dfrac{n}{V}$ we get:

$$n = C \times V$$

$$n_{NaOH(aq)} = \frac{0.100 \ mol}{L} \times 0.500 \ L$$

$$= 0.0500 \ mol$$

Step 2. Find the mass of $NaOH_{(aq)}$ using the formula,

$$m = n \times M$$

$$m_{NaOH(aq)} = 0.050 \ mol \times \frac{40.0 \ g}{mol}$$

$$= 2.00 \ g$$

To prepare this solution, 2.00 g of $NaOH_{(s)}$ is weighed out and placed in a 500 mL volumetric flask. It is then dissolved using distilled water until the 500 mL mark is reached.

Exercise 13.3

Find the mass of solute is required to prepare each of the following solutions: T/I

(a)   400.0 mL of 0.200 mol/L $H_2SO_{4(aq)}$

(b)   4.00 L of 0.100 mol/L $NaCl_{(aq)}$

(c)   200.0 mL of 0.500 mol/L $HCl_{(aq)}$

(d)   500.0 mL of 0.050 mol/L $KMnO_{4(aq)}$

(e)   1.00 L of 0.250 mol/L $Na_2CO_{3(aq)}$

13.5   Preparing Solutions using Hydrated Salts

As can be recalled, when calculating the molar mass of hydrated salts, it is important to include the mass for the number of moles of water present in the compound.

Sample problem 3:

What mass of $CuSO_4 \bullet 5H_2O_{(s)}$ is required to prepare 500.0 mL of a 0.250 mol/L $CuSO_{4(aq)}$ solution?

Solution:

Provided quantities:
Volume (V) = 0.500 L
Concentration (C) = 0.250 mol/L
Required quantity:
Mass of $CuSO_4 5H_2O_{(s)}$

Step 1. Find the number of moles of $CuSO_4 \cdot 5H_2O_{(s)}$ using the formula

$$n = C \times V$$

$$n_{CuSO_4.5H_2O_{(s)}} = \frac{0.250\ mol}{\cancel{L}} \times 0.500\ \cancel{L}$$

$$= 0.125\ mol$$

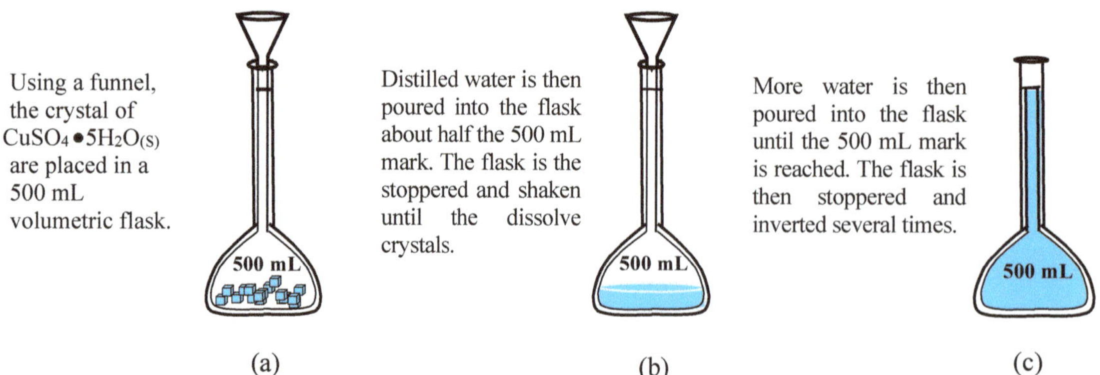

Step 2. Find the molar mass of $CuSO_4 \cdot 5H_2O_{(s)}$.

Finding the mass of the solute in this sample problem is different from the previous one, as the mass of water of crystallization must be included here.

Recall that the molar mass of $CuSO_4 \cdot 5H_2O_{(s)}$ is found by adding the molar mass of 1 mole of $CuSO_4$ and 5 moles of H_2O.

$$M_{CuSO_4.5H_2O} = M_{CuSO_4} + M_{5H_2O}$$

$$= \{\ 1\ mol(63.55\ g + 32.06\ g + 64.00\ g)\ /\ mol\ +\ 5\ mol(16.0\ g + 2.02\ g)\ /\ mol\ \}\ mol$$

$$= \{\ 159.6\ g\ +\ 90.1\ g\ \}\ mol$$

$$= 249.7\ g\ /\ mol$$

Step 3. Find the mass of $CuSO_4 \cdot 5H_2O_{(s)}$.

$$m_{CuSO_4 \cdot 5H_2O} = n \times M$$

$$= 0.125\ mol \times 249.7\ g/mol$$

$$= 31.2\ g$$

To prepare this solution, **31.2 g** $CuSO_4 \cdot 5H_2O_{(s)}$ is measured and placed in a 500 mL volumetric flask. It is then dissolved in distilled water until a volume of 500 mL is reached. The procedure is depicted by the sequence of figures below.

Figure 13.6. The preparation of copper (II) sulphate solution.

(a) What mass of $CaCl_2 \cdot 2H_2O_{(s)}$ is required to prepare 400.0 mL of a 0.200 mol/L $CaCl_{2(aq)}$ solution? **T/I**
(b) What mass of $Na_2SO_4 \cdot 10H_2O_{(s)}$ is required to prepare 800.0 mL of a 0.100 mol/L $Na_2SO_{4(aq)}$ solution? **T/I**

Summary of Concepts

13.6   **Calculating Molar Concentration, given Mass and Volume.**

a)   Finding the concentration of a solution produced by dissolving 6.72 g of $CuCl_{2(s)}$ in 200 mL of solution.

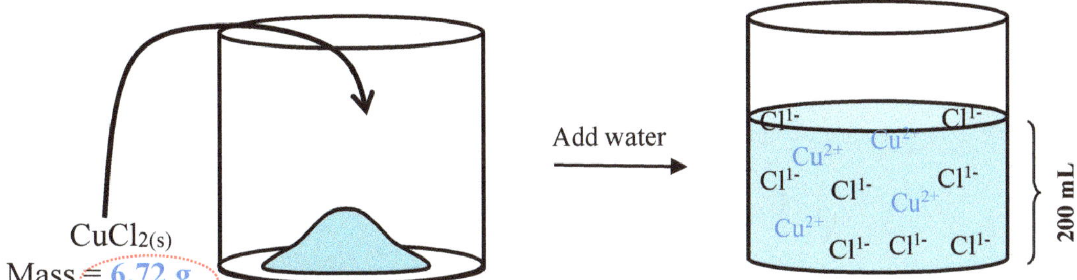

Figure 13.7(a). 6.72g $CuCl_{2(s)}$ **Figure 13.7(b).** 200.0 mL of $CuCl_{2(aq}$

1.   Find the number of moles of solute

$$n = \frac{m}{M}$$

$$= \frac{6.72\ g}{134.45 g/mol}$$

$$= 0.0500\ mol$$

2. Find the concentration of the solution

$$C = \frac{n}{V}$$

$$= \frac{0.0500\ mol}{0.200\ L}$$

$$= 0.250\ mol\ /\ L$$

Calculating Mass, given Volume and Concentration.

b)   Calculating the mass of $CuCl_{2(s)}$ that is present in or required to make 200.0 mL of 0.250 mol/L of $CuCl_{2(aq)}$. ***This problem is the reverse of the one above, see relationships.***

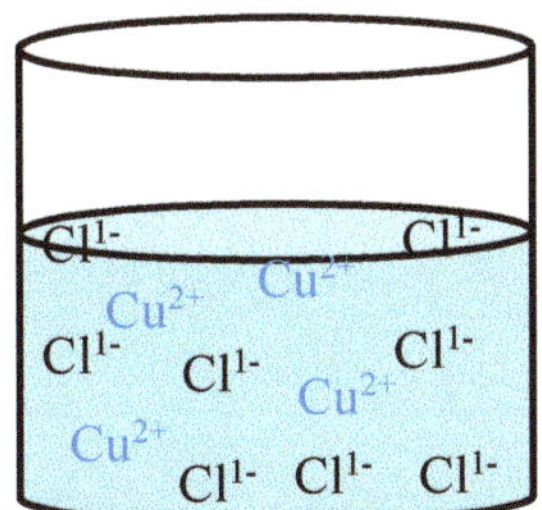

=

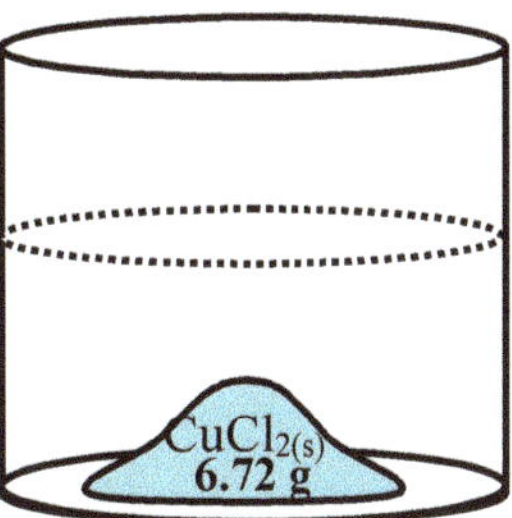

Figure 13.8(a). 200.0 mL of   **Figure 13.8(b).** 6.72 g of $CuCl_{2(s)}$ in
0.250 mol/L $CuCl_{2(aq)}$.   200.0 mL of solution.

1.   Find the number of moles of solute

$$n_{CuCl_2} = C \times V$$

$$= \frac{0.250\ mol}{L} \times 0.200\ L$$

$$= 0.0500\ mol$$

2.   Find the mass of the solute

$$m_{CuCl_2} = n \times M$$

$$= 0.0500\ mol \times \frac{134.45\ g}{mol}$$

$$= 6.72\ g$$

13.7 Equations for Dissociation of Ionic Compounds

As can be recalled, ionic compounds when dissolved in water, dissociate into their constituent ions. The following examples illustrate how some ionic compounds dissociate in water to produce their respective mole-ion ratios:

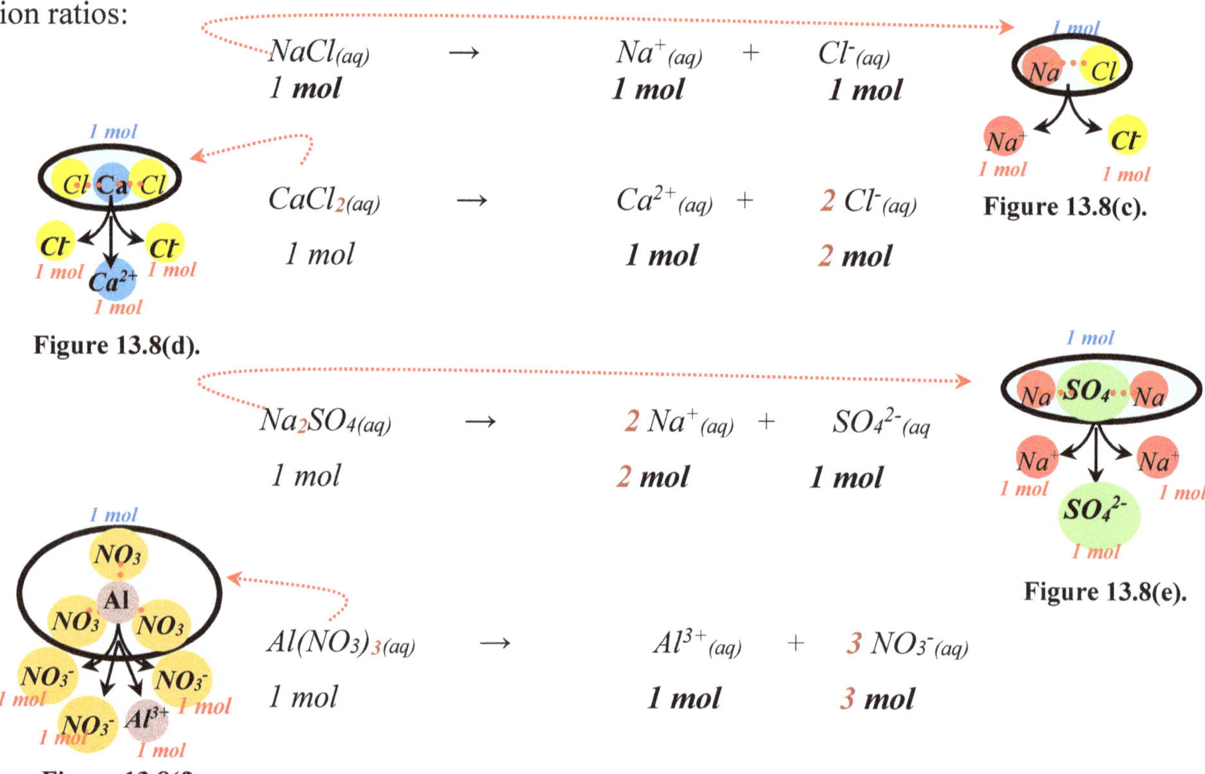

It will now be attempted to show how the concentrations of ions are calculated when ionic compounds dissolve in water. Two of the compounds in examples above are used below to illustrate how the mole ratios of their ion concentrations are established when one mole of each is dissolved in one liter of solution.

$$n_{CaCl_2} = 1.0\ mol$$
$$Volume = 1.0\ L$$
$$C_{CaCl_2} = \frac{1.0\ mol}{1.0\ L}$$
$$= 1.0\ mol/L$$
$$CaCl_{2(aq)} \rightarrow Ca^{2+}_{(aq)} + 2Cl^-_{(aq)}$$
$$1\ mol \rightarrow 1\ mol + 2\ mol$$
$$1\ mol/L \rightarrow 1\ mol/L + 2\ mol/L$$

Solution A

Figure 13.9(a). The dissociation of a 1.0 mol/L CaCl$_{2(aq)}$.

$$n_{Na_2SO_4} = 1.0\ mol$$
$$Volume = 1.0\ L$$
$$C_{Na_2SO_4} = \frac{1.0\ mol}{1.0\ L}$$
$$= 1.0\ mol/L$$
$$Na_2SO_{4(aq)} \rightarrow 2Na^+_{(aq)} + SO_4^{2-}_{(aq)}$$
$$1\ mol \rightarrow 2\ mol + 1\ mol$$
$$1\ mol/L \rightarrow 2\ mol/L + 1\ mol/L$$

Solution B

Figure 13.9(b). The dissociation of a 1.0 mol/L Na$_2$SO$_{4(aq)}$.

*Note that **when ionic compounds with radicals dissociate, the radicals do not break up, but move as one group**. This is so because the atoms that make up the radical are held together by covalent bonds.

Each of the illustrations before shows how one mole of an ionic compound dissociates to produce its specific mole ratio of ions. Each also shows how a compound having a concentration of 1.0 mol/L dissociates to produce its own specific *mole ratios of ion concentrations.*

Complete the dissociation equations for the following ionic compounds: K/U

a) $NaOH_{(aq)}$ $\rightarrow$

   1.0 mol/L

b) $ZnSO_{4(aq)}$ $\rightarrow$

   1.0 mol/L

c) $CaCl_{2(aq)}$ $\rightarrow$

   1.0 mol/L

d) $Mg(OH)_{2(aq)}$ $\rightarrow$

   1.0 mol/L

f) $FeCl_{3(aq)}$ $\rightarrow$

   1.0 mol/L

g) $Al_2(SO_4)_{3(aq)}$ $\rightarrow$

   1.0 mol/L

Ionization

As can be recalled, ionization refers to the process by which **molecular** compounds **form ions** when they dissolve in water. The following equation exemplifies this for the ionization of sulphuric acid.

$$H_2SO_{4(aq)} \quad \rightarrow \quad 2\,H^+_{(aq)} \quad + \quad SO_4^{2-}_{(aq)}$$

1 mol/L   *2 mol/L*   *1 mol/L*   (*Note again: radicals move as a group*)

Complete the ionization equations for the following molecular compounds: K/U

a) $HCl_{(aq)}$ $\rightarrow$
   1 mol/L

b) $HNO_{3(aq)}$ $\rightarrow$
   1 mol/L

c) $HI_{(aq)}$ $\rightarrow$
   1 mol/L

d) $HCN_{(aq)}$ $\rightarrow$
   1 mol/L

e) $HClO_{3(aq)}$ $\rightarrow$
   1 mol/L

Calculating Molar Concentration of Ions

Calculate the concentrations of $H^+_{(aq)}$ and $SO_4^{2-}_{(aq)}$ ions in a solution having 12.01 g of $H_2SO_{4(aq)}$ dissolved in 200.0 mL of solution. *(Assume complete ionization of $H_2SO_{4(aq)}$)*

Solution:

To solve these types of problems, it is useful to follow the steps below:

Step 1. Write a balanced equation to show how the compound dissociates or ionizes to give its specific mole ratios of ion concentrations.

$$H_2SO_{4(aq)} \quad \longrightarrow \quad 2\,H^+_{(aq)} \quad + \quad SO_4^{2-}_{(aq)}$$

$$1\ mol/L \qquad\qquad 2\ mol/L \qquad 1\ mol/L$$

Step 2. Change amount of solute from amount in mass to amount in moles.

$$n = \frac{m}{M}$$

$$n_{H_2SO_4} = \frac{12.01\,g}{98.06\,g/mol}$$

$$= 0.1220\,mol$$

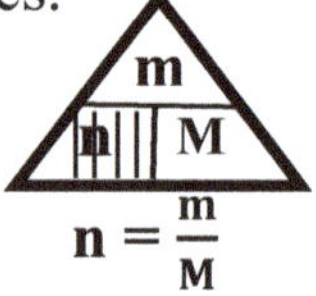

Step 3. Change the volume from milliliters to liters.

$$V = \frac{200.0\,mL}{1000\,mL/L}$$

$$= 0.200\,L$$

Step 4. Calculate the concentration of the solute, H_2SO_4, using the formula.

$$C = \frac{n}{V}.$$

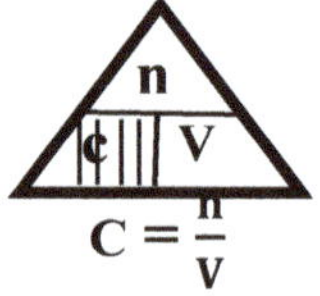

Note* **Concentration** of an entity **is** represented by **[ ]** parenthesis.

$$\left[H_2SO_{4(aq)} \right] = \frac{0.1220\,mol}{0.200\,L}$$

$$= 0.600\,mol/L$$

Step 5. Calculate the molar concentrations of each ion using the mole ratios from the equation in **step1.**

$$\left[H^+_{(aq)} \right] = 0.600\ \cancel{mol/L\ H_2SO}_{4(aq)} \times \frac{2\ mol/L\ H^+_{(aq)}}{1\ \cancel{mol/L\ H_2SO}_{4(aq)}}$$

$$= 1.20\,mol/L$$

$$\left[SO_4^{2-}_{(aq)} \right] = 0.600\ \cancel{mol/L\ H_2SO}_{4(aq)} \times \frac{1\,mol/L\ SO_4^{2-}_{(aq)}}{1\ \cancel{mol/L\ H_2SO}_{4(aq)}}$$

$$= 0.600\,mol/L$$

Assuming complete dissociation, calculate the molar concentration of all ions in the following solutions that have: **K/U T/I**

(a) 2.49 g of $CuSO_4 \cdot 5H_2O_{(s)}$ dissolved in 400.0 mL of solution
(b) 1.74 g of $Na_2SO_{4(s)}$ dissolved in 500.0 mL of solution
(c) 11.10 g of $CaCl_{2(s)}$ dissolved in 500.0 mL of solution
(d) 1.71 g of $Ba(OH)_{2(s)}$ dissolved in 200.0 mL of solution

13.8 Diluting a Stock Solution

Sometimes reagents are manufactured or stored in concentrations that are too high for safe use. For example, hydrochloric acid is sometimes purchased at a concentration of 10.0 mol/L **(stock solution)**, but students are required to use it only at lower safe concentrations such as 0.10 mol/L. It is therefore necessary to dilute the stock solution to whatever lower concentrations are required.

To understand how dilution happens, please observe what happens when water is added to a fishbowl in the following figures.

Aquarium A

Figure 13.10(a). Ten fish in aquarium of **V** volume of water.

Aquarium B

Figure 13.10(b). Ten fish in aquarium of **2 V** volume of water.

The aquarium A has ten fish in it. When water is added to, it a new situation (B) arises where the number of fish remained the same but the volume of water in which they can swim doubles. The aquarium B is now less concentrated with fish.

This is analogous to what happens when a concentrated aqueous solution is diluted. The following figures illustrate this.

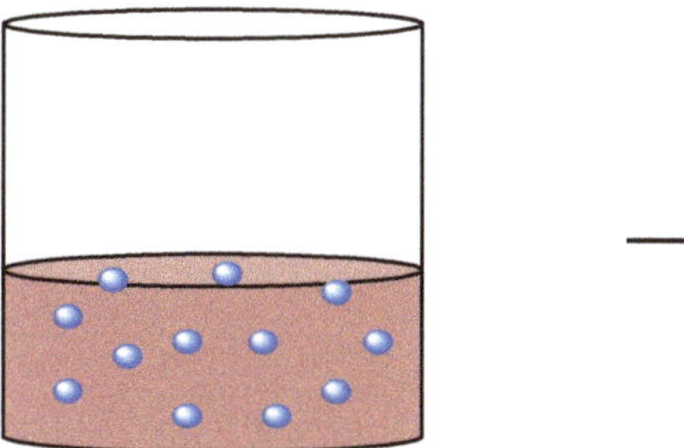
Solution A
Figure 13.11(a). Beaker with **twelve** particles in volume of solution **V**.

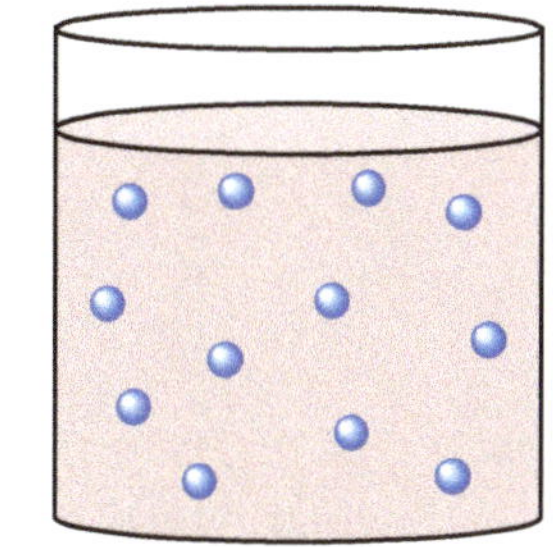
Solution B
Figure 13.11(b). Beaker with **twelve** particles in twice the volume of solution, **2V.**

As in the case with the fishbowl, where the number of fish remained the same after water is added, *the number of moles of particles remains the same when an aqueous solution is diluted.*

Calculations Involving Dilution of Solution

200.0 mL of a 4.00 mol/L $H_2SO_{4(aq)}$, solution A is diluted to produce 800.0 mL of solution B. What would be its final concentration?

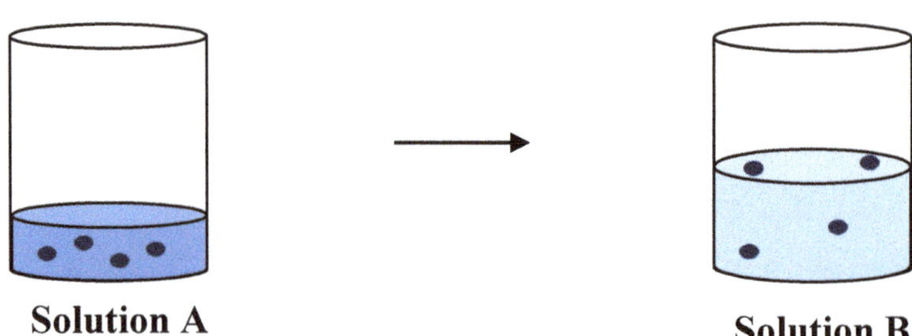

Solution A

Figure 13.12(a) C = 4.00 mol/L

Let V_1 = Volume in A
= 0.200 L

Let C_1 = Concentration in A

Solution B

Figure 13.12(b) V = 800.0 mL

Let V_2 = Volume in B
= 0.800 L

Let C_2 = Concentration in B

Note* When an aqueous solution is diluted, the number of moles of solute in the final solution remains the same as the original solution.

Since the number of moles is the same in both solutions, then:

$$n_{H_2SO_4} = C_1 \times V_1$$

For solution A

$$n_{H_2SO_4} = C_2 \times V_2$$

For solution B

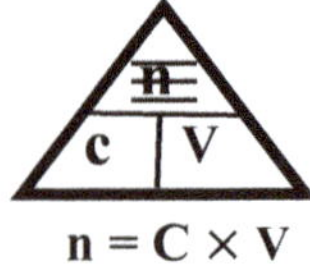

$$n = C \times V$$

Since n is common in both equations, equating these two equations we have:

$$C_1 \times V_1 = C_2 \times V_2$$

Solving for C2, we get: $$C_2 = \frac{C_1 \times V_1}{V_2}$$

Substituting values for C1, V1 and V2 in this equation we have:

$$C_2 = \frac{4.00 \, mol/L \times 0.200 \, L}{0.800 \, L} = 1.00 \, mol/L$$

Suppose that a sample of 2.00 L of 0.100 mol/L $HCl_{(aq)}$ is required and is to be prepared by diluting a sample from a 10.0 mol/L $HCl_{(aq)}$ stock solution. What volume of the stock solution must be diluted to obtain this dilute solution?

Provided quantities: **Required quantity:** V_1

$C_1 = 10.0$ mol/L
$V_2 = 2.00$ L
$C_2 = 0.100$ mol/L

Using the equation $C_1V_1 = C_2V_2$ to solve for V_1, we get

$$V_1 = \frac{C_2 \times V_2}{C_1}$$

Substituting values for C_2, V_2 and C_1 in this equation, we get :

$$V_1 = \frac{0.100 \; mol/L \times 2.00 \; L}{10.0 \; mol/L}$$

$$= 0.020 \; L$$

$$= 20.0 \; mL \; of \; 10.0 \; mol/L \; HCl_{(aq)}$$

**Note that when a concentrated acid is to be diluted, water must not be added to it. Instead, the acid is added to water. This is done by placing the container of distilled water in cold water bath and then gently pouring the calculated volume of acid into it. *All safety equipment must be worn (safety goggles, latex gloves and apron)*

Sample problem 7:

An aqueous solution has 16.42 g of Ca $(NO_3)_{2(s)}$ dissolved in 200.0 mL of solution. What would be the final concentration of $Ca^{2+}_{(aq)}$ and $NO_3^-_{(aq)}$ if this solution is diluted to 2.00 L?

Solution:

To solve this problem, it is useful to follow the steps below:

Step 1. Calculate the number of moles of Ca $(NO_3)_{2(s)}$.

$$n_{Ca(NO_3)_2} = \frac{m}{M} = \frac{16.42 \; g}{164.14 \; g/mol}$$

$$= 0.1000 \; mol$$

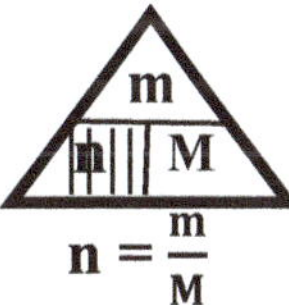

Step 2. Calculate the initial concentration of the Ca $(NO_3)_{2(aq)}$.

$$C_{Ca(NO_3)_{2(aq)}} = \frac{n}{V}$$

$$= \frac{0.1000 \; mol}{0.200 \; L}$$

$$= 0.500 \; mol/L$$

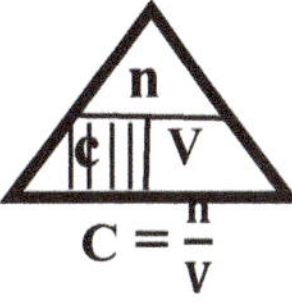

Step 3. Find the new concentration of the Ca $(NO_3)_{2(aq)}$ after dilution.

Using the equation, $C_1 \times V_1 = C_2 \times V_2$

$$C_2 = \frac{C_1 \times V_1}{V_2}$$

$$= \frac{0.500 \; mol/L \times 0.200 \; L}{2.00 \; L}$$

$$= 0.0500 \; mol/L$$

Step 4. Write the balanced equation to show how $Ca(NO_3)_{2(aq)}$ dissociates in water.

$$Ca(NO_3)_{2(aq)} \longrightarrow Ca^{2+} + 2\,NO_3^-{}_{(aq)}$$

$$1\ mol/L \qquad\qquad 1\ mol/L \qquad 2\ mol/L$$

Step 5. Calculate the concentration of each ion using the mole ratios from the equation before.

$$\left[Ca^{2+}{}_{(aq)}\right] = 0.0500\ mol/L\,Ca(NO_3)_{2(aq)} \times \frac{1\ mol/L\,Ca^{2+}{}_{(aq)}}{1.00\ mol/L\,Ca(NO_3^-)_{2(aq)}}$$

$$= 0.050\ mol/L$$

$$\left[NO_3^-{}_{(aq)}\right] = 0.050\ mol/L\,Ca(NO_3)_{2(aq)} \times \frac{2\ mol/L\,NO_3^-{}_{(aq)}}{1.00\ mol/L\,Ca(NO_3)_{2(aq)}}$$

$$= 0.100\ mol/L$$

Exercise 13.8

a) What is the final concentration of a 200 mL of 0.100 mol/L $HCl_{(aq)}$, if it is diluted to a new volume of 800 mL? **T/I**

b) A 100 mL of a solution of $NaOH_{(aq)}$ of was diluted to a new volume of 1 L. If its final concentration is 0.01 mol/L, determine its initial concentration. **T/I**

c) A student needs to make 2.0 L of 0.10 mol/L of $H_2SO_{4(aq)}$ by diluting a sample of a stock solution of 10.0 mol/L C. Determine the volume of the stock solution must be diluted. What precautions must be taken when diluting concentrated acids? **T/I**

d) To what final volume must 10.0 mL of 10.0 mol/L $HCl_{(aq)}$ be diluted for its concentration to change to 0.50 mol/L? **T/I**

e) An aqueous solution has 12.26 g $H_2SO_{4(aq)}$ dissolved in 500.0 mL of solution. What is the concentration of the solution? What will be the final concentration of $H^+{}_{(aq)}$ ions and $SO_4^{2-}{}_{(aq)}$ ions if the solution is diluted to 2.00 L? **T/I**

f) An aqueous solution has 2.92 g $Mg(OH)_{2(s)}$ dissolved in 200.0 mL of solution. What is the concentration of this solution? What will be the final concentration of $Mg^{2+}{}_{(aq)}$ ions and $OH^-{}_{(aq)}$ ions if the solution is diluted to 1.00 L? **T/I**

g) What mass of $Na_2CO_{3(s)}$ must be dissolved in 200.0 mL of solution to produce a solution having $CO_3^{2-}{}_{(aq)}$ ions of 0.500 mol/L? **T/I**

h) The concentration of $H^+{}_{(aq)}$ ions is important in determining the pH of solutions. What mass of $HCl_{(aq)}$ must be dissolved in 500.0 mL of solution to produce a concentration of 0.100 mol/L $H^+{}_{(aq)}$ ions? **T/I**

 Other Measurements of Solution Concentrations

Percentage Concentration

Unlike molar concentrations, where concentration is expressed as moles per litre of solution, many consumer products such as vinegar are sold as percentage concentrations.

Volume-Volume Concentration (%V/V)

In this type of measurement, a certain volume of the product is dissolved to make 100.0 mL of solution. Vinegar, for example, is sold as 5.0% acetic acid. This means that 5.0 mL of glacial (pure) acetic acid is dissolved in 95.0 mL of distilled water to make 100.0 mL of vinegar solution. This is otherwise expressed as 5.0% V/V. Since volumes of products are not limited only to 100.0 mL of solution, percentage by volume concentration of the solute is calculated using the general formula:

$$C = \frac{V_{solute}}{V_{solution}} \times 100\%$$

Sample problem 8.

A solution of rubbing alcohol has 3.50 L of isopropyl alcohol (C_3H_7OH) dissolved in 5.00 L tub of solution. What is the percentage by volume concentration of isopropyl alcohol?

Solution:

$$V_{C3H7OH(l)} = 3.50 \text{ L}$$
$$V_{C3H7OH(aq)} = 5.00 \text{ L}$$
$$C_{C3H7OH(aq)} = \frac{3.50 \text{ L}}{5.00 \text{ L}} \times 100\%$$
$$= 70 \text{ % V/V}$$

Sample problem 9.

The concentration of hydrochloric acid in a sample is 9.470 mol/L. If the density of the solution is 1.150 g/mL, calculate the volume percent of hydrochloric acid in the solution.

Solution:

Step 1: Change density to g/L since the concentration is given in mol/L.

$$d = \frac{1.150 \text{ g}}{mL} \times \frac{1000.0 \text{ mL}}{L}$$
$$= 1150 \text{ g / L}$$

Step 2: Find the mass of hydrochloric acid in 1.0 L of solution.
Using the formula: $m = n \times M$,

We get: $$m = 9.470 \text{ mol} \times \frac{35.45 \text{ g}}{mol}$$
$$= 335.7 \text{ g}$$

Step 3: Now that we know its mass and density, rearranging the formula for density gives the following:

$$d = \frac{m}{v}, \quad \therefore \ v = \frac{m}{d}$$

The volume of the hydrochloric acid in 1.0 L or 1000.0 mL of solution is

$$v = \frac{335.7 \ \cancel{g}}{1150 \ \cancel{g}/L}$$

$$= 2.92 \times 10^{-1} \ L$$

Step 4: Find the volume percentage of HCl.

$$= \frac{2.92 \times 10^{-1} \ \cancel{L}}{1.00 \ \cancel{L}} \times 100\%$$

$$= 29.2\%$$

Mass-Volume Concentration (% M/V)

In this type of measurement, the solute is a solid and the solvent is a liquid. The concentration of the solute is calculated using the general formula:

$$C = \frac{M_{solute(g)}}{V_{solution(mL)}} \times 100\%$$

Sample problem 10:

Your Chemistry teacher dissolved 1.50 g of iodine in ethyl alcohol to make 50.0 mL of solution. What is the percentage by mass concentration of iodine?

Solution:

$$m_{iodine} = 1.50 \ g$$
$$V_{solution} = 50.0 \ mL$$

$$C_{iodine} = \frac{1.50 \ g}{50.0 \ mL} \times 100\%$$

$$= 3.00 \ \% \ M/V$$

Mass-Mass Concentration (% M/M)

Alloys are made by melting two or more metals and then allowing the liquid solution to cool and solidify. They are thus considered to be solid solutions. Brass, for example, is an alloy made by melting copper and zinc in an 85:15 ratio, by mass. The concentration of the solute is calculated using the general formula:

$$C = \frac{M_{solute(g)}}{M_{solution(g)}} \times 100\%$$

A 10.0 g nickel coin contains 2.50 g of nickel and 7.50 g of copper. What is the percentage mass-mass concentration of nickel in the coin?

$m_{Ni} = 2.50$ g

$m_{alloy} = 7.50$ g

$$C_{Ni} = \frac{2.50 \text{ g}}{10.0 \text{ g}} \times 100\%$$

*Note that the mass of the solid solution is 10.0 g and not 7.50 g

$$= 25.0\% \text{ M/M}$$

Exercise 13.9

1. Vinegar is sold as a 5.00 % V/V solution of pure acetic acid in water. **T/I**
   (a) What volume of acetic acid and water must be mixed to make 10.0 L of a 5.00 %   V/V solution of vinegar?
   (b) If the density of acetic acid is 1.049 g/mL, find the mass of the acetic acid used.
   (c) How many moles of acetic acid used does your answer in part (b) correspond to?

2. The concentration of sulphuric acid in a sample is found to be 1.2 mol/L. If the density of the solution is 1.06 g/mL, calculate the volume percent of the sulphuric acid in the solution. **T/I**

3. A 100.0 mL box of mixed fruit juice has a fructose ($C_6H_{12}O_6$) concentration of 12.0% mass-volume. What mass of fructose would be found in 1.00 L of fruit juice? **T/I**

4. Coffee, brewed with pure water contains approximately 0.045 % M/VVT concentration of caffeine. What mass of caffeine will be found in a 200-mL cup of unsweetened coffee? **T/I**

5. Plumber's solder is 33.0% tin and 67.0% lead M/M percent. How many moles of tin are present in a 500.0 g sample of solder? **T/I**

Very Low Concentrations

The concentration of a particular substance in living organisms or the environment may be so low that it is not convenient to express it as either percent or molar concentration. Instead, it is expressed in **parts per million (ppm)** and **parts per billion (ppb).** In sea water, for example, the level of mercury is 1.80 parts per million. This means that there is 1.80 g of mercury for every 1.00×10^6 g of sea water.

The following formulas are used to calculate ppm and ppb:

$$ppm = \frac{Mass\ of\ solute}{Mass\ of\ solvent} \times 10^6$$

$$ppb = \frac{Mass\ of\ solute}{Mass\ of\ solvent} \times 10^9$$

If the level of mercury in farm raised rainbow trout is measured to be 0.0140 ppm. What mass of mercury will be in 10.0 t of trout?   (1.00 ton $= 1.00 \times 10^6$ g)

Solution:

Step 1: Convert mass in tones to mass in grams.

$$10.0\,t \times 1.00x\,10^6\,g/t = 1.00x\,10^7\,g$$

Step 2: Use the appropriate ratios to find the mass of mercury.

$$\frac{0.014\,g}{1.00x10^6\,g} = \frac{x\,g\,Hg}{1.00x10^7\,g}$$
$$x \quad = 0.140\,g\,Hg$$

The mass of mercury in 10.0 t of trout is 0.140 g.

Exercise 13.11

1. In an attempt to investigate the bioaccumulation of DDT up the food chain in a particular **lake with a DDT concentration of 5.00 ppb**, the tissues of some birds that feed on fish from the lake were analyzed. If 1.00 kg of bird tissues contained $3.00x10^{-2}$ kg of DDT, what is its concentration in pbb? **T/I A**

2. If 1.0 kg of polluted air contains 0.015 g of carbon monoxide, what is its concentration in ppm? **T/I**

3. If the concentration of carbon dioxide in an early morning outdoor air is 350 ppm, what mass of it will be in 2.0 kg of air? **T/I**

Finding Molar Concentration, given Density and Percentage by Volume

To solve these types of problems, the following sample solution is used:

Sample Problem 13:

A solution of $H_2SO_{4(aq)}$ is 80.0% V/V and has a density of 1.73g/mL 20° C, what is its molar concentration?

Solution:

Step 1: Establish that the volume of the acid in a1000.0 mL is $= 800.0$ ml since V/V $= 80.0\%$. The volume is found by changing the percentage to mL and multiplying by 10.

Step 2: Find the mass of the acid using the formula m $= $ D xV, where D $= $ density.

Mass $= 1.73$g/mL x 800.0 mL$= 1,384$ g

Step 3: Find the number of moles of H_2SO_4, n $= 1,384$ g /98.06g/mol $= 14.11$mol.

Step 4: Find the concentration of the acid**,** C by dividing by 1.000 L $= $ **14.11 mol /L**.

1. A solution of $H_2SO_{4(aq)}$ is 83.0% V/V and has a density of 1.22 g/mL at 20° C, what is its molar concentration? **T/I**

2. A solution of $HCl_{(aq)}$ is 26.0% by mass, and has a density of 1.23 g/mL 20° C, what is its molar concentration? **T/I**

3. A solution of $HNO_{3(aq)}$ is 50.0% by mass, and has a density of 1.31 g/mL 20° C, what is its molar concentration? **T/I**

Chapter Content:

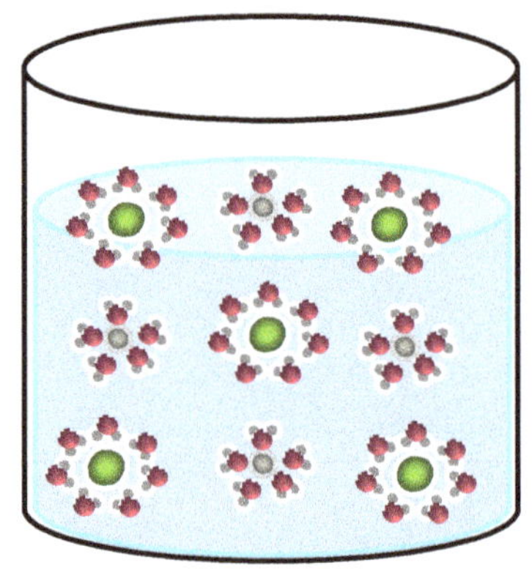

14.1 Solubility

If every solvent on Earth were to dissolve every solute, then everything would be in the form of one solution or another. But the natural order of the world, in terms of solubility, dictates that only specific solvent dissolve specific solutes (solids and gases).

The selective nature of water, for example, in dissolving only some types of substances is attributed to unique properties; it being a polar molecule and a liquid. This allows water to be in our bodies to transport dissolved wastes and nutrients without dissolving structures such as our cell membranes and bones. The relative abundance of water imposes special physiological adaptations on organisms, in terms of what types of excretory substances to produce. Aquatic animals which enjoy an unlimited supply of water, produce the very water-soluble ammonia as their main excretory product. Ammonia, being very soluble in water, dissolves well in the tissues and tissue fluids of organisms; lots of water is therefore needed to wash it away from their tissues. Terrestrial organisms produce less soluble urea and uric acid, in order to contend with less available water; these being less soluble, requires less water to be excreted These facts show that not all substances have the same solubility in water. Kidney stones and gall stones form because the compounds from which they are formed are sparingly soluble in aqueous solutions. These compounds form saturated solutions and crystallize out of solution. This chapter explains why some solutes are more soluble than others in water, and how temperature affects their solubility.

Factors That Affect Solubility

As can be recalled from Chapter 12, there must be some type of attraction between the particles of a solute and those of a solvent for mixing to take place. Ionic compounds, for example, dissolve in water because of the attraction between the polar water molecules and the ions present in the compounds. The relative ease with which the particles of a solute separate from each other and mix with the solvent particles determines the solubility of the solute in that particular solvent. At a particular temperature, the solubility of solute would depend on:

14.2 Force of Attraction Between the Particles of the Solute

In molecular crystals such as sugars, the particles are attracted by primarily weak van der Waals forces. It is thus easy for the polar water molecules to separate the slightly polar sugar molecules away from each other in a crystal of sugar. Sugar thus dissolves easily in water.

In ionic crystals, the particles are held by relatively stronger electrostatic force of attraction. The magnitude of this force is governed by Coulomb's law which is represented as follows:

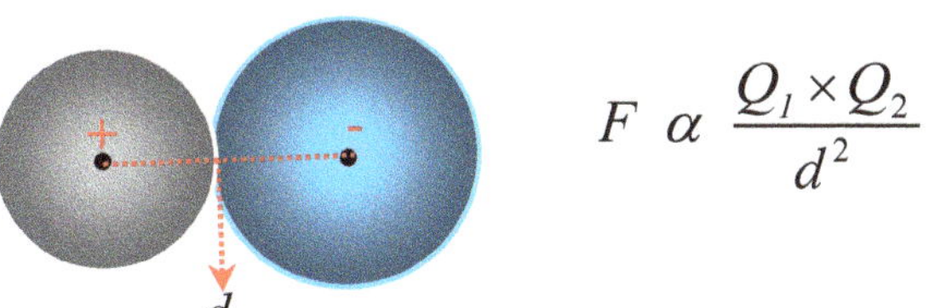

$$F \; \alpha \; \frac{Q_1 \times Q_2}{d^2}$$

Q_1 and Q_2 are the charges on the ions in Coulombs, d is interionic distance between the two ions. **Interionic distance, d,** *is the sum of the of the ionic radii of the two ions.* **F** *is the force of attraction between the ions.*

Figure 14.1. The interionic distance for an ionic compound.

The solubility of ionic compounds in water depends on the relative forces of attraction between the oppositely charged ions in the crystal. *If the ionic charges are small and the distances between them are large, then according to Coulomb's law, their forces of attraction are relatively weak and the compound dissolves relatively easily in water.* Cesium chloride, for example, is more soluble in water than sodium chloride is, at the same temperature. The ions in both compounds have the same magnitude of charge, but because the ionic radius of cesium is larger than that of sodium, the forces of attraction between the cesium ion and chloride ion are weaker than that between the sodium ion and the chloride ion. The cesium chloride **crystal lattice** is thus less stable than the sodium chloride **crystal lattice**, making it more soluble in water.

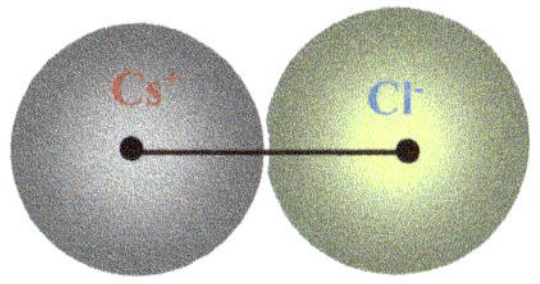

Figure 14.2(a). The relative sizes of Cs^+ and Cl^- ions in cesium chloride.

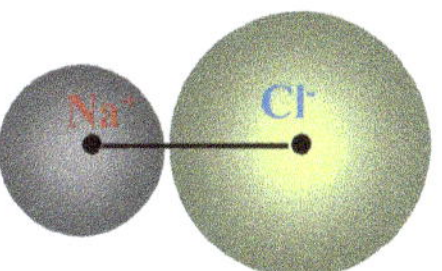

Figure 14.2(b). The relative sizes of Na^+ and Cl^- ions in sodium chloride.

Conversely, if their ionic charges are large and the distance between them are small **then** *their forces of attraction are relatively strong.* **In this case the ionic compound does not dissolve easily and may even remain insoluble depending on the magnitude of the force of attraction between their oppositely charged ions.** Calcium oxide, for example, is less soluble in water than sodium chloride at the same temperature. This is attributed to the greater magnitude of charges on the calcium and oxide ions, coupled with their smaller inter ionic distance, compared to the ions in sodium chloride which have smaller ionic charges and greater inter-ionic distances. The calcium oxide **crystal lattice** is thus more stable than sodium chloride **crystal lattice,** making it less soluble.

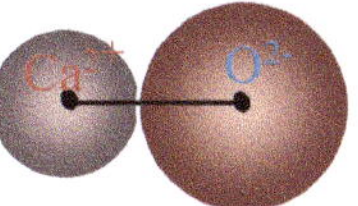

Figure 14.3(a) The double charges on the Ca^{2+} and O^{2-} ions with small inter-ionic radius

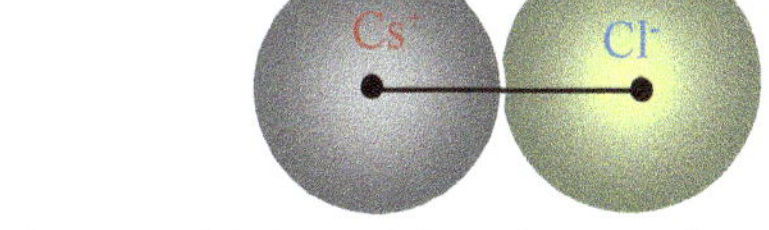

Figure 14.3(a) The single charges on the Cs^+ and Cl^- ions with large inter-ionic radius.

14.3 Force of Attraction between Solute particles and Solvent Particles

For the particles of solute to separate from each other and mix with the solvent particles, there must be a significant force of attraction between the particles of the solute and those of the solvent. Iodine, for example, a molecular crystal, dissolves in ethanol but ionic compounds do not. The presence of relatively weak van der Waals forces in both iodine and ethanol molecules are responsible for the solubility of iodine in ethanol. The ethanol molecules are able to remove the molecules of iodine from its crystal lattice with relative ease. But there is little or no attraction between the ethanol molecules and the ions present in ionic crystal to cause the ions to separate from each other and mix with the ethanol molecules. However, ionic compounds dissolve in water because of strong attraction between the highly polar water molecules and the ions present in their crystals. In the latter case, the water molecules are able to exert a stronger force of attraction to the ions of the crystal, than the ions in the crystals are able to exert on each other.

It can thus be concluded that for any solute to dissolve in a solvent, the force of attraction between the solvent and solute particles must be greater than the force of attraction between solute particles.

14.4 Solubility and Saturation

The solubility of a solute is the greatest mass of the solute that can dissolve in a given quantity of solvent at a particular temperature. For example, if small quantities of potassium nitrate are measured and dissolved sequentially in a 10.0 mL sample of water maintained at a constant temperature of 25.0 °C, it would be found that after a time, no more of the solute would dissolve. At this point a **saturated solution** of potassium nitrate is formed; no more potassium nitrate can dissolve in the 10.0 mL of water at 25.0 °C. In one experiment the mass of potassium nitrate was found to be 4.0 g. This means that any additional mass of potassium nitrate greater than 4.0 g, if added to 10.0 mL of water at 25.0 °C, would remain undissolved as crystals and would sink to the bottom of the solution. Also, any mass of potassium nitrate less than 4.0 g, if added to 10.0 mL of water at 25.0 °C, would dissolve completely but an **unsaturated solution** would result.

14.5 Temperature and Solubility

If the temperature of this **saturated** solution (the 10.0 mL of water at 25.0 °C having 4.0 g of dissolved of potassium nitrate) is raised, it then becomes **unsaturated**. This is so because the 10.0 mL of water would be at a temperature higher than 25.0 °C could dissolve more than 4.0 g of potassium nitrate before it becomes saturated. On the other hand, if the temperature of the saturated solution is cooled, crystallization will occur. This is so because the 10.0 mL of water at a temperature lower than 25.0 °C would not dissolve 4.0 g of potassium nitrate, but less. In the latter case, crystals would precipitate out of the solution, but the resulting solution would still remain saturated.

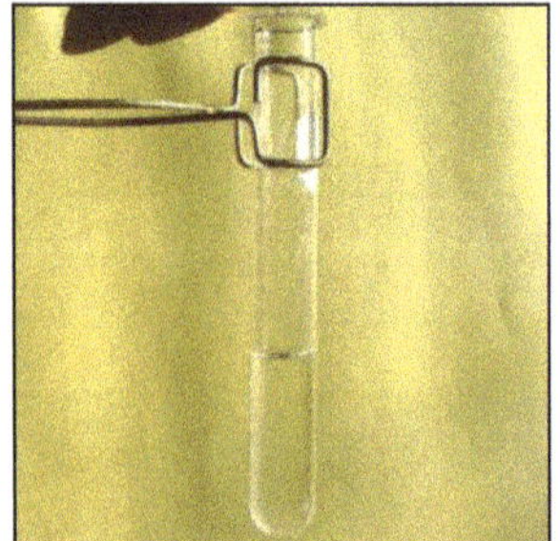

Figure 14.4(a).

Saturated Solution

10 mL of $H_2O_{(l)}$ at **25 °C** with 4.0 g KNO_3.

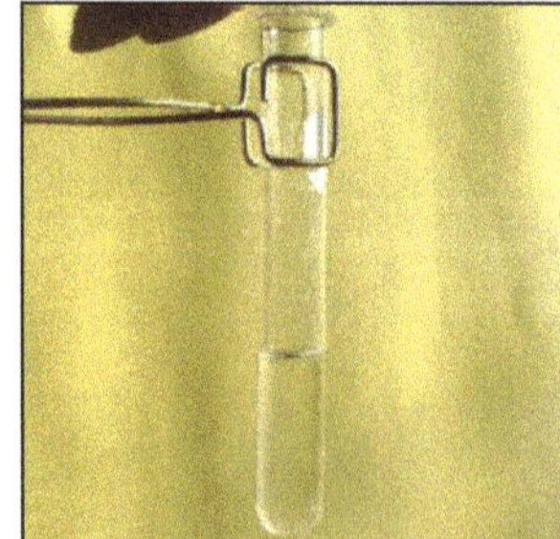

Figure 14.4(b).

Unsaturated Solution

10 mL of $H_2O_{(l)}$ at **30 °C** with 4.0 g KNO_3.

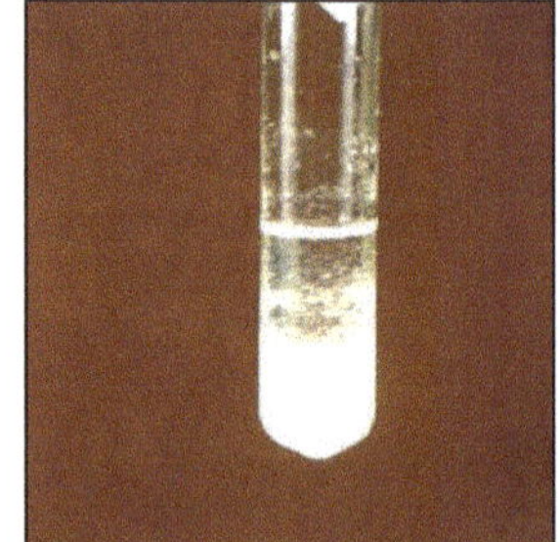

Figure 14.4(c).

Saturated Solution
(Crystallization occurs).

10 mL of $H_2O_{(l)}$ at **20 °C** with less than 4.0 g KNO_3

With few exceptions, an increase in temperature increases the solubility of solid solutes. Plausible explanations for increase in solubility of ionic compounds in water with rise in temperature could be:

 ➢ Any increase in temperature causes *the particles of the solid to gain more energy and separate farther away from each other* making their inter-particle force of attraction less. This allows the solvent particles to separate them more easily. Refer to chapter twelve (**Figure 12.7**).

 ➢ An increase in temperature causes *the particles of the solvent to gain more energy and separate from each other more easily.* In water, for example, more hydrogen bonds will be broken, and this allows more of the solvent molecules become available for the hydration process. Refer to chapter twelve (**Figure 12.6**).

➢ An increase in temperature *offers less chance for the solute particles getting closer together to attract each other* after the hydration process. *The farther apart the oppositely charged hydrated particles are, the less is the chance for them to recrystallize.* (**Figure 14.5b**).

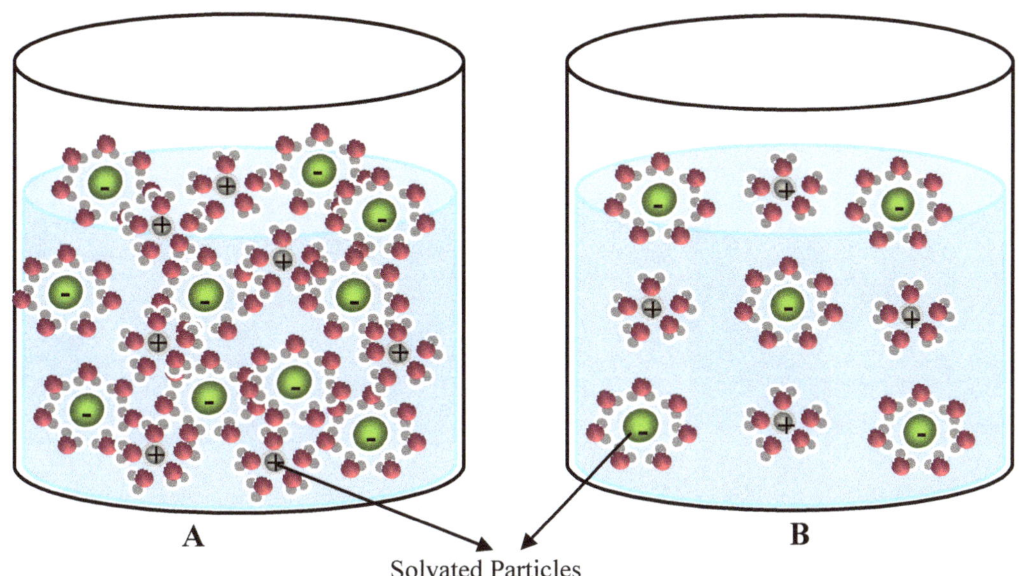

Figure 14.5(a). At a lower temperature, the dissolved particles are closer together to allow for faster re-crystallization.

Figure 14.5(b). At a higher temperature, the dissolved particles are farther away; the chance for re-crystallization is less.

14.6 Solubility of Gases in Water

When lid of a warm can of carbonated soda is opened and the sound of escaping gas is compared with that when the lid of a cold can of carbonated soda is opened, it is observed that the sound of the escaping gas from the warm can is louder than that of the cold can. At the end of the following activity, you should be able to explain this observation.

Activity:

Investigating the effect of temperature on solubility of a gas in water.

Experimental Design:

In this activity, two sets of carbonated soda will be subjected to different temperatures and the rate at which the dissolved gas evolves from them will be monitored.

Materials:
   *Two 5 cm long candles, on stands
   *Two 250 mL beakers
   *Two 350 mL glass bowls
   *Matches

   *Two 355 mL cans of carbonated pop soda; one cold and the other at room temperature
   *Approximately 200 mL of hot water
   *Approximately 200 mL of cold water

Procedure:
- Pour 110 mL of the cold and warm soda into one of the 250 mL beaker.
- Light the candles and place each to stand in the beakers with soda. The candles must be of the exact length and long enough to allow the burning wicks to be above the level of the soda.
- Place each beaker with soda and burning candle separately into the bowls. The bowls must be big enough to hold the two beakers and 200 mL water.
- Pour the 200 mL cold water into the bowl with cold soda and 200 mL hot water into the other bowl with warm soda.

Hypothesis:

Before carrying out this procedure, predict what will be observed in these two scenarios.

Observation:

The candle burnt for a longer time in the cold bowl.

Figure 14.6(a). Burning of candle in cold pop soda.

The candle burnt for a shorter time in the hot bowl.

Figure 14.6(b). Burning of candle in hot pop soda.

Table 14.1 Results from observing burning candles and the evolution of gases in warm and cold pop soda.

Scenarios	Results
Burning candle in soda placed in the bowl of hot water (Fig.14.6b).	There was vigorous effervescence with the liberation of a colorless gas of large bubbles. The flame of the candle became smaller until it was extinguished.
Burning candle in soda placed in the bowl of cold water (Fig.14.6a).	There was moderate effervescence with the liberation of a colorless gas of smaller bubbles. The flame of the candle took a much longer time to become extinguished.

Analysis 1

1. What gas is present in carbonated soda? **K/U**

2. What gas was produced by the carbonated soda during the effervescence that occurred?**T/I**

3. Does the gas that produced support combustion? **T/I**

4. Why was the candle in the hot bowl extinguished much faster than in the cold bowl? **T/I**

5. What are the implications of thermal pollution, (heating of bodies of water) on aquatic life forms? **A**

6. How would you account for the difference in sounds made by opening of the lids of the cold and warm cans of carbonated soda? **CT/I**

7. **Conclusion:**

What can you conclude about the effect temperature on solubility of gases in aqueous

Before carrying out this procedure, predict what will be observed in these two scenarios. **CT/I**

Activity: The effect of temperature on the solubility of potassium nitrate in water.
In an attempt to investigate the effect of temperature on the solubility of potassium nitrate in water, the following activity was carried out by a group of students.

Procedure:
1. Seven small test tubes were obtained.
2. 3.00 g samples of potassium nitrate were weighed out and placed in each test tube.
3. 3.0 mL, 4.0 mL, 5.0 mL, 6.0 mL 7.0 mL, 8.0 mL and 9.0 mL samples of distilled water were measured out and poured into each test tube respectively.
4. Using a water bath, each test tube was heated until the solid potassium nitrate dissolved completely.
5. A thermometer was placed in each test tube which was gradually cooled and stirred until crystallization occurred.
6. The temperature was taken in each test tube, just when crystallization occurred.

Evidence:

Table 14.2. Saturation temperatures for solutions of concentrations.

Test	#1	#2	#3	#4	#5	#6	#7
Mass of KNO₃	3.00 g	3.00 g	3.00 g	3.00 g	3.00 g	3.00 g	3.00 g
Volume of H₂O	3.00 mL	4.00 mL	5.00 mL	6.00 mL	7.00 mL	8.00 mL	9.00 mL
Saturation Temperature	56.0 °C	45.0 °C	38.0 °C	32.0 °C	28.0 °C	24.0 °C	20.0 °C

Solubility is normally expressed as the amount of solute dissolved in 100.0 mL at a particular temperature. Changing the solubility in the previous table to amount dissolved in 100.0 mL water, we have:

Test tubes: #1

$$\frac{3.00 \text{ g}}{3.00 \text{ mL}} = \frac{x \text{ g}}{100.0 \text{ mL}}$$
$$= 100.0 \text{ g at } 56\,°C$$

#2

$$\frac{3.00 \text{ g}}{4.00 \text{ mL}} = \frac{x \text{ g}}{100.0 \text{ mL}}$$
$$= 75.0 \text{ g at } 45\,°C$$

#3

$$\frac{3.00 \text{ g}}{5.00 \text{ mL}} = \frac{x \text{ g}}{100.0 \text{ mL}}$$
$$= 60.0 \text{ g at } 38\,°C$$

#4

$$\frac{3.00 \text{ g}}{6.00 \text{ mL}} = \frac{x \text{ g}}{100.0 \text{ mL}}$$
$$= 50.0 \text{ g at } 32\,°C$$

Test tubes: #5

$$\frac{3.00 \text{ g}}{7.00 \text{ mL}} = \frac{x \text{ g}}{100.0 \text{ mL}}$$
$$= 42.9 \text{ g at } 28\,°C$$

#6

$$\frac{3.00 \text{ g}}{8.00 \text{ mL}} = \frac{x \text{ g}}{100.0 \text{ mL}}$$
$$= 37.5 \text{ g at } 24\,°C$$

#7

$$\frac{3.00 \text{ g}}{9.00 \text{ mL}} = \frac{x \text{ g}}{100.0 \text{ mL}}$$
$$= 33.3 \text{ g at } 20\,°C$$

If these values are plotted on a graph, with solubility on the vertical axis and temperature on the horizontal axis, the following graph is obtained.

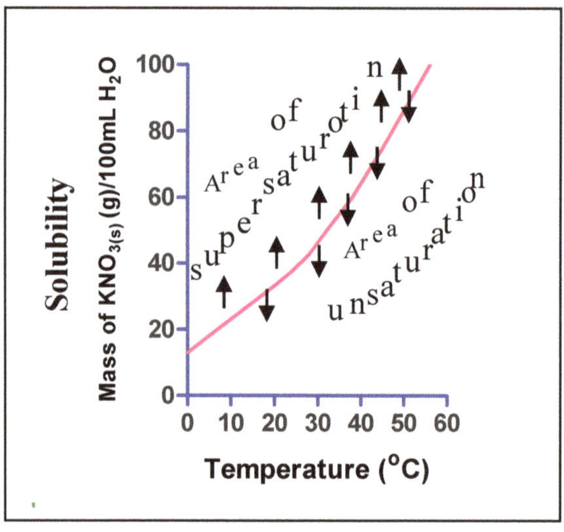

Figure 14.7. Solubility curve for KNO$_{3(aq)}$.

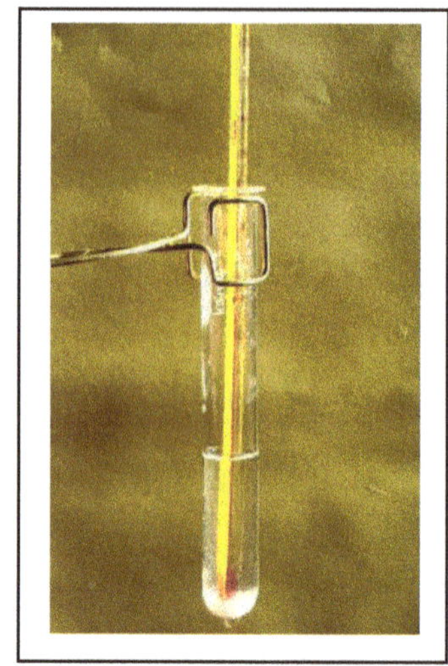

Figure 14.8. Crystallization of KNO$_3$ at saturation temperature.

The arrows above the curve in Figure 14.7, indicate the region of **supersaturation**. This region is where more solute has dissolved at any particular temperature than normally happens. The arrows below the curve indicate the region of **unsaturation**. Anywhere in this region the solution is still unsaturated and can dissolve more solute until saturation point is reached.

Analysis 2

1. What types of solutions were obtained in procedure 4? **T/I**

2. What types of solutions were obtained in procedure 5? **T/I**

3. What is the relationship between the solubility of potassium nitrate and temperature? **T/I**

Conclusion:

1. Write a concluding statement about the solubility of KNO$_3$ with temperature changes **T/I C**

The following graphs are obtained when the same is done to show how the solubilities of other substances change with temperature changes. The solubility curves obtained for each are shown in table 14.3.

Table 14.3. Solubility Curves for a few Ionic Compounds in Water

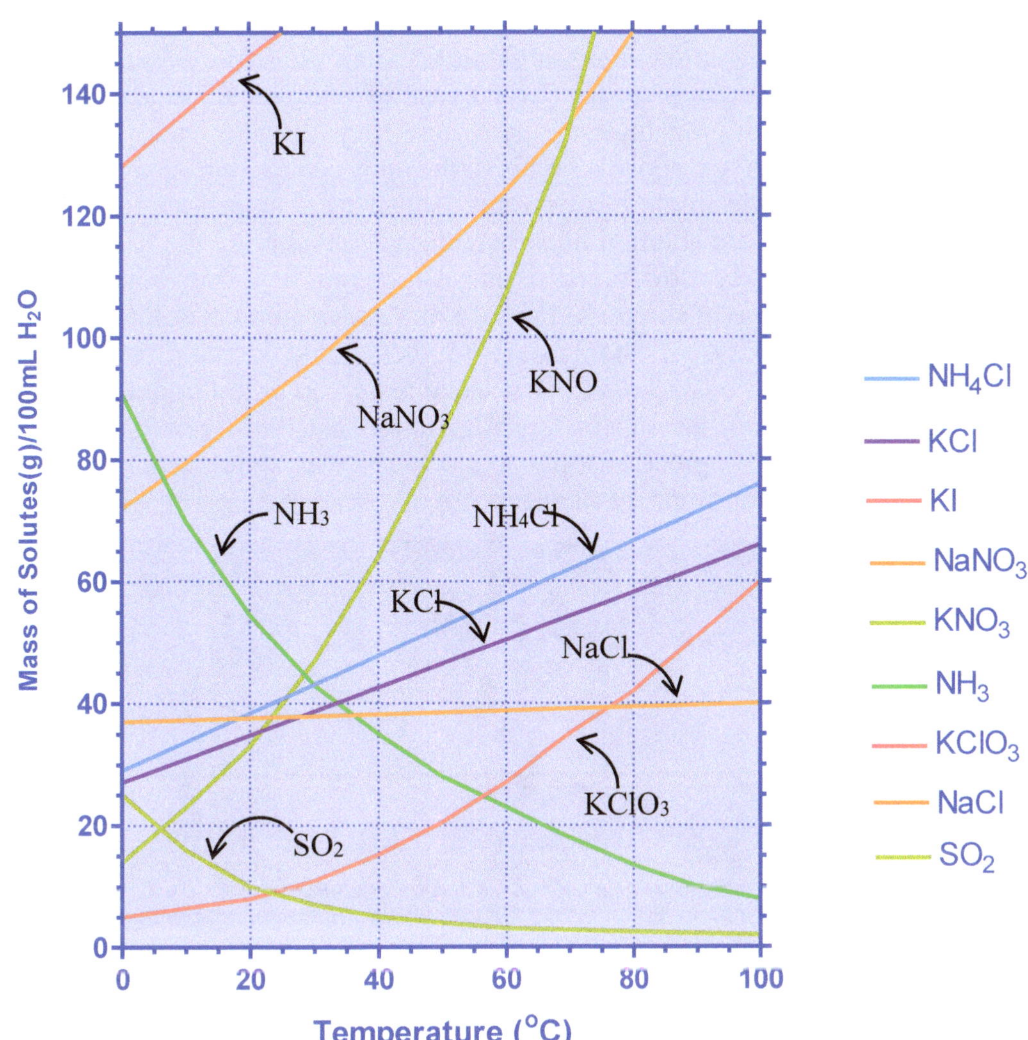

Analysis 3

Use the solubility curves in the table above to answer the following questions:

1. Which substance is most soluble at 0.0 °C? **T/I**

2. At what temperature do $NaNO_3$ and KNO_3 have the same solubility? **T/I**

3. If $KNO_{3(aq)}$ is cooled from 50.0 °C to 25.0 °C, what mass of crystals will precipitate? **T/I**

4. What substances exhibit decrease solubility with temperatures rise? **T/I**

5. Why is NH_3 more soluble than SO_2 at any temperature? **T/I**

6. If a saturated solution of $NaNO_{3(aq)}$ at 0.0 °C is raised to 80.0 °C, how much more solid is required to make the solution saturated at 80.0 °C? **T/I**

7. Can molar the concentration of any of the substances, at a particular temperature, be calculated from the line graphs provided? Explain your answer. **T/I C**

235

14.8 Supersaturated Solutions

A supersaturated solution contains more dissolved solutes than could be normally dissolved by that particular solvent at a particular temperature. In the case of aqueous solutions of some solid solutes, when their solutions are raised to very high temperatures, they dissolve more solutes than at lower temperatures. When these solutions are cooled, no crystallization occurs, but supersaturated solutions are formed. Sodium thiosulphate is an example of one solid that forms a supersaturated solution. Carbonated water is an example of a supersaturated solution of carbon dioxide in water.

In supersaturated solutions, the excess dissolved solutes can be removed from solution by the addition of certain "nuclei" such as a small piece of solute. In the case of a supersaturated solution formed from a solid solute, the addition of a small piece of **that solute** to the cooled solution, results in a spectacular precipitation of solute. The addition of Mentos to Diet Coke results in eruption due to the extremely rapid rate of carbon dioxide liberation from the supersaturated solution. Such observations are classical tests for supersaturation; the addition of a small piece of solute to a solution with spontaneous massive crystallization resulting. The following figure demonstrates this concept.

(a)   (b)   (c)   (d)

Figure 14.9. The crystallization of a supersaturated sodium thiosulphate solution in stages after a single crystal of sodium thiosulphate was added.

At temperatures around 25 °C, honey is a supersaturated solution of various sugars in water. It is composed of more than 70 % sugars and less 20 % water, depending on the source of the nectar. As in other supersaturated solutions, the water in honey dissolves more sugars than it would normally do. This makes honey relatively unstable. When honey is purchased during summer time it a viscous liquid. However, if it is left standing, it crystallizes as the temperature falls; more so, during winter time.

Exercise 14.1

1. Why does honey crystallize more during winter times than in others. T/I C

2. What should someone do to the honey to remove the crystals formed, without wasting any honey? What precaution must be taken not to destroy the properties of honey such as some useful enzymes, flavors, anti-oxidants and aroma? T/I A C

3. Raw, unfiltered honey that has bits of wax materials and pollens crystallizes faster than mildly heated and filtered honey. Explain this observation. T/I A

4. Does the crystallization of honey change its qualities? C T/I

5. Honey has variable combinations of glucose-fructose ratios. Fructose has a higher solubility than glucose. If you have a bottle of partially crystallized honey where would you most likely find some fructose? Explain. T/I C

CHAPTER 15
Soaps and Detergents

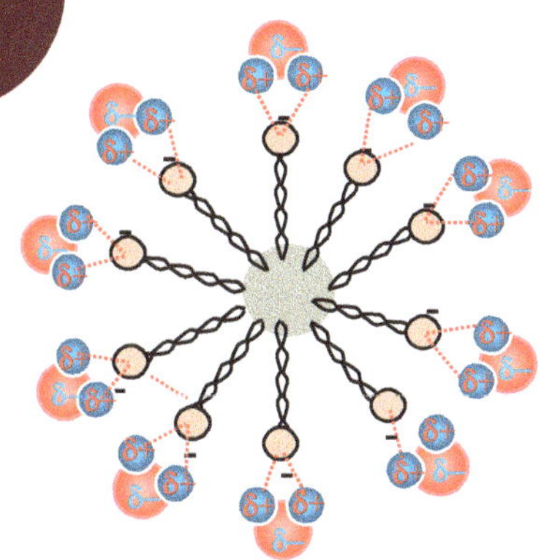

Chapter Content:
15.1 Soaps and Detergents
15.2 Hardness of water
15.3 Softening of hard water
15.4 **Activity:** Investigating the hardness of water

Have you ever thought of what our condition would be like without soap? Developed by Medieval Arabs, it benefited our world in immeasurable ways in terms of health and sanitation.

As a cleaning agent, water by itself would only be able to dissolve some substances from the materials that have to be cleansed. These include polar compounds such as acids, bases, and salts; others that do not dissolve in water will be left on the fabric. Washing a worn T-shirt for, example, with water alone, would only remove the substances such the urea and the ions that got on it from our sweat. The fatty sebum and other non-polar oily or greasy compounds would still remain on the fabric. The use of soap along with water allows for these polar as well as non-polar substances to be washed away from fabrics. Soap molecules bond with both these non-polar particles on fabrics as well as with water molecules. In this way, water is able to wash away non-polar compounds and other substances that have to be cleansed from fabrics.

Nevertheless, the cleansing action of soap can be negatively affected by some dissolved ions in water, such as $Ca^{2+}_{(aq)}$ and $Mg^{2+}_{(aq)}$. These ions react with the soap molecules precipitating them out of solution to form soap scum. This process not only wastes the soap, but the soap scum formed on it sticks to the clothes and blocks drainage pipes. Repairs to pipes and removal of scums from clothes can be very costly. Detergents molecules are not affected by these dissolved ions since they do not form precipitates with them. However, non-biodegradable detergents can cause problems for the environment by accumulating there.

This chapter explains how soaps and detergent molecules are able to cleanse fabrics of both polar and non-polar substances. It also investigates how $Ca^{2+}_{(aq)}$ and $Mg^{2+}_{(aq)}$ ions affect the actions of soap and detergents and how water is treated to remove these ions.

15.1 Soaps and Detergents (Surfactants)

Soaps and detergents are compounds that have been synthesized for cleaning. The following are examples of the formulas for a soap and a detergent molecule respectively.

$C_{17}H_{35}COO^-\ Na^+$
Sodium stearate (**Soap**)

Soaps are the sodium or potassium salts formed from various fatty acids such as stearic acid, linoleic acid and palmitic acid. etc.

Examination of these two compounds shows that each has one end with a long, neutral hydrocarbon chain, while the other end has a charge. Since the hydrocarbon chain is non-polar, it is described as **hydrophobic** (water-repelling).

$C_{12}H_{25}$ —⬡— $SO_3^-\ Na^+$

Sodium p-dodecyl benzene sulphonate

(Detergent)

Detergents are the sodium or potassium compounds of modified alkyl groups.

The other end which is ionic in nature is described as **hydrophilic** (water-loving). The Na$^+$ ion is only a spectator and thus plays no active part in the cleaning process.

How do Soaps and Detergents Clean?

When detergent or soap molecules are placed in water, the figure to the right illustrates what happens. Their hydrophobic hydrocarbon chains attract each other while their polar hydrophilic ends repel each other sideways but are attracted to water molecules by dipole-dipole forces. This cluster of hundreds of these molecules in this way forms **spherical** particles called micelles that dissolve in water.

If non-polar substances such as grease or oil, present on our skin or clothing they can be cleansed by both soaps and detergents. They do so by having the non-polar ends of their molecules dissolve the non-polar grease or oil particle, and their polar ends attract the water molecules. In this way, the non-polar oil or grease particles connect to the soap molecules which in turn connect to the water molecules. As the water molecules move they take away the non-polar subtances with them. These processes are illustrated by the following figures:

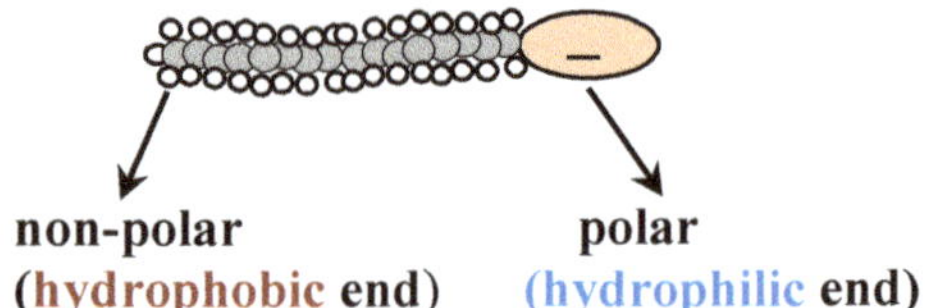

Figure 15.1. Generic model of a soap or detergent molecule.

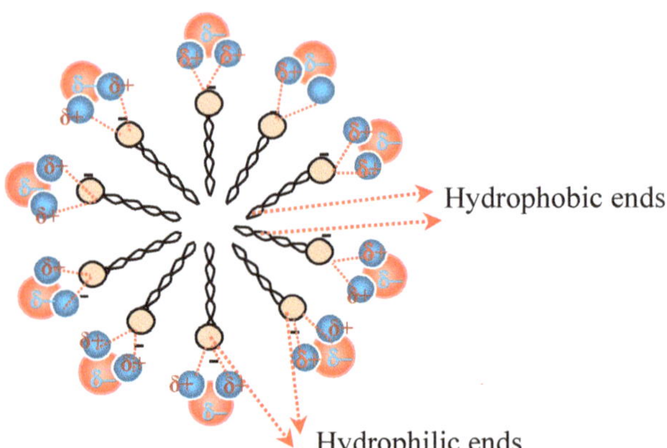

Figure 15.2. A detergent micelle in water.

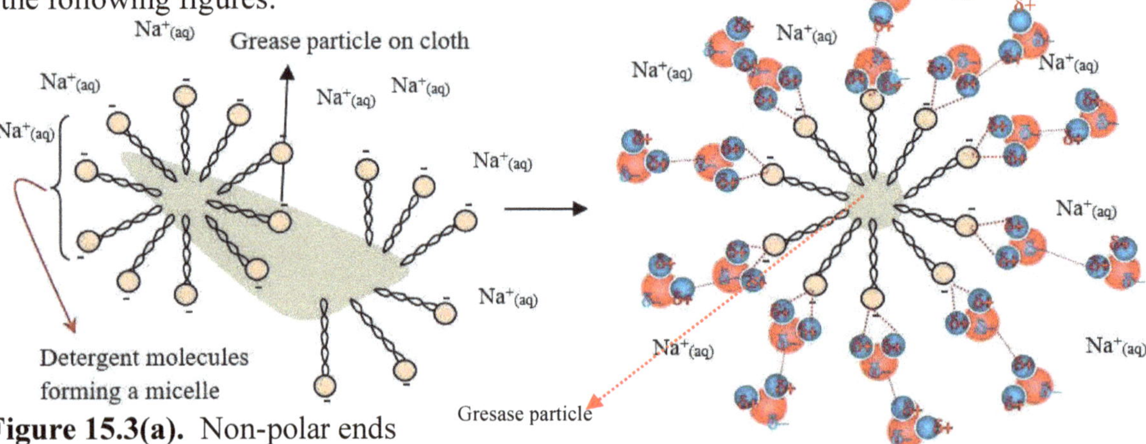

Figure 15.3(a). Non-polar ends of detergent molecules dissolving grease a particle on cloth.

Figure 15.3(b). A micelle with dissolved grease particle attracted to water molecules by dipole-dipole forces.

15.2 Hard and Soft Water

One of the number of ways that water becomes impure is by the presence of dissolved mineral ions such as Mg^{2+}(aq), Ca^{2+}(aq), Fe^{2+}(aq), Fe^{3+}(aq), Mn^{2+}(aq), etc. These ions invariably dissolve in ground water as it travels through different types of soils and rocks. One of the tasks of people using ground water, is to remove these ions. Inevitably, even after municipalities have treated their water in their endeavor to purify it, small quantities of these ions escape removal. The presence of these dissolved ions can cause health and industrial problems for people who use such type of water for drinking and washing. Hardness that is caused to water is one such type of problems. **Temporary hardness** is caused by the presence of Ca^{2+}(aq) in the form of Ca(HCO$_3$)$_{2}$(aq).

Causes of Hardness

In the **absence** of $Mg^{2+}{}_{(aq)}$ and $Ca^{2+}{}_{(aq)}$ ions, a soap solution will form a lather easily. Such water is described as **soft**. The amount of lather formed is an indication of the soap's capability to cleanse. If water contains dissolved $Mg^{2+}{}_{(aq)}$ and $Ca^{2+}{}_{(aq)}$ ions, they react with the soap molecules to form precipitates. This removal of the soap molecules from solution makes it difficult for the solution to lather. Such water is said to be **hard**. These reactions not only represent a waste of soap, but the precipitates formed are nuisances in many ways; they cling on to the clothes, stick to bathtubs and are responsible for pipe scales deposits. $Mg^{2+}{}_{(aq)}$ and $Ca^{2+}{}_{(aq)}$ ions react with soap molecules as follows:

$$Mg^{2+}{}_{(aq)} + 2C_{17}H_{35}COO^{-}{}_{(aq)} \longrightarrow Mg(C_{17}H_{35}COO)_{2(s)}\ (Scum)$$

$$Ca^{2+}{}_{(aq)} + 2C_{17}H_{35}COO^{-}{}_{(aq)} \longrightarrow Ca(C_{17}H_{35}COO)_{2(s)}\ (Scum)$$

15.3 Water Softening

Water softening involves the removal of $Mg^{2+}{}_{(aq)}$ and $Ca^{2+}{}_{(aq)}$ ions from water. There are a number of ways that this is done. These are as follows:

1. Use of **ion exchange** softener.

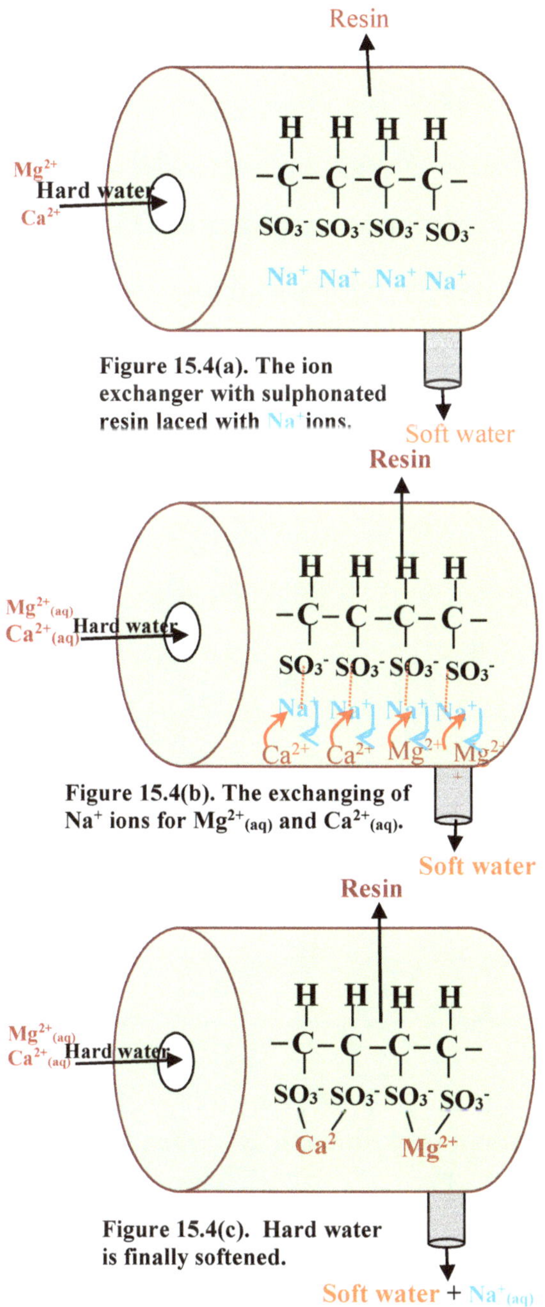

Figure 15.4(a). The ion exchanger with sulphonated resin laced with Na⁺ ions.

Figure 15.4(b). The exchanging of Na⁺ ions for $Mg^{2+}{}_{(aq)}$ and $Ca^{2+}{}_{(aq)}$.

Figure 15.4(c). Hard water is finally softened.

In this system, the hard water is passed through a tank filled with a non-soluble sulphonated resin which is in the form of beads. Before use, the resin is laced with a sodium salt, normally NaCl. This process allows $Na^{+}{}_{(aq)}$ ions to attach to the sulphonate groups (**Figure15.4a**).

As the hard water passes through the resin, the $Mg^{2+}{}_{(aq)}$ and $Ca^{2+}{}_{(aq)}$ are exchanged for the $Na^{+}{}_{(aq)}$ ions (**Figure15.4b**).

The water that exits the resin is 'softened' as it is now free of $Mg^{2+}{}_{(aq)}$ and $Ca^{2+}{}_{(aq)}$ ions, which are now stuck to the resin. However, the softened water is now higher in $Na^{+}{}_{(aq)}$ ion concentration (**Figure15.4c**). People who are sensitive to high $Na^{+}{}_{(aq)}$ ion concentration should be aware of this fact.

Periodically, the ion exchange softener must be cleansed of the absorbed $Mg^{2+}{}_{(aq)}$ and $Ca^{2+}{}_{(aq)}$ ions. This is done by flushing the ion exchanger with a concentrated solution of sodium chloride which allows the $Na^{+}{}_{(aq)}$ ions to reclaim their place on the resin by expelling the $Mg^{2+}{}_{(aq)}$ and $Ca^{2+}{}_{(aq)}$ ions out.

2. Addition of **sodium carbonate** to the hard water.

Since sodium carbonate is soluble in water, it dissociates as follows:
$$Na_2CO_{3(aq)} \rightarrow 2\,Na^+_{(aq)} + CO_3^{2-}_{(aq)}$$
Since the carbonates of calcium and magnesium are sparingly soluble in water, when the $Mg^{2+}_{(aq)}$ and $Ca^{2+}_{(aq)}$ ions present in hard water encounter the $CO_3^{2-}_{(aq)}$ ions, reactions take place that precipitate each of them out of solution. These reactions are as follows: $Mg^{2+}_{(aq)} + CO_3^{2-}_{(aq)} \rightarrow$ **$MgCO_{3(s)}$**
$$Ca^{2+}_{(aq)} + CO_3^{2-}_{(aq)} \rightarrow \mathbf{CaCO_{3(s)}}$$
The water is now 'softened' as it is now free of $Mg^{2+}_{(aq)}$ and $Ca^{2+}_{(aq)}$ ions. The left-over $Na^+_{(aq)}$ ions are spectator ions and have no effect on the soap.

Permanent Hardness: this refers to water that has $Mg^{2+}_{(aq)}$ and $Ca^{2+}_{(aq)}$ ions from sources other than $Ca(HCO_3)_{2(aq)}$. This type of hardness can only be softened by the methods described before.

15.4 Activity: **Investigating the hardness of water**

Question: What ions cause hardness of water?

Materials:

- distilled water
- 1.0 mol/L $MgCl_{2(aq)}$
- 1.0 mol/L $CaCl_{2(aq)}$
- 1.0 mol/L $Na_2CO_{3(aq)}$
- 1.0 mol/L $NaCl_{(aq)}$

* eye droppers
* test tubes
* test tube rack
* soap solution
* detergent solution

Procedure A:

1. Label four 10 mL test tubes: **1** distilled water, **2** $MgCl_{2(aq)}$, **3** $CaCl_{2(aq)}$ and **4** $NaCl_{(aq)}$.
2. Pour 5.0 mL of distilled water in the test tube labeled distilled water.
3. Pour 4.0 mL of distilled water into each of the other three labeled test tubes.
4. Pour 1.0 mL each of 1.0 mol/L $MgCl_{2(aq)}$, 1.0 mol/L $CaCl_{2(aq)}$ and 1.0 mol/L $NaCl_{(aq)}$ into the test tubes labeled $MgCl_{2(aq)}$, $CaCl_{2(aq)}$ and $NaCl_{(aq)}$, respectively.
5. Using the eye dropper, pour 10 drops of **soap solution** into each of the four test tubes.
6. Stopper each test tube, shake them vigorously and observe.

Evidence:

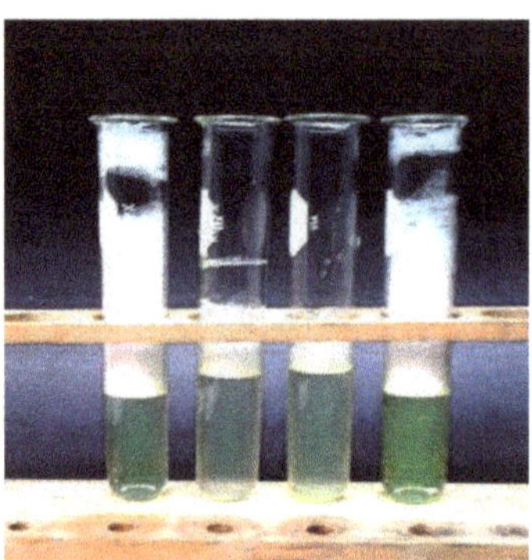

Figure 15.5. Test tubes showing results of reaction between soap solutions with distilled water, $MgCl_{2(aq)}$, $CaCl_{2(aq)}$ and $NaCl_{(aq)}$ after shaking.

Observation:

Table 15.1. Reaction of soap solutions with distilled water, $MgCl_{2(aq)}$, $CaCl_{2(aq)}$ and $NaCl_{(aq)}$.

Water samples	Observation after shaking
1 Distilled water + soap	
2 Distilled water + soap + $MgCl_{2(aq)}$	
3 Distilled water + soap + $CaCl_{2(aq)}$	
4 Distilled water + soap + $NaCl_{(aq)}$	

Analysis 1   **Having completed Table 15.1., use it to answer the questions that follow:**

1. In which test tubes was the most lather formed? T/I
2. In which test tubes was the most lather formed? T/I

3. In which test tubes were precipitates formed, if any, and why? T/I C

4. In which test tubes was hard water present? T/I

Conclusion:
1. Write a concluding statement about the ions responsible for hardness of water. T/I C A

2. If the soap was made from stearic acid, write the net ionic equations for the precipitates formed. T/I C

Procedure B:

1. Label four 10 mL test tubes: **1** distilled water, **2** $MgCl_{2(aq)}$, **3** $CaCl_{2(aq)}$ and **4** $NaCl_{(aq)}$.
2.   Pour 5.0 mL of distilled water in the test tube labeled distilled water.
3.   Pour 4.0 mL of distilled water into each of the other three labeled test tubes.
4. Pour 1.0 mL each of 1 mol/L $MgCl_{2(aq)}$, 1 mol/L $CaCl_{2(aq)}$ and 1 mol/L $NaCl_{(aq)}$ into the test tubes labeled $MgCl_{2(aq)}$, $CaCl_{2(aq)}$ and $NaCl_{(aq)}$, respectively.
5. Pour 2.0 mL of 1 mol/L $Na_2CO_{3(aq)}$ into each test tube.

Evidence:

Figure 15.6. Test tubes showing results of reaction between $Na_2CO_{3(aq)}$ with distilled water, $MgCl_{2(aq)}$, $CaCl_{2(aq)}$ and $NaCl_{(aq)}$.

Table 15.2. Reaction of $Na_2CO_{3(aq)}$ with distilled water, $MgCl_{2(aq)}$, $CaCl_{2(aq)}$ and $NaCl_{(aq)}$.

Observation:

Water samples	Observation
1 Distilled water + $Na_2CO_{3(aq)}$	
2 Distilled water + $Na_2CO_{3(aq)}$ + $MgCl_{2(aq)}$	
3 Distilled water + $Na_2CO_{3(aq)}$ + $CaCl_{2(aq)}$	
4 Distilled water + $Na_2CO_{3(aq)}$ + $NaCl_{(aq)}$	

Analysis 2

1. (a) In which test tubes were precipitates formed, if any. T/I

   (b) Write the net ionic equations for the reactions that produced precipitates. T/I C

Procedure B (cont'd):

6. Using the eye dropper, pour **10 drops of soap solution** into each of the four test tubes.

7. Stopper each test tube, shake them vigorously and observe.

Evidence:

Figure 15.7. Test tubes showing results of reaction between soap solution with distilled water, $MgCl_{2(aq)}$, $CaCl_{2(aq)}$ and $NaCl_{(aq)}$ after reaction with $Na_2CO_{3(aq)}$ and shaking.

Table 15.3 Reaction between soap solution with distilled water $MgCl_{2(aq)}$, $CaCl_{2(aq)}$ and $NaCl_{(aq)}$ after reaction with $Na_2CO_{3(aq)}$ and shaking.

Observation:

Water samples	Observation after shaking
1. Distilled water + soap + $Na_2CO_{3(aq)}$	
2. Distilled water + $Na_2CO_{3(aq)}$ + soap + $MgCl_{2(aq)}$	
3. Distilled water + $Na_2CO_{3(aq)}$ + soap + $CaCl_{2(aq)}$	
4. Distilled water + $Na_2CO_{3(aq)}$ + soap + $NaCl_{(aq)}$	

Analysis 3

1. Is there a difference in the pattern of lather formed compared to that of Procedure A? If so, in which test tubes? T/I
2. Was there any hard water present after the addition of $Na_2CO_{3(aq)}$? T/I
3. Why was the water softened? T/I C

Conclusion:

1. Write a concluding statement with an explanation concerning how hard water can be softened based on the results of procedure B, above. A C

Procedure C:

1. Label four 10 mL test tubes: 1 distilled water, 2 $MgCl_{2(aq)}$, 3 $CaCl_{2(aq)}$ and 4 $NaCl_{(aq)}$.
2. Pour 5.0 mL of distilled water in the test tube labeled distilled water.
3. Pour 4.0 mL of distilled water into each of the other three labeled test tubes.
4. Pour 1 mL each of 1 mol/L $MgCl_{2(aq)}$, 1 mol/L $CaCl_{2(aq)}$ and 1 mol/L $NaCl_{(aq)}$ into the test tubes labeled $MgCl_{2(aq)}$, $CaCl_{2(aq)}$ and $NaCl_{(aq)}$, respectively
5. Using the eye dropper, pour 10 drops of **detergent solution** into each of the four test tubes.
6. Stopper each test tube, shake them vigorously and observe.

Evidence:

Figure 15.8. Test tubes showing results of reaction between a detergent solution with distilled water, $MgCl_{2(aq)}$, $CaCl_{2(aq)}$ and $NaCl_{(aq)}$ after shaking.

Observation:

Table 15.4. Reaction of detergent solution distilled water, $MgCl_{2(aq)}$, $CaCl_{2(aq)}$ and $NaCl_{(aq)}$.

Water samples	Observation after shaking
1. Distilled water + detergent	
2. Distilled water + detergent + $MgCl_{2(aq)}$	
3. Distilled water + detergent + $CaCl_{2(aq)}$	
4. Distilled water + detergent + $NaCl_{(aq)}$	

Analysis 4

1. Was there any difference in the amount of lather formed in the various test tubes? T/I C
2. Was any precipitate formed? C
3. Are the detergent molecules affected by hard water? T/I C

Conclusion:

1. Write a concluding statement explaining the ability of detergent to work with hard water. C AT/I

Extension:

Synthetic detergents were manufactured to offset the problems that soap encountered with hard water. The detergent that was used in this experiment was an example of an alkyl benzene sulphonate. These are called anionic because of the negative charge on the polar end. There are other detergents that are classified as:

- Cationic
- Non-ionic
  Research the structures and chemical compositions of these, and how they work with hard water. Also find out what are their other related uses. T/I C A

Temporary Hardness

Temporary hardness is caused by the presence of calcium hydrogen carbonate in water. This type of hardness can be softened by boiling water that contains this compound. By boiling the water gets rid of the calcium ions in the form of calcium carbonate that precipitates out of solution. The following equation describes how this happens.

$$Ca(HCO_3)_{2(aq)} \xrightarrow{heat} CaCO_{3(s)} + H_2O_{(l)}$$

CHAPTER REVIEW SOLUTION AND SOLUBILITY (CHAPTER 12-15)

Matching

Match each term in the table below with the correct statements that follow

A	Water	G	Solvent
B	Solute	H	mol/L
C	Polar	I	Hydrogen bonds
D	Ionization	J	Saturated
E	Water of crystallization	K	Volumetric flask
F	Inter molecular	L	Electrostatic

1. This is used to dissolve the solute during the solution process.
2. This is called a universal solvent.
3. These hold the water molecules together in its liquid form.
4. This is dissolved by the solvent during the solution process.
5. The unit used to describe molar concentration.
6. These types of molecules attract each other by dipole-dipole forces.
7. The process by which a molecule breaks up into ions.
8. A solution that cannot dissolve any more solute at a specific temperature.
9. Hydrated salts have these.
10. Equipment used to accurately prepare solutions.
11. Force of attraction between molecules.
12. Force of attraction between the ions in an ionic compound.

True or False

Read each of the following statements and then decide if it is *True* or *False*.

1. Water is a bent polar molecule.
2. Grease can be dissolved by water.
3. The mixing of water and a concentrated acid can be highly exothermic.
4. Evaporating water from a solution changes its concentration.
5. Adding water to a solution decreases the solution's concentration.
6. In any solution of $ZnCl_{2(aq)}$, the concentration of $Zn^{2+}_{(aq)}$ and $Cl^-_{(aq)}$ will always be the same.
7. Hydrogen bonding is the major intermolecular force of attraction between water molecules.
8. When a negative ion is solvated by water molecules, it is surrounded by the oxygen ends of the water molecules.
9. The solvation of ions is by water molecules is an endothermic process.

10.	Water molecules auto ionize to produce $H^+_{(aq)}$ and $O^{2-}_{(aq)}$ ion.
11.	The process whereby the ions of an ionic compound separate is called ionization.
12.	The concentration of $NO_3^-{}_{(aq)}$ in 0.1 mol/L $KNO_{3(aq)}$ is half of that present in 0.1 mol/L $Ba(NO_3)_{2(aq)}$.
13.	The concentration of $Cl^-_{(aq)}$ in 0.2 mol/L $NaCl_{(aq)}$ is the same as in 0.1 mol/L $BaCl_{2(aq)}$.
14	If a solution of salt is heated for some time, the amount of salt in it increases.

Multiple Choice

Choose the letter that best answers the question.

1. Which of the following is **not** a solution?

a	carbon dioxide in water	d.	salt in water
b	copper in gold	e.	alcohol in water
c	sand in water		

2. Water molecules are attracted to each other by

a	London dispersion forces	d.	hydrogen bonds
b	ionic bonds	e.	both a and d
c	magnetic forces		

3. Which of the following processes are involved in dissolving an ionic compound?

a	breaking of the hydrogen bonds in water	d.	none of the above
b	breaking of the ionic bonds	e.	a, b and c
c	solvation of the ions		

4. Water dissolves ethyl alcohol because they are both

a	colourless molecules	d.	non-polar molecules
b	small molecules	e.	none of the above
c	polar molecules		

5. .In aqueous solutions, the solvent is always

a	air	d.	oil
b	alcohol	e.	none of the above
c	water		

6. A solution that has a relatively large quantity of solute dissolved in it is said to be

a	concentrated	d.	Strong
b	dilute	e.	a and c only
c	Saturated		

7. Which of the following factors determine the concentration of a solution?

a	the volume of solvent	d	the size of the particles
b	the temperature of the solution	e	a and c only
c	the number of moles particles dissolved		

8. The mass of sodium hydroxide dissolved in 500 mL of 0.20 mol/L $NaOH_{(aq)}$ is

a	4.0 g	d	1.0 g
b	0.20 g	e	none of the above
c	2.0 g		

9. A solution having one mole of a solute dissolved in 500 mL of solution will have a concentration of

a	1.00 mol/L	d	0.25 mol/L
b	2.0 mol/L	e	none of the above
c	0.50 mol/L		

10. What volume of water should be added to 250 mL of a 2.0 mol/L $HCl_{(aq)}$ solution to dilute it to 0.50 mol/L?

a	500 mL	d	750 mL
b	1.0 L	e	375 mL
c	250 mL		

11. If 4.00 g of $NaOH_{(s)}$ is dissolved in water to make 200 mL of solution, then its concentration will be

a	2.0 mol/L	d	0.50 mol/L
b	4.0 mol/L	e	none of the above
c	1.0 mol/L		

12. Barium hydroxide dissociates in water as follows: $Ba(OH)_{2(aq)} \rightarrow Ba^{2+}_{(aq)} + 2\,OH^{-}_{(aq)}$. If the concentration of the $Ba(OH)_{2(aq)}$ is 0.50 mol/L, the $[Ba^{2+}_{(aq)}]$ and $[OH^{-}_{(aq)}]$ will be

a	$[Ba^{2+}_{(aq)}] = 0.50$ mol/L	d	$[OH^{-}_{(aq)}] = 1.0$ mol/L
b	$[OH^{-}_{(aq)}] = 0.50$ mol/L	e	only a and d
c	$[Ba^{2+}_{(aq)}] = 1.0$ mol/L		

13. If 9.52 g of $MgCl_{2(s)}$ is dissolved in water to make 400 mL of solution, the $[Cl^{-}_{(aq)}]$ will be

a	0.50 mol/L	d	4.0 mol/L
b	1.0 mol/L	e	none of the above
c	2.0 mol/L		

14.

Hardness of water is caused by what ions?				
a	Na^+ and K^+		*d*	Mg^{2+} and Al^{3+}
b	Ca^{2+} and Mg^{2+}		*e*	none of the above
c	Cl^- and I^-			

15.

Copper(II) sulphate pentahydrate dissociates in water as follows: $CuSO_4.5H_2O_{(s)} + H_2O_{(l)} \rightarrow Cu^{2+}_{(aq)} + SO_4^{2-}_{(aq)}$. What mass of $CuSO_4.5H_2O_{(s)}$ is needed to prepare 250 mL of 0.200 mol/L of $CuSO_{4(aq)}$?

a	24.96 g		*d*	15.96 g
b	12.48		*e*	none of the above
c	7.98 g			

16.

In what volume of solution must 11.10 g of $CaCl_{2(s)}$ be dissolved in to produce 0.20 mol/L$Cl^-_{(aq)}$?

a	2.0 L		*d*	1.0 L
b	0.50 L		*e*	none of the above
c	4.0 L			

17.

Water does not dissolve grease because				
a	they are colourless molecules		*d*	they are both non-polar molecules
b	they are small molecules		*e*	water is polar but grease is non-polar
c	they are both polar molecules			

18.

It is dangerous to dump untreated sewage and other effluent in lakes and rivers because				
a	they contain poisons		*d*	all of the above
b	they can cause diseases		*e*	none of the reasons given
c	they can cause algal blooms			

CHAPTER 16
Water Chemistry

Chapter Content:

16.1 Importance of Clean Water

Of all the available water on the Earth's surface, less than 1% is fresh and suitable for drinking; approximately 97.5% is saline and is unsuitable for use, while the rest is stored in the polar icecaps. Since drinking water is one of our most important resources, it is important that it be used conservatively, recycled where possible, and that its natural sources remain unpolluted. Sewage and industrial wastes pose the most serious threats to drinking water supplies.

Raw sewage contains a host of substances which if discarded into the environment could cause catastrophic health and environmental problems. These include organic solids, grease and oils, pathogenic micro-organisms, nitrogenous wastes and heavy metals such as copper, mercury, lead and chromium; with each of these having their specific effects. These can enter the human body by ingestion of contaminated water or food. It is essential therefore that sewage be treated to remove those substances that are harmful to human health if ingested, or which could cause adverse environmental effects.

There are some notable cases of people dying or becoming sick by drinking water contaminated by different substances, or where the environment has been seriously compromised. In May of 2000, seven people died and hundreds more became very sick in Walkertown, Ontario, by drinking water that became contaminated by a deadly strain of the *E. coli* bacteria. It was discovered that the *E. coli* seeped from a farm runoff into a well; the source of water for the residents of Walkertown. The bacteria remained undetected only after it wreaked havoc on the residents of Walkertown. In 2016, in Flint, Michigan over 100 000 residents there, experienced elevated levels of lead in their drinking water. It was learnt that this problem started in 2014, when the water source to the city was switched from one source to another, in an effort to save the city millions of dollars. It was later discovered that some substance in the new source of water caused corrosion of the old lead pipes resulting in lead being leached from them into drinking water. For this reason, municipalities now treat water with anti-corrosion agents during its purification process. The new water source was also probably responsible for the 10 deaths and the 77 other people who suffered from Legionnaires disease. This disease is caused by drinking water contaminated with the *Legionella* bacteria. In August of 1995, in my native country of Guyana, the walls of a huge man-made reservoir containing millions of liters of sludge, laced with waste cyanide from OMAI gold mine, broke and polluted 80 miles of river water. Cyanide is used to extract gold from its ore and is stored afterwards in these reservoirs for its natural decomposition. This caused immeasurable destruction to fish and other wildlife; the lives of over 40 000 indigenous people who depended on the river for drinking water and wildlife were severely affected. This represented was one of the worst man-made disasters, in recent history in South America.

As the human population keeps increasing, so does the need for drinking water. It thus becomes incumbent upon us all develop a more caring attitude in terms of how much water we use to avoid wasting it, and to minimize its contamination from industries and other sources. Nations may eventually have to invest into costly technologies to recycle wastewater or to use solar energy to distil water from saline sources. One of the leading nations at this time in this technology is Israel that recycles up to, at least, 85% of its wastewater.

16.2 Sewage Treatment

Water that goes down our sinks, flushed down our toilets and from storms, together is called sewage. Sewage consists of a whole range of substances that include organic compounds, inorganic compounds and a host of pathogenic microbes. The aim of municipalities is to treat the sewage and the ensuing water to destroy the pathogens and to reduce the levels of the other dissolved substances to safe limits for ultimate disposal into the environment or use.

Sewage enters a network of underground pipes called sewers which finally collects into a large pipe from which it is pumped to the treatment plant. The following diagram depicts a schematic view of a typical sewage treatment plant.

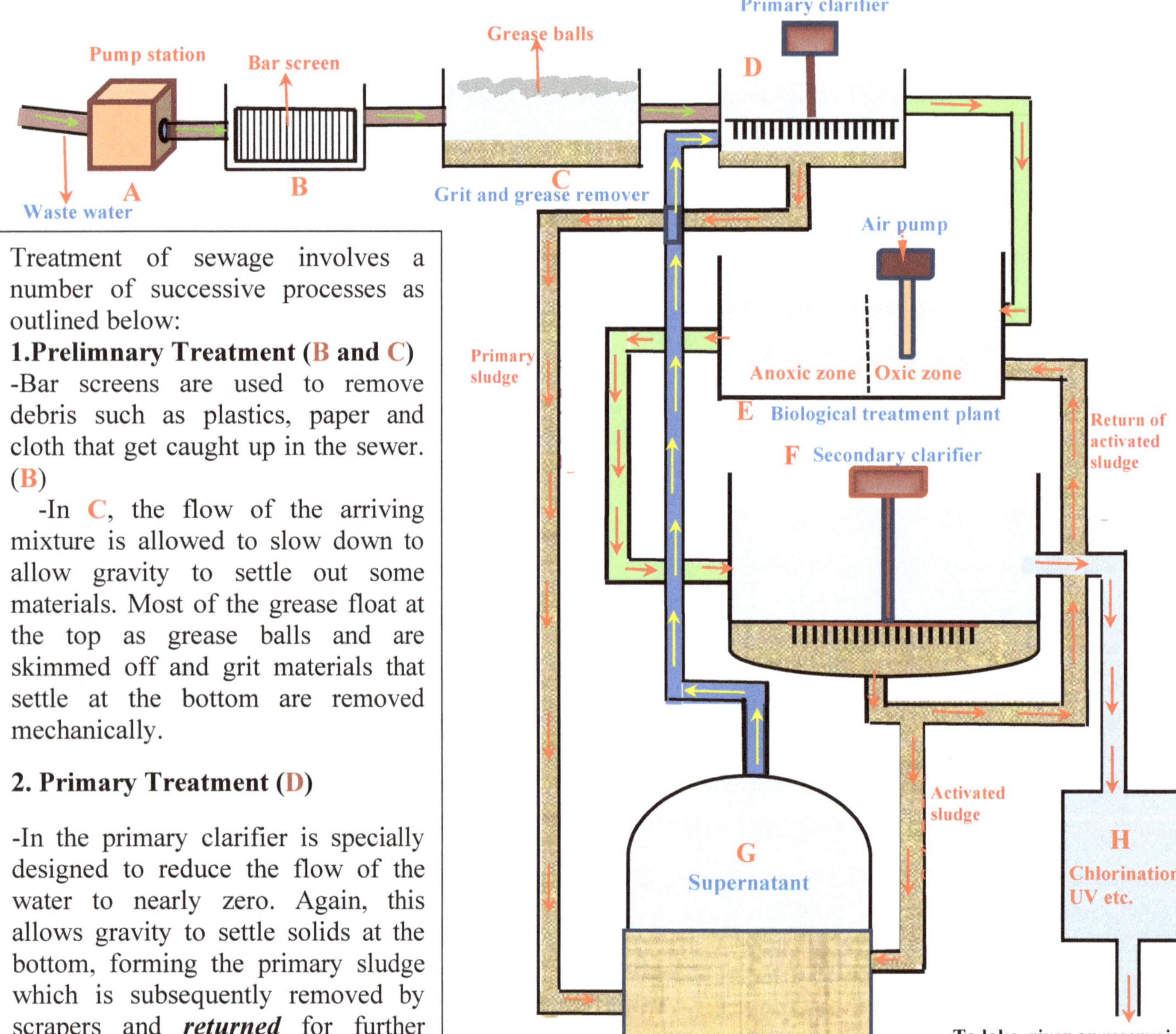

Treatment of sewage involves a number of successive processes as outlined below:

1. Prelimnary Treatment (B and C)

-Bar screens are used to remove debris such as plastics, paper and cloth that get caught up in the sewer. (B)

-In C, the flow of the arriving mixture is allowed to slow down to allow gravity to settle out some materials. Most of the grease float at the top as grease balls and are skimmed off and grit materials that settle at the bottom are removed mechanically.

2. Primary Treatment (D)

-In the primary clarifier is specially designed to reduce the flow of the water to nearly zero. Again, this allows gravity to settle solids at the bottom, forming the primary sludge which is subsequently removed by scrapers and **returned** for further treatment at point, G. Other lighter materials are skimmed off at the top. More than 50% of solids are removed at this point.

Figure 16.1. Schematic diagram of a sewage treatment plant.

3. Secondary Treatment (E and F).

Secondary treatment utilizes a host of specially selected *bugs* (bacteria and protozoans) to degrade organic wastes by feeding on them, converting them to other substances or by absorbing some inorganic substances.

The effluent from the primary clarifier(s) enters a huge circular basin which is inoculated with the bugs. This water still has lots of organic carbon, pathogens and inorganic compounds. To reduce these, ***dissolved oxygen gradients*** are established in the basin to facilitate the activities of microbes that are oxygen sensitive; some microbes use oxygen while others do not.

In the first zone (*oxic*) of the circular basin, warm air under high pressure is pumped in to encourage the proliferation of the ***aerobic bugs***. Warm air is used to create a conducive temperature and the oxygen necessary for uptimal microbial activities. With these suitable conditions, the bugs are then able to feed on the organic matter, thereby reducing their concentrations. In this zone also, the main nitrogenous animal waste, urea is first oxidized to ammonia, which is then changed, through a series of nitrifying aerobic bacteria, into nitrates ions.

$$\text{Urea} \longrightarrow NH_{3(aq)} \xrightarrow[\text{Nitrifying bacteria}]{O_2} NH_4^+ \xrightarrow[\text{Nitrifying bacteria}]{O_2} NO_2^- \xrightarrow[\text{Nitrifying bacteria}]{O_2} NO_3^-$$

Dissolved oxygen

Nitrate in water is still dangerous to human health and the environment. In infants, it can cause '***blue baby syndrome***' and in bodies of water it can cause algal bloom. More treatment of the water is therefore required to reduce the concentration of these and other undesirable substances.

The second zone is deliberately deprived of oxygen, and is called the (*anoxic zone*). This allows for the proliferation of ***anaerobic bugs*** which are able to feed on the organic carbon by using the **oxygen bonded in the NO$_3^-$ ions**, but **not from any dissolved oxygen.** As a result of this, the unhealthy nitrate ions are degraded, and the nitrogen and other gases produced in this process are released into the air.

$$\text{Food} + \text{Bugs} + NO_3^- \longrightarrow N_{2(g)} + CO_{2(g)} + \text{Energy}$$

Anaerobic bacteria ⟶ These bonded oxygen are used by the bacteria

After the water has spent about six hours in this basin, to allow the bugs to cleanse the water of organic carbon and some inorganic substances, it is then sent into the **secondary clarifier**. In this basin, the water flow is slowed down to allow any other insoluble materials to settle out, as what is called the ***activated sludge***.

The activated sludge which is composed of aggregates of bugs and other colloidal materials, together is called ***floc.*** In some water treatment plants, aluminum sulphate or iron sulphate are used to coagulate colloidal materials by reducing the mutual repulsion between their particles, making them clump together and settle out of the mixture. As the colloids coagulate and settle, they carry away with them some bugs and other contaminants that adhere to them; the removal of these substances from the water increases its transparency. The activated sludge is scraped out and sent to the sludge treatment plant, **G** where water is pressed out of it as a **supernatant.** The **supernatant** is subsequently *returned to the primary clarifier for purification.* In **G,** the sludge is further digested by anaerobic bugs, resulting in the production of other useful substances such as methane gas and fertilizer materials. *Some of the activated sludge, which is still rich in nutrients and microbes, is periodically returned to basin, E, to replenish any depleted bug population and to provide food for their sustenance. It is essential for the latter to happen for sustained biological degradation to occur in basin E.*

Tertiary treatment G

The water that leaves the secondary clarifier still has a load of pathogens that must be destroyed before it is released into reservoirs such as lakes or rivers. This is done by treating the water with pathogen- destroying agents such as ozone, chlorine, or UV; pH correction also takes place here.

    The Canadian Guidelines for Drinking Water Quality

Since water can be contaminated by various types of pathogen organisms, organic and inorganic substances which are harmful to human health, Health Canada, in its guidelines for *Canadian Drinking Water Quality* ***(https://www.canada.ca/en/health-canada/services/environmental-workplace-health/water-quality/drinking-water/canadian-drinking-water-guidelines.html)*** has set certain criteria that decide what contaminants are safe to be in drinking water and what are not. The guidelines have established strict controls on the levels of contaminants permitted to be in water, based on the known health effects associated with each contaminant, exposure levels, and the availability of treatment. The level for each contaminant that is deemed to be safe in drinking water is referred to as its maximum acceptable concentration (MAC). Water having any contaminant that exceeds this maximum concentration is deemed unsafe to drink. The guidelines may be revised from time to time to address any issues or problems that may arise, or from the findings of ongoing research with respect to the various contaminants. The following table gives a few chemicals, their MACs and their sources.

Table 16.1. MACs for a few chemicals normally found in Canadian drinking water

Chemicals	Barium	Lead	Fluoride	Nitrate	Benzene
MACs	1.0	0.010	1.5	45.0	0.005
Sources	Industrial wastes and spills	Leaching from lead pipes and other lead appliances	Erosion of rocks and soil	Run off from farms, fertilizer use and sewage	Industrial spills

Exercise 16.1.

1. Referring to the diagram of the sewage treatment plant, answer the following:

a) What could be concluded about the plant operation if the water collected after treatment at point **H** is still found to have a high population of the *E.Coli* bacteria and a high concentration of nitrate? How could this problem be remedied? T/I CA

b) What might happen to the water quality if an inadequate supply of air is pumped into the oxic zone? CT/I

c) If the water stays too long in the primary clarifier, what might happen with dissolved oxygen? How might this impact on the cost for pumping air into the oxic zone? T/I AC

d) What might happen in basin E if enough activated sludge is not returned? A C

2. The following table shows the relative level of a few substances when water samples from two different sites were analyzed.

Table 16.2. Comparison of substances in two sample of water with contaminants.

Sites	Nitrates	*E.Coli*	Dissolved O$_2$	Ammonia	Turbidity
A	high	High	low	high	high
B	low	Low	high	low	low

Based on the relative concentrations of the substances, what can be concluded about the sources of the water? Justify your answer. AC

3. The following table shows the levels of a few chemicals and bacteria found in two sample of drinking water; one of which is not properly purified.

Table 16.3. Levels of some substances in two samples of drinking water.

Substances	Lead	Nitrates	Barium	E.coli
Sample A	0.015	120	2.0	high
Sample B	0.005	30	0.8	none

a) Compare their values to their MACs in table 16.2 and then decide which sample of water is unsafe to drink. AT/I

b) Described how you would identify each of the *chemicals* present in the water. T/IC

c) Research the health effects if any, if water containing of each of the four components is consumed over an extended period of time. T/IA

4. Imagine that you were living in a place where there is no source of drinking water except that from a running river. Its water is turbid but has no dangerous dissolved chemicals such as lead, mercury, and other harmful organic compounds such as benzene, phenols etc, in it. All you have is some aluminum sulphate, materials to make a filter bed and a source of heat. Describe how you could obtain a relatively safe sample of water to drink. Explain the significance of each step in your procedure. K/U T/I C

16.4 Water Purification

Communities obtain their water from two sources: **ground water** and **surface water**. *Ground water* is what percolates through the soil and accumulates below the water table. It is obtained by drilling wells and then pumping the water out. **Surface water** is that which collects in lakes, oceans, rivers and streams or reservoirs.

Before water from either source can be used for drinking, it must be must be treated to remove any unwanted ions or suspended solids and to destroy any pathogenic microbes that might be present in it. This involves a number of processes as shown in Figure 16.2.

Coagulation and flocculation

In this process, coagulants are added to allow dissolved solids to clump and form larger particles called *floc.* The coagulants commonly used are sulphates of iron and aluminum.

Sedimentation

The neutralized heavy particles from the coagulation process, settle out of solution as a sludge by the force of gravity. This process is called *sedimentation.* The sediments (sludge) are remove the sedimentation tanks mechanically.

Filtration

After all the **floc** has settled to the bottom of the sedimentation tank, the water is then passed through a filter bed to remove any remaining dissolved solids, odors, some chemicals and some microscopic organisms. This filter bed is composed of charcoal, gravel and sand particles of selected sizes to ensure that most of the contaminants are removed. The filtered water is the sent for further treatment.

Disinfection

To ensure that there are no remaining harmful microorganisms present in the filtered water, it is treated with one of many types of disinfectants. These include chlorine, chloramines, UV, chlorine dioxide or ozone. Some anticorrosion agents may be added also to prevent corrosion of lead pipes which may lead to leaching of harmful lead into the water. In affluent countries, lead pipes are gradually being replaced by ones that are safer.

Quality Control

Turbidity, Temperature and Pathogens check.
 The water is monitored to ensure that it is free of pathogens and suspended solids, and that its temperature is suitable for consumption.

pH Correction
It is important to balance the pH of water as any imbalance could have adverse effects on human health and the environment. If the water is acidic it is neutralized by calcium carbonate or if it is basic it neutralized by adding acids. After this stage the water is now safe for drinking and it can be pumped into homes or where ever it is required.

All of these checks take place in reservoirs F and G (Fig 6.2 p.265).

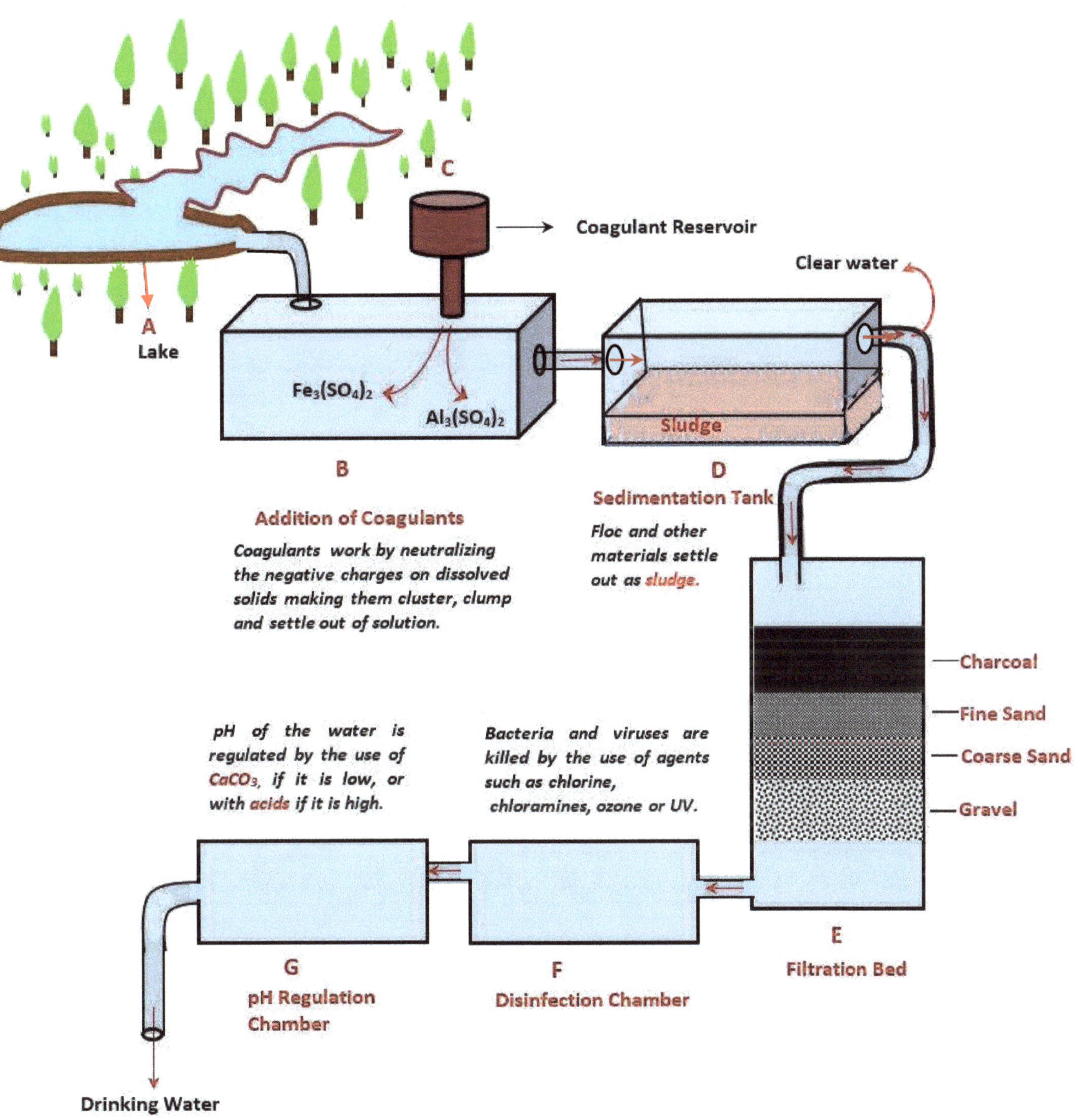

Figure 16.2 Schematic diagram of a water treatment plant

Water being a universal solvent, has the ability to dissolve a wide range of substances. Because of this, naturally pure water does not exist. The majority of **naturally dissolved substances** pose little health risk to humans and the danger to the environment if their concentrations are maintained within certain acceptable range. However, it is human activities coupled with usage of the wide range of synthetic substances that are mostly responsible for the introduction of harmful substances (pollutants) to bodies of water, the land and air. In spite of this, some pollutants can be tolerated by humans with minimum health risks if their concentrations are contained within certain range, but some of these can be deleterious in very minute concentrations. It therefore incumbent upon us, that we become more aware of the nature of these pollutants, and to take the necessary steps to control their usage and disposal, in the safest possible ways. Undesirable substances find their way into water in a number of ways; these are outlined as follows.

- As water flows over the earth's surface or percolates through it, it dissolves (leaches) any substances that are soluble in it. These may include carbonates, sulphates, bicarbonates, calcium, magnesium, iron, potassium, etc.

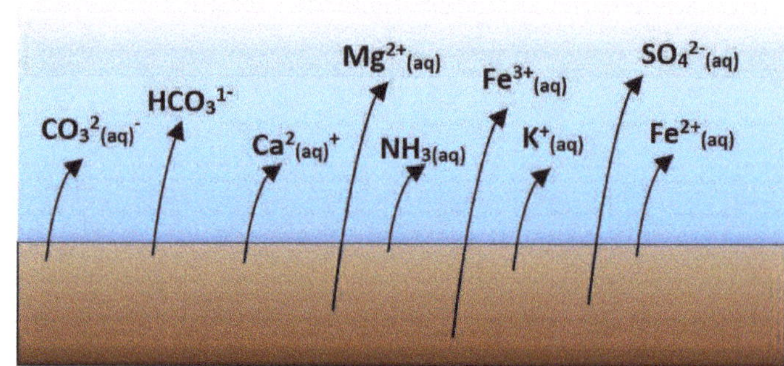

Figure 16.2. Ions being leached out of soil by water

- From industrial wastes and untreated sewage that are deliberately or inadvertently dumped into lakes and rivers and other bodies. These substances may include pesticides, organic wastes, micro-organisms (bacteria, viruses) mercury, lead, phosphates and nitrates, arsenic, ammonia, lye, chloride, sodium, barium, hydrocarbons (benzene, waste oils (etc).

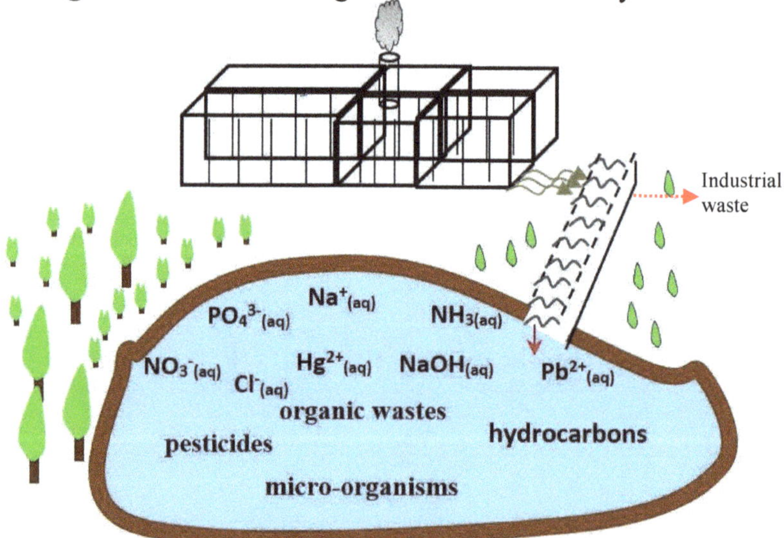

Figure 16.3. Industrial wastes being dumped into a lake.

- From runoff from agricultural farms and feeding lots. These substances include primarily phosphates, nitrates, pesticides, dissolved solids, urea and ammonia from artificial fertilizers and from the breakdown of faeces.

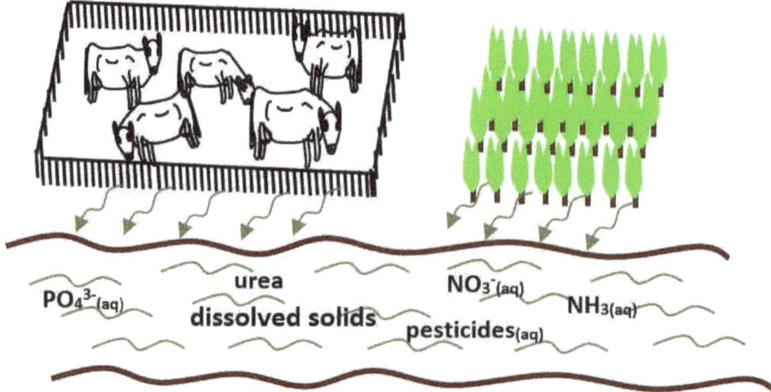

Figure 16.4. Runoffs from a feeding lot and a farm, entering a river.

- From precipitations such as acid rain, hail and snow. These include carbonates, bicarbonates, sulphates, sulphites, nitrites, nitrates and hydrogen ions.

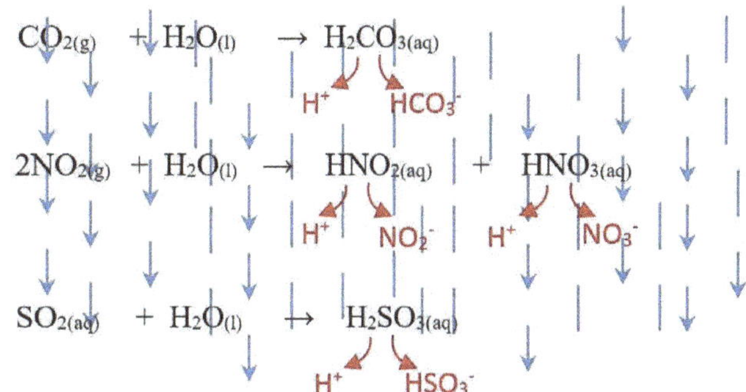

Figure 16.5. Ions from acid precipitation entering the environment.

All of these substances affect water quality in a number of ways:

- Calcium and magnesium ions cause hardness of water.
- Carbonates, bicarbonates and lye increase alkalinity of water.
- Sodium can affect hypertensive patients, raising their blood pressure.
- Pesticides, lead, mercury, cadmium, arsenic and thalium are poisonous substances.
- Micro-organisms such as coliform bacteria are infectious and can be deadly.
- Dissolved solids affect water transparency and photosynthesis.
- Hydrogen ions increase the acidity of water and soil.
- Nitrates and phosphates cause algal blooms.

16.6 Organic Wastes and Fertilizers on Dissolved Oxygen

Biodegradable organic wastes, such as those found in sewage, affect the amount of dissolved oxygen in water, directly or indirectly. Organic wastes serve as food for some microbes such as bacteria. The presence of organic wastes in water thus quickens the growth of these bacteria. As bacteria act on organic wastes, they break them down (decompose) into simpler compounds and minerals, that they use for respiration and growth. Some of these are recycled to the environment.

Decomposition of organic wastes has a direct impact on the amount of dissolved oxygen in water. As aerobic bacteria feed on organic wastes, they inevitably use some of the dissolved oxygen, that is normally used by fish and other beneficial aquatic organisms, for cellular respiration. The presence of organic wastes in water thus increases the demand for dissolved oxygen. This concept is referred to as **BOD** (*biological oxygen demand*). Scientists use the amount of organic carbon in water as a measure of BOD since the breakdown of these is commensurate with oxygen usage by microbes.

The released phosphates and nitrates from the decomposition process, along with any added manufactured fertilizers to water, may lead to excessive growth of algae and other aquatic plants. This rapid excessive growth is referred to as "**algal bloom**". Apart from blocking light for photosynthesis in the body of water, the algae eventually die because of competition for space and nutrients. The dead algae create more organic matter for decomposition; bacterial population and their actions skyrocket, further increasing the BOD of the water. In severe cases, dissolved oxygen concentration may become so low, that it leads to extensive fish death.

16.7 Activity: The Effects of Biodegradable Organic Matter on Dissolved Gases ($O_{2(g)}$, $CO_{2(g)}$)

Background Information

Aerobic bacteria undergo cellular respiration according to the following equation:

$$C_6H_{12}O_{6(aq)} + 6\,O_{2(aq)} \longrightarrow 6\,H_2O_{(l)} + 6\,CO_{2(g)} + \text{Energy}$$

During cellular respiration in aquatic organisms, the carbon dioxide gas that is produced dissolves in water to produce the weak carbonic acid. Carbonic acid is detected by bromothymol blue indicator (**BTB**). Depending on the concentration of carbonic acid formed, the BTB undergoes a series of colours changes, ranging from blue to light green to light yellow to yellow. This can be demonstrated by exhaling gently through a solution of B*TB, using a straw.*

In normal oxygenated water, methylene blue has a blue colour. However, if the oxygen content of water diminishes, some methylene blue molecules decolourise. Its final colour in water would thus depend on how much oxygen is removed from the water; if all the oxygen is removed, all the methylene blue molecules go colourless and so does the entire solution. The degree of decolourisation of methylene blue will be used to monitor the amount of the amount of dissolved oxygen in this activity.

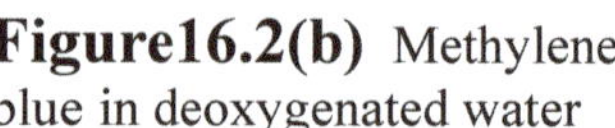
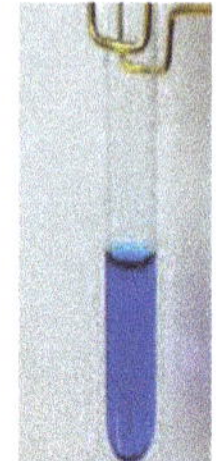

Figure16.2(b) Methylene blue in deoxygenated water

Figure 16.2(a) Methylene blue in oxygenated water

In this activity, decaying rice will be used as the organic material and the bacteria present on the rice will be the decomposers.

Materials:

- Ten test tubes
- Two test tube racks
- Ten stoppers
- Decaying cooked rice
- Methylene blue indicator with dropper
- Bromothymol blue indicator with dropper
- Distilled water

Procedure:

1. Place five of the test tubes in one rack and the rest in the other rack
2. Label these 1, 2, 3, 4, 5, 6, 7, 8, 9 and10.
3. Put two drops of methylene blue into each of the first *five* tubes.
4. Leave test tube 1 alone.
5. In test tube 2 place 2 grains of decaying rice.
6. In test tube 3 place 4 grains of decaying rice.
7. In test tube 4 place 6 grains of decaying rice.
8. In test tube 5 place 8 grains of decaying rice.
9. Pour approximately 8 mL of distilled water into each of the *five* test tubes (1-5); adjust the water to make all volumes equal.
10. Stopper each of these five test tubes firmly and observe over the next few days.
11. Put four drops of bromothymol blue into each of the other five test tubes (6-10).
12. Leave test tube 6 alone.
13. In test tube 7 place 2 grains of decaying rice.
14. In test tube 8 place 4 grains of decaying rice.
15. In test tube 9 place 6 grains of decaying rice.
16. In test tube 10 place 8 grains of decaying rice.
17. Pour 8 mL of distilled water into each of these test tubes; adjust the water to make all the volumes equal in each test tube.
18. Stopper each of these five test tubes firmly and observe them over the next few days.

Do not shake the test tubes as it would affect the amount of dissolved gases and give false results.

Hypotheses:

Make hypotheses as to what you would expect to observe in each of the test ten tubes.

Results:

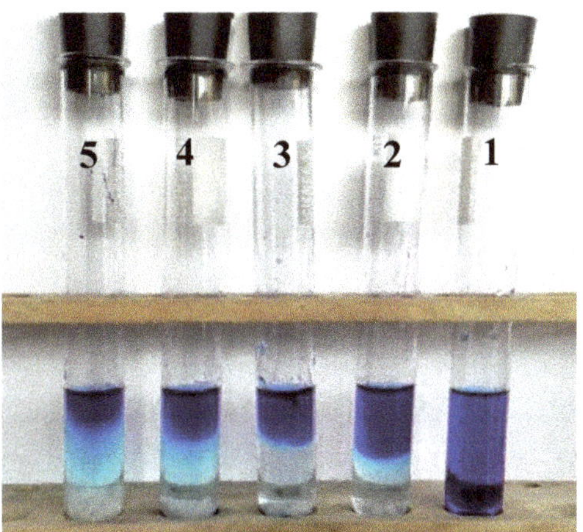

Figure 16.3(a) Methylene blue solution with decaying rice after a few days.

Figure 16.3(b) Bromothymol blue solution with decaying rice after a few days.

Analysis 4   *(With reference to test tubes 2,3,4,5,7,8, 9 and 10)*

1. a) In which test tube was the least amount of carbon dioxide produced? T/I

    b) In which test tube was the most amount of carbon dioxide produced? T/I

    c) In which test tube was the least amount of oxygen used? T/I

    d) In which test tube was the most amount of oxygen used? T/I

    e) Why was it important to stopper the test tubes firmly? K/UC

Conclusion:

Make a concluding statement about the relationship between the amount of organic matter and the amount and dissolved oxygen. K/UC

Evaluation:

Do the results agree with your hypotheses? K/U

Additional questions:

2. a) What was the purpose of test tubes 1 and 6? K/U

    b) What would eventually be observed if the test tubes were observed for over an extended period of time? T/I

    c) Why did the changes start from the bottom of the mixtures? T/I K/U

    d) Was there any correlation between the amount of oxygen used and the amount of carbon dioxide produced? Explain your answer. K/UC

    e) What might be observed if the all the stoppers were removed from the test tubes? T/I

    f) What relationship exists between the amount of organic matter and the number of bacteria? T/I

3. Why people should refrain from dumping organic wastes into bodies of water?

   What may happen if they do not refrain from this practice? K/UA

4. Why do municipalities construct spill weirs along the courses of rivers polluted with organic wastes? T/I

5. Composting is one of the various processes by which organic wastes can dealt with that minimises the risk of water pollution. How would you convince the people in your neighbourhood about having composts for their homes? K/UCA

   5. Gasification is another process of using carbonaceous organic wastes. Research how this process works and lists the disadvantages and benefits of this method of using wastes. T/I A

   6. People use air pumps to continuously aerate their fish tanks. What are the benefits of this practice, and what might happen if the they discontinue this practice? A T/I

CHAPTER REVIEW: SEWAGE TREATMENT (CHAPTER 16)

Matching

Match the item in the table below with the statements that follows

A	Sewage	G	Aerobic
B	Oxic zone	H	Anaerobic
C	Anoxic zone	I	Algal bloom
D	BOD	J	Eutrophic
E	Bugs	K	Biodegradable
F	Activated sludge	L	MAC

1. A condition that results when dissolved oxygen in a body of water is severely reduced.
2. A mixture of faeces and wastewater from domestic and industrial use.
3. In sewage treatment, this intermediate waste consists of organic material wastes and living   microbes.
4. These substances can be broken down by bacteria and fungi.
5. This term is used to collectively describe all the microbes utilized to feed on sewage.
6. The maximum acceptable concentration of any substance permitted to be in drinking water.
7. These types of microbes use dissolved oxygen to feed.
8. These types of microbes extract the oxygen bonded in nitrate ions to feed.
9. The zone in the biological treatment basin that is well aerated.
10. The zone in the biological treatment basin that is deprived of oxygen.
11. A measure of the amount of dissolved oxygen needed by microbes to feed on organic carbon.
12. This type of condition is normally produced when lots of fertilizer enters bodies of water such as rivers and lakes.

True or False

Read each of the following statements and then decide if it is *True* or *False*.

1. Too much fertilizer in water cannot affect the amount of dissolved oxygen.
2. Too much sodium in drinking water is not healthy for hypertensive people.
3. Raw sewage is laced with a host of pathogens.
4. All inorganic substances are removed from water after treatment.
5. Sewage consists mainly of water.
6. Acid rain raises the pH of water.
7. The use of UV is less eco-friendly than chlorine for water treatment.
8. Coagulants allow colloidal materials to clump together and settle.
9. A fish tank operating without an air pump could become anoxic.
10. Turbid water has less dissolved solids than transparent water.

Multiple Choice

Choose the letter that best answers the questions.

1. Which of the following are examples of coagulants?

a	Zinc Chloride	d	Ferric sulphate
b	Aluminum sulphate	e	b and c
c	Calcium Nitrate		

2. In the absence of oxygen methylene blue goes

a	blue	d	colourless
b	yellow	e	to none of the above
c	pink		

3. Which of the following substances are non-biodegradable?

a	plastics	d	oil spills
b	metals	e	all except c
c	wood		

4. In the presence of lot of carbon dioxide bromothymol blue goes

a	blue	d	colourless
b	yellow	e	to none of the above
c	pink		

5. Which of the following substances are biodegradable?

a	Fruits	d	Organic wastes
b	Fallen leaves	e	All of the above
c	Dead animals		

6. Which of the following are examples of heavy metals found in sewage?

a	Lead	d	Cadmium
b	Mercury	e	All of the above
c	Copper		

7. Drinking contaminated water can cause

a	cholera	d	hepatitis A
b	dysentery	e	all of the above
c	guinea worms		

8. Which of the following would **not** affect the amount of dissolved oxygen?

a	rise in temperature	d	fertilizers
b	organic wastes	e	windy conditions
c	arsenic		

9. Dumping of Lye in water

a	reduces dissolved oxygen	d	lowers the pH
b	increases dissolved carbon dioxide	e	has no effect
c	raises the pH		

10. An increase in the turbidity of water

a	lowers the rate of photosynthesis	d	changes pH
b	increases transparency	e	a and d
c	decreases transparency		

11. Which of the following incidents was associated with the *E. Coli* bacteria?

a	Flint drinking water	d	The Chernobyl disaster
b	OMAI gold mine	e	None of the above
c	Walkerton drinking water		

12. Water contaminated with heavy metals such as mercury enter our bodies

a	by drinking the contaminated water	d	by inhalation
b	ingestion via food chains	e	a and b
c	by bathing in the water		

13. Aluminum sulphate and ferric sulphate are used to purify water because they

a	kill pathogens	d	regulate the pH
b	remove odours	e	do all of the above
c	coagulate colloidal materials		

14. Allowing polluted water in rivers to flow along spill weirs facilitates purification by

a	allowing the water to flow fast	d	allowing dissolve solids to settle
b	allowing more aeration to take place	e	none of the above
c	allowing more photosynthesis to happen		

15. Activated charcoal is used in filters to purify water by removing

a	carbon dioxide	d	heavy metals
b	odours	e	acidity
c	pathogens		

CHAPTER 17
Acids and Bases

Chapter Content:

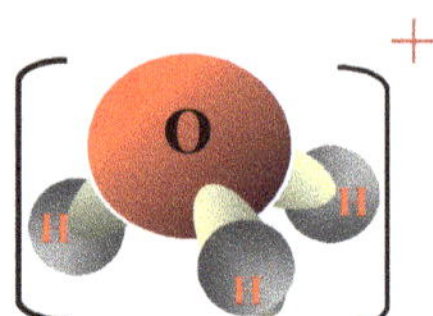

Do you know that if the pH of your blood changes by a value of 0.2 above and below its normal value of 7.4, it could be fatal? To maintain a constant blood pH, our bodies have a number of buffering systems present in the kidneys and lungs. A number of acid-base theories have been proposed to better our understanding of acids and bases. Through experimentation, acids and bases have been categorized as being either weak or strong. Weak acids and bases have different properties from strong ones. Knowledge of these makes people become aware of the potential benefits and dangers when dealing with these chemicals. For example, to treat acid reflux or heart burns medicines containing weak bases are administered to patients: strong bases would be too caustic to the stomach. For similar reasons, only weak acetic acids are used to make pickles; strong acids would cause stomach burns. Acid-base chemistry has far-reaching applications; acids are used industrially to produce fertilizers, explosives, detergents, batteries, and as food additives to enhance their flavors or act as preservatives. Bases are used for making soaps, antacids, cement and fertilizers. Many household commodities contain either acids or bases as their active ingredients. These include drain cleaners, bleach cleaners and glass cleaners, among others. Both acids and bases are corrosive and extreme care must be taken when using these alone, or when they are reacting with other substances. Appropriate protective equipment such as gloves, goggles and aprons must be used in these cases.

17.1 Acid-Base Theories

While there are a number of acid-base theories, only two of these will be briefly described in this chapter.

The Arrhenius Theory of Acids and Bases

In 1887, a Swedish chemist, Svante Arrhenius formulated his acid-base theory. For this theory to be understood, it is first important to understand how electrolytes conduct electricity, as this concept was the basis on which Arrhenius formulated his theory.

Electrolyte:

An electrolyte is any substance that conducts electricity in its molten state or in its aqueous form.
How does an electrolyte conduct electricity?

For any electrolyte to conduct electricity, it must allow a continuous flow of electrons from the cathode to the anode, **within the solution,** as shown in Figure 17.1 below. Salt solutions are examples of electrolytes. As can be recalled, when ionic compounds dissolve in water, they **dissociate** into their respective ions. $CuCl_{2(s)}$, for example, dissociates in water according to the following equation:

$$CuCl_{2(s)} + H_2O_{(l)} \rightarrow Cu^{2+}_{(aq)} + 2\,Cl^{1-}_{(aq)}$$

The $Cu^{2+}_{(aq)}$ ions and $Cl^{1-}_{(aq)}$ ions, now freely mobile, travel to the elctrodes due to electrostatic forces of attraction; the $Cu^{2+}_{(aq)}$ ions move to the negative cathode, and the $Cl^{1-}_{(aq)}$ ions move to the negative anode. Since negatively charged ions, such as $Cl^-_{(aq)}$, are attracted to the **anode,** they are called **anions.** And since positively ions, such as $Cu^{2+}_{(aq)}$, are attracted to the **cathode,** they are called **cations.** Each different electrolyte will have its own specific type of cations and anions.

The following diagram illustrates how the movement of these ions enables this aqueous salt solution conduct to electricity.

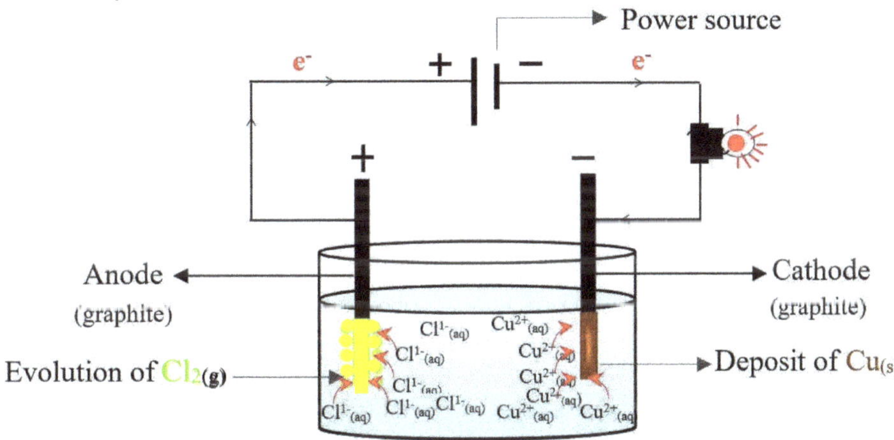

Figure 17.1. A copper (II) chloride solution conducting electricity.

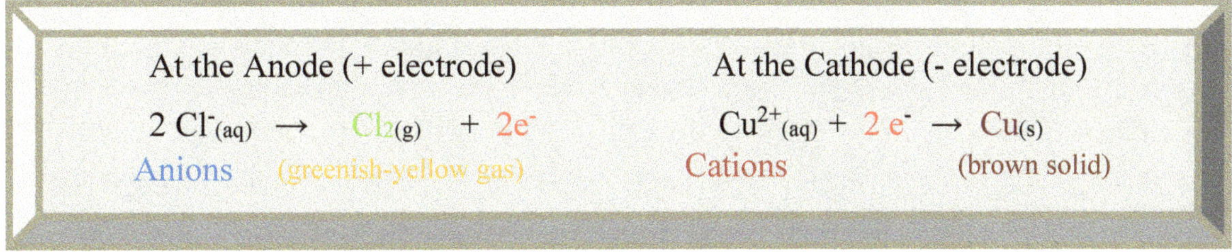

When the ions reach their respective electrodes, the following two reactions take place:

At the cathode, the cations ($Cu^{2+}_{(aq)}$) collect electrons from the cathode and become **_discharged_** as copper atoms; these are seen as a brown deposit on the cathode. Electroplating uses this concept.

At the anode the anions ($Cl^-_{(aq)}$) give up their electrons to the anode and also become **_discharged_** as chlorine molecules; these are seen as a greenish-yellow gas. These are the two processes that are involved in the passage of a current through any electrolyte; one set of ions ($Cu^{2+}_{(aq)}$) *accepting* the electrons **from the cathode,** and another set ($Cl^-_{(aq)}$) *releasing* electrons **to the anode.** These two processes are continuous since there is a large number of these ions present at any time in the aqueous solution; they eventually stop when all the ions discharge A wire placed across these two electrodes allows for the current to flow externally.

Arrhenius **observed** that **aqueous solutions of acids and bases also conduct electricity**. He therefore **concluded** that *acids and bases must also produce ions when they dissolve in water,* since it is aqueous ions that are involved in electrical conductivity. The following are examples of the formulas of a few acids and bases:

<table>
<tr><td><u>**Acids**</u></td><td><u>**Bases**</u></td></tr>
<tr><td>$HCl_{(aq)}$</td><td>$NaOH_{(aq)}$</td></tr>
<tr><td>$HNO_{3(aq)}$</td><td>$KOH_{(aq)}$</td></tr>
<tr><td>$H_2SO_{4(aq)}$</td><td>$Mg(OH)_{2(aq)}$</td></tr>
<tr><td>$CH_3COOH_{(aq)}$</td><td>$CsOH_{(aq)}$</td></tr>
<tr><td>$HClO_{3(aq)}$</td><td>$Ba(OH)_{2(aq)}$</td></tr>
</table>

Furthermore, from his observation of the formulas of acids and bases, Arrhenius theorized that:

1. Acids produce $H^+_{(aq)}$ ions in aqueous solutions. Hydrochloric acid, for example, ionizes according to the following equation:
$$HCl_{(aq)} \rightarrow H^+_{(aq)} + Cl^-_{(aq)}$$
(Hydrochloric acid)

2. Bases produce $OH^-_{(aq)}$ ions in aqueous solutions. Sodium hydroxide, for example, dissociates according to the following equation:
$$NaOH_{(aq)} \rightarrow Na^+_{(aq)} + OH^-_{(aq)}$$
(Sodium hydroxide)

The Arrhenius theory, however, had a few short commings. The are outlined as follows:

- Empirical evidence shows that acidic solutions contain primarily hydronium ions, $H_3O^+_{(aq)}$ rather than $H^+_{(aq)}$ ions. Scientist believe that $H^+_{(aq)}$ ions, being positively charged, are attracted to the highly polar water molecules to which they bond and form hydronium ions.
- Empirical evidence also shows that aqueous solutions of some salts have pH values indicating that they are either acids or bases even though their formulas do not contain any hydrogen ions or hydroxyl groups. The following table shows a few examples:

Table 17.1. pH of a few aqueous compounds without any $H^+_{(aq)}$ ions or $OH^-_{(aq)}$ ions.

Compounds	pH	Acidic	Basic
$CaO_{(aq)}$	10		√
$Na_2CO_{3(aq)}$	9		√
$NH_4Cl_{(aq)}$	6	√	

- Also, according to the Arrhenius theory, acid-base reactions are only limited to taking place in water. In fact, some acid base reactions can happen in solvents other than water.
  A number of other acid-base theories were also proposed; another one is described below.

The Brönsted-Lowry Theory

In 1923, Johannes Brönsted of Denmark and Thomas Lowry of England formulated another acid-base theory. This theory states that:

- **An acid is a proton donor**
- **A base is a proton acceptor**

Note* A proton is a hydrogen atom that has lost its electron.

The following equations will illustrate the essence of this theory.

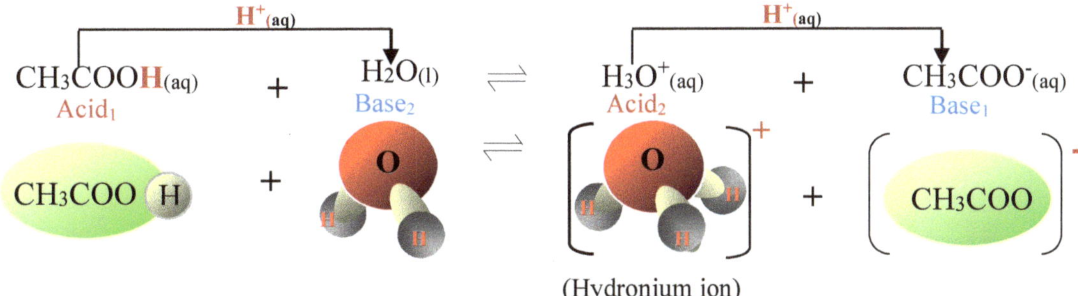

Figure 17.2. Conjugate acid-base pairs formation between acetic acid and water.

In the above illustrations, $CH_3COOH_{(aq)}$ is donating a proton to $H_2O_{(l)}$ in the forward reaction. According to the Brönsted-Lowry theory, the **$CH_3COOH_{(aq)}$** molecule is the **acid** and the **$H_2O_{(l)}$** molecule is the **base.** In the reverse reaction, the $H_3O^+_{(aq)}$ (hydronium ion) is donating a proton to the $CH_3COO^-_{(aq)}$ ion. In this reaction, the **$H_3O^+_{(aq)}$** ion is the **acid** and the $CH_3COO^-_{(aq)}$ ion is the **base.**

17.2   Conjugate Acid-Base Pair Formation

When any molecule acts as an acid by donating an $H^+_{(aq)}$ ion, what is left of it is called its **conjugate base.** For example, when acetic acid acts as an acid, its conjugate base is formed as follows:

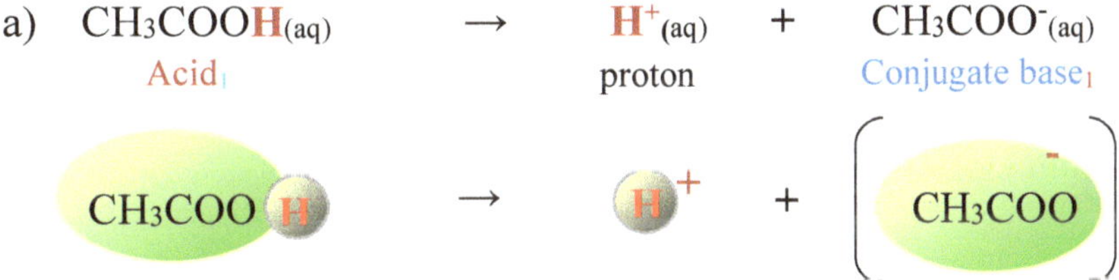

Figure 17.3. The formation of the conjugate base from acetic acid.

In this example, $CH_3COOH_{(aq)}$ is donating the $H^+_{(aq)}$ ion, and the $CH_3COO^-_{(aq)}$ ion is what is left of the acid; it is now called the **conjugate base** of the acid, $CH_3COOH_{(aq)}$.
For the reverse reaction, we get:

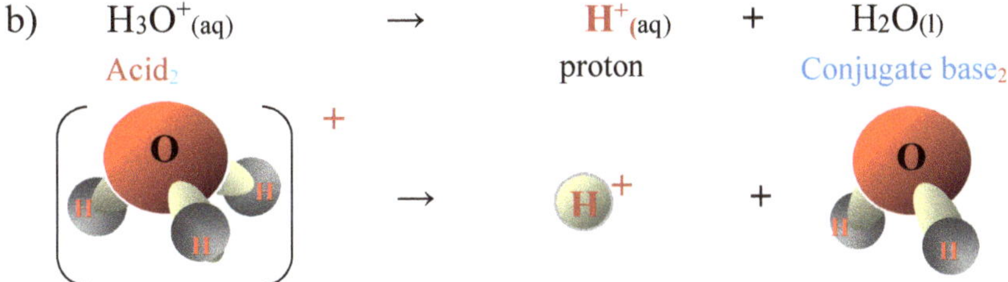

Figure 17.4.   The formation of the conjugate base from a hydronium ion.

In this other example, $H_3O^+_{(aq)}$ is the acid. After it donates the $H^+_{(aq)}$ ion, the $H_2O_{(l)}$ is what is left of the acid; it is now called the **conjugate base** of the acid $H_3O^+_{(aq)}$.
Returning to our illustrations, the following conjugate acid-base pairs can be identified:

1. $CH_3COOH_{(aq)}$: $CH_3COO^-_{(aq)}$
2. $H_3O^+_{(aq)}$: $H_2O_{(l)}$

*Based on these examples, it can be concluded that **an acid differs from its conjugate base by a proton.***

Use the Brönsted-Lowry theory to identify the conjugate acid-base pairs in the following acid-base reactions. K/U

(a)　$HCO_3^-{}_{(aq)} + S^{2-}{}_{(aq)} \rightleftharpoons HS^-{}_{(aq)} + CO_3^{2-}{}_{(aq)}$
(b)　$CO_3^{2-}{}_{(aq)} + HC_2H_3O_2 \rightleftharpoons HCO_3^-{}_{(aq)} + C_2H_3O_2^-{}_{(aq)}$
(c)　$H_3PO_4{}_{(aq)} + OCl^-{}_{(aq)} \rightleftharpoons HClO_{(aq)} + H_2PO_4^-{}_{(aq)}$
(d)　$HSO_4^-{}_{(aq)} + HPO_4^{2-}{}_{(aq)} \rightleftharpoons SO_4^{2-}{}_{(aq)} + H_2PO_4^-{}_{(aq)}$

17.3　Strong Acids versus Weak Acids.

Acids are molecular compounds and are therefore formed by covalent bonds. Normally, molecular compounds do not ionize when dissolved in water. But, if a certain bond within the molecule is highly polar, it is possible that the molecule may ionize when it dissolves in water.

Strong acids and bond polarity

As can be recalled from Chapter 6, whether or not a covalent bond is polar, depends solely on the difference in electronegativity of the two bonded atoms. The following examples illustrate this concept.

$$H\text{———}H \qquad\qquad Cl\text{———}Cl$$

Electronegativity values:　　=　　2.1　　2.1　　　　　3.5　　3.5
Electronegativity difference =　　　　0　　　　　　　　0

The bond in each of molecules above is non-polar because there is an even distribution of the shared electron cloud between the two atomic nuclei; their nuclei exert equal pull on the electron cloud.

Figure 17.5. The even electron cloud distributions in the H_2 and Cl_2 molecules.

As can be recalled from Chapter 6, also, if there is an electronegativity difference greater than 0.4 between the two bonded atoms, then the covalent bond becomes polar. The following example illustrates this concept:

$$\overset{\delta+}{H}\text{———}\overset{\delta-}{Cl}$$

　　Electronegativity values　　=　2.1　　3.5
　　Electronegativity difference =　　1.4

As already explained in Chapter 6, the bond formed between the hydrogen and chlorine atoms would be polar, because the chlorine atom, which is significantly more electronegative, exerts a greater pull on the shared electron cloud. Because of this stronger pull, more of the electron cloud stays with the chlorine atom making it acquire a small negative charge, symbolized as $\delta-$, and the hydrogen atom acquiring a small positive charge, symbolized as $\delta+$. The small positive charge, $\delta+$, that is acquired by the hydrogen atom is due to the due to its electron cloud being shifted away from its nucleus; it is not fully neutralized now.

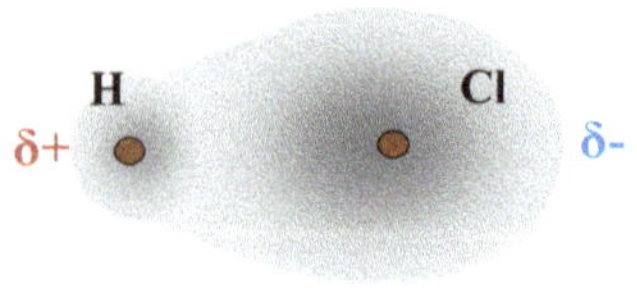

Figure 17.6. The electron cloud distribution for the hydrogen chloride molecule.

Again, as can be recalled from chapter 6, the greater the electronegativity difference, the more polar and weaker the covalent bond becomes. However, if conditions become favourable, the more electronegative atom may even take complete possession of the shared pair of electrons from the covalent bond. When this happens, the molecule is said to ionize. This is exactly what happens when $HCl_{(g)}$, hydrogen chloride gas dissolves in water. The following figure illustrates this concept.

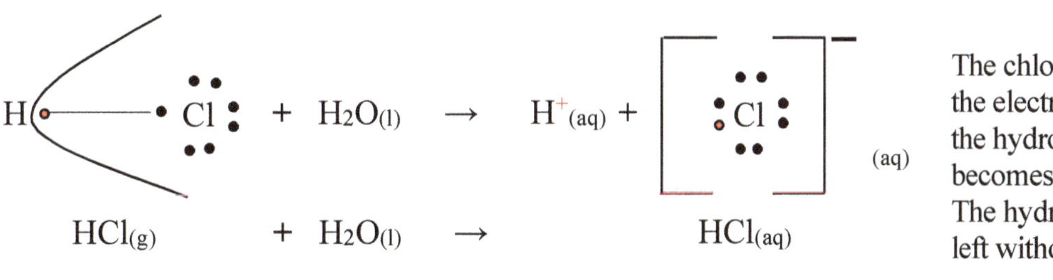

Figure 17.7. The ionization of hydrogen chloride in water to form hydrochloric acid.

Arrhenius observed that acids such as $HCl_{(aq)}$ are very good conductors of electricity, and he concluded that for this to happen, they must ionize completely in water. These acids are called **strong acids**. **A strong acid** is one that **completely ionizes in water.** This is portrayed in the table below. The generic formula for acids, HA is used in the illustrations beside the table.

Table 17.2.   The complete ionization of hydrochloric acid

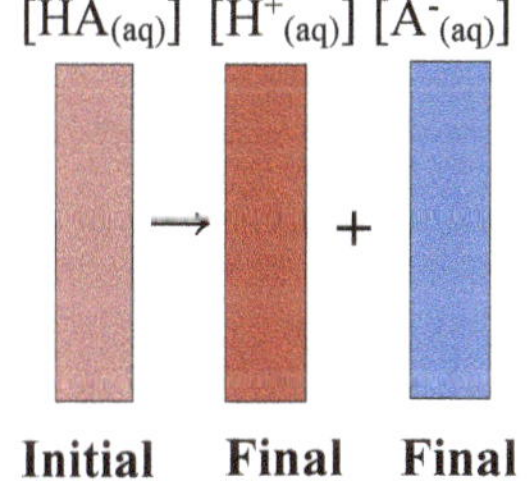

Concentration	$HCl_{(aq)}$	$\rightarrow$ $Cl^-_{(aq)}$	$+$ $H^+_{(aq)}$
Initial	1.0 mol/L	0.0 mol/L	0.0 mol/L
Final	0.0 mol/L	1.0 mol/L	1.0 mol/L

An aqueous solution of hydrochloric acid, (HCl) would therefore have only $H^+_{(aq)}$ and $Cl^-_{(aq)}$ ions, but no $HCl_{(aq)}$ molecules. Note that in this type of ionization equation, a single arrow symbol ($\rightarrow$) is used, and it symbolizes complete ionization.

Arrhenius also observed that solutions of some acids, such as ethanoic acid (vinegar) and formic acid, conduct electricity poorly. He theorized that for poor electrical conductivity, there must be only few ions present in these solutions. This would happen if the acids only partially ionize. **These acids which ionize only partially in water are said to be weak.** The ionization of ethanoic (acetic acid) is used to exemplify this concept.

Table 17.3.   The partial ionization of ethanoic acid

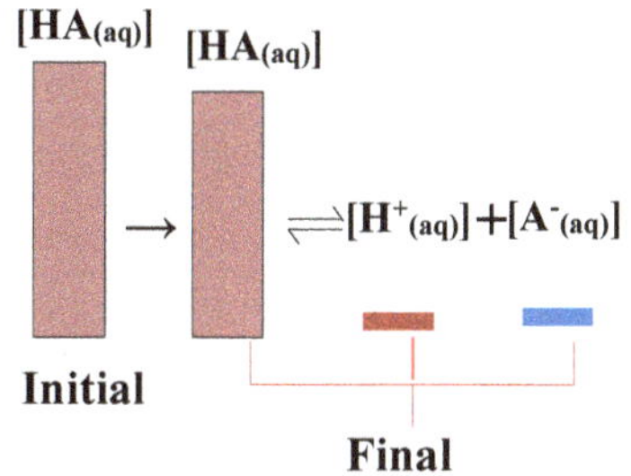

CONCENTRATION	$CH_3COOH_{(aq)}$	$\rightleftharpoons$ $CH_3COO^-_{(aq)} +$	$H^+_{(aq)}$
Initial	1 mol/L	0 mol/L	0 mol/L
Final	0.987 mol/L	0.013 mol/L	0.013 mol/L

In this example, the acid is only 1.3% ionized. An aqueous solution of acetic acid would therefore have 987 molecules of $CH_3COOH_{(aq)}$, 13 $CH_3COO^-_{(aq)}$ ions and 13 $H^+_{(aq)}$ ions respectively, per every 1000 molecules of $CH_3COOH_{(aq)}$ dissolved in water. Low $[H^+_{(aq)}]$ ions, means weak acidity. Note that in this type of ionization equations, a double arrow symbol ($\rightleftharpoons$) is used, and it symbolizes partial ionization. This sign also indicates that the reaction is reversible.

Figure17.8(a). Hydrochloric acid conducting electricity very well.

Figure17.8(b). Acetic acid conducting electricity poorly.

Exercise 17.2

What are the differences of between 1.0 mol/L $CH_3COOH_{(aq)}$ and 0.10 mol/L $HCl_{(aq)}$?

Solution

a) 1.0 mol/L $CH_3COOH_{(aq)}$ is 10 times more concentrated than 0.10 mol/L $HCl_{(aq)}$.
b) 1.0 mol/L $CH_3COOH_{(aq)}$ is weaker than the 0.10 mol/L $HCl_{(aq)}$ even though it is 10 times more concentrated since it is very poorly ionized.
c) 1.0 mol/L $CH_3COOH_{(aq)}$ has a higher pH than 0.10 mol/L $HCl_{(aq)}$.

Conjugate Base Strength

Hydrochloric acid is said to be a strong acid as it undergoes **complete ionization.** This is exemplified by the following equation:

$$HCl_{(aq)} \ + \ H_2O_{(l)} \ \longrightarrow \ H_3O^+_{(aq)} \ + \ Cl^-_{(aq)} \qquad 100\% \ \text{ionization}$$

Acid   Conjugate base

(Strong)   **(Weak)**

Recall that according to the Brönsted-Lowry theory, **a base is a proton acceptor.** In this case, the $Cl^-_{(aq)}$ ion, which is the conjugate base of the acid $HCl_{(aq)}$, is **reluctant** to accept the proton from the $H_3O^+_{(aq)}$ ion to reform $HCl_{(aq)}$. It is thus said to be a **weak conjugate base**. For the same reason, it lets go of the $H^+_{(aq)}$ ion from the $HCl_{(aq)}$ so easily that it results in complete ionization. A **strong acid** thus has a **weak conjugate base**.

Ethanoic acid is said to be a weak acid as it undergoes **partial ionization.** This is exemplified by the following equation:

$$CH_3COOH_{(aq)} + \ H_2O_{(l)} \ \rightleftharpoons \ H_3O^+_{(aq)} + CH_3COO^-_{(aq)} \quad \textbf{(1.3\% ionization)}$$

Acid   Conjugate base

(Weak)   **(Strong)**

In this case, the $CH_3COO^-_{(aq)}$ ion, which is the conjugate base of the acid, $CH_3COOH_{(aq)}$, is very willing to accept the proton from the $H_3O^+_{(aq)}$ ion to reform $CH_3COOH_{(aq)}$. It is thus said to be a **strong conjugate base**. For the same reason, it is very reluctant to let go of $H^+_{(aq)}$ ion from the $CH_3COOH_{(aq)}$ easily, resulting in partial ionization. A **weak acid** thus has a **strong conjugate base.**

The following table summarizes the difference between a strong acid and a weak acid.

Table 17.4. Some properties of strong acids and weak acids. (See Fig. 17.07 on p.300)

Properties	Strong Acids Example: $HCl_{(aq)}$	Weak Acids Example: $CH_3COOH_{(aq)}$
Electrical conductivity	High	Low
Degree of ionization	High	Low
pH	Lower	Higher
Conjugate base	Weak	Strong

17.4 Strong Bases versus Weak Bases

The theory that explains the concepts of strong acids and weak acids is similar to that for strong bases and weak bases.

By definition, a strong base is one that completely ionizes in water. The following example illustrates this concept:

Table 17.5. The complete ionization of potassium hydroxide.

Concentration	$KOH_{(aq)}$ $\rightarrow$	$K^+_{(aq)}$ $+$	$OH^-_{(aq)}$
Initial	1.0 mol/L	0 mol/L	0 mol/L
Final	0.0 mol/L	1.0 mol/L	1.0 mol/L

According to the information provided in this table, an aqueous solution of potassium hydroxide would have only $K^+_{(aq)}$ and $OH^-_{(aq)}$ ions, but no $KOH_{(aq)}$ formula units.

By definition, a weak base is one that **ionizes only partially** in water. This is exemplified by the reaction between ammonia and water molecules.

$$NH_{3(aq)} \ + \ H_2O_{(l)} \ \rightleftharpoons \ NH_4^+_{(aq)} \ + \ OH^-_{(aq)}$$

Ammonia molecules   Ammonium ions

Here again the double arrow symbol ($\rightleftharpoons$) is used to indicate partial ionization as defined earlier. In this example, an aqueous solution of ammonia will have lots of ammonia molecules and only few ammonium and hydroxide ions.

Acidic hydrogen in acids

Monoprotic **acids**: Since the $H^+_{(aq)}$ ion in reality is a **proton,** acids that are capable of producing only one mole of $H^+_{(aq)}$ ions per mole of acid are called *mono**protic** acids*.
The following equations illustrate how a few of these acids ionize:

$$HI_{(aq)} \ \rightarrow \ H^+_{(aq)} + I^-_{(aq)}$$
$$HClO_{3(aq)} \ \rightarrow \ H^+_{(aq)} + ClO_3^-_{(aq)}$$
$$CH_3COOH_{(aq)} \ \rightleftharpoons \ H^+_{(aq)} + CH_3COO^-_{(aq)}$$

Diprotic **acids:** These are acids that are of capable producing two moles of $H^+_{(aq)}$ ions per mole of the acid. The following equations show how an example of these acids, sulphuric acid, ionizes:

$$H_2SO_{4(aq)} \ \rightarrow \ H^+_{(aq)} + HSO_4^-_{(aq)}$$
$$HSO_4^-_{(aq)} \ \rightarrow \ H^+_{(aq)} + SO_4^{2-}_{(aq)}$$

***Triprotic* acids:** These are acids that are capable of producing three moles of $H^+_{(aq)}$ ions per mole of the acid. The following equations show how an example of these acids, phosphoric acid, ionizes:

$$H_3PO_{4(aq)} \rightarrow \quad H^+_{(aq)} + H_2PO_4^-{}_{(aq)}$$
$$H_2PO_4^-{}_{(aq)} \rightarrow \quad H^+_{(aq)} + HPO_4^{-2}{}_{(aq)}$$
$$HPO_4^{-2}{}_{(aq)} \rightarrow \quad H^+_{(aq)} + PO_4^{-3}{}_{(aq)}$$

Notice that diprotic and triprotic acids ionize in sequential steps. In these, only after the first mole of $H^+_{(aq)}$ ions is produced and it is completely neutralized by a base, does any further ionization occurs. The reaction between sulphuric acid and sodium hydroxide is used illustrates this concept.

$$H_2SO_{4(aq)} + NaOH_{(aq)} \rightarrow \quad NaHSO_{4(aq)} + H_2O_{(l)}$$
$$1 \text{ mol} \qquad\quad 1 \text{ mol} \qquad\qquad 1 \text{ mol (acid salt)}$$

The equation above shows that there are two moles of acidic $H^+_{(aq)}$ ions present per mole of $H_2SO_{4(aq)}$, but only one mole is neutralized by the one mole of $NaOH_{(aq)}$.

If neutralization was stopped at this point, the salt produced will be acidic since it is still capable of producing one more mole $H^+_{(aq)}$ ions. This is illustrated by the following equations:

$$NaHSO_{4(aq)} \rightarrow \quad Na^+_{(aq)} + HSO_4^-{}_{(aq)}$$
$$1 \text{ mol} \qquad\qquad\qquad 1 \text{ mol}$$
$$HSO_4^-{}_{(aq)} \rightleftharpoons \quad SO_4^{2-}{}_{(aq)} + H^+_{(aq)} \rightarrow \quad \text{(the cause of acidity)}$$
$$1 \text{ mol} \qquad\qquad\qquad 1 \text{ mol}$$

For complete neutralization, another mole of the base must be used as illustrated below.
$$NaHSO_{4(aq)} + NaOH_{(aq)} \rightarrow Na_2SO_{4(aq)} + H_2O_{(l)h}$$
$$1 \text{ mol} \qquad\qquad 1 \text{ mol}$$

Hydroxide Ions in Bases

***Monoprotic* bases:** These bases are capable of producing one mole of hydroxide ions per mole after ionization. They are thus each capable of neutralizing only one mole of a monoprotic acid. The equation below illustrates this concept:
$$NaOH_{(aq)} + CH_3COOH_{(aq)} \rightarrow CH_3COONa_{(aq)} + H_2O_{(l)}$$
$$1 \text{ mol} \qquad\qquad 1 \text{ mol}$$
$$\text{Monoprotic base} \qquad \text{Monoprotic acid}$$

***Diprotic* bases:** These bases are capable of neutralizing two moles of a monoprotic acid or one mole of a diprotic acid. The equations below illustrate this concept.

$$Mg(OH)_{2(aq)} + 2\ HNO_{3(aq)} \rightarrow Mg(NO_3)_{2(aq)} + 2\ H_2O_{(l)}$$
$$1 \text{ mol} \qquad\qquad 2 \text{ mol}$$
$$\text{Diprotic base} \qquad \text{Monoprotic acid}$$

$$Ca(OH)_{2(aq)} + H_2SO_{4(aq)} \rightarrow CaSO_{4(s)} + 2\ H_2O_{(l)}$$
$$1 \text{ mol} \qquad\qquad 1 \text{ mol}$$
$$\text{Diprotic base} \qquad \text{Diprotic acid}$$

1. Ascorbic acid (vitamin C), found in citrus fruits is essential in many metabolic pathways, plays roles in maintaining healthy tissues and acts as an antioxidant. An aqueous solution of it conducts electricity extremely poorly. Its molecular formula is $HC_6H_7O_6$ and it is monoprotic.

    a) Would this be a strong or weak acid? T/I

    b) Would the pH of this acid be close to 7 or much below it? T/I

    c) Write the ionization equation for this acid with water. K/U C

    d) Identify the conjugate acid-base pairs formed in (c), above. T/I

    e) How many moles of NaOH can this acid neutralize? Write an equation for this reaction. T/I K/U C

2. Potassium hydroxide ionizes in water according to the following equation:

$$KOH_{(s)} + H_2O_{(l)} \rightarrow K^+_{(aq)} + OH^-_{(aq)}$$

    a) Is this a strong or weak base? Use the Arrhenius's theory to explain your answer. T/I C

    b) How would an aqueous solution of this base conduct electricity? Explain. T/I C

    c) Strong bases have pH high above 7, while weak ones have pH well above 7. Where would the pH of this base be? T/I

    d) One mole of this base is mixed with one mole of $H_2SO_{4(aq)}$.

        i) Write a balanced equation for the reaction that takes place here. K/U C

        ii) If a pH meter is placed in an aqueous solution of the salt formed, what would it indicate? T/I

3. A solution of an aqueous substance, X, has a pH of about 11, but conducts electricity poorly.

    a) What type of substance could X be? T/I

    b) Choose from the following list of substances which one would best fit the description of X: $KOH_{(aq)}$, $HCl_{(aq)}$, $NH_4Cl_{(aq)}$ or $NH_{3(aq)}$. K/U

    c) Write an equation for the reaction of substance you have chosen above with water, and then identify the conjugate acid-base pairs formed. K/U C

17.5 Acid- Base Neutralization Reactions

As learnt in the previous chapter, the acidity of a substance is due to its ability to produce $H^+_{(aq)}$ ions and the basicity of a substance is due to its ability to produce $OH^-_{(aq)}$ ions. An acid-base neutralization would therefore involve $OH^-_{(aq)}$ ions reacting with $H^+_{(aq)}$ ions forming water, which is a neutral compound. The following word equation represents a general neutralization reaction:

$$Acid \quad + \quad Base \quad \rightarrow \quad Salt \quad + \quad Water$$

The following is an example of a chemical equation for a specific acid-base neutralization reaction:

$$HCl_{(aq)} \quad + \quad NaOH_{(aq)} \quad \rightarrow \quad NaCl_{(aq)} \quad + \quad H_2O_{(l)}$$

    Hydrochloric acid   Sodium hydroxide   Sodium chloride   Water

Whenever there is complete neutralization that involves strong bases and strong acids, all the products **are neutral**. Not all acid-base neutralization gives neutral products; if a weak acid neutralizes a strong base, the resulting solution is basic, and if a strong acid neutralizes a weak base, an acidic solution is formed. The latter two situations result because the salts formed are hydrolysed by water. Hydrolysis of salts will be learnt in a higher level Chemistry course.

Ionic Equation and Net Ionic Equation.

Any acid-base neutralization reaction can be represented by a net ionic equation. To derive a net ionic equation, an **ionic equation** must first be created. For the above reaction, we have the following ionic equation:

$$H^+_{(aq)} + Cl^-_{(aq)} + Na^+_{(aq)} + OH^-_{(aq)} \rightarrow Na^+_{(aq)} + Cl^-_{(aq)} + H_2O_{(l)}$$

After the spectator ions are removed from the ionic equation, the following **net ionic equation** remains:

$$H^+_{(aq)} + OH^-_{(aq)} \rightarrow H_2O_{(l)}$$

The equation above shows how the net effect of an acid-base neutralization reaction really is $H^+_{(aq)}$ ions neutralizing $OH^-_{(aq)}$ ions in a 1: 1 mole ratio, forming water.

Balancing acid-base neutralization equations.

The key to balancing acid-base neutralization equations is to know the number of moles of ionizable $H^+_{(aq)}$ ions that the acid has, and the number of moles of ionizable $OH^-_{(aq)}$ ions that the base has. How could this information be used to complete and balance the following equation?

$$H_2SO_{4(aq)} + NaOH_{(aq)} \rightarrow Products?$$

$$\textit{# of mol?} \qquad \textit{# of mol?}$$

Since one mole $H_2SO_{4(aq)}$ has **two** moles of ionizable $H^+_{(aq)}$ ions and one mole of $NaOH_{(aq)}$ has only **one** mole of ionizable $OH^-_{(aq)}$ ions, then it necessary to use **two** moles of $NaOH_{(aq)}$ for every **one** mole $H_2SO_{4(aq)}$ for neutralization. Completing this equation, we get the following:

$$H_2SO_{4(aq)} + 2\,NaOH_{(aq)} \rightarrow Na_2SO_{4(aq)} + 2\,H_2O_{(l)}$$

$$\textit{1 mol} \qquad \textit{2 mol}$$

Using the same reasoning, we have the following balanced equation for the reaction between $Mg(OH)_{2(aq)}$ and $HNO_{3(aq)}$.

$$Mg(OH)_{2(aq)} + 2\,HNO_{3(aq)} \rightarrow Mg(NO_3)_{2(aq)} + 2\,H_2O_{(l)}$$

$$\textit{1 mol} \qquad \textit{2 mol}$$

Exercise 17.4

Complete and balance the following equations: K/U C

1. $HCl_{(aq)} + Ca(OH)_{2(aq)} \rightarrow$
2. $H_2SO_{4(aq)} + KOH_{(aq)} \rightarrow$
3. $Mg(OH)_{2(aq)} + H_2SO_{4(aq)} \rightarrow$
4. $H_3PO_{4(aq)} + NaOH_{(aq)} \rightarrow$
5. $CH_3COOH_{(aq)} + KOH_{(aq)} \rightarrow$
6. $H_3PO_{4(aq)} + Ba(OH)_{2(aq)} \rightarrow$
7. $LiOH_{(aq)} + H_2SO_{4(aq)} \rightarrow$
8. $H_2SO_{3(aq)} + Ca(OH)_{2(aq)} \rightarrow$
9. $HClO_{3(aq)} + Mg(OH)_{2(aq)} \rightarrow$
10. $HBrO_{4(aq)} + Al(OH)_3 \rightarrow$

Monoprotic Acid versus Monoprotic Base

Sample problem 1:

What volume of 0.100 mol/L $HCl_{(aq)}$ is required to neutralize 25.0 mL of 0.200 mol/L $NaOH_{(aq)}$?

Solution:

Provided quantities:
Concentration of $HCl_{(aq)}$ = 0.100 mol/L
Volume of $NaOH_{(aq)}$ = 25.0 mL or 0.025 L
Concentration of $NaOH_{(aq)}$ = 0.200 mol/L

Required quantity:

Volume of 0.10 mol/L $HCl_{(aq)}$

Step 1. Write a balanced equation for the reaction between the acid and base.

$$\underset{1\ mol}{\underset{(Given)}{NaOH_{(aq)}}} \quad + \quad \underset{1\ mol}{\underset{(Required)}{HCl_{(aq)}}} \quad \longrightarrow \quad NaCl_{(aq)} \quad + \quad H_2O_{(l)}$$

Here, we see that $NaOH_{(aq)}$ and $HCl_{(aq)}$ react proportionately in a 1:1 mole ratio.

Step 2. Calculate the number of moles of the **given reactant**.

*Since the **concentration** and **volume** of the $NaOH_{(aq)}$ are given, it becomes the **given quantity**, as its number of moles can be found using the formula:*

$$n_{(NaOH)} = C \times V$$
$$= \frac{0.200\ mol}{L} \times 0.025\ L$$
$$= 0.00500\ mol$$

$$n = C \times V$$

Step 3. Calculate the number of moles of the **required quantity,** $HCl_{(aq)}$.

$$\frac{Given}{Required} = \frac{1\ mol}{1\ mol} = \frac{0.00500\ mol\ NaOH_{(aq)}}{X\ mol\ HCl_{(aq)}}$$
$$X = 0.00500\ mol\ HCl_{(aq)}$$

Step 4. Using the equation, $V = \dfrac{n}{C}$, find the volume of the **$HCl_{(aq)}$.**

$$v = \frac{n}{c}$$

$$V_{HCl(aq)} = \frac{0.00500\ mol}{0.100\ mol\ /\ L}$$
$$= 0.0500\ L$$
$$= 50.0\ mL$$

Diprotic Acid versus Monoprotic Base

Sample problem 2:

Find the concentration of $NaOH_{(aq)}$ if 50.0 mL of it is required to neutralize 50.0 mL of 0.100 mol/L $H_2SO_{4(aq)}$.

Solution:

Provided quantities:
Concentration of $H_2SO_{4(aq)}$ = 0.100 mol/L
Volume of $H_2SO_{4(aq)}$ = 0.0500 L
Volume of $NaOH_{(aq)}$ = 0.0500 L

Required quantity:
Concentration of $NaOH_{(aq)}$

Step 1. Write a balanced equation for the reaction between the acid and base.

     (Given)   *(Required)*

$$H_2SO_{4(aq)} \;+\; 2\,NaOH_{(aq)} \;\longrightarrow\; Na_2SO_{4(aq)} \;+\; 2\,H_2O_{(l)}$$

    1 mol   *2 mol*

Here, we see that $H_2SO_{4(aq)}$ and $NaOH_{(aq)}$ react proportionately in a 1:2 mol ratio.

Step 2. Calculate the number of moles of the ***given reactant.***
    *Since the **concentration** and **volume** of the $H_2SO_{4(aq)}$ are given, it becomes the **given quantity** as its number of moles can be found using the formula:*

$$n = C \times V$$

$$n_{H_2SO_4\,(aq)} = C \times V$$
$$= 0.100\ mol\,/\,L \times 0.0500\ L$$
$$= 0.00500\ mol$$

Step 3. Calculate the number of moles of the **required quantity,** $NaOH_{(aq)}$.

$$\frac{Given}{Required} = \frac{1\ mol}{2\ mol} = \frac{0.00500\ mol\ H_2SO_4}{X\ mol\ NaOH}$$

$$X = 0.0100\ mol\ NaOH$$

Step 4. Use the equation, $C = \dfrac{n}{V}$ to find the concentration of **NaOH**$_{(aq)}$.

$$C = \frac{0.0100\ mol\ NaOH}{0.0500\ L}$$
$$= 0.200\ mol\,/\,L$$

Diprotic Base versus Monoprotic Base

Sample problem 3:

What is the concentration of a $Mg(OH)_{2(aq)}$ solution, if 200.0 mL of it is neutralized by 50.0 mL of 0.20 mol/L $HNO_{3(aq)}$?

Solution:

Provided quantities:
Concentration of $HNO_{3(aq)}$ = 0.20 mol/L
Volume of $HNO_{3(aq)}$ = 0.0500 L
Volume of $Mg(OH)_{2(aq)}$ = 0.200 L

Required quantity:
Concentration of $Mg(OH)_{2(aq)}$

Step 1. Write a balanced equation for the reaction between the acid and base.

(Required)   *(Given)*

$$Mg(OH)_{2(aq)} \ + \ 2\,HNO_{3(aq)} \ \longrightarrow \ Mg(NO_3)_{2(aq)} \ + \ 2\,H_2O_{(l)}$$

$$1\ mol \qquad\qquad 2\ mol$$

Here, we see that $Mg(OH)_{2(aq)}$ and $HNO_{3(aq)}$ react proportionately in a 1:2 mole ratio.

Step 2. Calculate the number of moles of the **given reactant**.

*Since the **concentration** and **volume** of the $HNO_{3(aq)}$ are given, it becomes the **given quantity,** as its number of moles can be found using the formula:*

$$n_{HNO_{3(aq)}} = 0.20\ mol/L \times 0.0500\ L$$

$$= 0.010\ mol$$

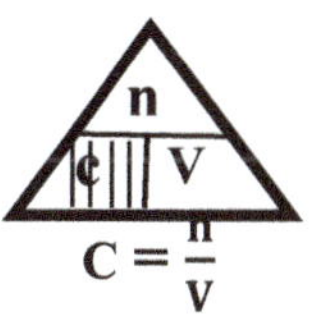

Step 3. Calculate the number of moles of the **required reactant, $Mg(OH)_{2(aq)}$.**

$$\frac{Given}{Required} = \frac{2\ mol}{1\ mol} = \frac{0.010\ mol\ HNO_{3(aq)}}{X\ mol\ Mg(OH)_{2(aq)}}$$

$$2X = 0.010\ mol\ Mg(OH)_{2(aq)}$$

$$X = 0.0050\ mol\ Mg(OH)_{2(aq)}$$

Step 4. Use the equation $C = \dfrac{n}{V}$ to find the concentration of $Mg(OH)_{2(aq)}$.

$$C = \frac{0.0050\ mol\ Mg(OH)_{2(aq)}}{0.200\ L}$$

$$= 0.025\ mol/L\ Mg(OH)_{2(aq)}$$

Sample problem 4:

What mass of $Ca(OH)_{2(s)}$ is required to neutralize 100.0 mL of 0.500 mol/L $HClO_{3(aq)}$?

Solution:

Provided quantities:
  Volume of $HClO_{3(aq)} = 0.100L$
  Concentration of $HClO_{3(aq)} = 0.500$ mol/L

Required quantity:
  Mass of $Ca(OH)_{2(s)}$

Step 1. Write a balanced equation for the reaction between the acid and base.

(Required)   *(Given)*
$$Ca(OH)_{2(s)} \ + \ 2\,HClO_{3(aq)} \ \longrightarrow \ Ca(ClO_3)_{2(aq)} \ + \ 2\,H_2O_{(l)}$$
$$1\,mol \qquad\quad 2\,mol$$

Here, we see that $Ca(OH)_{2(s)}$ and $HClO_{3(aq)}$ react proportionately in a 1:2 mole ratio.

Step 2. Calculate the number of moles of the **given reactant**.

  Since the **concentration** and **volume** of the $HClO_{3(aq)}$ are given, it becomes the **given quantity** as its number of moles can be found using the formula:
$$n = C \times V$$
$$n_{HClO_3\,(aq)} = C \times V$$
$$n = C \times V$$
$$= \frac{0.500\,mol}{\cancel{L}} \times 0.100\,\cancel{L}$$
$$= 0.0500\,mol$$

Step 3. Calculate the number of moles of the required reactant, **$Ca(OH)_{2(s)}$**.

$$\frac{Given}{Required} = \frac{2\,mol}{1\,mol} = \frac{0.0500\,mol\ HClO_{3(aq)}}{X\,mol\ Ca(OH)_{2(s)}}$$
$$2X = 0.0500\,mol\ Ca(OH)_{2(s)}$$
$$X = 0.0250\,mol\ Ca(OH)_{2(s)}$$

Step 4. Use the equation, $m = M \times n$ to find the mass of $Ca(OH)_{2(s)}$.
  In this equation, M is the molar mass of $Ca(OH)_{2(s)}$.
$$M_{Ca(OH)_2} = 74.1\ g/mol$$

$$m = 74.1\,g/mol \times 0.0250\,mol$$
$$= 1.86\,g$$

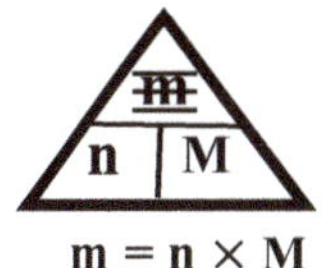

Student difficulty:

Students sometimes do not know which formula to use to find the number of moles of a given substance when solving problems in stoichiometry.

- When working with **pure solids, pure liquids and pure gases**, quantities in moles are calculated using the formula: $n = \dfrac{mass}{molar\ mass}$ or $n = \dfrac{m}{M}$

For example, if a quantity of sodium hydroxide has a mass of **10.0 g**, then the number of moles in this quantity will be calculated as follows:

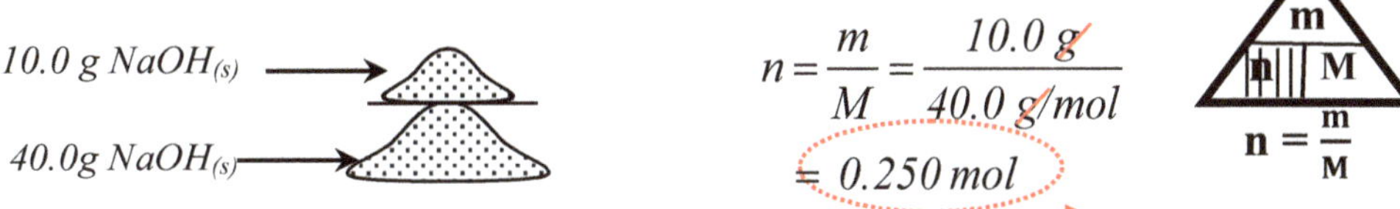

$$n = \frac{m}{M} = \frac{10.0\,g}{40.0\,g/mol}$$
$$= 0.250\,mol$$

If this same mass of sodium hydroxide (10.0 g) is now dissolved in water to make a certain volume of solution, *its number of moles remains unchanged*. For example, if it dissolved in water to make 500.0 mL **of solution, we will have the following scenario:**

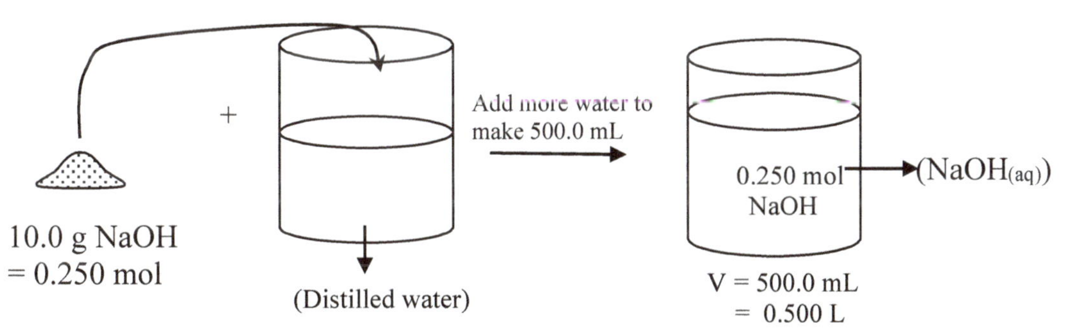

Figure 16.8(a). Distilled Water.   **Figure 16.8(b).** NaOH(aq).

The concentration of the NaOH(aq) formed is calculated as follows:

Using the formula, $C = \dfrac{n}{V}$, we get: $C = \dfrac{0.250\,mol}{0.500\,L} = $ **0.500 mol/L**

If a student was given this same solution, **0.500 L** of **0.500 mol/L NaOH(aq)**, and was required to find the number of moles in the solution, then he/she **must** get a value of **0.250 mole**, since this is quantity of NaOH that was used to make the solution. This is calculated using the formula,

$$n = C \times V$$

$$n_{NaOH} = \frac{0.500\,mol}{L} \times 0.500\,L$$
$$= 0.250\,mol$$

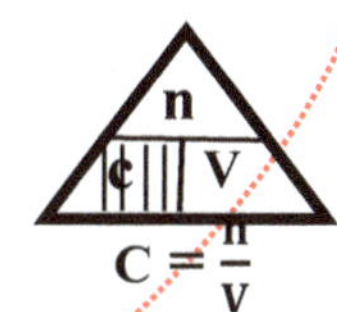

* Notice how using the formula, $n = C \times V$ gives the number of moles of solute in an **aqueous solution** while the formula, $n = \dfrac{m}{M}$, gives the number of moles for **pure** **solids, liquids and gases.**

17.7

Summary for Calculating Number of Moles in Aqueous Solutions

For pure solids, liquids and gases	**For aqueous solutions** (salts, acids and bases)

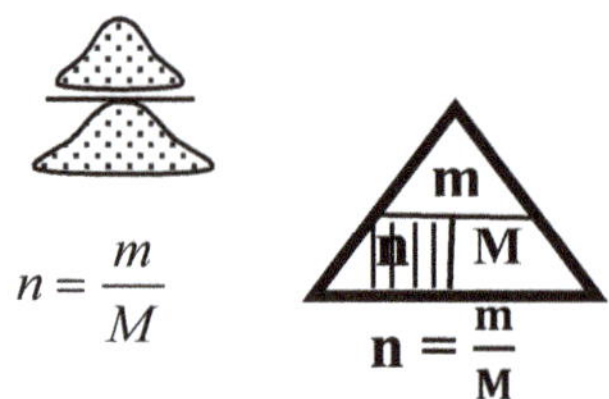

$$n = \frac{m}{M}$$

$$n = \frac{m}{M}$$

For pure solids, liquid and gases, use only the formula, $n = \dfrac{m}{M}$ to find number of moles.

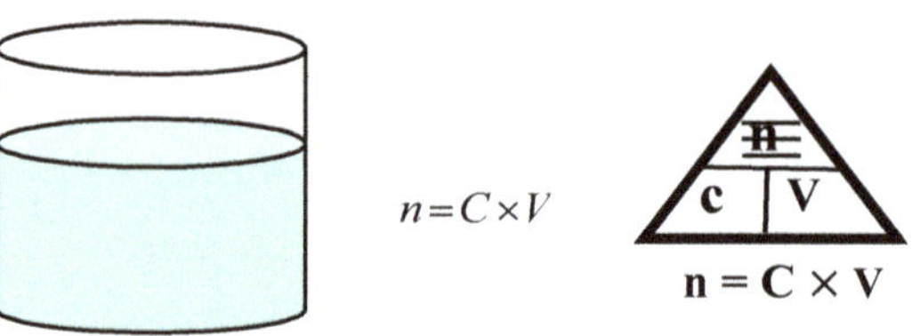

$n = C \times V$

$$n = C \times V$$

For aqueous solutions use only the formula, $n = C \times V$ to find number of moles.

Exercise 17.5

1. What volume of 0.200 mol/L of $HCl_{(aq)}$ is needed to neutralize 20.0 mL of 0.100 mol/L $Ca(OH)_{2(aq)}$? T/I

2. What volume of 0.10 mol/L $Na_2CO_{3(aq)}$ must be used to neutralize a 100.0 mL of 1.0 mol/L $HCl_{(aq)}$ spill? T/I

3. What mass of antacid tablet made of $Mg(OH)_2$ is needed to 100.0 mL of stomach acid having $HCl_{(aq)}$ of concentration 2.00 mol/L? T/I

4. If acid rain has 0.01 mol/L $H_2SO_{3(aq)}$ as its main ingredient, what volume of 0.05 mol/L of $Na_2CO_{3(aq)}$ is needed to neutralize 10.0 L of this type of rainwater? T/I

5. Acetic acid, $CH_3COOH_{(aq)}$ is sold under name vinegar. If 10.0 mL of 0.10 mol/L $NaOH_{(aq)}$ is required to neutralize 8.0 mL of vinegar, what is the concentration of the vinegar? T/I

6. In an experiment, 10.0 mL of 0.10 mol/L $NaOH_{(aq)}$ was needed to neutralize 10.0 mL of a $H_2SO_{4(aq)}$ solution. Calculate the concentration of the acid $H_2SO_{4(aq)}$ solution. T/I

7. A student mixed 100.0 mL of 0.20 mol/L $H_2SO_{4(aq)}$ with 400.0 mL of 0.20 mol/L $NaOH_{(aq)}$. What volume of 0.1 mol/L $HCl_{(aq)}$ is required to neutralize the excess base? T/I

8. What mass of $Mg_{(s)}$ would be required to neutralize the excess acid from a mixture of 200.0 mL of 0.10 mol/L $NaOH_{(aq)}$ and 250.0 mL of 0.10 mol/L $H_2SO_{4(aq)}$? Also, find the mass of hydrogen gas that would be produced. T/I

9. In an effort to neutralize an acid spill of 500.0 mL of 0.10 mol/L $HNO_{3(aq)}$, a student poured 200.0 mL of 0.20 mol/L $KOH_{(aq)}$ onto it. What mass of $NaHCO_{3(s)}$ is required to react with any excess acid? T/I

10. 100.0 mL of 0.10 mol/L $HCl_{(aq)}$ is mixed with 140.0 mL of a solution of $Ba(OH)_{2(aq)}$. A pH meter placed in the mixture shows that the mixture is basic and that it required a further 40 mL of 0.05 mol/L $HCl_{(aq)}$ for neutralization. Calculate the concentration of the $Ba(OH)_{2(aq)}$. T/I

11. Figure A below shows how a solution of $HCl_{(aq)}$ can be reacted with a mass of powdered $CaCO_{3(s)}$ and any gas produced be collected in a balloon. This done by first placing the $CaCO_{3(s)}$ in a deflated balloon which is then secured over the mouth of an erlenmeyer flask. The $CaCO_{3(s)}$ is the poured into the acid. T/I
   a) Write a balanced equation for the reaction between $CaCO_{3(s)}$ and $HCl_{(aq)}$.
   b) What mass of $CO_{2(g)}$ is collected in each balloon?
   c) In which flask is the acid completely neutralized?
   d) How do you account for the differences in volumes of the balloons?

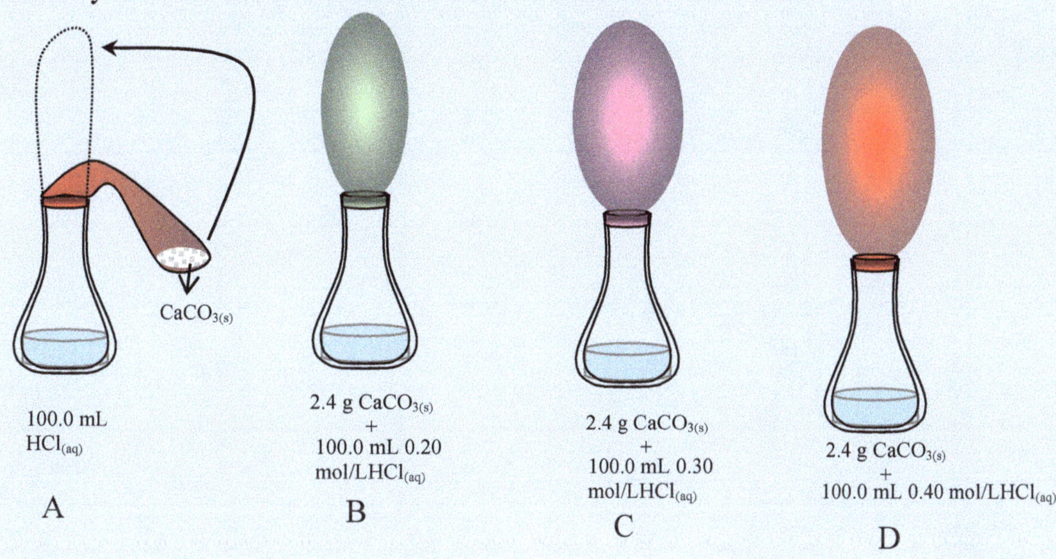

17.8 Acid-Base Titration

Suppose that a bottle containing hydrochloric acid lost its label and its concentration is to be determined, what technique could be used to achieve this? The technique that is commonly used is called titration. In this procedure, the substance to be analyzed, $HCl_{(aq)}$, is reacted with a basic solution, such as $NaOH_{(aq)}$, of known concentration. This reactant, $NaOH_{(aq)}$, of known concentration is called a *standard solution*. A known volume of the acid, *aliquot,* is measured out using a pipette, and is placed in an Erlenmeyer flask. The base is placed in a graduated burette and is released gradually (*titrated*), into the Erlenmeyer flask containing the acid until neutralization exactly occurs. The substance in the burette, in this case the **$NaOH_{(aq)}$**, is called the *titrant.* To know when complete neutralization exactly occurs, *indicators* are used. An **indicator** is a substance that changes colour in acids and in bases. The following table shows how a few indicators change colour when placed in solutions of acids and bases.

Table 16.6. Indicators and their colours in solutions of acids and bases.

Indicators	Acids	Bases
Phenolphthalein	colourless	pink
Litmus	red	blue
Bromothymol blue	yellow	blue
Methyl orange	red	yellow
Phenol red	yellow	red

Figure 16.9. Pipette and bulb.

As the base is titrated into the acid, it gradually neutralizes the acid, forming salt and water until all of the acid is just neutralized. At this point in the titration, only water and salt are present. One more drop of the base will make the solution turn slightly basic enough to change the colour of the indicator placed in the flask. This point is called the *endpoint* of the titration. If phenolphthalein indicator is used in an acid-base titration, it will change from colourless to pink if it is placed in the flask with the acid when the end point is reached. The volume of the $NaOH_{(aq)}$ used is read off from the burette. In most cases, a number of trials are performed for greater accuracy. The following figures are used to illustrate the titration of an acid with a base using phenolphthalein indicator.

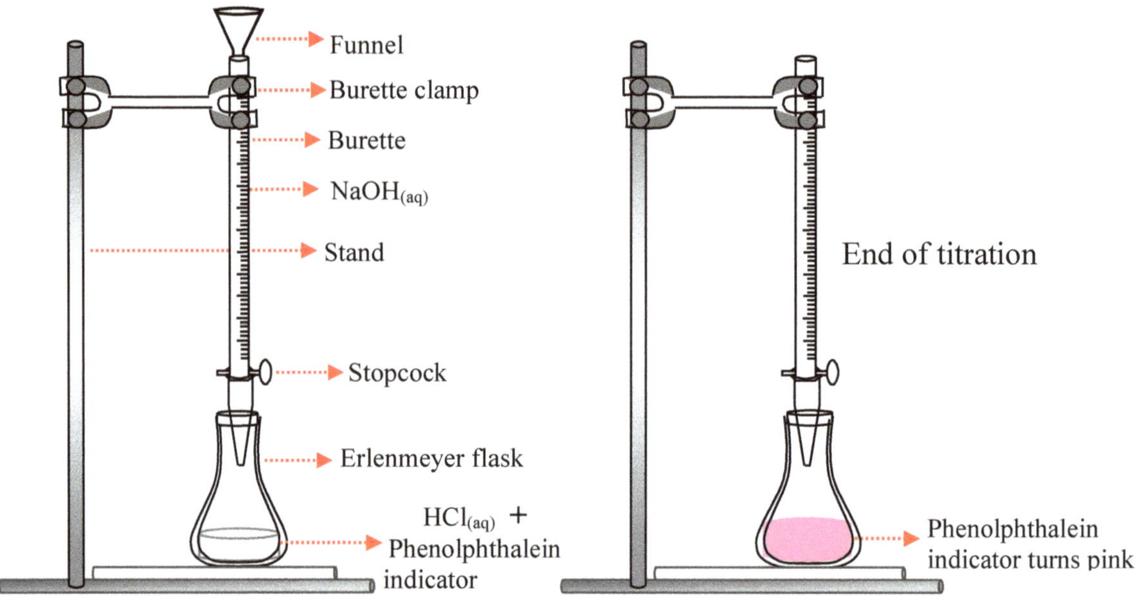

Figure 16.10(a). Titration of $HCl_{(aq)}$ with $NaOH_{(aq)}$. **Figure 16.10(b).** End point of titration

The following sample problem serves to outline how a solution of $HCl_{(aq)}$ of unknown concentration is determined by titrating it with a base of known concentration.

Monoprotic Acid versus Monoprotic Base

Sample problem5:

The following results were obtained when 10.0 mL aliquots of the $HCl_{(aq)}$ of unknown concentration were titrated with 0.100 mol/L $NaOH_{(aq)}$. Calculate the concentration of the $HCl_{(aq)}$.

Table 16.7. Volume of 0.100 mol/L $NaOH_{(aq)}$ used.

Trial	1	2	3	4
Final burette reading (mL)	11.0	21.5	31.7	41.9
Initial burette reading (mL)	0	11.0	21.5	31.7
Volume of $NaOH_{(aq)}$ used (mL)	11.0	10.5	10.2	10.2

Since the first trial is inconsistent with the others, its result can be discarded. The average volume of the $NaOH_{(aq)}$ used is found by using only the last three trials.

Step 1. Find the average of the titrant $NaOH_{(aq)}$.

Average volume of the $NaOH_{(aq)}$ used:

$$V_{av} = \frac{10.5\ mL + 10.2\ mL + 10.2\ mL}{3}$$
$$= 10.3\ mL$$
$$= 0.0103\ L$$

Step 2. Find the number of moles of the given quantity, **$NaOH_{(aq)}$**.

In this titration, since the volume and the concentration of the $NaOH_{(aq)}$ are known, its number of moles can be calculated using the formula, $n = C \times V$.

$$n_{NaOH} = \frac{0.100\ mol}{L} \times 0.0103\ L$$
$$= 0.00103\ mol$$

Note* The **given reagent** is the one of known **C** and **V** values as these are used to find **n**.

Step 3. Find the number of moles of the required quantity from the mole ratios in the balanced equation:

$$\underset{1\ mol}{\underset{(Given)}{NaOH_{(aq)}}} + \underset{1\ mol}{\underset{(Required)}{HCl_{(aq)}}} \rightarrow NaCl_{(aq)} + H_2O_{(l)}$$

$$\frac{Given}{Required} = \frac{1\ mol}{1\ mol} = \frac{0.00103\ mol\ NaOH}{X\ mol\ HCl}$$
$$X = 0.00103\ mol\ HCl$$

Step 4. Find the concentration of the $HCl_{(aq)}$ using the formula, $C = \dfrac{n}{V}$.

$$C_{HCl_{(aq)}} = \frac{0.00103\ mol}{0.0100\ L}$$
$$= 0.100\ mol/L$$

Diprotic Acid versus Monoprotic Base

The following results were obtained when 0.100 mol/L $NaOH_{(aq)}$ was titrated against 20.0 mL aliquots of $H_2SO_{4(aq)}$ of unknown concentration. Calculate the concentration of the $H_2SO_{4(aq)}$.

Table 16.8. Volume of 0.100 mol/L $NaOH_{(aq)}$ used.

Trial	1	2	3	4
Final burette reading (mL)	13.0	24.6	37.1	49.6
Initial burette reading (mL)	0	12.0	24.6	37.1
Volume of $NaOH_{(aq)}$ used (mL)	13.0	12.6	12.5	12.5

Solution:

Since the first trial is inconsistent with the others, its result can be discarded. The average volume is found by taking the last three trials.

Step 1. Find the average of the titrant, $NaOH_{(aq)}$.

Average volume of the $NaOH_{(aq)}$ used:

$$V_{av} = \frac{12.6\ mL + 12.5\ mL + 12.5\ mL}{3}$$

$$= 12.5\ mL$$

Step 2. Find the number of moles of the given quantity, **$NaOH_{(aq)}$**.

In this titration, since the volume and the concentration of the $NaOH_{(aq)}$ are known, its number of moles can be calculated using the formula, $n = C \times V$.

$$n_{NaOH(aq)} = 0.100\ mol\ /\ L \times 0.0125\ L$$

$$= 0.00125\ mol$$

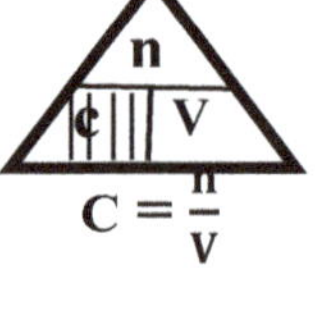

$$n = C \times V$$

Step 3. Find the number of moles of the required quantity, **$H_2SO_{4(aq)}$**, from the mole ratios in the balanced equation:

(Given) *(Required)*

$2\ NaOH_{(aq)}\ +\ H_2SO_{4(aq)}\ \rightarrow\ Na_2SO_{4(aq)}\ +\ 2\ H_2O_{(l)}$

$2\ mol$ $1\ mol$

$$\frac{Given}{Required} = \frac{2\ mol}{1\ mol} = \frac{0.00125\ mol\ NaOH}{X\ mol\ H_2SO_4}$$

$$2X = 0.00125\ mol\ H_2SO_4$$

$$X = 0.000625\ mol\ H_2SO_4$$

Step 4. Find the concentration of the $H_2SO_{4(aq)}$ using the formula, $C = \frac{n}{V}$.

$$C_{H_2SO_4\ (aq)} = \frac{0.000625\ mol}{0.0200\ L}$$

$$= 0.0310\ mol\ /\ L$$

Diprotic Acid versus Diprotic Base

A student obtained the following results when 0.100 mol/L $Ba(OH)_{2(aq)}$ was titrated against 10.0 mL aliquots of $H_2SO_{4\,(aq)}$ of unknown concentration. Calculate the concentration of the $H_2SO_{4(aq)}$.

Sample problem 7:

Table 16.9. Volume of 0.100 mol/L $Ba(OH)_{2(aq)}$ used.

Trial	1	2	3	4
Final burette reading (mL)	12.0	23.2	34.3	35.4
Initial burette reading (mL)	0	12.0	23.2	34.3
Volume of $Ba(OH)_{2(aq)}$ used (mL)	12.0	11.2	11.1	11.1

Solution:

Since the first trial is inconsistent with the others, its result can be discarded. The average volume is found by taking the last three trials.

Step 1. Find the average volume of the titrant, $Ba(OH)_{2(aq)}$.

Average volume of the $Ba(OH)_{2(aq)}$ used:

$$V_{av} = \frac{11.2\,mL + 11.1\,mL + 11.1\,mL}{3}$$
$$= 11.1\,mL$$

Step 2. Find the number of moles of the given quantity, **$Ba(OH)_{2(aq)}$**.

In this titration, since the volume and the concentration of the $Ba(OH)_{2(aq)}$ are known, its number of moles can be calculated using the formula, $n_{Ba(OH)_2(aq)} = C \times V$.

$$n_{Ba(OH)_2(aq)} = 0.100\,mol\,/\,L \times 0.0111\,L$$
$$= 0.00111\,mol$$

$n = C \times V$

Step 3. Find the number of moles of the required quantity, **$H_2SO_{4(aq)}$**, from the mole ratios in the balanced equation:

(Given)   (Required)

$$Ba(OH)_{2(aq)} \; + \; H_2SO_{4(aq)} \; \rightarrow \; BaSO_{4(s)} \; + \; 2H_2O_{(l)}$$

$$1\,mol \qquad\qquad 1\,mol$$

$$\frac{Given}{Required} = \frac{1\,mol}{1\,mol} = \frac{0.00111\,mol\;Ba(OH)_2}{X\,mol\;H_2SO_4}$$

$$X = 0.00111\,mol\;H_2SO_4$$

Step 4. Find the concentration of the $H_2SO_{4(aq)}$ using the formula, $C = \dfrac{n}{V}$.

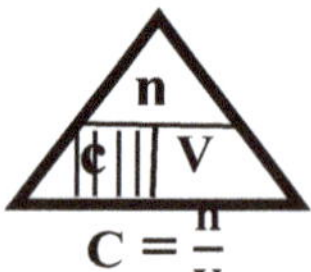

$$C_{H_2SO_4(aq)} = \frac{0.00111\ mol}{0.0100\ L}$$

$$= 0.110\ mol\,/\,L$$

Monoprotic Acid versus Solution

Sample problem 8:

The results were obtained when 10.0 mL aliquots of the $HCl_{(aq)}$ of unknown concentration were titrated with a 0.200 mol/L $Na_2CO_{3(aq)}$. Find the concentration of the $HCl_{(aq)}$.
Note* $Na_2CO_{3(aq)}$ is a weak base.

Table 16.10. Volume of 0.200 mol/L $Na_2CO_{3(aq)}$ used.

Trial	1	2	3	4
Final burette reading (mL)	21.0	41.4	20.5	40.9
Initial burette reading (mL)	0	21.0	0	20.5
Volume of $Na_2CO_{3(aq)}$ used (mL)	21.0	20.4	20.5	20.4

Solution:

Since the first trial is inconsistent with the others, its result can be discarded. The average *volume is found by taking the last three trials.*

Step 1. Find the average volume of the **titrant,** $Na_2CO_{3(aq)}$, used.

Average volume of the $Na_2CO_{3(aq)}$ used:

$$V_{av} = \frac{20.4\ mL + 20.5\ mL + 20.4\ mL}{3}$$

$$= 20.4\ mL$$

Step 2. Find the number of moles of the given quantity, **$Na_2CO_{3(aq)}$.**

In this titration, since the volume and the concentration of the $NaOH_{(aq)}$ are known, its number of moles can be calculated using the formula,

$$n_{Na_2CO_3(aq)} = C \times V.$$

$$n_{Na_2CO_3(aq)} = 0.200\ mol\,/\,L \times 0.0.0204\ L$$

$$= 0.00408\ mol$$

286

Step 3. Find the number of moles of the required quantity, $HCl_{(aq)}$, from the mole ratios in the balanced equation:

$$Na_2CO_{3(aq)} + 2\,HCl_{(aq)} \rightarrow 2\,NaCl_{(aq)} + H_2O_{(l)} +$$

$$\text{1 mol} \qquad \text{2 mol}$$

$$\frac{Given}{Required} = \frac{1\,mol}{2\,mol} \frac{0.00408\,mol\,Na_2CO_3\,(aq)}{X\,mol\,HCl}$$

$$X = 0.00816\,mol\,HCl$$

Step 4. Find the concentration of the $HCl_{(aq)}$ using the formula,

$$C = \frac{n}{V}.$$

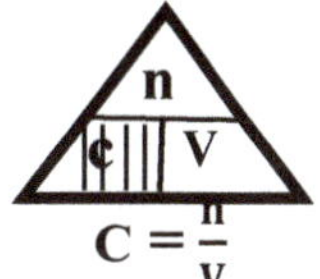

$$C_{HCl(aq)} = \frac{0.00816\,mol}{0.0100\,L}$$

$$= 0.816\,mol\,/\,L$$

Exercise 17.6   T/I

a) The table next shows the results obtained when a 0.100 mol/L $LiOH_{(aq)}$ is titrated against 10.0 mL aliquots of $H_2SO_{4\,(aq)}$ of unknown concentration:

Table 16.11. Volume of 0.100 mol/L $LiOH_{(aq)}$ used.

Trial	1	2	3	4
Final burette reading (mL)	10.2	20.3	30.4	30.5
Initial burette reading (mL)	0	10.2	20.3	30.4
Volume of $LiOH_{(aq)}$ used (mL)				

Use the evidence provided in the table to calculate the concentration of the $H_2SO_{4\,(aq)}$.

b) The table below shows the results obtained when a 0.200 mol/L $Ba(OH)_{2(aq)}$ is titrated against 10.0 mL aliquots of $HCl_{(aq)}$ of unknown concentration:

Table 16.12. Volume of 0.200mol/L $Ba(OH)_{2(aq)}$ used.

Trial	1	2	3	4
Final burette reading (mL)	12.4	24.7	37.1	49.5
Initial burette reading (mL)	0	12.4	24.7	37.1
Volume of $Ba(OH)_{2(aq)}$ used (mL)				

Use the evidence provided in the table to calculate the concentration of the HCl$_{(aq)}$.

c) The table next shows the results obtained when a solution of Ca(OH)$_{2(aq)}$ of unknown concentration is titrated against 10.0 mL aliquots of 0.100 mol/L HCl$_{(aq)}$

Table 16.13. Volume of Ca(OH)$_{2(aq)}$ of unknown concentration used.

Trial	1	2	3	4
Final burette reading (mL)	13.4	25.7	38.1	50.5
Initial burette reading (mL)	0	12.4	24.7	37.1
Volume of Ba(OH)$_{2(aq)}$ used (mL)				

Use the evidence provided in the table to calculate the concentration of the Ca(OH)$_{2(aq)}$.

d) The table below shows the results obtained when a solution of Na$_2$CO$_{3(aq)}$ of unknown concentration is titrated against 10.0 mL aliquots of 0.100 mol/L H$_2$SO$_{4(aq)}$

Table 16.14. Volume of Na$_2$CO$_{3(aq)}$ of unknown concentration used.

Trial	1	2	3	4
Final burette reading (mL)	15.2	30.3	45.5	50.7
Initial burette reading (mL)	0	15.2	30.3	35.5
Volume of Na$_2$CO$_{3(aq)}$ used (mL)				

i) Use the evidence provided in the table to calculate the concentration of the   Na$_2$CO$_{3(aq)}$.

ii) What mass of Na$_2$CO$_3$ must have been used to make 500.0 mL of Na$_2$CO$_{3(aq)}$?

17.9 pH of Solutions

Auto-Ionization of Water

Even though water is a molecular compound, the use of very sensitive conductivity instruments shows that it does conduct electricity extremely slightly. Scientists attribute this to a small degree of auto-ionization of water molecules, that produces H$^+_{(aq)}$ and   OH$^-_{(aq)}$ ions. The following equation illustrates how water auto-ionizes:

$$H_2O_{(l)} \rightleftharpoons H^+{}_{(aq)} + OH^-{}_{(aq)}$$

Equilibrium concentration: 1.0×10^{-7} *mol/L* 1.0×10^{-7} *mol/L*

The concentrations of the ions shown in the equation above were obtained experimentally for water at 25 ^{0}C. As can be seen, the concentrations of the $H^+{}_{(aq)}$ and $OH^-{}_{(aq)}$ produced are extremely small. Despite the fact that water undergoes auto-ionization, it still remains a neutral compound. This is due to the fact that when water molecules auto-ionize, they produce equal concentrations of H$^+_{(aq)}$ ions and OH$^-_{(aq)}$ ions.

However, the addition of any substance to water that changes this equality in concentration between $H^+_{(aq)}$ ions and $OH^-_{(aq)}$ ions will produce an aqueous solution that is either *acidic* or *basic.*

It should be noted that when $H^+_{(aq)}$ ions are produced, they quickly bond with water molecules forming hydronium ions. This is illustrated as follows:

$$H^+_{(aq)} \;+\; H_2O_{(l)} \quad \rightarrow \quad H_3O^+_{(aq)}$$
$$\text{(hydronium ion)}$$

The $H_3O^+_{(aq)}$ ions have the same acidic property as free $H^+_{(aq)}$ ions, as they can quickly decompose to produce $H^+_{(aq)}$ ions when required for neutralization. ***For the purpose of this text****, it shall be assumed that the $H^+_{(aq)}$ ions are representative of the $H_3O^+_{(aq)}$ ions.*

Ionic Product of Water: K_w

This concept will be further developed in your Grade 12 chemistry course. But, for the sake of clarity it will be briefly introduced here.

The product of the $H^+_{(aq)}$ and $OH^-_{(aq)}$ ion concentrations in water is called the ionic product of water, symbolized K_w, and is represented by the equation below:

$$K_w \;=\; [H^+_{(aq)}] \times [OH^-_{(aq)}]$$

$$\textbf{At 25 °C, } K_w \;=\; (1.0 \times 10^{-7}) \times (1.0 \times 10^{-7}) = 1.0 \times 10^{-14}$$

The $[H^+_{(aq)}]$ and $[OH^-_{(aq)}]$ are both 1.0×10^{-7} mol/L, which is a very small value.

It must be emphasized that at 25° C, for any **aqueous solution**; neutral, acidic or basic, the ionic product of water **will always be 1.0×10^{-14}**. Based on this understanding, the following can be inferred that for *any aqueous solution:*

➢ If the *$[H^+_{(aq)}]$* **increases,** then the *$[OH^-_{(aq)}]$* must **decrease** proportionally such that their ionic product, 1.0×10^{-14}, remains the same at 25 °C.

➢ If the *$[OH^-_{(aq)}]$* **increases,** then the *$[H^+_{(aq)}]$* the must **decrease** proportionally such that their ionic product, 1.0×10^{-14}, remains the same at 25 °C.

Based on the above inferences, the following equations can be derived for water.

$$[H^+_{(aq)}] \;=\; \frac{K_w}{[OH^-_{(aq)}]} \qquad \text{and} \qquad [OH^-_{(aq)}] \; \frac{K_w}{[H^+_{(aq)}]}$$

Addition of an Acid to water.

Recall that acids produce $H^+_{(aq)}$ ions when they dissolve in water. Therefore, **when an acid is added to water, the *$[H^+_{(aq)}]$* ion increases above the 1.0×10^{-7} mol/L concentration. To maintain the ionic product of water, the *$[OH^-_{(aq)}]$* ion must decrease proportionately below its 1.0×10^{-7} mol/L concentration.** How this happens, is illustrated and explained below.

Before the addition of acid

$$K_w \;=\; [H^+_{(aq)}] = [OH^-_{(aq)}]$$
$$= (1.0 \times 10^{-7}) \times (1.0 \times 10^{-7})$$
$$= 1.0 \times 10^{-14}$$

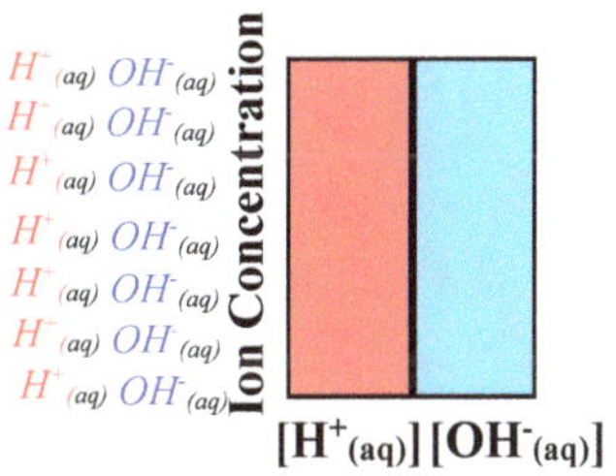

Figure 17.02. Equal $H^+_{(aq)}$ and $OH^-_{(aq)}$ concentrations.

According to the following illustrations, the $H^+_{(aq)}$ ions added from the acid neutralize some of the $OH^-_{(aq)}$ ions present in water to form neutral water molecules. The result is that the $[H^+_{(aq)}]$ ion increases and the $[OH^-_{(aq)}]$ ion decreases, making the solution acidic.

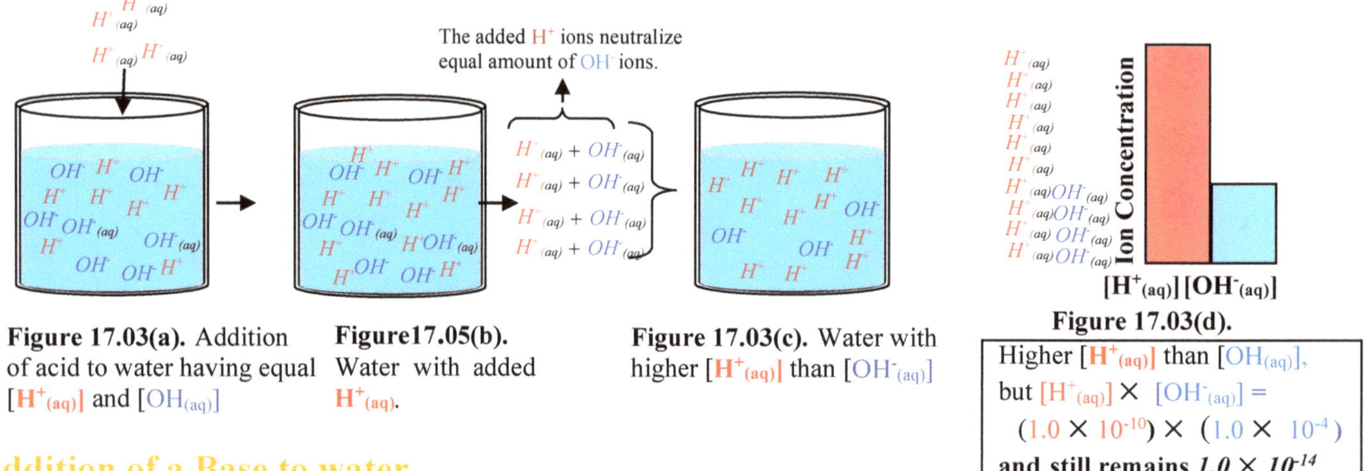

Figure 17.03(a). Addition of acid to water having equal [**H⁺**(aq)] and [OH(aq)]

Figure 17.05(b). Water with added **H⁺**(aq).

Figure 17.03(c). Water with higher [**H⁺**(aq)] than [OH⁻(aq)]

Figure 17.03(d).
Higher [**H⁺**(aq)] than [OH(aq)], but [H⁺(aq)] × [OH⁻(aq)] =
$(1.0 \times 10^{-10}) \times (1.0 \times 10^{-4})$
and still remains 1.0×10^{-14}

Addition of a Base to water.

Recall that bases produce $OH^-_{(aq)}$ ions when they dissolve in water. Therefore, **when a base is added to water, the $[OH^-_{(aq)}]$ ion increases above the 1.0×10^{-7} mol/L concentration. To maintain the ionic product of water, the $[H^+_{(aq)}]$ must decrease proportionally below its 1.0×10^{-7} mol/L concentration.** How this happens is illustrated and explained below.

$$K_w = [H^+_{(aq)}] \times [OH^-_{(aq)}]$$
$$(1.0 \, x10^{-7}) \times (1.0 \, x10^{-7})$$
$$= 1.0 \times 10^{-14}$$

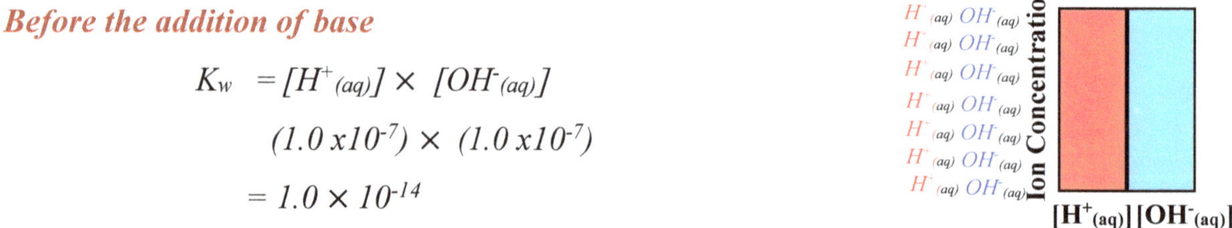

Figure 17.04. Equal H⁺(aq) and OH⁻(aq) concentrations.

According to the following illustrations, the $OH^-_{(aq)}$ ions added from the base neutralize some of the $H^+_{(aq)}$ ions present in water to form neutral water molecules. The result is that the $[H^+_{(aq)}]$ decreases and the $[OH^-_{(aq)}]$ increases making the solution basic.

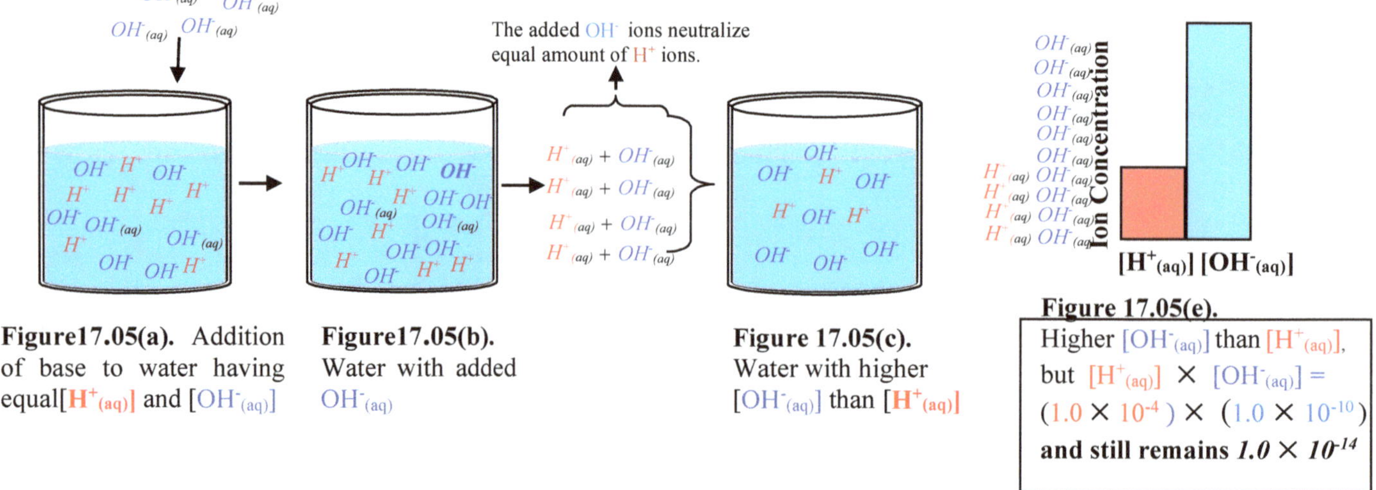

Figure17.05(a). Addition of base to water having equal[**H⁺**(aq)] and [OH⁻(aq)]

Figure17.05(b). Water with added OH⁻(aq)

Figure 17.05(c). Water with higher [OH⁻(aq)] than [**H⁺**(aq)]

Figure 17.05(e).
Higher [OH⁻(aq)] than [H⁺(aq)], but [H⁺(aq)] × [OH⁻(aq)] =
$(1.0 \times 10^{-4}) \times (1.0 \times 10^{-10})$
and still remains 1.0×10^{-14}

Summary:
- ➢ **In acidic solutions:**

  $[H^+_{(aq)}] > [OH^-_{(aq)}]$

- ➢ **In basic solutions:**

  $[OH^-_{(aq)}] > [H^+_{(aq)}]$

- ➢ **In neutral solutions:**

  $[H^+_{(aq)}] = [OH^-_{(aq)}]$

* **Ionic product of water at 25 ^{0}C**

  $K_w = [H^+_{(aq)}]\,[OH^-_{(aq)}] = 1.0 \times 10^{-14}$

* **Hydogen ion concentration:**

  $[H^+_{(aq)}] = \dfrac{K_w}{[OH^-_{(aq)}]}$

* **Hydroxyl ion concentration:**

  $[OH^-_{(aq)}] = \dfrac{K_w}{[H^+_{(aq)}]}$

> *In all aqueous solutions; salts, acids and bases, the product of $[OH^-_{(aq)}]$ and $[H^+_{(aq)}]$ must be 1.0×10^{-14} at 25 ^{0}C.*

pH (Power of hydrogen) Calculations

The expression, pH has been associated with the way of indicating the concentration of hydrogen ions $[H^+_{(aq)}]$ in aqueous solutions. Developed about 100 years ago by Danish scientist Sørenson, is expressed mathematically as:

$$pH = -\log [H^+_{(aq)}]$$

Sample problem 9:

What is the pH of water at 25°C?

Solution:

Provided: $[H^+_{(aq)}] = 1.0 \times 10^{-7}$ mol/L (the concentration of $H^+_{(aq)}$ at 25 °C is 1.0×10^{-7} mol/L).

Required quantity: pH

Using the formula, $pH = -\log [H^+_{(aq)}]$, we get:

$$pH = -\log 1.0 \times 10^{-7} \; mol/L$$
$$= 7.0$$

Sample problem 10:

Calculate the pH of a 0.10 mol/L $HCl_{(aq)}$.

Solution:

To solve this problem, it is useful to follow the step below.

Step 1. Complete an equation to show the number of moles of $H^+_{(aq)}$ ions produced per mole of the acid as it ionizes.

$$HCl_{(aq)} \longrightarrow H^+_{(aq)} + Cl^-_{(aq)}$$
$$1.0 \; mol/L \qquad\quad 1.0 \; mol/L \quad\; 1.0 \; mol/L$$

Step 2. Calculate the $[H^+_{(aq)}]$ in the solution.

$$[H^+_{(aq)}] = \frac{0.10 \text{ mol/L HCl}_{(aq)} \times \dfrac{1 \text{ mol/L H}^+_{(aq)}}{1 \text{ mol/L HCl}_{(aq)}}}{}$$

$$= 0.10 \text{ mol/L H}^+_{(aq)}$$

Step 3. Calculate the pH using the expression:

$$pH = -log\ [H^+_{(aq)}]$$

$$pH = -log\ 0.10 \text{ mol/L } H^+_{(aq)}$$

$$= 1.0$$

Calculating the pH of a Strong Diprotic Acids

When strong diprotic acids such as $H_2SO_{4(aq)}$ ionize in water, they do so in stages as the following equations show:

$$H_2SO_{4(aq)} \longrightarrow H^+_{(aq)} + HSO_4^-{}_{(aq)}$$

$$1 mol/L \qquad\qquad 1\ mol/L \quad 1\ mol/L$$

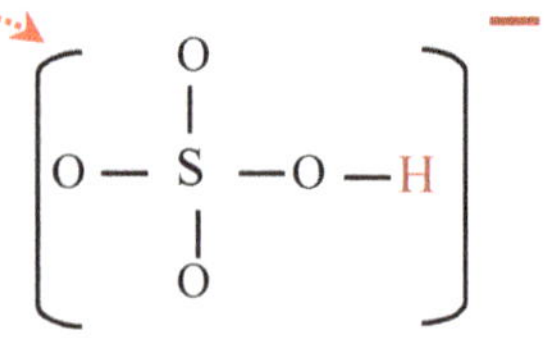

Figure 17.05(d). HSO4⁻ ion with negative charge

Since these acids are strong, ***there is complete ionization of the first mole of $H^+_{(aq)}$ ions** per mole of the acid.* However, for the second mole of $H^+_{(aq)}$ ions still present in the hydrogen sulphate ions, ***HSO_4^-{}_{(aq)}***, (figure besides equation above) this is not the case.

Note that in the case of $H_2SO_{4(aq)}$, there is no negative (-) charge present on the molecule, so the first mole of hydrogen atoms was relatively easy to ionize. But, because of the negative charge (-) now present on the ***HSO_4^-{}_{(aq)}*** ion, it exerts a force of attraction on the positively charged $H^+_{(aq)}$ ion that wants to leave. The result is that there is only partial ionization of the $HSO_4^-{}_{(aq)}$ ions as indicated by the following equation:

$$HSO_4^-{}_{(aq)} \rightleftharpoons H^+_{(aq)} + SO_4^{2-}{}_{(aq)}$$

A book of data shows that $HSO_4^-{}_{(aq)}$ ions undergo only 20% ionization.

To calculate the pH of diprotic acids, it is necessary to determine the sum of the $[H^+_{(aq)}]$ produced in both ionization steps. The following sample illustrates how this is done for $H_2SO_{4(aq)}$.

Sample problem 11:

Calculate the pH of a 0.010 mol/L $H_2SO_{4(aq)}$.

Solution:

Step 1. Complete an equation to show the number of moles of $H^+_{(aq)}$ ions produced per mole of acid as it undergoes its first ionization process.

$$H_2SO_{4(aq)} \longrightarrow H^+_{(aq)} + HSO_4^-{}_{(aq)}$$

$$1\ mol/L \qquad\qquad 1\ mol/L$$

Step 2. Calculate the $[H^+_{(aq)}]$ in the solution.

$$[H^+_{(aq)}] = 0.010 \; mol/L \; H_2SO_{4(aq)} \times \frac{1 \; mol/L \; H^+_{(aq)}}{1 \; mol/L \; H_2SO_{4(aq)}}$$

$$= 0.010 \; mol/L \; H^+_{(aq)}$$

Step 3. Calculate the $[H^+_{(aq)}]$ produced from the ionization of $HSO_4^-{}_{(aq)}$ during the second ionization process, using the formula:

$$[H^+_{(aq)}] = \frac{\%\; ionization \times initial \; acid \; concentration}{100\%}$$

$$[H^+_{(aq)}] = \frac{20\% \times 0.010 \; mol/L}{100\%}$$

$$= 0.0020 \; mol/L$$

Step 4. Add the $[H^+_{(aq)}]$ ion concentrations obtained from the two ionization processes.

$$= \; 0.010 \; mol/L + 0.002 \; mol/L$$

$$= \; 0.012 \; mol/L$$

Step 5. Calculate the pH of the solution.

$$pH \;\; = -log \; 0.012$$

$$= 1.92$$

Note that if the $[H^+_{(aq)}]$ ions for the first ionization were only used to calculate the pH, we would have:

$$pH = -log \; 0.010 = \textbf{2.0} \; \text{(for first ionization process only)}$$

The difference between this value and that obtained in **Step 5** is very small, **0.08**. *So, using the $[H^+_{(aq)}]$ ions for the first ionization process only, **it may suffice** to calculate the final pH of strong diprotic acids.*

What pH would result if 6.30 g of $HNO_{3(aq)}$ is dissolved in water to make 200.0 mL of $HNO_{3(aq)}$?

To do this, it is useful to follow the steps below:

Step 1. Find the number of moles of $HNO_{3(aq)}$.

$$n_{HNO_3(aq)} = \frac{m}{M}$$

$$= \frac{6.30 \; g}{63.02 \; g/mol}$$

$$= 0.0100 \; mol$$

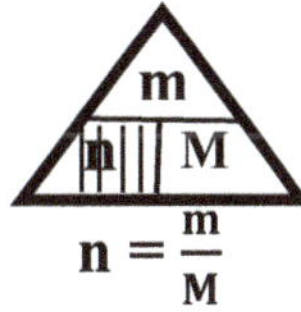

Step 2. Calculate the concentration of $HNO_{3(aq)}$, using the formula:

$$C_{HNO_3(aq)} = \frac{n}{V} \times$$

$$= \frac{0.0100 \ mol}{0.200 \ L}$$

$$= 0.0500 \ mol \ / \ L$$

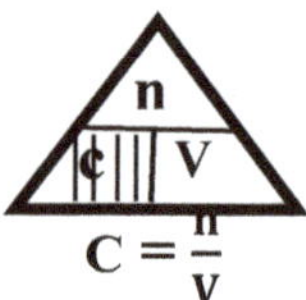

Step 3. Complete an equation to show the number of moles of $H^+_{(aq)}$ ions produced per mole of acid as it ionizes.

$$HNO_{3(aq)} \quad \longrightarrow \quad H^+_{(aq)} \quad + \quad NO_3^-_{(aq)}$$

$$1 \ mol/L \qquad\qquad\qquad 1 \ mol/L$$

Step 4. Calculate the $[H^+_{(aq)}]$ in the solution.

$$[H^+_{(aq)}] = 0.0500 \ mol \ / \ L \ HNO_{3(aq)} \times \frac{1 \ mol \ / \ L \ H^+_{(aq)}}{1 \ mol \ / \ L \ HNO_{3(aq)}}$$

$$= 0.0500 \ mol \ / \ L \ H^+_{(aq)}$$

Step 5. Calculate the pH using the expression, $pH = -log \ [H^+_{(aq)}]$.

$$pH = -log \ 0.0500$$

$$= 1.30$$

Calculating $[H^+_{(aq)}]$ from given pH

If the $[H^+_{(aq)}]$ in an acidic solution is found to be **0.10 mol/L**, then its calculated **pH** will be **1.0**. It is only logical then, that if a solution has a **pH** of **1.0**, its $[H^+_{(aq)}]$ must be **0.10 mol/L**. Based on these relations, the following mathematical expression was established to calculate $[H^+_{(aq)}]$ ions for any given pH:

$$[H^+_{(aq)}] = 10^{-pH} \ mol/L$$

Sample problem 13:

What is the $[H^+_{(aq)}]$ of an aqueous solution that has a pH of 2.0?

Solution:

Using the formula: $[H^+_{(aq)}] = 10^{-pH}$ *we get :*

$$[H^+_{(aq)}] = 1.0 \times 10^{-2} \ mol/L$$

Exercise 17.7

Calculate the pH of the following solutions: T/I

a) 1.0×10^{-3} mol/L $HNO_{3(aq)}$

b) 1.2×10^{-2} mol/L $HCl_{(aq)}$

c) 2.0×10^{-4} mol/L $HClO_{3(aq)}$

Exercise 17.8

Determine the pH of following solutions that has: T/I

a) 6.30 g $HNO_{3(aq)}$ dissolved in 1.0 L of solution

b) 18.25 g $HCl_{(aq)}$ dissolved in 2.0 L of solution

c) 8.45 g $HClO_{3(aq)}$ dissolved in 4.0 L of solution

d) 1.28 g $HI_{(aq)}$ dissolved in 400.0 mL of solution

e) 9.11 g $HCl_{(aq)}$ dissolved in 4.0 L of solution

Exercise 17.9

Calculate the $[H^+_{(aq)}]$ in the following solution: T/I

a) $H_2SO_{4(aq)}$ of pH 2.0

b) $HCl_{(aq)}$ of pH 4.0

c) $HClO_{3(aq)}$ of pH 4.4

d) $HI_{(aq)}$ of pH 3.0

e) $CH_3COOH_{(aq)}$ of pH 5.0

17.10   pH and Dilution

To understand how the pH of an acidic solution changes as it is diluted, the following sample problem is first solved:

Sample problem 1:

What will be the change in pH if 10.00 mL of 0.0100 mol/L $HNO_{3(aq)}$ is diluted to 100.0 mL?

Solution:

To see how the pH of an acidic solution changes with dilution, its original pH is first found and then this is compared with its pH after dilution.

Step 1. Complete the equation to show the mole ratio of $H^+_{(aq)}$ per moles of acid as it ionizes.

$$HNO_{3(aq)} \quad \rightarrow \quad H^+_{(aq)} \quad + \quad NO_3^-{}_{(aq)}$$

$$1 \; mol/L \qquad\qquad 1 \; mol/L$$

Step 2. Calculate the $[H^+_{(aq)}]$ in the solution of $HNO_{3(aq)}$.

$$[H^+_{(aq)}] = \; 0.0100 \; mol/L \; HNO_{3(aq)} \times \frac{1 \; mol/L \; H^+_{(aq)}}{1 \; mol/L \; HNO_{3(aq)}}$$

$$= \; 0.0100 \; mol/L \; H^+$$

Step 3. Calculate the original pH of the $HNO_{3(aq)}$ solution.

$$pH \; = \; -\log 0.0100$$
$$= \; 2.00$$

Step 4. Find the final concentration of the $[H^+_{(aq)}]$ in the diluted solution.

Since this involves dilution of a solution the formula, $C_1 \times V_1 = C_2 \times V_2$ is used to find the final concentration C_2, of $[H^+_{(aq)}]$.

$$C_{2[H^+_{(aq)}]} = \frac{C_1 \times V_1}{V_2}$$

$$= \frac{0.0100 \; mol/L \times 0.0100 \; L}{0.100 \; L}$$

$$= 0.00100 \; mol/L$$

Step 5. Calculate the final pH of the solution.

$$pH = -\log 0.00100$$
$$= 3.00$$

> • Based on these calculations, it can be concluded that **as an acidic solution is diluted, *its pH increases.***

The pH scale

Different aqueous solutions have different $[H^+_{(aq)}]$ and thus different pH. Grouping all of these possible values together produces a pH scale from which the $[H^+_{(aq)}]$ for various aqueous solutions can be compared. The following pH scale gives the pH for a few substances:

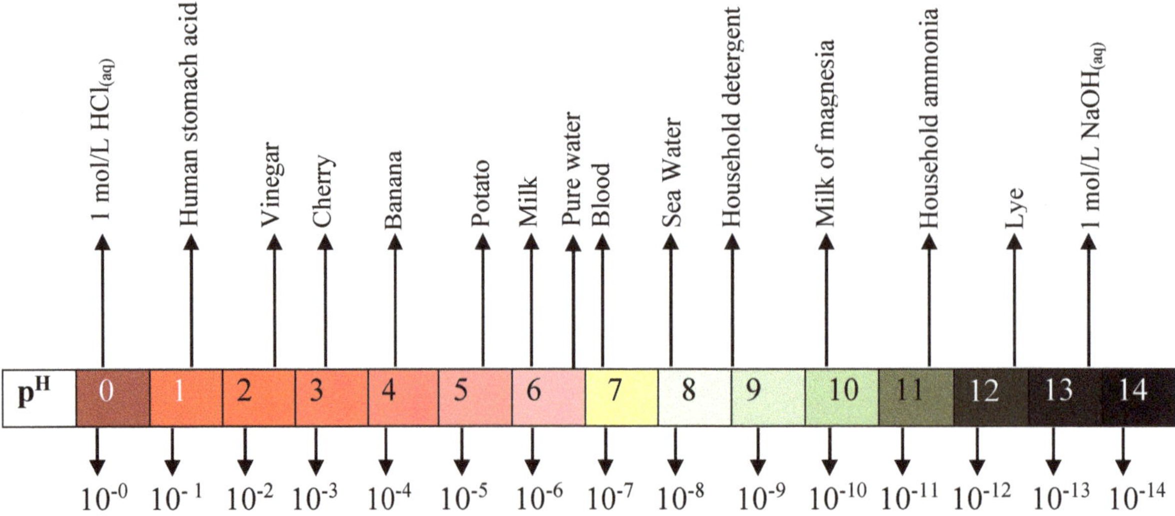

Figure 17.06. The pH scale with corresponding $[H^+_{(aq)}]$ with some substances.

From the pH scale above, a number of relationships can be formulated between pH and $[H^+_{(aq)}]$ changes. The following table shows some of these as they relate to the dilution of an acid.

Table 17.06. Dilution of an acid and corresponding pH changes.

Original pH and $[H^+_{(aq)}]$	Final pH and $[H^+_{(aq)}]$	Change in pH (final – original)	Number of times $[H^+_{(aq)}]$ is less in final than in original
$[H^+_{(aq)}] = 1.0 \times 10^{-1}$ mol/L pH = 1.0	$[H^+_{(aq)}] = 1.0 \times 10^{-2}$ mol/L pH = 2.0	1	**10** times less
$[H^+_{(aq)}] = 1.0 \times 10^{-1}$ mol/L pH = 1.0	$[H^+_{(aq)}] = 1.0 \times 10^{-3}$ mol/L pH = 3.0	2	**100** times less
$[H^+_{(aq)}] = 1.0 \times 10^{-1}$ mol/L pH = 1.0	$[H^+_{(aq)}] = 1.0 \times 10^{-4}$ mol/L pH = 4.0	3	**1 000** times less
$[H^+_{(aq)}] = 1.0 \times 10^{-2}$ mol/L pH = 2.0	$[H^+_{(aq)}] = 1.0 \times 10^{-6}$ mol/L pH = 6.0	4	**10 000** times less

Student Misconception:

Because there is a tenfold increase in $[H^+_{(aq)}]$ for every pH change of 1, students commonly make the mistake, when trying to find the change in $[H^+_{(aq)}]$, by multiplying the changes in pH by a factor of 10. For example, if there was a change in pH of 3 and students were required to find the change in $[H^+_{(aq)}]$, some of them would multiply the 3 by 10. The result here would 30, which is totally different from the correct answer of 1000 which is 10^3. The table above gives useful information how to find change in hydrogen ion concentration $[H^+_{(aq)}]$ for changes in pH.

17.11 pOH of Solutions

Like pH, which is a way of indicating the concentration of $[H^+_{(aq)}]$ in an aqueous solution, pOH is used to indicate the $[OH^-_{(aq)}]$ in aqueous solutions. The mathematical principle used to calculate the pOH of solutions is similar to that used to calculate pH.

$$pOH = -\log [OH^-_{(aq)}]$$

The following sample problem will illustrate how the pOH of an aqueous solution of strong base is calculated.

Sample problem 2:

What is the pOH of 0.0100 mol/L $NaOH_{(aq)}$ solution?

Solution:

To solve this problem, it is useful to follow the steps below:

Step 1. Complete the equation to show the mole ratio of [OH⁻$_{(aq)}$] and [NaOH$_{(aq)}$] during ionization:

$$NaOH_{(aq)} \quad \rightarrow \quad Na^+{}_{(aq)} \quad + \quad OH^-{}_{(aq)}$$

$$1 \ mol/L \hspace{6cm} 1 \ mol/L$$

Step 2. Calculate the [OH⁻$_{(aq)}$] in the solution of NaOH$_{(aq)}$.

$$\left[OH^-{}_{(aq)} \right] = 0.0100 \ \cancel{mol/L} \ \cancel{NaOH}_{(aq)} \times \frac{1 \ mol/L \ OH^-{}_{(aq)}}{1 \ \cancel{mol/L} \ \cancel{NaOH}_{(aq)}}$$

$$= 0.0100 \ mol/L$$

Step 3. Calculate the pOH of the solution using the formula:

$$pOH = -\log [OH^-{}_{(aq)}]$$

$$pOH = -\log 0.0100$$

$$= 2.00$$

pKw and Calculations

$$\text{Recall that } K_w = 1.0 \times 10^{-14}$$

$$\mathbf{pKw = -\log kw}$$

$$= -\log 1.0 \times 10^{-14}$$

$$= 14$$

$$\text{Recall also that } K_w = [H^+{}_{(aq)}] \times [OH^-{}_{(aq)}]$$

$$\therefore \left[OH^-{}_{(aq)} \right] = \frac{K_w}{\left[H^+{}_{(aq)} \right]}$$

$$\text{and, } \left[H^+{}_{(aq)} \right] = \frac{K_w}{\left[OH^-{}_{(aq)} \right]}$$

Calculate the pH of the solution in the *sample problem above*, 0.010 mol/L NaOH$_{(aq)}$.

Step 1. Calculate the [H⁺$_{(aq)}$] using the formula:

$$\left[H^+{}_{(aq)} \right] = \frac{K_w}{\left[OH^-{}_{(aq)} \right]}$$

$$\left[H^+{}_{(aq)} \right] = \frac{1.0 \times 10^{-14}}{0.010}$$

$$= 1.00 \times 10^{-12} \ mol/L$$

Step 2. Calculate the pH of the solution using the formula:

$$pH = -\log [H^+{}_{(aq)}]$$

$$pH = -\log 1.00 \times 10^{-12}$$

$$= 12.0$$

Conclusion:

From the previous calculations it can be seen that the same solution; 0.01 mol/L $NaOH_{(aq)}$ has a **pH of 2.0** and a **pOH of 12.0** The sum of its pH and pOH is thus **14.0.**

Since pKw of any aqueous solution is also **14.0,** it would appear that a relationship can be formulated between pH, pOH and pKw as follows:

$$pKw = pH + pOH$$

From this the following the other relationships can also be established:

$$pH = pKw - pOH$$

$$pOH = pKw - pH$$

Sample problem 4:

What is the pH of a 0.00200 mol/L $KOH_{(aq)}$ solution?

Solution:

Since this is a base, it is first useful to find the pOH of the solution.

Step 1. Complete the equation to show the mole ratio of $[OH^-_{(aq)}]$ and $[KOH_{(aq)}]$ during ionization:

$$KOH_{(aq)} \quad \rightarrow \quad K^+_{(aq)} \quad | \quad OH^-_{(aq)}$$
$$1 \ mol/L \qquad\qquad\qquad 1 \ mol/L$$

Step 2. Calculate the $[OH^-_{(aq)}]$ in the solution of $KOH_{(aq)}$.

$$\left[OH^-_{(aq)}\right] = 0.00200 \ mol/L \ KOH_{(aq)} \times \frac{1 \ mol/L \ OH^-_{(aq)}}{1 \ mol/L \ KOH_{(aq)}}$$

$$= 0.00200 \ mol/L$$

Step 3. Calculate the pOH of the solution using the formula:

$$pOH = -log \ [OH^-_{(aq)}]$$

$$pOH = -log \ 0.00200 = \textbf{2.70}$$

$$pH = pKw - pOH$$

$$= 14.0 - 2.7 = \textbf{11.3}$$

The following table summarizes the interchanges among $[H^+_{(aq)}]$, $[OH^-_{(aq)}]$, pH, pOH, K_w and pKw.

Table 17.07. Relationships among $[H^+_{(aq)}]$, $[OH^-_{(aq)}]$, pH, pOH, K_w and pKw.

$[H^+_{(aq)}]$ mol/L	pH	pOH	$[OH^-_{(aq)}]$ mol/L	Kw	pH + pOH = pKw
0	0	14	10^{-14}	10^{-14}	14
10^{-1}	1	13	10^{-13}	10^{-14}	14
10^{-2}	2	12	10^{-12}	10^{-14}	14
10^{-3}	3	11	10^{-11}	10^{-14}	14
10^{-4}	4	10	10^{-10}	10^{-14}	14
10^{-5}	5	9	10^{-9}	10^{-14}	14
10^{-6}	6	8	10^{-8}	10^{-14}	14
10^{-7}	7	7	10^{-7}	10^{-14}	14
10^{-8}	8	6	10^{-6}	10^{-14}	14
10^{-9}	9	5	10^{-5}	10^{-14}	14
10^{-10}	10	4	10^{-4}	10^{-14}	14
10^{-11}	11	3	10^{-3}	10^{-14}	14
10^{-12}	12	2	10^{-2}	10^{-14}	14
10^{-13}	13	1	10^{-1}	10^{-14}	14
10^{-14}	14	0	0	10^{-14}	14

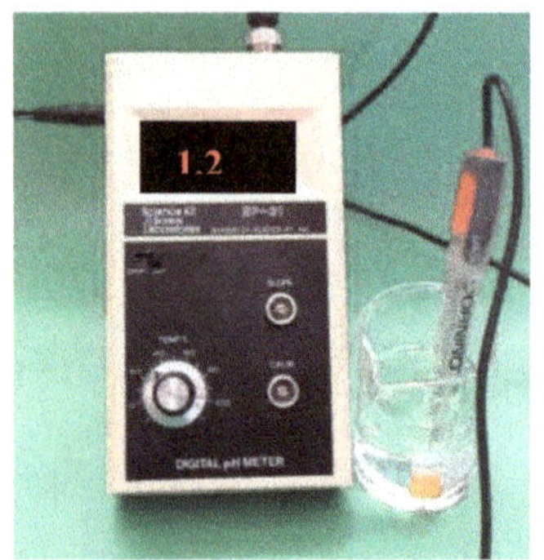

Figure 17.07(a).
pH of $HCl_{(aq)}$.

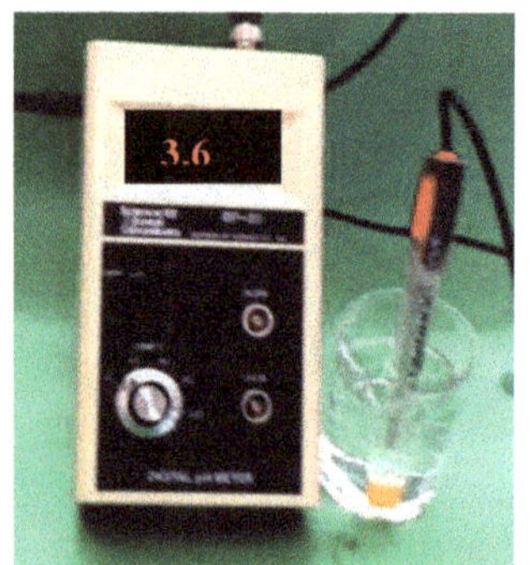

Figure 17.07(b).
pH of $CH_3COOH_{(aq)}$.

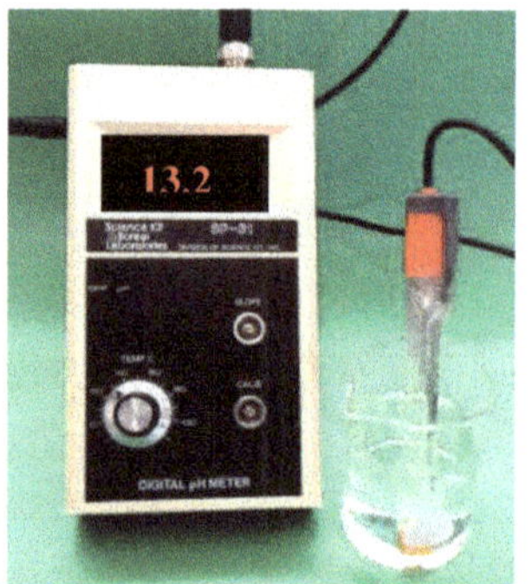

Figure 17.07(c).
pH of $NaOH_{(aq)}$.

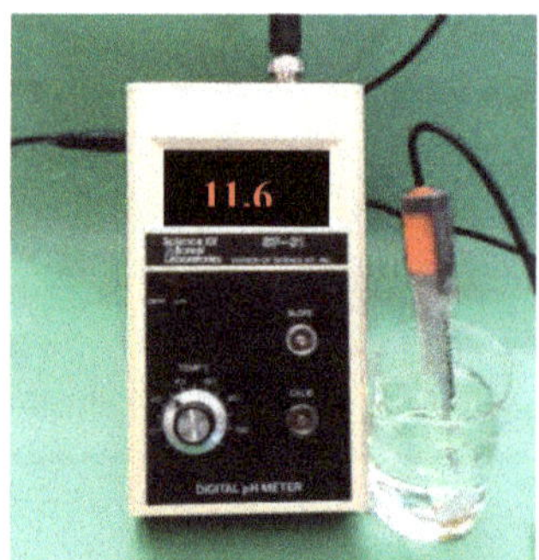

Figure 17.07(d).
pH of $NH_{3(aq)}$.

 ## pH and pOH Changes during an Acid-Base Titration

To understand how the pH of any aqueous solution changes as its $[H^+_{(aq)}]$ and $[OH^-]$ change, the following problem shall be examined.

Sample problem 5:

What will be the final pH and pOH of a solution produced from the mixing of 300.0 mL, *(0.3000 L)* of 0.100 mol/L $HCl_{(aq)}$ with 200.0 mL *(0.200 L)* of 0.100 mol/L $NaOH_{(aq)}$?

Solution:

To solve this problem, *the key is to find the final $[H^+_{(aq)}]$ or $[OH^-_{(aq)}]$ in the resulting mixture*. To do this, it is useful to follow the steps below:

Step 1. Calculate the number of moles of $HCl_{(aq)}$ in 0.100 mol/L $HCl_{(aq)}$ as this will lead to finding the number of moles of $H^+_{(aq)}$.

$$n_{HCl_{(aq)}} = 0.100 \ mol \ / \ L \times 0.3000 \ L$$

$$= 0.03000 \ mol$$

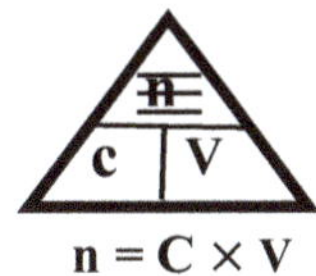

Step 2. Complete the equation to show the mole ratio of H$^+$$_{(aq)}$ ions produced per mole of HCl$_{(aq)}$ that ionizes.

$$HCl_{(aq)} \quad \rightarrow \quad H^+{}_{(aq)} \quad + \quad Cl^-{}_{(aq)}$$

$$1 \ mol \qquad\qquad 1 \ mol$$

Step 3. Calculate the number of moles of H$^+$$_{(aq)}$ in 0.030 L of 0.10 mol/L HCl$_{(aq)}$.

$$n_{H^+{}_{(aq)}} = 0.03000 \ mol \ HCl_{(aq)} \times \frac{1 \ mol \ H^+{}_{(aq)}}{1 \ mol \ HCl_{(aq)}}$$

$$= 0.03000 \ mol$$

Step 4. Calculate the number of moles of NaOH$_{(aq)}$ in 0.200 L of 0.100 mol/L NaOH$_{(aq)}$, as this will lead to finding the number of moles of OH$^-$$_{(aq)}$.

Using the equation, $n = C \times V$, we get:

$$n_{NaOH_{(aq)}} = 0.100 \ mol \ / \ L \times 0.200 \ L$$

$$= 0.0200 \ mol$$

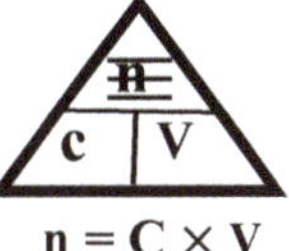

Step 5. Complete the equation to show the mole ratio of OH$^-$$_{(aq)}$ produced per mole of NaOH$_{(aq)}$ that dissociates:

$$NaOH_{(aq)} \quad \rightarrow \quad Na^+{}_{(aq)} \quad + \quad OH^-{}_{(aq)}$$

$$1 \ mol \qquad\qquad\qquad 1 \ mol$$

Step 6. Calculate the number of moles of OH$^-$$_{(aq)}$ in the solution.

$$n_{OH^-{}_{(aq)}} = 0.0200 \ mol \ NaOH_{(aq)} \times \frac{1 \ mol \ OH^-{}_{(aq)}}{1 \ mol \ NaOH_{(aq)}}$$

$$= 0.0200 \ mol$$

Step 7.
As can be recalled from acid-base neutralization reaction, effectively what is happening is the H$^+$$_{(aq)}$ ions, from the acid reacting with OH$^-$$_{(aq)}$ ions from the base in a 1:1 mole ratio according the equation below:

$$H^+{}_{(aq)} \quad + \quad OH^-{}_{(aq)} \quad \rightarrow \quad H_2O_{(l)}$$

$$1 \ mol \qquad\quad 1 mol \qquad\qquad 1 \ mol$$

The next step would be to compare the number of moles of H$^+$$_{(aq)}$ ions with the number moles of OH$^-$$_{(aq)}$ ions in the mixture, to see which of the two will remain in excess after the neutralization process. In this situation, we have the following:

$$n_{H^+{}_{(aq)}} = 0.0300 \ mol$$

$$n_{OH^-{}_{(aq)}} = 0.0200 \ mol$$

$$Excess \ \ n_{H^+{}_{(aq)}} = 0.0100 \ mol$$

Subtract the smaller quantity from the larger quantity to find the excess reagent.

Since equal number of moles of OH$^-$$_{(aq)}$ ions neutralize equal number of moles of H$^+$$_{(aq)}$ ions from inspection of the above numbers, it can be concluded that **0.0200 mole of OH$^-$$_{(aq)}$** ions will neutralize exactly 0.0200 mole of H$^+$$_{(aq)}$ ions, **leaving 0.0100 mole of H$^+$$_{(aq)}$ ions in excess.**

NOTE So far, only the **number of moles** of $H^+_{(aq)}$ and $OH^-_{(aq)}$ ions are calculated to ascertain which of the two will be in excess and by how much. The next step calculates the **concentration of ions that leads to pH calculations.***
Step 8. Find the concentration of the excess ion, $\mathbf{H^+_{(aq)}}$.

To better understand how this is done, the following figures are used to illustrate this.

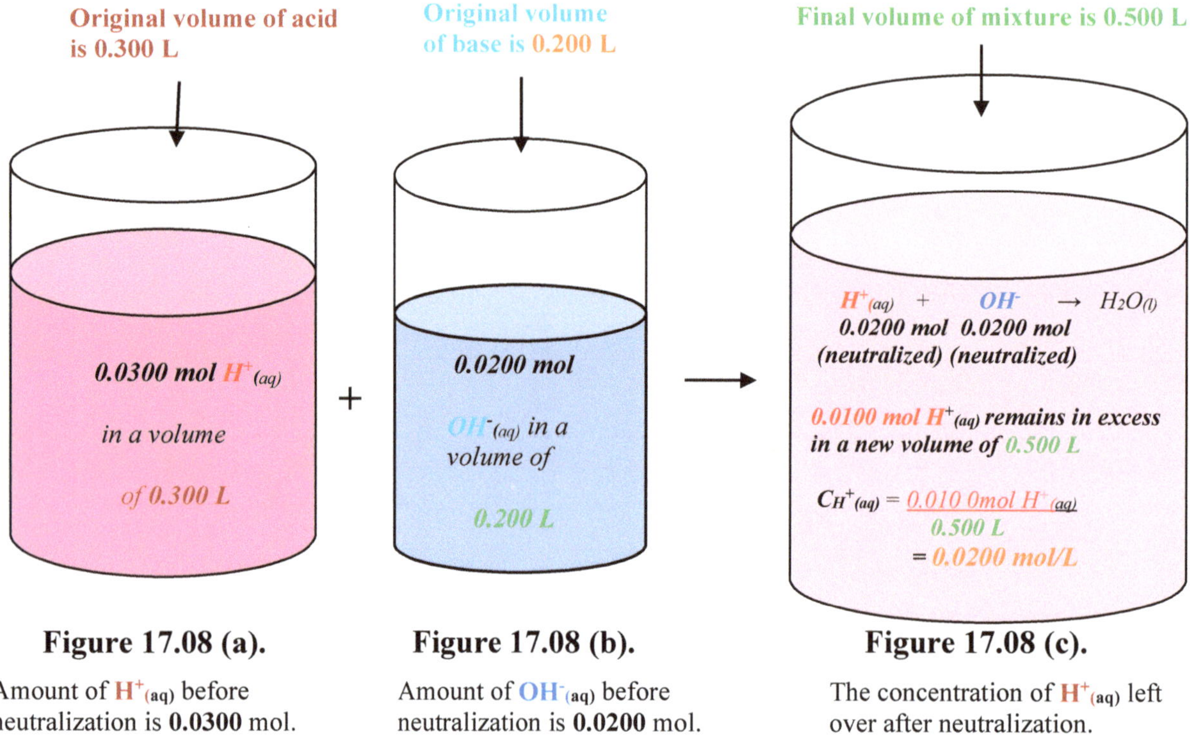

Figure 17.08 (a).	**Figure 17.08 (b).**	**Figure 17.08 (c).**
Amount of $\mathbf{H^+_{(aq)}}$ before neutralization is **0.0300** mol.	Amount of $\mathbf{OH^-_{(aq)}}$ before neutralization is **0.0200** mol.	The concentration of $\mathbf{H^+_{(aq)}}$ left over after neutralization.

 Note* Colours are only filled in to create an effect and they do not represent the actual colours of the reagents.

Figures 17.08 (a) and 17.08 (b) above show the number of moles of $H^+_{(aq)}$ and $OH^-_{(aq)}$ originally present in the acid and base, respectively. Figure 17.08 (c) shows how **0.0200 mol** of $H^+_{(aq)}$ ions **neutralizes** exactly **0.0200 mol** of $OH^-_{(aq)}$ ions with **0.0100 mol** of $H^+_{(aq)}$ ions left over in a new volume of **0.500 L,** instead of the **0.300 L** from which it originally came. This new volume of **0.500 L** was obtained from the mixing of the original volumes of the acid and base:

$$0.300\ L \quad + \quad 0.200\ L \quad = \quad 0.500\ L$$

Original volume of acid	**Original volume of base**	**Final volume of mixture**

To calculate the concentration of excess $H^+_{(aq)}$ ions, the new volume of 0.50 L is now used, using the formula:

$$C_{H^+(aq)} = \frac{n}{V}.$$

$$C_{H^+(aq)} = \frac{0.0100\ mol}{0.500\ L}$$

$$= 0.0200\ mol\,/\,L$$

Step 9. Calculate the pH of the mixture.

Using the formula pH = -log $[H^+_{(aq)}]$ we get,
$$\mathbf{pH} = -log\ 0.0200\ mol/L$$
$$= \mathbf{1.69}$$

Step 10. Calculate the pOH of the mixture

$$Recall\ that\ \textbf{\textit{pOH}} = \textit{pKw} - \textit{pH}$$
$$\textbf{\textit{pOH}} = \textbf{\textit{14.00}} - \textit{1.69} = \textbf{\textit{12.31}}$$

These same steps are to be followed when finding the resulting pH and pOH changes during an acid base titration.

Exercise 17.11

a). A solution of $HCl_{(aq)}$ of pH 2.0 is diluted until its new pH is 5.0. What is the change in $[H^+_{(aq)}]$? T/I

b). A solution of $HNO_{3(aq)}$ of pH 1.0 is diluted until its new pH is 6.0. What is the change in $[H^+_{(aq)}]$? T/I

c) To what new volume must a 100.0 mL of 1.0×10^{-2} mol/L $HCl_{(aq)}$ be diluted to achieve a final pH of 4.0? T/I

d) A solution is having 12.79 g of $HI_{(aq)}$ dissolved in 2.0 mL is diluted to 4.0 L. What is its T/I

    i) Original pH?

    ii) Final pH?

    ii) Change in $[H^+_{(aq)}]$?

e) If a solution has 2.0 g of $HCl_{(aq)}$ dissolved in 1.0 L of solution, determine the following: T/I

    i) Its pOH

    ii) Its pH

f) What mass of $NaOH_{(s)}$ is required to neutralize a 200.0 mL $HCl_{(aq)}$ of pH 2? T/I

g) If 200.0 mL of 0.10 mol/L $HCl_{(aq)}$ is mixed with 300.0 mL of 0.20 mol/L $NaOH_{(aq)}$. Determine the pH of the resulting mixture. T/I

h) 200.0 mL of 0.10 mol/L $HClO_{3(aq)}$ is mixed with 100.0 mL of 0.10 mol/L $NaOH_{(aq)}$. Determine the pH of the resulting mixture. T/I

i) If 2.0 g of $Mg_{(s)}$ is added to 200.0 mL of 1.0 mol/L $HCl_{(aq)}$, determine what volume of 1.0 mol/L $NaOH_{(aq)}$ is required to neutralize any excess acid. T/I

j) Only 200.0 mL of 0.50 mol/L $Na_2CO_{3(aq)}$ was available to neutralize 500.0 ml of 0.50 mol/L $HCl_{(aq)}$ spill. Was the spill neutralized? If not, determine what mass of $Na_2CO_{3(s)}$ powder must be used to react with it for complete neutralization. T/I

k) During a titration, 2.00 mL of 0.10 mol/L of $H_2SO_{4(aq)}$ was added to a 10.0 mL aliquot of 0.10 mol/L $KOH_{(aq)}$ in an Erlenmeyer flask. Calculate the final pH of the mixture in the flask. T/I

17.14   Calculations of pH changes during an Acid- Base Titration

The following calculations show how the pH of a **10.0 mL** of 0.100 mol/L $HCl_{(aq)}$ changes as   portions of a 0.100 mol/L $NaOH_{(aq)}$ are added to it sequentially.

Initial pH of the acid without any addition of the base.

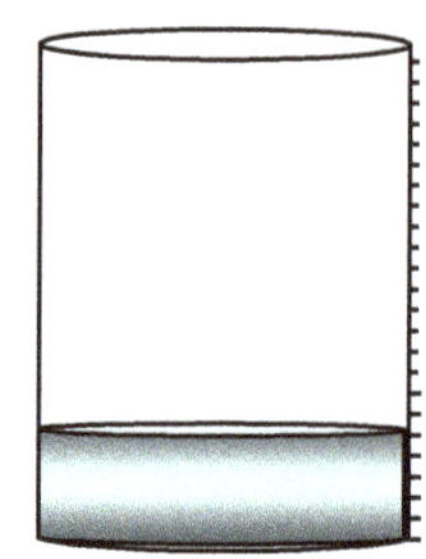

$$HCl_{(aq)} \longrightarrow H^+_{(aq)} + Cl^-_{(aq)}$$

$$0.10 \text{ mol/L} \qquad\qquad 0.10 \text{ mol/L}$$

$$pH = -\log 0.10 = \mathbf{1.0}$$

Fig. 17.09 (a).
$V_{HCl(aq)} = 0.0100 \text{ L}$

A) Addition of 2.00 mL of 0.100 mol/L NaOH$_{(aq)}$ to **10.0 mL acid.**

Step 1. Find the initial **amount** of H$^+_{(aq)}$ in **10.0 mL** of 0.100 mol/L HCl$_{(aq)}$.

$$n_{HCl} = n_{H^+}$$

$$n = c \times v = 0.100 \; \frac{mol}{L} \times 0.0100 \; L = \underline{1.00 \times 10^{-3} \, mol}$$

Step 2. Find the amount of OH$^-_{(aq)}$ added

$$n_{NaOH_{(aq)}} = n_{OH^-_{(aq)}}$$

$$n = c \times v = 0.100 \frac{mol}{L} \times 0.0200 \; L = 2.00 \times 10^{-4} \; mol$$

Step 3. Find the amount of **excess** H$^+_{(aq)}$ left in mixture after neutralization.

$$n_{H^+_{(aq)}} - n_{OH^-_{(aq)}} = \underline{1.00 \times 10^{-3} mol} - 2.00 \times 10^{-4} \, mol = \underline{8.00 \times 10^{-4} mol}$$

Step 4. Find the new volume of the mixture

$$V_{new} = 10.0 \; mL + 2.0 \; mL = 12.0 \; mL = 0.0120 \; L$$

Step 5. Find the new concentration of H$^+_{(aq)}$

$$C = \frac{n}{v} = \underline{8.00 \times 10^{-4} mol} \div 0.0120 \; L = 6.66 \times 10^{-2} mol \, / \, L$$

Step 6. Find the new pH of the mixture = -log [H$^+_{(aq)}$]

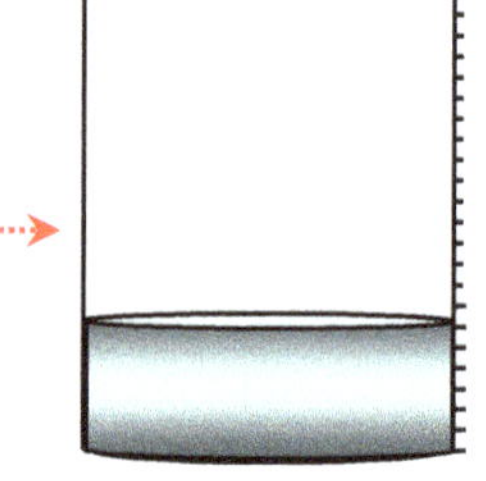

Fig. 17.09 (b). $V_{mixture} = 0.0120 \text{ L}$

$$= -\log 6.66 \times 10^{-2} = \mathbf{1.18}$$

B) Addition of **another** 2.00 mL of 0.100 mol/L NaOH$_{(aq)}$ to **12.0 mL of mixture.**

Step 1. Find the **amount** of H$^+_{(aq)}$ in **12.0 mL** of mixture.

$$= 8.00 \times 10^{-4} \, mol \quad \text{* Already calculated in part A}$$

Step 2. Find the amount of OH$^-_{(aq)}$ added

$$n = c \times v = 0.100 \frac{mol}{L} \times 0.0200 \; L = 2.00 \times 10^{-4} \; mol$$

Step 3. Find the amount of **excess** H$^+_{(aq)}$ left in mixture after neutralization.

$$n_{H^+_{(aq)}} - n_{OH^-_{(aq)}} = \underline{8.00 \times 10^{-4}} mol - 2.00 \times 10^{-4} \; mol = \underline{6.00 \times 10^{-4}} \; mol$$

Step 4. Find the new volume of the mixture

$$V_{new} = 12.0 \; mL + 2.0 \; mL = 14.0 \; mL = 0.0140 \; L$$

Step 5. Find the new concentration of H$^+_{(aq)}$

$$C = \frac{n}{v} = 6.00 \times 10^{-4} mol \div 0.0140 \; L = 4.28 \times 10^{-2} mol \, / \, L$$

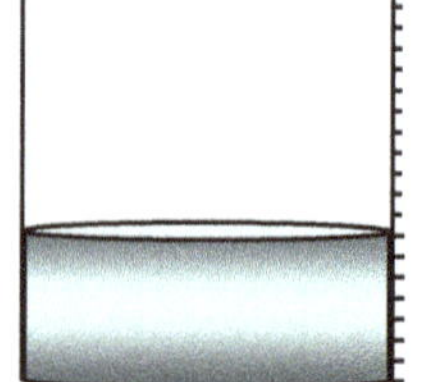

Fig. 17.09 (c).
$V_{mixture} = 0.0140 \text{ L}$

Step 6. Find the new pH of the mixture = -log [H$^+$$_{(aq)}$]

$$= \text{-log } 4.28 \text{ x} 10^{-2} = 1.37$$

C) Addition of another **2.00 mL of 0.100 mol/L NaOH**$_{(aq)}$ to **14.0 mL of mixture.**

Step 1. Find the **amount** of H$^+$$_{(aq)}$ in **14.0 mL** of mixture.

$$= 6.00 \times 10^{-4} \, mol \quad \text{* Already calculated in part B}$$

Step 2. Find the amount of OH$^-$$_{(aq)}$ added

$$n = c \times v = 0.100 \frac{mol}{L} \times 0.0200 \, L = 2.00 \times 10^{-4} \ mol$$

Step 3. Find the amount of **excess** H$^+$$_{(aq)}$ left in mixture after neutralization.

$$n_{H^+_{(aq)}} - n_{OH^-_{(aq)}} = 6.00 \times 10^{-4} \, mol - 2.00 \times 10^{-4} \, mol = 4.00 \times 10^{-4} \ mol$$

Step 4. Find the new volume of the mixture

$$V_{new} = 14.0 \ mL + 2.0 \ mL = 16.0 \ mL = 0.0160 \ L$$

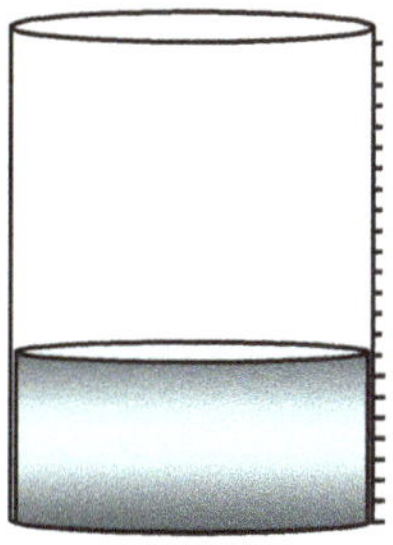

Step 5. Find the new concentration of H$^+$$_{(aq)}$

$$C = \frac{n}{v} = 4.00 \times 10^{-4} \, mol \div 0.0160 \ L = 2.5 \times 10^{-2} \, mol \, / \, L$$

Step 6. Find the new pH of the mixture = -log [H$^+$$_{(aq)}$]

$$- \text{-log } 2.50 \text{ x} 10^{-2} - 1.60$$

Fig. 17.09 (d).
V$_{mixture}$ = 0.0160 L

D) Addition of another **2.00 mL of 0.100 mol/L NaOH**$_{(aq)}$ to **16.0 mL of mixture.**

Step 1. Find the **amount** of H$^+$$_{(aq)}$ in **16.0 mL** of mixture.

$$= 4.00 \times 10^{-4} \, mol \quad \text{* Already calculated in part C}$$

Step 2. Find the amount of OH$^-$$_{(aq)}$ added

$$n = c \times v = 0.100 \frac{mol}{L} \times 0.0200 \, L = 2.00 \times 10^{-4} \ mol$$

Step 3. Find the amount of **excess** H$^+$$_{(aq)}$ left in mixture after neutralization.

$$n_{H^+_{(aq)}} - n_{OH^-_{(aq)}} = 4.00 \times 10^{-4} \, mol - 2.00 \times 10^{-4} \ mol = 2.00 \times 10^{-4} \ mol$$

Step 4. Find the new volume of the mixture

$$V_{new} = 16.0 \ mL + 2.0 \ mL = 18.0 \ mL = 0.0180 \ L$$

Step 5. Find the new concentration of H$^+$$_{(aq)}$

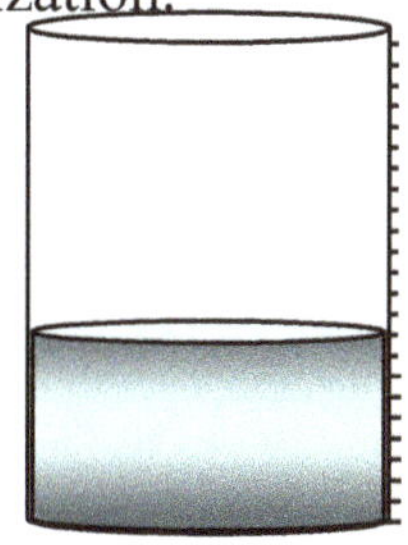

$$C = \frac{n}{v} = 2.00 \times 10^{-4} \ mol \div 0.0180 \ L = 1.1 \times 10^{-2} \ mol \, / \, L$$

Fig. 17.09 (e).
V$_{mixture}$ = 0.0180 L

Step 6. Find the new pH of the mixture = -log [H$^+$$_{(aq)}$]

$$= \text{-log } 2.50 \text{ x} 10^{-2} = 1.94$$

E) Addition of another **2.00 mL of 0,100 mol/L NaOH**$_{(aq)}$ to **18.0 mL of mixture.**

Step 1. Find the initial **amount** of $H^+_{(aq)}$ in **10.0 mL** of 0.100 mol/L $HCl_{(aq)}$

$$= \underline{2.00 \times 10^{-4} \; mol} \quad \text{* Already calculated in part D}$$

Step 2. Find the amount of $OH^-_{(aq)}$ added

$$n = c \times v = 0.100 \frac{mol}{L} \times 0.0200 \; L = 2.00 \times 10^{-4} \; mol$$

Step 3. Find the amount of excess $H^+_{(aq)}$ left in mixture after neutralization

$$n_{H^+_{(aq)}} - n_{OH^-_{(aq)}} = \underline{2.00 \times 10^{-4} mol} - 2.00 \times 10^{-4} \; mol = 0 \; mol$$

> **Equivalence point has been reached; equal amounts of $H^+_{(aq)}$ and $OH^-_{(aq)}$ have reacted.**

- **Since the acid is strong and the base is strong, the mixture is now neutral. The concentration of the $H^+_{(aq)}$ is that of water at 25 °C, which is 1.0×10^{-7} mol/L.**

The pH of the mixture = -log $[H^+_{(aq)}]$

$$= \text{-log } 1.0 \times 10^{-7} = 7.0$$

The new volume of the mixture is

$$0.002 \; L + 0.018 \; L = 0.020 \; L$$

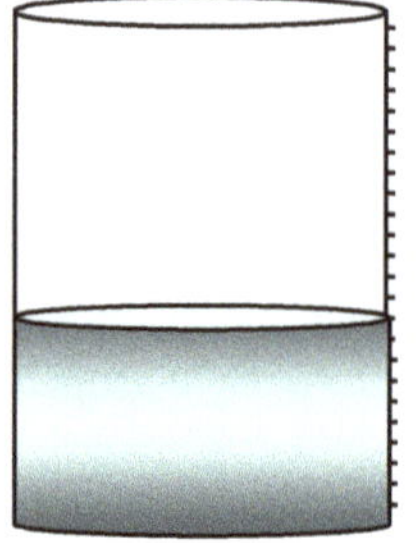

Fig. 17.09 (f).
$V_{mixture} = 0.020 \; L$

F) Addition of **2.00 mL of 0.100 mol/L $NaOH_{(aq)}$** to the **20.0 mL mixture.**

- From henceforth $OH^-_{(aq)}$ will in excess

Step 1. Find the amount of $OH^-_{(aq)}$ added

$$n = c \times v = 0.100 \frac{mol}{L} \times 0.00200 \; L = \underline{2.00 \times 10^{-4} \; mol}$$

Step 2. Find the amount of **excess $OH^-_{(aq)}$** left in mixture **after neutralization.**

$$= \underline{2.00 \times 10^{-4}} \; mol$$

Step 3. Find the new volume of the mixture

$$V_{new} = 2.00 \; mL + 20.0 \; mL = 22.0 \; mL = 0.0220 \; L$$

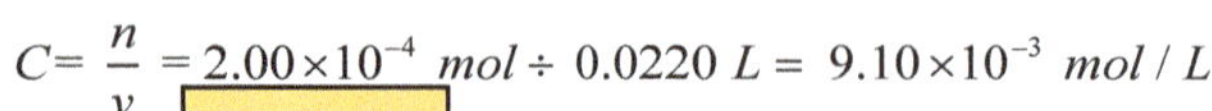

Step 4. Find the new concentration of **excess $OH^-_{(aq)}$.**

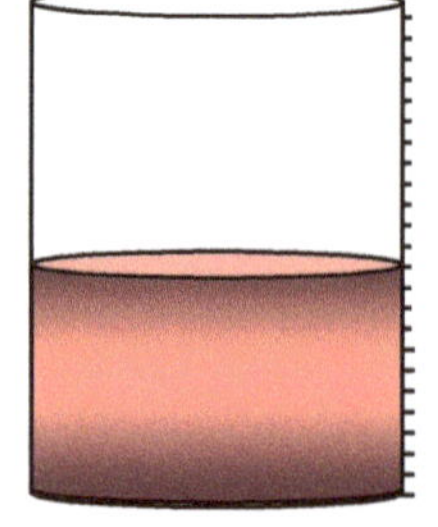

Fig. 17.09 (g).
$V_{mixture} = 0.022 \; L$

$$C = \frac{n}{v} = \underline{2.00 \times 10^{-4} \; mol} \div 0.0220 \; L = 9.10 \times 10^{-3} \; mol / L$$

Step 5. Find the new pH of the mixture

Recall that $pK_w = pH + pOH$

$$14 = pH + pOH$$

$$\therefore pH = 14 - pOH$$

$$pOH = -log[OH^-_{(aq)}] = -log\ 9.10 \times 10^{-3}\ mol/L$$

$$= 2.04$$

$$pH = 14.0 - 2.04 = 11.96$$

G) Addition of **2.00 mL of 0.100 mol/L NaOH$_{(aq)}$** to the **22.0 mL mixture.**

Step 6. Find the amount of OH$^-_{(aq)}$ added.

$$n = c \times v = 0.100\ \frac{mol}{L} \times 0.00200\ L = 2.00 \times 10^{-4}\ mol$$

Step 7. Find the amount of **excess OH$^-_{(aq)}$** left in mixture **after neutralization.**

$$= 2.00 \times 10^{-4}\ mol + 2.00 \times 10^{-4}\ mol$$

$$= 4.00 \times 10^{-4}\ mol$$

Step 8. Find the new volume of the mixture.

$$V_{new} = 2.00\ mL + 22.0\ mL = 24.0\ mL = 0.0240\ L$$

Step 9. Find the new concentration of **excess OH$^-_{(aq)}$.**

$$C = \frac{n}{v} = 4.00 \times 10^{-4}\ mol \div 0.0240\ L = 1.67 \times 10^{-2}\ mol/L$$

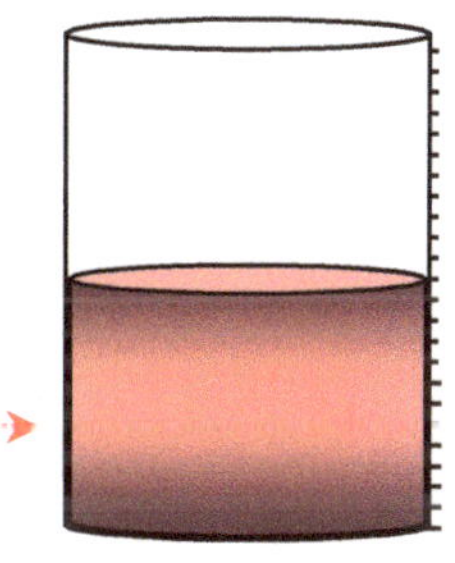

Fig. 17.09 (h).
$V_{mixture} = 0.024\ L$

Step 10. Find the new pH of the mixture.

$$pH = 14.0 - pOH$$

$$pOH = -log[OH^-_{(aq)}] = -log\ 1.67 \times 10^{-3}$$

$$= 1.77$$

$$pH = 14.0 - 1.77 = 12.23$$

As an exercise, find the pH of the mixture if an additional 2.00 mL of the NaOH$_{(aq)}$ was added to the mixture above. The following figure shows the titration curve that is obtained when a plots of the volumes of the bases added and the pH obtained are made.

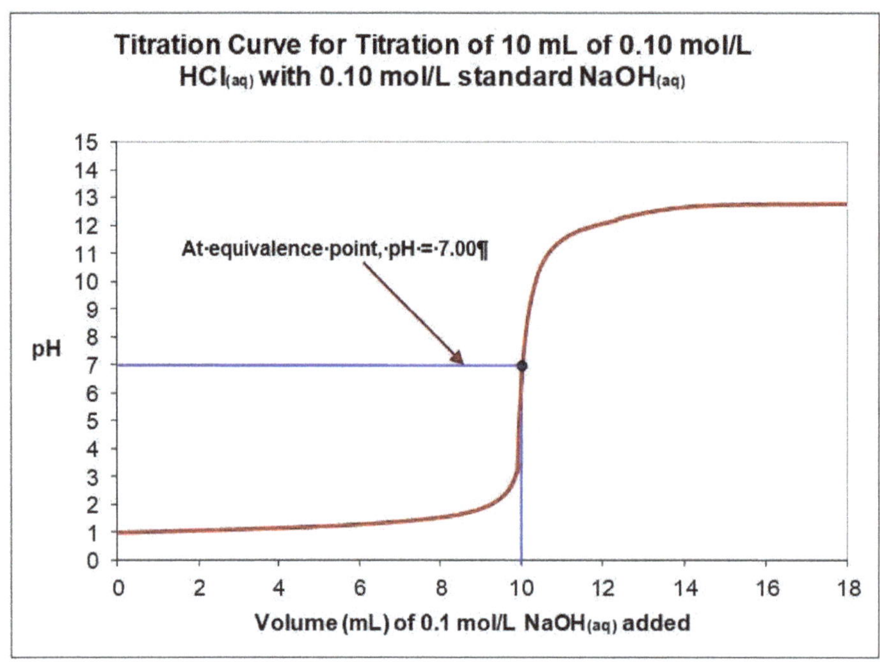

Figure 17.10. The titration curve for a strong base and strong acid.

Notice how high above 7.00 the pH of the solution rises after only one drop of $NaOH_{(aq)}$ is added. The following calculations show how this happens, assuming that one drop of the $NaOH_{(aq)}$ added has a volume of 0.050 mL = 5.0×10^{-5} L.

Solution:

At pH 7.00, the total volume of the solution is 20.00 mL.

Since the pH of the solution is 7.00, then $[OH^-_{(aq)}]$ is 1.00×10^{-7} mol/L (as in the case of water).

The number of moles of $OH^-_{(aq)}$ in 20.0 mL of mixture is found using the formula, $n = C \times V$.

$$n_{OH^-} = 1.00 \times 10^{-7} \, mol \, / \, L \times 0.0200 \, L$$
$$= 2.00 \times 10^{-9} \, mol$$

Neutral mixture

The number of moles in one drop of 0.100 mol/L $NaOH_{(aq)}$.

$$n_{OH^-} = 0.100 \, mol \, / \, L \times 5.0 \times 10^{-5} \, L = 5.00 \times 10^{-6} \, mol$$

Total amount of $OH^-_{(aq)}$ in the mixture = 5.00×10^{-6} mol + 2.00×10^{-9} mol $\approx$ 5.00×10^{-6} mol

The concentration of $OH^-_{(aq)}$, $\quad c = \dfrac{n}{v} = \dfrac{5.0 \times 10^{-6} \, mol}{0.020 L} = 2.50 \times 10^{-4} \, mol \, / \, L$

*Assuming that the volume of a drop does contribute to the final volume

$$pOH = - \log 2.50 \, x10^{-4} = 3.60 \qquad pH = 14.0 - 3.60 = 10.4$$

 Naming Acids

Writing Formulas and Names of Binary Acids.

A **binary acid** is formed when the hydride of any one of the following non-metals: fluorine, chlorine, bromine, iodine and sulphur, is dissolved in water. For example, when the compound hydrogen chloride $HCl_{(g)}$ is dissolved in water, it is changed to the acid hydrochloric acid or aqueous hydrogen chloride ($HCl_{(aq)}$).

There are two systems of naming binary acids, *the Classical* and *IUPAC* systems as outlined in the following table.

Table 17.07. Naming the Binary Acids

Hydride + Water	$HF_{(g)}$ hydrogen fluoride + $H_2O_{(l)}$	$HCl_{(g)}$ hydrogen chloride + $H_2O_{(l)}$	$HBr_{(g)}$ hydrogen bromide + $H_2O_{(l)}$	$HI_{(g)}$ hydrogen iodide + $H_2O_{(l)}$	$HS_{(g)}$ hydrogen sulphide + $H_2O_{(l)}$
Formula of binary acid	$HF_{(aq)}$	$HCl_{(aq)}$	$HBr_{(aq)}$	$HI_{(aq)}$	$H_2S_{(aq)}$
IUPAC name	*aqueous* hydrogen fluoride	*aqueous* hydrogen chloride	*aqueous* hydrogen bromide	*aqueous* hydrogen iodide	*aqueous* hydrogen sulphide
Classical name	hydrofluoric acid	hydrochloric acid	hydrobromic acid	hydroiodic acid	hydrosulphuric acid
Prefix	hydro-	hydro-	hydro-	hydro-	hydro-
Root	-fluor	-chlor	-brom	-iod	-sulphur
Suffix / End	ic acid	ic acid	ic acid	ic acid	ic acid

In naming the binary acids, using the **classical method** it useful to note the following:
1.) The name **hydrogen** is replaced by the prefix, ***hydro-***
2.) The next part of the name is derived from the other non-metal present. For example, in $HCl_{(aq)}$ the chlorine is modified to *–chlor*. This then followed by the suffix – *ic* and then the word "acid." $HCl_{(aq)}$ thus becomes hydrochloric acid.

In naming the binary acids, using the **IUPAC** method it useful to note the following:
3.) The name of the hydride is retained and is preceded by the word, *aqueous*. For example, $HCl_{(aq)}$ is named aqueous hydrogen chloride.

Note*. When writing the formulas of acids, the subscript *(aq)* must be used in the formula.

Writing Formulas and Names of Ternary or Oxyacids.

Ternary acids, also called **oxyacids** acids, contain **oxyanions** and hydrogen. The name of the oxyanion and the acid that it forms with hydrogen depend on the nature of the non-metal and the number of oxygen atoms present in the oxyanion. For example, the oxyanion, NO_3^- is called nitrate because of the nitrogen and three oxygen atoms. The acid formed from it, $HNO_{3(aq)}$ is called nitric acid. The following table provides useful information about some common oxyanions and the corresponding acids that they form.

Non-metal in acid			Oxygen atoms 2 less = 1 Prefix: hypo- Suffix: -ous	Oxygen atoms 1 less = 2 Prefix: ——— Suffix: -ous	Oxygen atoms Standard = 3 Prefix: ——— Suffix: -ic	Oxygen atoms 1 more = 4 Prefix: Per- Suffix: -ic
Cl	Acid	→	$HClO_{(aq)}$ hypochlorous acid	$HClO_{2(aq)}$ chlorous acid	$HClO_{3(aq)}$ chloric acid	$HClO_{4(aq)}$ perchloric acid
	Radical	→	$ClO^-_{(aq)}$ hypochlorite	$ClO_2^-{}_{(aq)}$ chlorite	$ClO_3^-{}_{(aq)}$ chlorate	$ClO_4^-{}_{(aq)}$ perchlorate
Br	Acid	→	$HBrO_{(aq)}$ hypobromous acid	$HBrO_{2(aq)}$ bromous acid	$HBrO_{3(aq)}$ bromic acid	$HBrO_{4(aq)}$ perbromic acid
	Radical	→	$BrO^-_{(aq)}$ hypobromite	$BrO_2^-{}_{(aq)}$ bromite	$BrO_3^-{}_{(aq)}$ bromate	$BrO_4^-{}_{(aq)}$ perbromate
I	Acid	→	$HIO_{(aq)}$ hypoiodous acid	$HIO_{2(aq)}$ iodous acid	$HIO_{3(aq)}$ iodic acid	$HIO_{4(aq)}$ periodic acid
	Radical	→	$IO^-_{(aq)}$ hypoiodite	$IO_2^-{}_{(aq)}$ iodite	$IO_3^-{}_{(aq)}$ iodate	$IO_4^-{}_{(aq)}$ periodate
N	Acid	→	X	$HNO_{2(aq)}$ nitrous acid	$HNO_{3(aq)}$ nitric acid	X
	Radical	→	X	$NO_2^-{}_{(aq)}$ nitrite	$NO_3^-{}_{(aq)}$ nitrate	X
C	Acid	→	X	$H_2CO_{2(aq)}$ carbonous acid	$H_2CO_{3(aq)}$ carbonic acid	X
	Radical	→	X	$CO_2^{2-}{}_{(aq)}$ carbonite	$CO_3^{2-}{}_{(aq)}$ carbonate	X

Exceptions with respect to number of oxygen atoms in some oxyanion

Non-metal in acid			Oxygen atoms 2 less = 2	Oxygen atoms 1 less = 3	Oxygen atoms Standard = 4	1 more = 5
S	Acid	→	$H_2SO_{2(aq)}$ hyposulphurous acid	$H_2SO_{3(aq)}$ sulphurous acid	$H_2SO_{4(aq)}$ sulphuric acid	$H_2SO_{5(aq)}$ persulphuric acid
	Radical	→	$SO_2^{2-}{}_{(aq)}$ hyposulphite	$SO_3^{2-}{}_{(aq)}$ sulphite	$SO_4^{2-}{}_{(aq)}$ sulphate	$SO_5^{2-}{}_{(aq)}$ persulphate
P	Acid	→	$H_3PO_{2(aq)}$ hypophosphorous acid	$H_3PO_{3(aq)}$ phosphorous acid	$H_3PO_{4(aq)}$ phosphoric acid	X
	Radical	→	$PO_2^{3-}{}_{(aq)}$ hypophosphite	$PO_3^{3-}{}_{(aq)}$ phosphite	$PO_4^{3-}{}_{(aq)}$ phosphate	X
			Hypo-			**Per-**

*The letter, **X** means that the compound does not exist.

Table 16.18 before gives the classical names and formulas for the oxyacids. The following table compares the classical and IUPAC systems for naming the various oxyacids formed by the non-metal, bromine.

Table 17.09. Comparison of the classical and IUPAC systems for naming oxyacids.

Formula of acid	HBrO$_{(aq)}$	HBrO$_{2(aq)}$	HBrO$_{3(aq)}$	HBrO$_{4(aq)}$
Classical name	hypobromous acid	bromous acid	bromic acid	perbromicic acid
IUPAC name	*aqueous* hydrogen hypobromite	*aqueous* hydrogen bromite	*aqueous* hydrogen bromate	*aqueous* hydrogen perbromate

The nonmetals: **F, Cl, Br, I, C, P, N and S** along with the number of oxygen atoms bonded to them, determine the name of the oxyanion. In the examples above, they are hypobromite (BrO$^-$), bromite (BrO$_2^-$), bromate (BrO$_3^-$), and perbromate (BrO$_4^-$) since the nonmetal is **bromine** and the number of oxygen atoms is different in each case. To complete the names of the acids, the words "*aqueous*" and "**hydrogen**" are placed in front of the oxyanions.

There are some common acids and polyatomic ions that do not follow a definite system of naming and these must be committed to memory. The following table gives a list of a few of them.

Table 17.01. List of compounds with non-systemic names.

Formula	Name
HCN$_{(aq)}$	hydrocyanic acid
CN$^-$	cyanide
CH$_3$COOH$_{(aq)}$	acetic acid
H$_2$C$_2$O$_{4(aq)}$	oxalic acid
Cr$_2$O$_7^{2-}$	dichromate
SCN$^-$	thiocyanate
S$_2$O$_3^{2-}$	thiosulphate
NH$_4^+$	ammonium

Naming and writing the formulae for hydrates.

Hydrates are salts that incorporate molecules of water in their crystals. For example, in copper(II) sulphate, there are five water molecules per formula unit of the salt. In writing the formula of a hydrated compound, *the formula of the compound is first written and then followed by a period and then the number of water molecules present.* For the salt copper (II) sulphate, for example, we have the formula, CuSO$_4$.5H$_2$O. The IUPAC name for this would be copper (II) sulphate pentahydrate. As can be seen from this example, Greek prefixes are used to indicate the number of water molecules in its formula unit.

Writing formulas for compounds containing polyatomic ions

The principle that is used to write binary ionic compound also applies to writing polyatomic ionic compounds; the sum of the positive valences and the negative valences add up to zero. For example, writing the formula for copper (II) nitrate can be done in two ways:

(a) 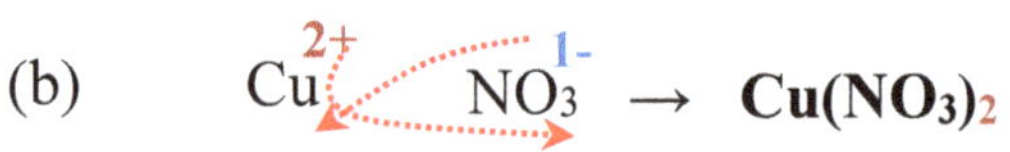 $\rightarrow$ **Cu(NO₃)₂**

In this method, the number of cations and anions required to add up to produce a sum of zero charge are written separately. In this case, one Cu^{2+} ion requires two NO_3^- ions for neutrality. The formula of the compound is then written according to the number of ions present. It should be noted that when a compound has more than one polyatomic ion in its formula, it must be placed in parenthesis with the appropriate subscript outside of the parenthesis.

(b) 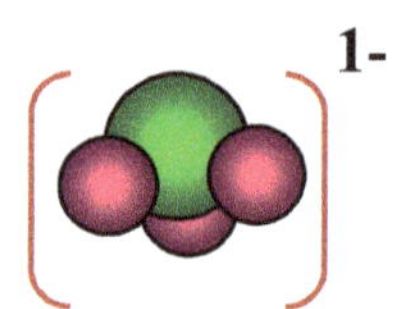 $\rightarrow$ **Cu(NO₃)₂**

The other method involves the crisscrossing of the valences of the cations and anions. All numerals except 1 are shown. Please note also, that the charges on the numerals are dropped.

For these types of compounds with polyatomic ions, the following point should be noted.
* The negative valence is not on the oxygen atoms, but on the whole ion.

$$\left[O - N - O \atop \quad \| \atop \quad O \right]^{1-}$$

Figure 17.17(a). The structural formula of the nitrate ion showing valence on the whole ion.

Figure 17.17(b). The space-filling model of the nitrate ion showing valence on the whole ion.

Exercise 17.12

Table 17.02. Write the formulas for the following compounds. K/U

(a)	magnesium bromate	(h)	ammonium sulphate
(b)	zinc iodite	(i)	potassium cyanide
(c)	Lead (II) sulphite	(j)	sodium periodate
(d)	ferric chlorate	(k)	plumbic nitrate
(e)	Tin (IV) carbonate	(l)	potassium hydrogensulphate
(f)	cupric acetate	(m)	lithium phosphate
(g)	calcium hypoiodite	(n)	calcium nitrite

Table 17.03. Write the classical and / or IUPAC names for the following compounds. K/U

(a)	$AgIO_2$	(j)	K_3PO_4
(b)	$MgCO_3$	(k)	$Be(CN)_2$
(c)	$Ca(HSO_4)_2$	(l)	$Ba(HCO_3)_2$
(d)	$Na_2SO_4 \cdot 7H_2O$	(m)	$SrSO_3$
(e)	$HBrO_{3(aq)}$	(n)	$KClO$
(f)	$Sn(NO_3)_2$	(o)	$KMnO_4$
(g)	$(NH_4)_2SO_4$	(p)	$HClO_{4(aq)}$
(h)	$Pb(NO_2)_2$	(q)	$Fe(IO_3)_3$
(i)	$Mg(CH_3COO)_2$	(r)	$Zn_3(PO_3)_2$

It is generally observed that students encounter more difficulty naming compounds than writing formulas. The following figure can be used to guide students in naming compounds accurately.

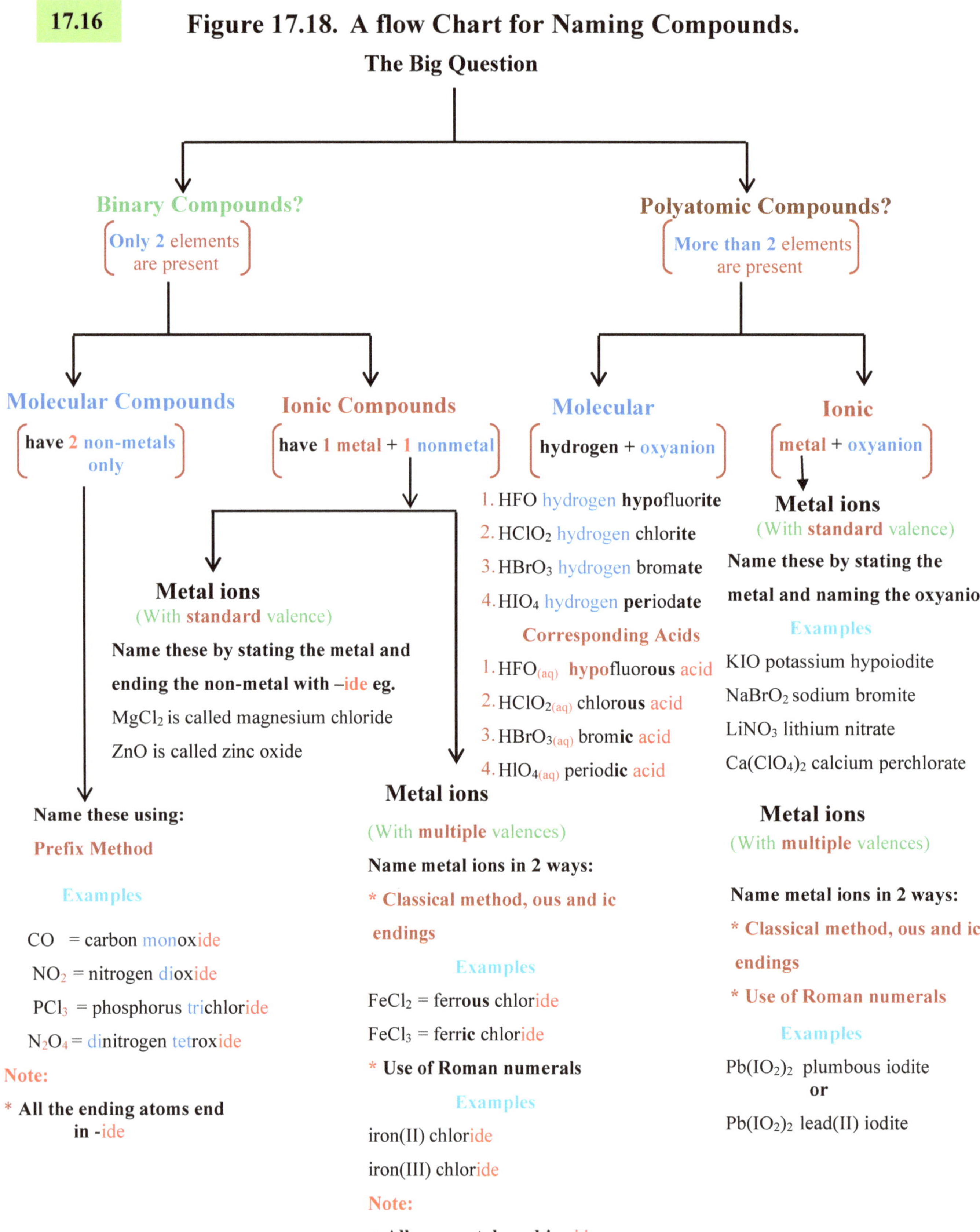

CHAPTER REVIEW ACIDS AND BASES (17)

Matching

Match each term in the table below with the correct statements that follow

A	$K_w = \left[H^+_{(aq)}\right] \times \left[OH^-_{(aq)}\right]$	H	Aqueous ions
B	Strong acids	I	Weak acids and weak bases
C	Electrolytes	J	Brönsted-Lowry
D	Diprotic acids	K	Conjugate base
E	Titrant	L	Arrhenius theory
F	End point	M	pH scale
G	$H^+_{(aq)} + OH^-_{(aq)} \rightarrow H_2O_{(l)}$	N	Standard solution

1.	His theory states that acids produce $H^+_{(aq)}$ ions and bases produce $OH^-_{(aq)}$ ions respectively when they dissolve in water.
2.	These species are responsible for the electrical conductivity in electrolytes.
3.	Their theory states that an acid is a proton donor and a base is a proton acceptor.
4.	These molecules ionize completely when they dissolve in water.
5.	These molecules ionize only partially when they dissolve in water.
6.	These conduct electricity when they dissolve in water.
7.	This is what is left of the acid after it donates a proton to an acceptor.
8.	These are capable of producing two moles of $H^+_{(aq)}$ ions per mole of molecules.
9.	The net ionic equation for an acid base neutralization.
10.	The reagent that is placed in the burette during titration.
11.	The solution of known concentration used during titration.
12.	The ionic product of water equation.
13.	Tells if a substance is neutral, basic or acidic and to what degree it is either basic or acidic.
14.	The point in titration when equal amount of $OH^-_{(aq)}$ is titrated with equal amount of $H^+_{(aq)}$ that causes the indicator to change colour.

True or False

Read each of the following statements and then decide if it is *True* or *False*.

1.	Strong acids ionize in water to produce a low concentration of $H^+_{(aq)}$.
2.	Weak acids have higher pH than stronger acids because they fully ionize when they dissolve in water.
3.	A pH scale measures the $[H^+_{(aq)}]$ in a solution.
4.	Unripe fruits taste sour because of the weak acids that they contain.
5.	A diprotic acid produces only one mole of $H^+_{(aq)}$ ions.
6.	Antacids are composed of strong bases.
7.	A Brönsted-Lowry acid is a proton donor.
8.	Bases have pH values lower than 7.
9.	In an acid solution, $[H^+_{(aq)}] > [OH^-_{(aq)}]$.
10.	A weak acid cannot neutralize a strong base.
11.	Dilution of an acid solution raises its pH.
12.	Pure water at temperatures above 25 ^{0}C may have a pH greater than 7
13.	A dilute solution of a strong acid has lower pH than a concentrated solution of a weak acid.
14.	If the sting of insects is caused by formic acid, the treatment for which is made of weak bases.

Multiple Choice

Choose the letter that best answers the questions.

1. A compound that *dissociates* in water and conducts electricity very well is most likely to be

a	a weak acid	*d.*	a soluble salt
b	a metal	*e.*	none of the above
c	a weak base		

2. Sodium chloride dissolves in water as follows: $NaCl_{(s)} + H_2O_{(l)} \rightarrow Na^+_{(aq)} + Cl^-_{(aq)}$. This change can be classified as

a	dissociation	*d.*	a and c only
b	melting	*e.*	none of the above
c	ionization		

3. Acetic acid (vinegar) is classified as a weak acid because

a	it is more than 99 % ionized	d.	it cannot neutralize strong bases
b	it ionizes about 50 %	e.	none of the above
c	it is only 1.3 % ionized		

4. Carbonic acid ionizes as follows in water: $H_2CO_{3(aq)} + H_2O_{(l)} \rightleftharpoons HCO_3^-{}_{(aq)} + H_3O^+{}_{(aq)}$

The conjugate acid-base pairs produced are

a	$H_2CO_{3\,(aq)}$ and $H_2O_{(l)}$	d	$H_2CO_{3(aq)}$ and $HCO_3^-{}_{(aq)}$
b	$HCO_3^-{}_{(aq)}$ and $H_3O^+{}_{(aq)}$	e	d and c only
c	$H_2O_{(l)}$ and $H_3O^+{}_{(aq)}$		

5. At 25 °C which of the following is true for pure water?

a	$[OH^-{}_{(aq)}] = 1 \times 10^{-7}$ mol/L	d.	Its pH = 7.0
b	$[H^+{}_{(aq)}] = 1 \times 10^{-7}$ mol/L	e.	all of the previous
c	$[H^+{}_{(aq)}] = [OH^-{}_{(aq)}]$		

6. If the pH of an aqueous solution is 9, then its pOH must be

a	9	d	1×10^{-9}
b	5	e	1×10^{-5}
c	14		

7. Which of the following is **not** true for a 0.10 mol/L $HCl_{(aq)}$ (a strong acid)?

a	It is > 99% ionized	d.	It has a pH of 1.0
b	It produces 0.1 mol/L $H^+{}_{(aq)}$	e.	It has a pH of 6.0
c	It is very sour		

8. If an aqueous solution has a pOH of 11, its $[H^+{}_{(aq)}]$ must be

a	1.0×10^{-3} mol/L	d	3.0×10^{-1} mol/L
b	11.0×10^{-3} mol/L	e	none of the above
c	14.0×10^{-3} mol/L		

9. If the $[OH^-{}_{(aq)}]$ of an aqueous solution is 1.0×10^{-9} mol/L, its $[H^+{}_{(aq)}]$ must be

a	1.0×10^{-14} mol/L	c	1.0×10^{-7} mol/L
b	1.0×10^{-9} mol/L	d	none of the above
c	1.0×10^{-5} mol/L		

10. Water is evaporated a basic solution of pH 10, which of the following will occur?

a	$[OH^-{}_{(aq)}]$ will increase	d	pOH will decrease
b	$[H^+{}_{(aq)}]$ will decrease	e	all of the above
c	pH will increase		

11. When $HI_{(aq)}$, a strong acid, is diluted with water, there would most likely be

a	an increase in pH	*d.*	an increase in $[H^+_{(aq)}]$
b	a decrease in pOH	*e.*	a, b, and c
c	a decrease in $[H^+_{(aq)}]$		

12. Hydrochloric acid of concentration 1.0×10^{-3} mol/L will have a pH of

a	11.0	*d.*	1.0
b	3.0	*e*	none of the above
c	10		

13. Water reacts in the following two ways: $H_2O_{(l)} + H_2O_{(l)} \rightleftharpoons H_3O^+_{(l)} + OH^-_{(aq)}$
$$H_2O_{(l)} + HCl_{(aq)} \rightarrow H_3O^+_{(l)} + Cl^-_{(aq)}$$
Based on these reactions, water can be described as

a	an acid	*d*	a, b and c
b	a base	*e*	None of the above
c	amphoteric		

14. If acid A has a pH of 3 and another acid, B has a pH of 5, which of the following is true?

a	Acid B is two times more acidic than acid A
b	Acid A is one hundred times more acidic than acid B
c	Acid A is one hundred times more concentrated with $[H^+_{(aq)}]$ than acid B
d	Acid B is two hundred times more concentrated with $[H^+_{(aq)}]$ than acid A
e	b and c

15. Sodium hydroxide reacts with chloric acid according to the equation below:
$$NaOH_{(aq)} + HClO_{3(aq)} \rightarrow NaClO_{3(aq)} + H_2O_{(l)}$$
Which of the following statements is **not** true?

a	25 mL of 0.1 mol/L $HClO_{3(aq)}$ will completely neutralize 50 mL of 0.05 mol/L $NaOH_{(aq)}$
b	100 mL of 0.05 mol/L $HClO_{3(aq)}$ will completely neutralize 50 mL of 0.1 mol/L $NaOH_{(aq)}$
c	One mole of $NaOH_{(aq)}$ neutralizes exactly one mole of $HClO_{3(aq)}$
d	A basic solution is produced if 50 mL of 0.2 mol/L $HClO_{3(aq)}$ is mixed with 40 mL of 0.1 mol/L $NaOH_{(aq)}$
e	An acidic solution is produced if 25 mL of 0.2 mol/L $HClO_{3(aq)}$ is mixed with 15 mL of 0.3 mol/L $NaOH_{(aq)}$

16. A 25 mL of 0.1 mol/L of $CH_3COOH_{(aq)}$ (a weak acid) will

a	completely neutralize 25 mL of 0.1 mol/L $NaOH_{(aq)}$ (a strong base).
b	**not** completely neutralize 25 mL of 0.1 mol/L $NaOH_{(aq)}$.
c	completely neutralize 50 mL of 0.05 mol/L $Ca(OH)_{2(aq)}$ (a strong base).
d	**not** completely neutralize 50 mL of 0.05 mol/L $Ca(OH)_{2(aq)}$.
e	a and c

17. Which of the following is **not** true of antacids?

a	They neutralize acids	d	They are weak bases
b	They have a pH greater than 7	e	They are used to treat heart burns
c	They are strong bases		

18. The bond in HF molecule is much more polar bond than HI molecule, yet $HI_{(aq)}$ is a very much stronger acid than $HF_{(aq)}$. The reason for this is due to the fact that

a the iodine atom is so much bigger than the fluorine atom that it allows the $HI_{(aq)}$ to ionize much easier than $HF_{(aq)}$.

b iodine is more electronegative than fluorine

c fluorine exerts a greater attractive force on hydrogen than iodine does

d the polarity of bonds do not affect how easily a molecules ionize in water

e a and c

19. Which of the following would not cause a change in pH, when added to water?

a	$Mg(OH)_2$	d	$C_2H_4O_2$
b	$NaCl$	e	NaH_2PO_4
c	$HClO$		

20. Which of the following when dissolved in water would produce a basic solution?

a	$Na_2O_{(s)}$	d	$NH_{3(g)}$
b	$SO_{2(g)}$	e	a and d
c	$NaCl_{(s)}$		

21. Which of the following is true of $NaHSO_4$?

a It is a basic salt

b It is an acidic salt

c one mole of it will completely neutralize one mole of NaOH

d It is a neutral salt

e a and c

UNIT 4
Gases

Atmospheric pressure
Atmospheric pressure
Atmospheric pressure
Atmospheric pressure
Atmospheric pressure
Atmospheric pressure
Increased volume
Movable plunger
Atmospheric pressure
Atmospheric pressure
Atmospheric pressure
Atmospheric pressure
Atmospheric pressure
Immovable plunger
Atmospheric pressure

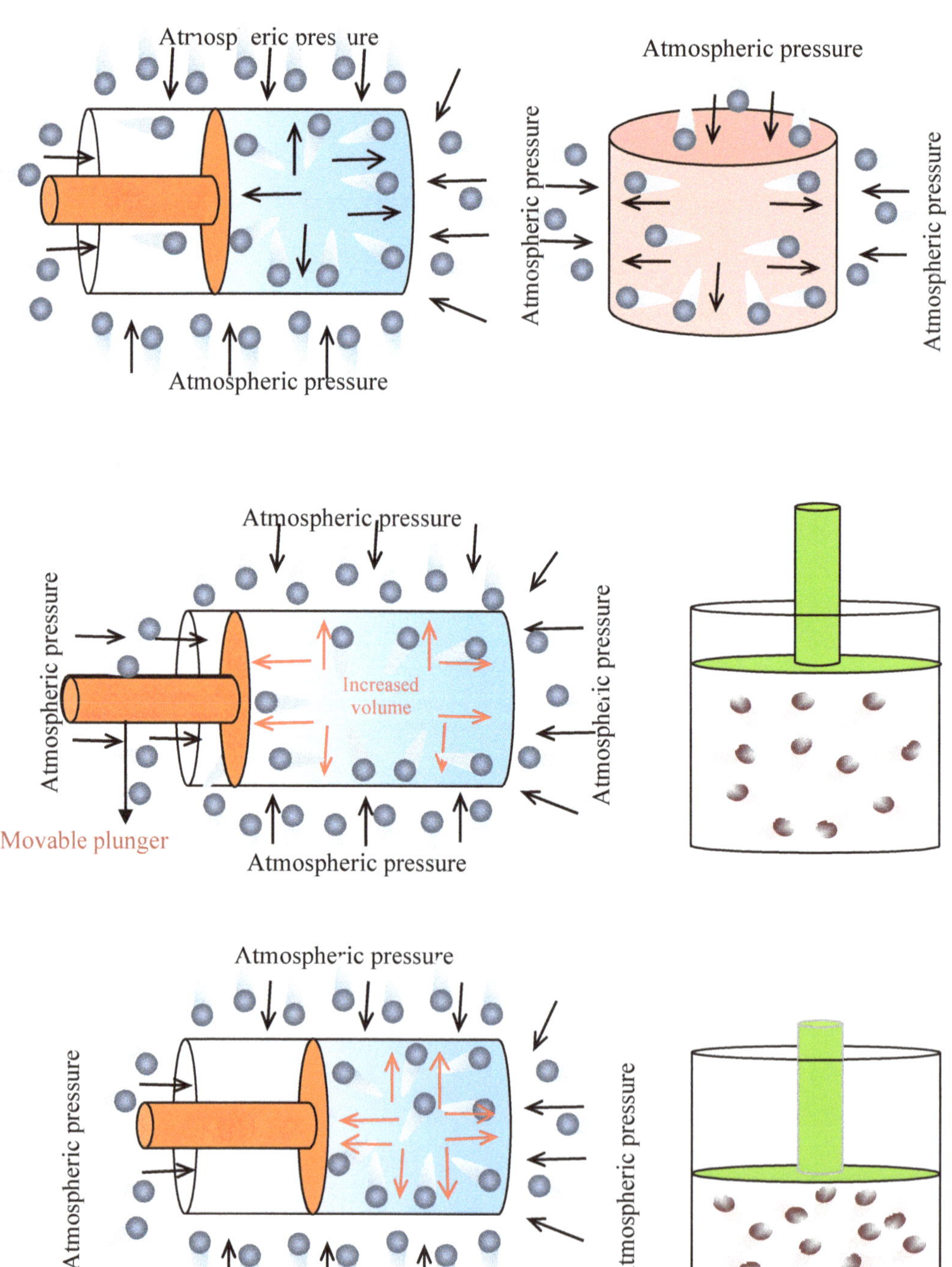

Chapter Content:

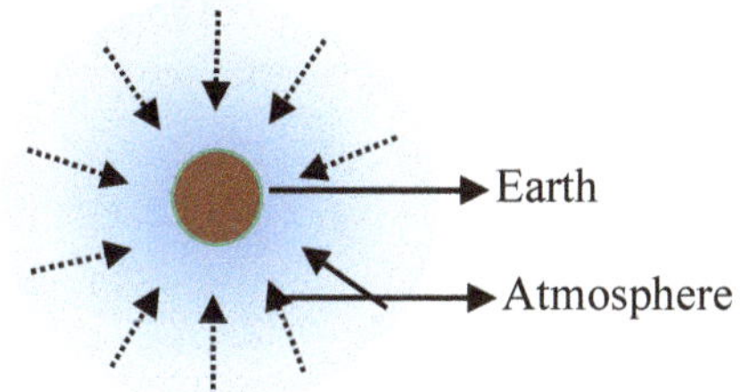

18.1 Atmospheric Pressure Effects

Have you ever wondered why your ears would sometimes 'pop' as you ascend or descend the atmosphere in an airplane? The answer to this has to do with the fluctuations in atmospheric pressure with respect to our stable blood pressure as we ascend or descend the atmosphere. At sea level, the pressure of our blood and tissues is the same as that exerted by the atmosphere on us. As we ascend into the atmosphere, its pressure decreases progressively, but our body's pressure remains the same. To prevent our bodies from exploding when we are ascending the atmosphere in an aircraft, the aircraft is pressurized to compensate for the decrease in atmospheric pressure. Initially, we feel this added pressure on our eardrums, but swallowing allows air to enter the middle ear, by way of the eustachian tube, to equate this pressure. The opposite is done as the aircraft descends. What must astronauts do to survive in space where there is no atmosphere?

Very often we hear of deadly accidents involving cylinders containing compressed gases. Yes, compressed gases exert pressures on the walls of their containers and if the cylinders are not strong enough to withstand these, they will inevitably explode. It is therefore vital that gas cylinders be tested to know their pressure capacity in order to know how much gas they can store. You might have noticed that the pressures in your car's tires are a bit higher after a long journey. The reason for this is due to the fact that friction between the tires and the road raises the temperature of the air in the tires which causes a higher pressure. Why should you not adjust the pressure in your tires **to their maximum safe capacity,** before a long summer journey? Why do you think pressure cookers have nozzles? Why are pressure cookers indispensable for people living in mountainous places? Why does air rush into and out of our lungs during inhalation and exhalation, respectively? Do you know that an entire aircraft is lifted off the ground because its specially designed wings create a lower air pressure above it during takeoff? This chapter deals with atmospheric pressure, describes and explains how the gas laws are derived.

18.2 The Gaseous State

The various states of matter can be attributed to the differences in the magnitude of the forces of attraction between the particles of the various types of matter. Some forms of matter are solid because the forces of attraction between their particles are strong to maintain their particles in relatively fixed orders. In this state, only vibration of their particles is permitted.

In substances that are liquids at room temperature, the forces of attraction between their particles are weaker than those of solids. The force of attraction is so weak that the particles have more freedom of movement to slip and slide over each other, taking the shape of the containers, in which they are placed. Some solids liquefy, only if they are heated to very high temperatures to significantly weaken the strong bonds that hold their particles in their fixed lattice. Ionic compounds are examples of these. Other solids, such as molecular crystals, will liquefy at relatively lower temperature when heated.

In gases, very weak forces of attraction exist between their particles. Because of this, their particles tend to diffuse to infinite distances from each other. Gases occupy the same volume as their containers.

18.3   Atmospheric Pressure

Before gas the laws are formulated and understood, it is important to study the following concepts:

Force: This is calculated by multiplying the mass of an object by its acceleration.
Anything that has mass exerts a force downwards due to the Earth's gravity on it. An object of mass 1.0 kg, for example, would exert a force as calculated by the equation:

$$Force = Mass \times Acceleration$$

The force exerted this mass: Force = 1.0 Kg × 9.8m/s^2 = 9.8 Newtons
(9.8 m/s^2 is the acceleration due to the Earth's gravity)

Pressure: This is defined as force per unit area.

$$Pressure = \frac{Force}{Area}$$

If a force of 1 Newton, for example, is applied over an area of 1.0 m^2, then the pressure exerted would be:

$$Pressure = \frac{1.0\,N}{1.0\,m^2} = 1.0\,pascal\,(Pa)$$

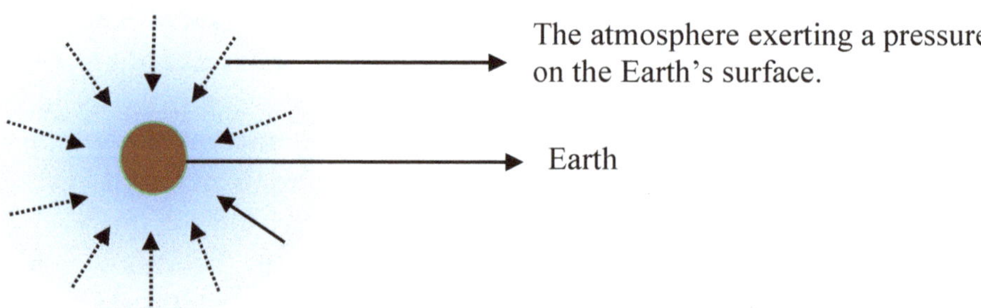

Figure 18.1. The Earth and its atmospheric pressure.

Since air has mass, then the column of air above the Earth's surface, the atmosphere, is exerting a force on its surface. At sea level, this force is measured to be 101.0 kN. The pressure exerted on an area of 1.0 m^2 by the atmosphere on Earth would thus be:

$$Average\ atmospheric\ pressure = \frac{101.0\,kN}{1.00\,m^2} = 101.0\,kPa\ at\ 0\,°C.$$ This pressure is the same pressure what a

mass of approximately 11.08 tons would exert, if spread over an area of 1.0 m^2. This is indeed an awesome mass, yet we do not feel the effect of its pressure on our bodies.

18.4   Demonstration of Atmospheric Pressure

One of the dramatic ways of demonstrating atmospheric pressure is that of it crushing an empty pop can. This is done by placing about 15 mL of water in an empty popcan and heating it on a hot plate until almost all the water is boiled off. Using a pair of tongs, the can is then quickly inverted in a large beaker of ice-cold water. Immediately following this, the can is suddenly crushed- **Figure 18.2(d).**

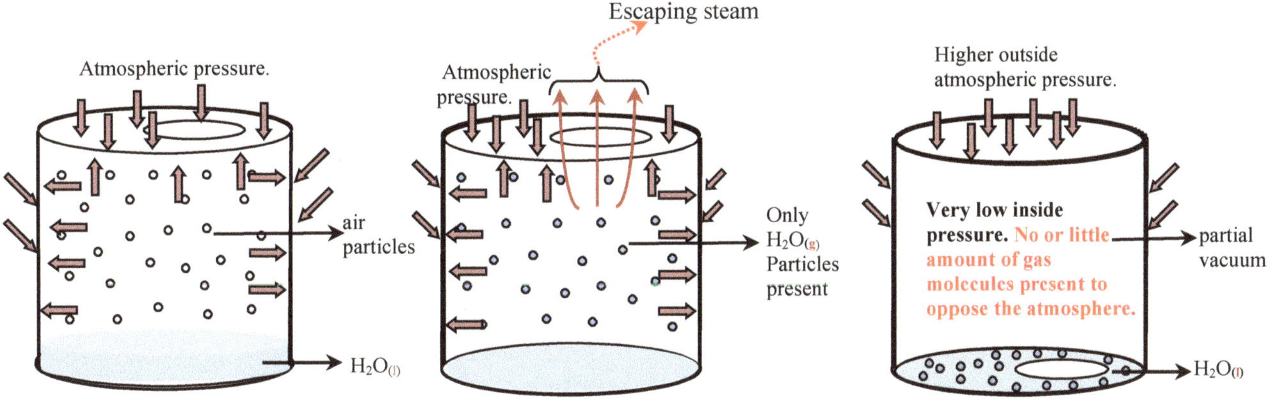

Figure 18.2(a). Pop can with water and air. Pressure inside is the same as outside.

Figure 18.2(b). Pop can with only water vapor. Pressure inside is still the same as outside.

Figure 18.2(c). Inverted Pop can with a partial vacuum.

Explanation of this observation:

Initially, the pop can contained air particles, along with the liquid water poured in at the bottom. Before heating, the pressure exerted by the air particles inside the pop can was the same as that exerted by the atmospheric- Figure **18.2(a).** As the can was heated and the water started to boil, a current of water vapor was created that eventually pushed all or most of the air out of the can- Figure **18.2(b).** The water vapor that was left in the can, like the air particles before, exerted an equal and opposite pressure as that of the outside atmospheric pressure.

However, as the can is inverted in the cold water, Figure **18.2.c,** the water vapor in it condensed, creating a partial vacuum. As this happened, the pressure inside the can became much less than the outside atmospheric pressure. The greater outside pressure crushed the can - **Figure 18.2(d).**

Figure 18.2(d). Pop can being crushed by the atmosphere.

18.5 Evangelista Torricelli

The first barometer, an instrument for measuring atmospheric pressure was designed by the Italian scientist, Evangelista Torricelli, in the seventeenth century. Torricelli discovered the atmosphere is capable of supporting a column of mercury in a vacuum tube. Figure **18.3(a)** shows how a vacuum tube is placed in a container of mercury. When this tube is unsealed with its lower end immersed in mercury, the atmosphere exerted a pressure on the mercury in the container and forced some of it up the tube. The height of the mercury was found to be 760.0 mm at sea level. A simpler way of doing this is to completely fill a long enough tube with mercury, stopper the open end, invert it, immerse it in a container of mercury and then open it. Some of the mercury will flow down the filled tube into the container until the height of what remains is 760 mm. A vacuum will remain at the top of the tube - Figure **18.3(b).**

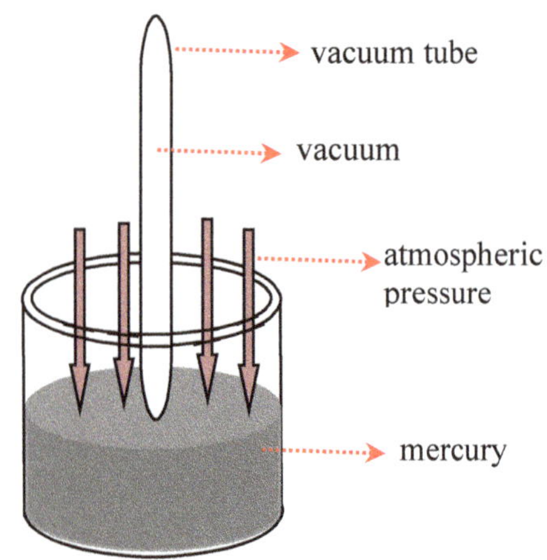

Figure 18.3(a). A vacuum tube
placed in a container of mercury.

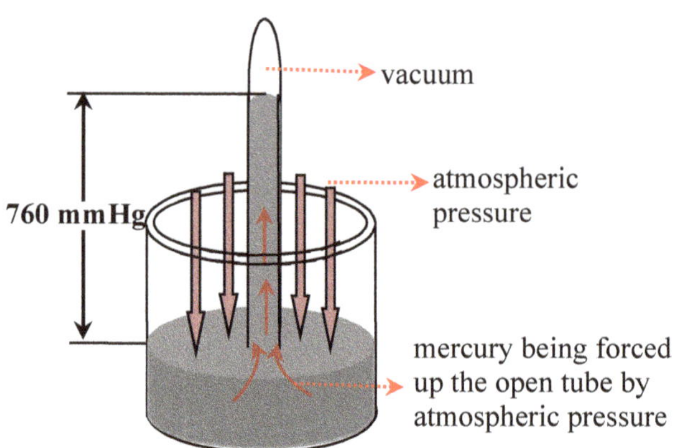

Figure 18.3(b). Vacuum tube is unsealed
at the end, immersed in mercury.

Why did the mercury not go up more than 760 mm? If your answer is that the atmosphere is capable of supporting a height of only 760 mm, then you are correct.

What measurements would be made if this activity is done on top of a mountain, or with water?

*Hint: On top of a mountain the atmospheric pressure is less than at sea level.
The pressure of a column of liquid is given by the equation, $P = h \rho g$
Where h is the height of the liquid, ρ is its density, and g is the acceleration due to gravity.
* Mercury has a density of 13.53 g/mL and water is 1.0 g/mL

Units of Atomospheric Pressure

Millimeters of Mercury: This was the original unit used by Torricelli, until it gave way to more suitable ones.

Torrs: This used in honor of Torricelli to simplify the usage of "millimeters of mercury".

Pascals: Named after the French scientist Blaise Pascal, it is the SI unit of pressure and is symbolized by Pa. It is the pressure exerted by a force of one Newton over one square meter of surface. Since the Pascal is very small, for most practical purposes, units such as kilo and mega Pascals are used.

Bars: This unit is used mostly by meteorologists to express pressure.

Atmosphere: This is the pressure exerted by the Earth's atmosphere measured at sea level. It is equal to 101.3 kPa or 760 mm Hg or 101.325 kPa. This unit is used mostly to express high pressures. For example, in the synthesis of ammonia by the Haber process, a pressure of 150 to 250 atm. is used, however this not an SI unit.

Standard Ambient Temperature and Pressure (SATP)

The earliest set of standard conditions that were used to measure gases were 0 °C and 760 mm Hg. This was known as **standard temperature and temperature (STP).**
These set of conditions were subsequently changed by IUPAC in the 1970s to values that were more applicable and reflective of conditions in which not only gases, but other substances are studied. These new standard conditions are called **standard ambient temperature and pressure (SATP),** which are **25 °C** (298 K) and **100 kPa.**

Summary:
- ➢ **STP**: *Standard temperature and pressure which are **0 °C and 101.325 kPa***

- ➢ **SATP**: *Standard ambient temperature and pressure which are **25.0 °C and 100.0 kPa**.*

Pressure in an Open Container:

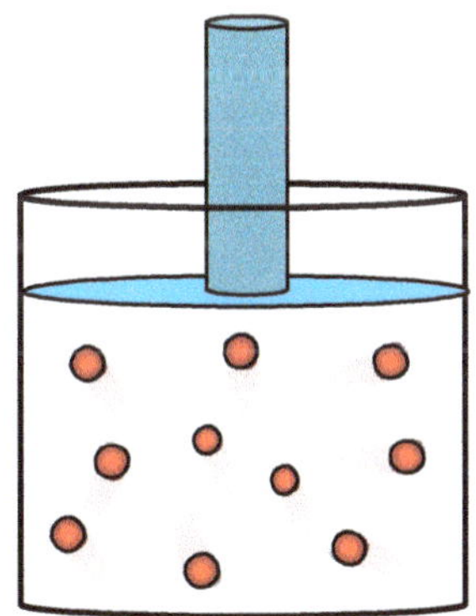

Figure 18.4 Atmospheric pressure in an open container.

The pressure of a gas in a container is due to the bombardment of its particles against the walls of its container. In an open container, the inside walls are equally bombarded by the gas particles as its outside walls are. Because of this, the pressure inside the container is the same as that outside. A concluding statement for the pressure in any open system would be that, ***the pressure inside is the same as that outside.***

18.6 **Boyle's Law: Volume and Pressure changes**

Suppose that we have equal masses of a gas (*same number of moles of particles*) at the same temperature in both of the containers below, what relationships between their volumes and pressures could be established?

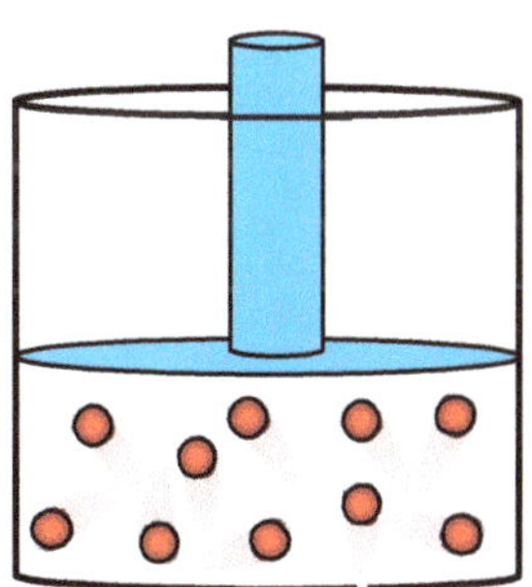

Figure 18.5(a). A Volume = 1.0 L
Pressure = 110.0 kPa *(Low).*
A

Figure 18.5(b). B Volume = 0.50 L
Pressure = 220.0 kPa *(High).*
B

Cylinder **A** with a fixed mass of gas having an initial volume of 1.0 L and pressure of 110.0 kPa is changed to cylinder **B** by reducing its volume to half at constant temperature. The new pressure of **B** was subsequently measured and found to be 220.0 kPa.

When the pressure and volume of the gases in these two cylinders are compared, it can be observed that as the volume is decreased its pressure increases. In fact, when a series of volume changes on a fixed mass of a gas at 25.0 °C are made, and the pressure for each change is noted and tabulated, the following results were obtained:

Table 18.1. Relationship between volume and pressure changes.

Experiment	1	2	3	4	5
Volume (L)	1.00	0.920	0.820	0.750	0.920
Pressure (kPa)	100.0	108.0	120.0	135.0	160.0
Pressure ×Volume (kPa.L)	100.	99.4	98.4	101.	102.

In the table above, neglecting the minor differences, it can be seen that for a fixed mass of gas at constant temperature, the product of its volume and pressure for the various experiments is constant.

Robert Boyle (1627-1691) was the first to discover the relationship between *the volume and pressure* of *a fixed mass of gas at constant temperature*. He concluded that **the volume of a fixed mass of gas at a constant temperature is inversely proportional to its pressure**. This relationship can be represented as:

$$V \propto \frac{1}{P}$$

From the preceding table, it can be seen that $PV = k$
Using the values in experiments 1 and 2 above, we get:
Experiment 1:   *100.0 kPa × 1.00 L = 100. kPa.L = (P_1×V_1)*
Experiment 2:   *108.0 kPa × 0.920 L = 99.4 kPa ($P_2 \times V_2$)*

Except for the very minute difference, the product of volume and pressure in these two experiments is the same, as they are for the other experiments in the table above.
From these results, Boyle's law can alternatively be expressed as:

$$P_1 \times V_1 = P_2 \times V_2$$

Sample problem 1:

What pressure must be applied to 2.00 L of a gas at 100.0 kPa to change its volume to 0.800 L?

Solution:

Provided quantities:
$V_1 = 2.00\,L$
$P_1 = 100.0\,kPa$
$V_2 = 0.800\,L$
Required quantity: P_2
Using the equation, $P_1 \times V_1 = P_2 \times V_2$

$$P_2 = \frac{P_1 \times V_1}{V_2}$$

$$= \frac{100.0\,kPa \times 2.00\,L}{0.800\,L}$$

$$= 2.50 \times 10^2\,kPa$$

What will be the final volume of air in a cylinder if 1.00 L of it at 110.0 kPa is pressurized to a 200.0 kPa?

Provided quantities:
$$V_1 = 1.00 L$$
$$P_1 = 110.0\,kPa$$
$$P_2 = 200.0\,kPa$$

Required quantity: V_2

Using the equation, $P_1 \times V_1 = P_2 \times V_2$

$$V_2 = \frac{P_1 \times V_1}{P_2}$$

$$= \frac{110.0\,kPa \times 1.00\,L}{200.0\,kPa}$$

$$= 0.550\,L$$

Summary:

$$P_1 \times V_1 = V_2 \times P_2$$

$$P_1 = \frac{V_2 \times P_2}{V_1}$$

$$V_1 = \frac{V_2 \times P_2}{P_1}$$

$$V_2 = \frac{P_1 \times V_1}{P_2}$$

$$P_2 = \frac{P_1 \times V_1}{V_2}$$

Exercise 18.1

1. A balloon filled with an inert gas has an initial volume of 2.0 L and pressure of 120.0 kPa. T/I

    a) What will happen to its volume if it ascends into the atmosphere at constant temperature?

    b) What will be its final pressure if its new volume was 3.0 L?

2. A cylinder containing 5.0 L helium gas is at 4.0 atm, is used to fill balloons to a volume of 500.0 mL at a pressure of 1.0 atm. How many balloons will get filled? T/I

3. A cylinder contains air of volume of 2.0 L at pressure 1.0 atm. If the walls of the cylinder can only withstand a maximum pressure of 4.0 atm, to what critical volume can the air be compressed before the cylinder bursts? T/I

4. A syringe has 40.0 mL of air at 101.0 kPa. What pressure must be applied to the gas for it to have a final volume of 30.0 mL. Assume that the mass and temperature of the gas stay constant. T/I

Charles's Law: Volume and Temperature changes

French physicist, *Jacques Charles (1746-1823) investigated the relationship between the volume of a **fixed mass of gas** at constant pressure and changes in temperature on it*. In a series of experiments, a volume of air in a closed syringe with a freely movable plunger is subjected to different temperatures. For each different temperature, the new volume of the air in the syringe is noted and recorded. Figure 18.6 illustrates how this is done.

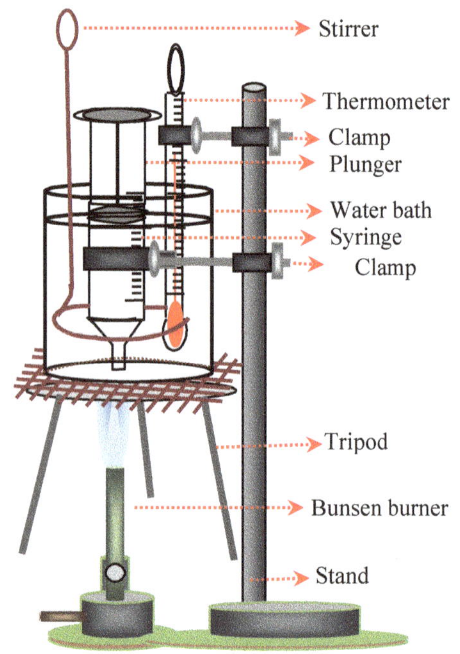

Figure 18.6. Bunsen burner heats up water bath having a gas syringe with movable plunger.

The Bunsen burner is used to heat up the water bath. The syringe with air absorbs the heat from the water bath and its plunger pushes out. The thermometer in the water bath monitors temperature changes. By regulating the flame of the burner, the temperature is changed every 5-10 °C up to about 90 °C. The new volume of the air at each new temperature change is noted and tabulated.

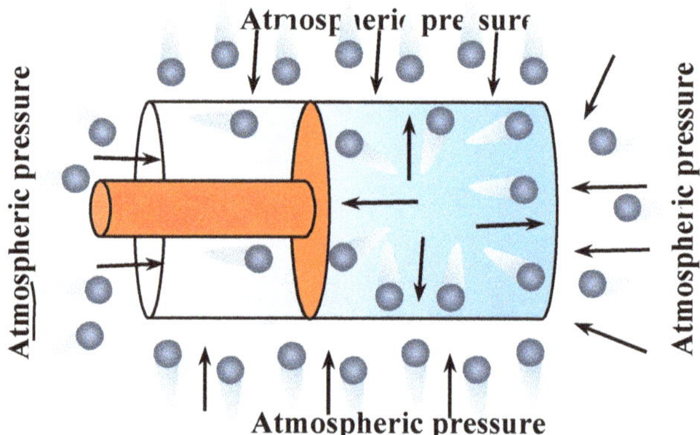

Figure 18.7(a). Syringe before heating with pressure inside being the same as that outside.

Why does a gas expand at constant pressure when it is heated?

In the Figure 18.7(a) above, the atmosphere is pushing on the plunger from the outside and the air particles in the syringe are pushing on it from the inside. If these two pressures are the same, the plunger does not move. Figure 18.7(b) below represents the case where the air in the syringe is being heated in a water bath.

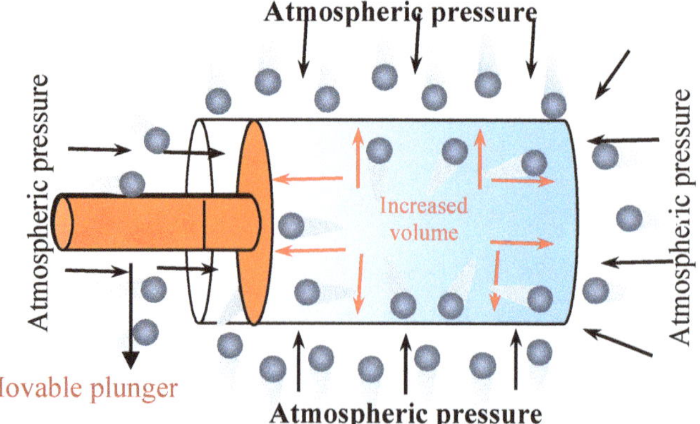

Figure 18.7(b). Plunger of syringe pushes out due to rise in temperature of gas.

As the air particles are heated up in a closed container, their kinetic energy increases. As a result of this, the particles collide against the walls of the container more forcefully thereby increasing the pressure in the container. The pressure within the syringe now becomes greater than the outside atmospheric pressure. However, because the plunger is freely mobile, it able to move outwards to negate any increase in pressure within the syringe, until the pressure inside it is the same as the outside atmospheric pressure. The gas is said to be heated at constant pressure because the pressure within the syringe is kept constant by allowing the plunger to move outwards. The net result is that there is a small increase in the volume of the air in the syringe. *If the air particles are kept at this constant temperature, the final volume would stay the same and the pressure inside would still be equal to that of the outside atmosphere.* The reverse will happen if the temperature of the air in the syringe is lowered; the pressure within the syringe would reduce and the atmosphere, being at a higher pressure, would push the plunger in producing a decrease in volume.

18.7   The Kelvin Temperature Scale

When plots of volume change with change temperatures are made on a fixed mass of gas, graphs of the following nature are obtained.

Figure 18.8. The plot of change in volume with change in temperature for a couple of gases.

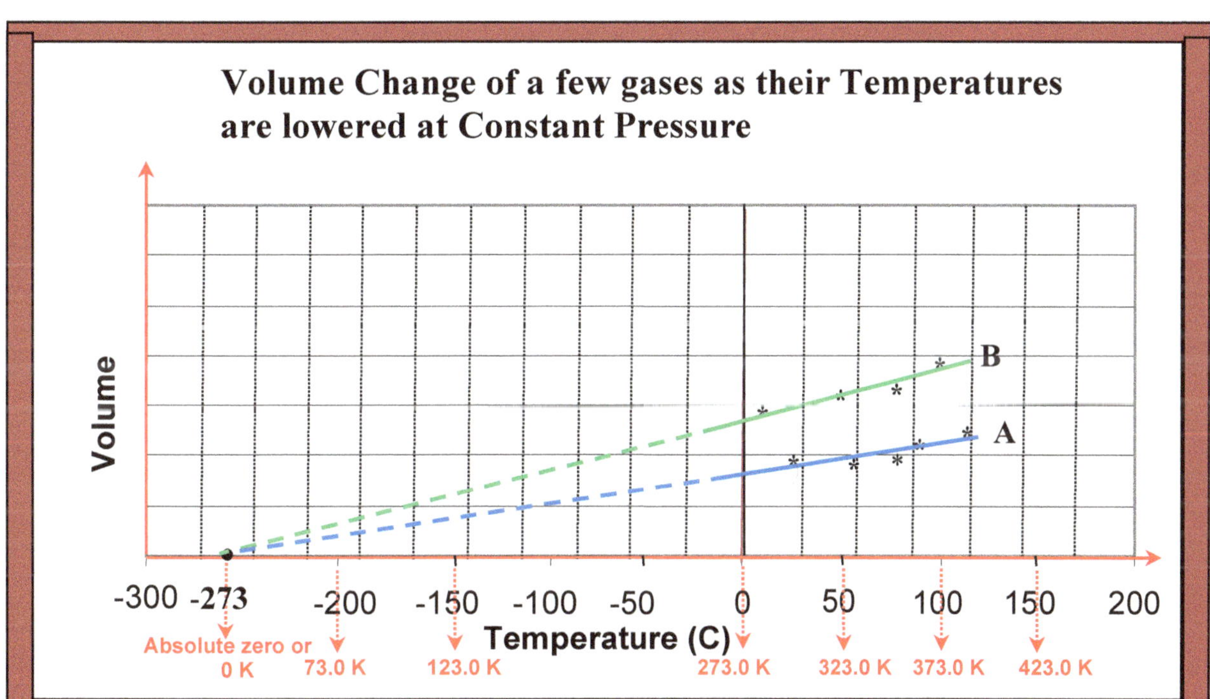

The graphs above show that as the temperature of any of the gases (at constant pressure) increases, its volume increases proportionately. The graph also shows that if the lines are extrapolated downwards, they touch the horizontal axis at -273.0 °C. The same would also be observed for all different types of gases. *In theory, this would mean that a gas would have zero volume at -273.0 °C.* However, in practice, this is not what is obtained; gases liquefy before they reach these low temperatures.

This temperature, **-273.0 °C**, is called **absolute zero,** and it is the lowest possible temperature that can be measured.

Scottish scientist, Lord Kelvin created another temperature scale. On this scale he makes -273.0°C, zero Kelvin. On the Celsius scale, 0 °C becomes 273.0 K. To make temperature conversions between these two temperature scales, it useful to follow the instructions below:

- To change from Kelvin to Celsius temperature, subtract 273.0 from the Kelvin value and rename it Celsius.
- To change from Celsius temperature to Kelvin, add 273 to the value and rename.

1. Convert the following Celsius temperature to Kelvin: K/U
   a) 10 °C
   b) 100 °C
   c) 120 °C
   d) − 50 °C
2. Convert the following Kelvin values to Celsius temperature: K/U
   a) 75 K
   b) 120 K
   c) 373 K

18.8   Charles' Law

From the graph obtained on the previous page, it can be concluded that *the volume of fixed mass gas at constant pressure is directly proportional to temperature.* Mathematically, this can be represented as:

$$V \propto T$$

$$or$$

$$V = kT \quad where\ k\ is\ a\ constant$$

$$k = \frac{V}{T} \quad {}^{*}Note\ that\ this\ is\ only\ possible\ if\ T\ is\ Kelvin\ and\ not\ Celsius\ temperature$$

The table next will illustrate how $\dfrac{V}{T}$ *is constant*

Table 18.2. The relationship between volume and temperature changes of a fixed mass of gas at constant pressure.

Experiments	1	2	3	4
Temperature T (°C)	40.00	70.00	100.0	120.0

Change to Kelvin ⟶   Add 273   Add 273   Add 273   Add 273

	1	2	3	4
Temperature, T (K)	313.0	343.0	373.0	393.0
Volume (L)	4.000	4.940	5.370	5.660
$k = \frac{V}{T}$ **(L/K)**	**0.0144**	**0.0144**	**0.0144**	**0.0144**

Based on the results obtained and the analyses made from his experiments, Charles formulated his law which states that *the volume of a fixed mass of gas at constant pressure, increases proportionally as its temperature increases.* Mathematically, this law can be expressed as:

$$\frac{V_1}{T_1} = \frac{V_2}{T_2} = a\ constant, k$$

Using values from experiments 1 and 2 in the Table 2, we get for K in both cases:

$$Experiment\ 1:\ k = \frac{4.000\ L}{313.0\ K} = 0.0144\ L/K \quad Experiment\ 2: k = \frac{4.940\ L}{343.0\ K} = 0.0144\ L/K$$

A balloon filled with neon gas having a volume of 2.00 L at 25.0 °C is placed in a refrigerator at 2.0 °C. What will be its final volume?

Solution:

Provided quantities:

$$V_1 = 2.00\,L$$
$$T_1 = 298.0\,K \quad * \text{Note that all temperatures are converted to Kelvin}$$
$$T_2 = 275.0\,K$$

Required quantity: V_2

Using the relationship, $\dfrac{V_1}{T_1} = \dfrac{V_2}{T_2}$

$$V_2 = \frac{V_1 \times T_2}{T_1} = \qquad V_2 = \frac{2.00\,L \times 275.0\,K}{298.0\,K}$$

$$= 1.85\,L$$

Exercise 18.3

a) Air in a cylinder with a movable piston has a volume of 4.00 L at 20.0 °C. To what temperature must the air be warmed to reach a new volume of 4.50 L at constant pressure? T/I

b) Inside a house, a balloon containing helium gas at 22.0 °C has a volume of 5.0 L. What will be its new volume if it is placed outside where temperature is lowered to 5 °C? Assume that the atmospheric pressure remains constant. T/I

c) 2.0 L of argon gas is heated at 20.0 °C until its final temperature is 32.0 °C at constant pressure. What is its final volume? T/I

d) A balloon filled with only helium gas had a volume of 4.00 L at 22.0 °C. When placed out doors for one cold night it attained a new volume of 3.75 L. Assuming that the atmospheric pressure remained constant, what was the highest temperature that balloon was subjected to? T/I

18.9　Pressure and Temperature Law or Gay-Lussac's Law

Suppose that we have a fixed mass of gas in a syringe *with an immovable plunger* and the gas is heated up, what will happen to its pressure?

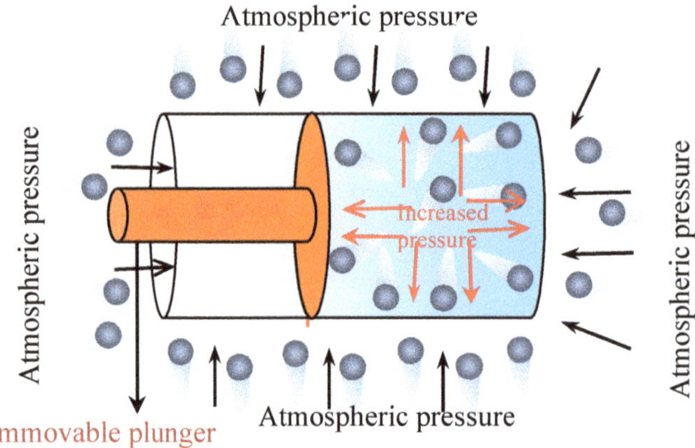

Figure 18.9. Pressure inside syringe of fixed volume increases with rise in temperature.

Like the previous case, as the temperature of the gas in the container increases, so does its pressure. However, unlike the previous case where the plunger is able to move back and forth to equate the pressure inside the syringe with that outside, this is not allowed here. In this case, because the plunger is fixed, the pressure is allowed to build up inside the syringe. As the temperature of the gas in the syringe is raised, the higher is the pressure that builds up there. In experiments to determine the relationship between pressure and temperature changes, the temperature of the gas in the syringe is gradually raised and its new pressure is noted and recorded for each new temperature change.

Table 18.3 below shows how the temperature and pressure law is derived after a series of experiments is carried out on a gas of fixed mass of and volume.

Table 18.3 Variation of pressure as temperature of a fixed volume of gas is changed.

Experiments	**1**	**2**	**3**
Temperature, T (^{o}C)	25.0	35.0	60.0
Change to Kelvin ⟶	Add 273	Add 273	Add 273
Temperature, T (K)	298.0	308.0	333.0
Pressure (kPa)	100.0	103.0	111.0
$K = \dfrac{P}{T}$ (kPa/K)	0.3340	0.3340	0.3330

Analysis of the results obtained in the table above shows that the pressure of a fixed volume of gas is directly proportional to its temperature. If the pressure of the gas for each experiment is divided by its corresponding Kelvin temperature, a constant is obtained. Since $k = \dfrac{P}{T}$, then the pressure and temperature law can be expressed as:

$$\frac{P_1}{T_1} = \frac{P_2}{T_2}$$ Knowing any three quantities, the fourth can be found.

Sample problem 4:

The walls of a sealed gas cylinder can withstand a maximum pressure of 1000.0 kPa.
If the gas in it has an initial pressure of 250.0 kPa and temperature of 25.0 °C, to what temperature must the cylinder be heated before its safety valve opens to prevent explosion of the cylinder?

Provided quantities:

P_1 = 250.0 kPa

T_1 = 298.0 K

P_2 = 1000.0 kPa

Required quantity: T_2

Using the relationship, $\dfrac{P_1}{T_1} = \dfrac{P_2}{T_2}$

$$T_2 = \frac{P_2 \times T_1}{P_2}$$

$$= \frac{1000.0 \, kPa \times 298.0 \, K}{250.0 \, kPa}$$

$$= 1192 \, K$$

$$= 919.0 \, {}^0C$$

Exercise 18.4

a) A sealed container has 2.00 L of methane gas at a pressure of 120.0 kPa and temperature of 20.0 °C. If the container is placed in a room where the temperature is 35.0 °C, what will be its new pressure? T/I

b) A cylinder of hydrogen gas at 22.0 °C has a pressure of 4.00 atm. If the maximum internal pressure that the cylinder can withstand is 12.00 atm, find the minimum temperature at which the cylinder would explode if it were in a building where a fire broke out? T/I

c) The volume of gas above a half-filled soda glass bottle is 500.0 mL at a temperature of 25.0 °C and pressure of 110.0 kPa. What will be the new pressure of the gas if the bottle is placed in a refrigerator where it is cooled to 5.0 °C and 10% of the gas particles dissolved in the soda? T/I

d) Assuming that after several hours of continuous driving at high speed the maximum temperature that tires reach is 50.0 °C. What must be the safe initial pressure of the tires at 20.0 °C before going on a long journey if the maximum pressure that the tires can withstand is 280.0 kPa? Assume that the volume of the tires stays constant. T/I

18.10   The Combined Gas Law

So far, none of the three gas laws that have been studied dealt with the combined relationships of volume, pressure and temperature changes for a fixed mass of gas. All of these laws can be combined as the following illustrations show.

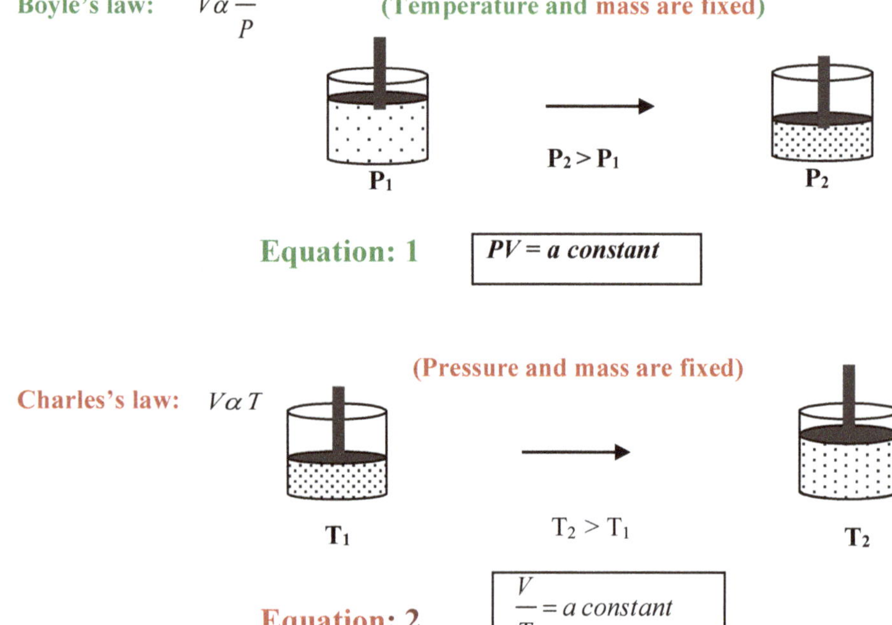

Pressure and temperature law: $P \alpha T$ (Volume and mass are fixed)

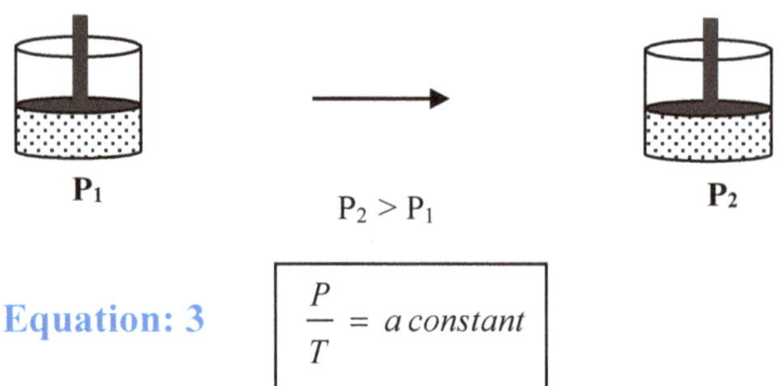

Summary:

$PV = a\ constant$ (Boyle's law)

$\dfrac{V}{T} = a\ constant$ (*Charles's law*)

$\dfrac{P}{T} = a\ constant$ (*Pressure and temperature law*)

A combination of these three laws gives the following equation:

$$\frac{PV}{T} = k$$

In this equation, the relationships of volume, temperature and pressure are expressed. Calculations can now be made for all combinations of pressure, temperature and volume changes on a ***fixed mass of any gas.***

Another way of expressing this **combined gas law** is:

$$\frac{P_1 V_1}{T_1} = \frac{P_2 V_2}{T_2}$$

In this combined gas law, all the other gas laws can be found. It therefore suffices for all the others when calculations involving changes in pressure, temperature and volume on a ***fixed mass of any gas are required.***

- In **Boyle's law**, *temperature is constant* so it would not change. **Removing temperature** from the combined gas law equation, we have:

$$\frac{P_1V_1}{\cancel{T_1}} = \frac{P_2V_2}{\cancel{T_2}} \quad \rightarrow \quad P_1V_1 = P_2V_2$$

- In **Charles's law**, *pressure is constant*, so it would not change. **Removing pressure** from the combined gas law equation, we have:

$$\frac{\cancel{P_1}V_1}{T_1} = \frac{\cancel{P_2}V_2}{T_2} \quad \rightarrow \quad \frac{V_1}{T_1} = \frac{V_2}{T_2}$$

- In **the Pressure and Temperature law**, *volume is constant* so it would not change. **Removing volume** from the combined gas law equation, we have:

$$\frac{P_1\cancel{V_1}}{T_1} = \frac{P_2\cancel{V_2}}{T_2} \quad \rightarrow \quad \frac{P_1}{T_1} = \frac{P_2}{T_2}$$

Sample problem 5:

A syringe containing 2.00 L of air at 25.0 °C and 100.0 kPa is placed in a refrigerator where the temperature is 5.0 °C. What will be the final pressure in the syringe if the plunger is fixed?

Solution:

Provided quantities:

$V_1 = 2.00$ L $\qquad P_1 = 2.00$ L

$V_2 = 2.00$ L $\qquad T_2 = 278.0$ K

$T_1 = 298.0$ K

Required quantity: P_2

Using the combined gas law equation,

$$\frac{P_1V_1}{T_1} = \frac{P_2V_2}{T_2}$$

We eliminate V_1 and V_2 **since the volume remains constant**. Doing this, we are left with the temperature and pressure law: $\dfrac{P_1}{T_1} = \dfrac{P_2}{T_2}$ Making P_2 the subject of the equation, we get:

$$P_2 = \frac{P_1T_2}{T_1} = \frac{100.0\ kPa \times 278.0\ K}{298.0\ K}$$

$$= 93.29\ kPa$$

A balloon filled with helium gas having a volume of 4.00 L at a temperature of -10 °C and a pressure of 40.00 kPa in the atmosphere is lowered to where the temperature is 20 °C and the pressure is 98.0 kPa. What will be its new volume if the mass of gas remains constant?

Solution:

Provided quantities:

$V_1 = 4.00$ L $T_2 = 278.0$ K

$T_1 = 263.0$ K $P_2 = 98.0$ kPa

$P_1 = 40.00$ kPa

Required quantity: V_2

Using the combined gas law equation,

$$\frac{P_1 V_1}{T_1} = \frac{P_2 V_2}{T_2}, \text{ and making } V_2 \text{ the unknown, we get:}$$

$$V_2 = \frac{P_1 V_1 T_2}{T_1 P_2} \quad \text{Substituting values in this equation, we get:}$$

$$V_2 = \frac{40.00 \, kPa \times 4.00 \, L \times 278.0 \, K}{263.0 \, K \times 98.0 \, kPa}$$

$$= 1.73 \, L$$

Exercise 18.5

a) Helium gas is stored in a 10.0 L steel cylinder at 20.0 °C and 1200.0 kPa pressure. How many balloons of capacity of 2.0 L can be filled with helium gas at 25.0 °C and 120.0 kPa from the cylinder? T/I

b) A weather balloon contains 6.00 L of an inert gas at 15.0 °C and 90.0 kPa pressure. It is brought to the surface of the Earth where the new temperature and pressure are 25.0 °C and 101.0 kPa respectively. What would be its new volume? T/I

c) A car tire filled with air has a pressure of 240.0 kPa, a volume of 12.0 L and temperature of 15.0 °C. If after several hours of high-speed driving, its temperature rose to 40.0 °C and its volume increased by 5.00%, what is the new pressure? T/I

d) Oxygen gas collected above water at 25 °C and 102.5 kPa has a volume of 200.0 mL. What would be the volume of dry oxygen at STP? (The vapour pressure of water at 25.0 °C is 3.17 kPa). T/I

e) At what new temperature must 10.0 L of $CO_{2(g)}$ at 20.0 °C and 100.2 kPa be heated to have a volume of 12.5 L at 100.0 kPa? T/I

f) 24.0 L of carbon dioxide gas is at a temperature of 25 °C and at a pressure of 100.0 kPa. To what temperature must the gas be cooled to attain a new volume of 14.0 L at 98.0 kPa? T/I

g) To what new pressure 10.4 L of nitrogen at 101.2 kPa and 25 °C be raised, if its temperature is raised to 60.0 °C. T/I C A

h) Scuba divers use air tanks for their supply of oxygen. Air has approximately 20% oxygen, approximately 80 % nitrogen. As the diver descends, the column of water above creates a pressure on him/her; the lower the decent the higher is the pressure experienced. Because of this high pressure, inevitably the gases that are inhaled liquefy in the tissues of the body. Unlike oxygen that is constantly consumed and removed by cellular respiration, liquefied nitrogen remains in the tissues. If the diver ascends suddenly, the reduced pressure on his tissues causes the liquid nitrogen to change into gaseous nitrogen that affects various organs such as the heart, brain and lungs (Caisson disease). Research this disease to find out how it can be avoided and what treatment is given to patients who suffer from this disease. A T/I C

i) Without the use of a pressure cooker it may take about 20 to 25 minutes to cook some peas. With a pressure cooker, it may take less than 10 minutes. Find out how a pressure cooker is able to achieve this. T/I C A

j) As an aircraft ascends higher and higher into the atmosphere, the atmospheric pressure gets progressively less and less and this has serious consequences for those in the aircraft; the pressure and concentration of oxygen present get too low to sustain life. To solve these problems, air is taken from the outside and is compressed to increase the oxygen concentration. The air is then cooled and circulated inside the aircraft; the used air is removed subsequently. What might happen to the passengers if a hole suddenly appears in one of the cabin walls of the aircraft? T/I A

Chapter Content:

Partial pressures of mixtures of non-reacting gases, and measured volumes of gases have wide applications, ranging from medical diagnosis to life-saving airbags of automobiles.

An airbag, in automobiles, works on the principle that when a severe head-on collision occurs, it causes a pre-calculated mass of sodium azide (NaN_3) solid to decompose spontaneously, producing a specific volume of nitrogen gas that inflates the air bag within 40 milliseconds of the impact. More or less than that specific volume of gas produced could be fatal to the driver or passengers; a greater volume could rupture the bag, while a smaller volume would fail to provide adequate cushioning. In marine salvaging operations, airbags are also used to refloat sunken ships or their important cargoes within. In this case, the volumes of gases used must be enough to create a magnitude of buoyancy large enough to raise the objects out of the water. The steam engine which works by the pressure created by water vapour (steam) depends on water being heated to a very high temperature to form a certain volume of steam every second. The partial pressures of oxygen and carbon dioxide in the blood are delicately balanced by some of our organ systems; any prolonged fluctuations in these can be diagnostic of some respiratory or metabolic disorders. In respirometry, the use of lung volumes, tidal volumes and vital capacities aid in differentiating between different types of respiratory abnormalities. In scuba diving, extreme care is taken to select the types of gas mixtures used for breathing to reduce complications, such as diver's bends and narcosis, while at the same time facilitating an adequate supply of oxygen. This chapter illustrates Dalton's law of partial pressure, the vapour pressure of water and Gay-Lussac's law of combining volumes. It also explains how Avogadro's hypothesis enables molar volumes of gases to be calculated.

Gas Mixtures

19.1   Dalton's Law of Partial Pressure

English scientist, John Dalton, hypothesised that if, in a mixture of gases, there is no reaction between them, and that if each gas behaves independently of each other, then each gas would contribute its own partial pressure to the total pressure of the gas mixture. He carried out a series of experiments to test his hypothesis. In these experiments, he placed a fixed number of moles of gas separately into a cylinder of known volume and he recorded its pressure. He repeated this for different non-reacting gases. Finally, he places all the gases into one cylinder and measured the new total pressure of the gas mixture. The results of these are illustrated in the following figures:

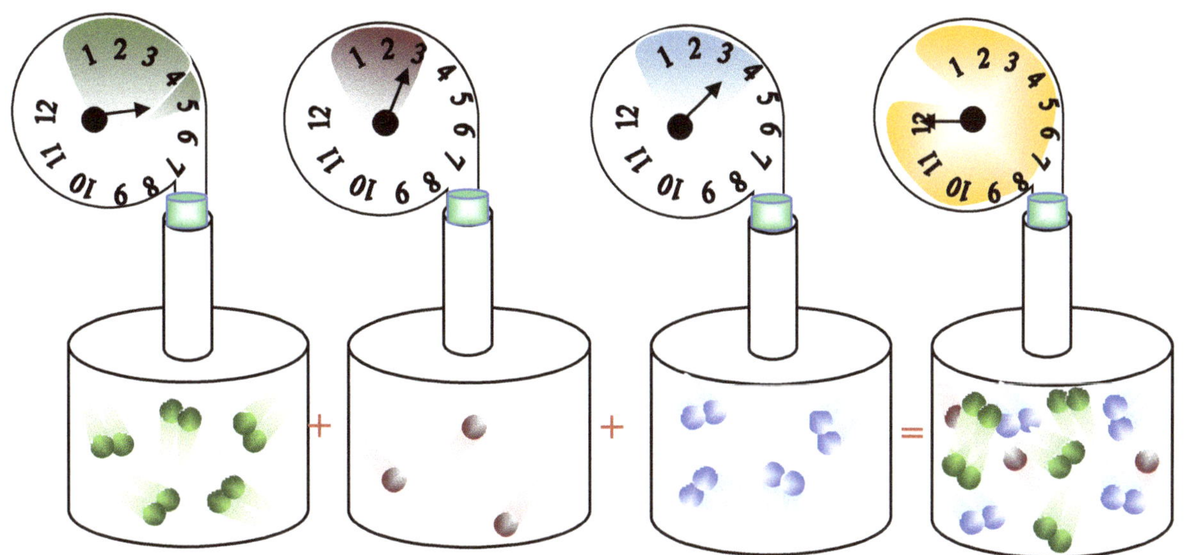

Figure 19.1(a). **5 mol** of $N_{2(g)}$ in volume V give **5 units** of pressure.

Figure 19.1(b). **3 mol** of $He_{(g)}$ in volume V give **3 units** of pressure.

Figure 19.1(c). **4 mol** of $O_{2(g)}$ in volume V give **4 units** of pressure.

Figure 19.1(d). mixture of all gases in volume V gives **12 units**. of pressure

Recall that there are two ways of increasing the pressure of a gas held at constant temperature in a container: by reducing its volume (Boyle's law) or by increasing its number of particles. The first three figures above show that the pressure obtained by each different sample of gas is related to its own number of moles of particles. The last figure shows the pressure obtained when all these gases are placed in a single container.

Table 19.1. Relationships between number of gas particles and pressure units.

Figures	Type of particles	Number of moles of particles	Number of pressure units
1	N_2 molecules	5	5
2	He atoms	3	3
3	O_2 molecules	4	4
4	Mixtures of N_2 and O_2 molecules and He atoms	12	12

Analysis of the results in the previous table shows that the total pressure obtained from a mixture of non-reacting gases is the sum of the pressures of the individual gases. From this, Dalton formulated his law of partial pressure which states that *the total pressure of a mixture of non-reacting gases is the sum of the partial pressures of the individual gases.*

The basic understanding here is that regardless of what type of non-reacting gas is placed in a container, whether it be in the form of molecules or atoms, each will have the same pressure effect of one unit.

According to the above results, the following mathematical deductions can be made for the various gases, in terms of their partial pressures:

Gases	$N_{2(g)}$	$He_{(g)}$	$O_{2(g)}$	**Mixture**

$$\textit{Partial pressure units}: \quad \frac{5}{12}\,total \;+\; \frac{3}{12}\,total \;+\; \frac{4}{12}\,total \;=\; \frac{12}{12}\,total$$

$$\textit{Total pressure}: \quad P_1 \quad + \quad P_2 \quad + \quad P_3 \quad = \quad P_{total}$$

A 2-L flask containing $N_{2(g)}$, $O_{2(g)}$ and $CO_{2(g)}$ with partial pressures of 78.00 kPa, 20.00 kPa and 1.00 kPa respectively. What is the total pressure in the flask?

Provuded quantities:

$$P_{N_2} = 78.00 \text{ kPa}$$
$$P_{O_2} = 20.00 \text{ kPa}$$
$$P_{CO_2} = 1.00 \text{ kPa}$$

Required quantity: P_{total}

$$P_{total} =$$

$$P_{N_2} + P_{O_2} + P_{CO_2}$$
$$78.00 \, kPa + 20.00 \, kPa + 1.00 \, kPa$$
$$99.00 \, kPa$$

Exercise 19.1

1. A gas cylinder has a mixture of $O_{2(g)}$ and $He_{(g)}$. If the total pressure inside is 10.0 atm, but the partial pressure of $He_{(g)}$ is 9.0 atm, what is the partial pressure of the $O_{2(g)}$? T/I
2. How might the partial pressures of inhaled $O_{2(g)}$, $CO_{2(g)}$ and $H_2O_{(g)}$ differ from those exhaled? T/I

19.2   Vapour Pressure of Water

In the laboratory, some gases such as oxygen are collected by the downward displacement of water. Inevitably, because the collected gas is in contact with water, some of the water molecules vaporize and mix with the collected gas. The total pressure of the collected gas is thus not due to the gas alone, but also due to the vapour pressure exerted by the water molecules at that temperature. To know what the partial pressure of the dry gas would be, the vapour pressure of water at that temperature is noted and then subtracted from the pressure of the gas mixture in the container. Note that the total pressure in the filled container is the same as the atmospheric pressure, at that time of collection. The table below gives the vapour pressure of water at a few different temperatures.

Table 19.2. Vapour pressure of water at different temperatures.

Temperature(°C)	0	5.0	10.0	15.0	20.0	25.0	30.0	35.0	40.0	45.0	50.0
Pressure(kPa)	0.61	0.87	1.23	1.71	2.34	3.17	4.24	5.62	7.38	9.58	12.33

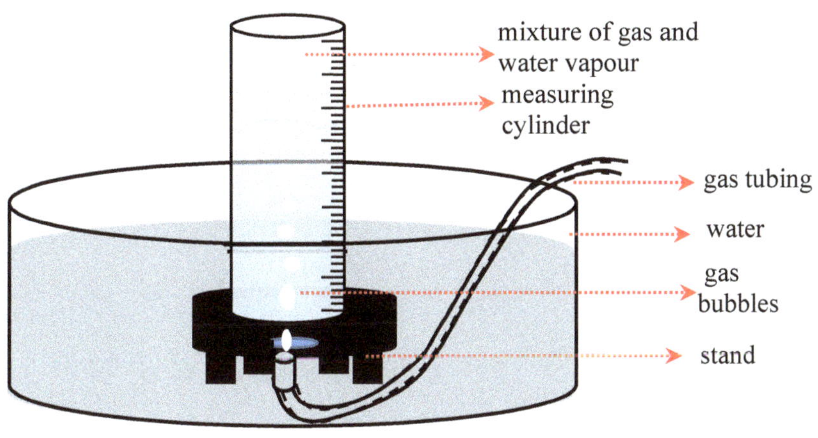

The figure to the left depicts the collection of a gas above water. The measuring cylinder is filled with water which is then inverted over a water-filled trough. As the gas rises up the cylinder, it displaces the water downwards.

Note* The combined pressure of the gas and water vapor becomes equal to atmospheric pressure **only** when the level of the water in the cylinder is the same as the water level in the trough.

Figure 19.2. The collection of a gas over water using a pneumatic trough.

Sample problem 2:

If oxygen gas is collected in the laboratory where the temperature is 25.0 °C and the atmospheric pressure is 100.0 kPa, what is the partial pressure of the dry oxygen gas?

Solution:

Provided quantities:

$$P_{total} = 100.00 \ kPa$$
$$P_{water\ vapour} = 3.17 \ kPa$$

Required quantity: P_{O_2}

$$P_{O_2} = \frac{P_{total} - P_{water\ vapour}}{100.0 \ kPa - 3.17 \ kPa}$$
$$96.83 \ kPa$$

19.3 Gay-Lussac's Law of Combining Gas Volumes

Joseph Louis Gay-Lussac (1778-1850) observed that, when gases at the same temperature and pressure **react** and their volumes are observed, they do so in ratios of small whole numbers. For example, in the reaction between carbon monoxide and oxygen, the following happens:

$$CO_{(g)} \quad + \quad CO_{(g)} \quad + \quad O_{2(g)} \quad \rightarrow \quad CO_{2(g)} \quad + \quad CO_{2(g)}$$

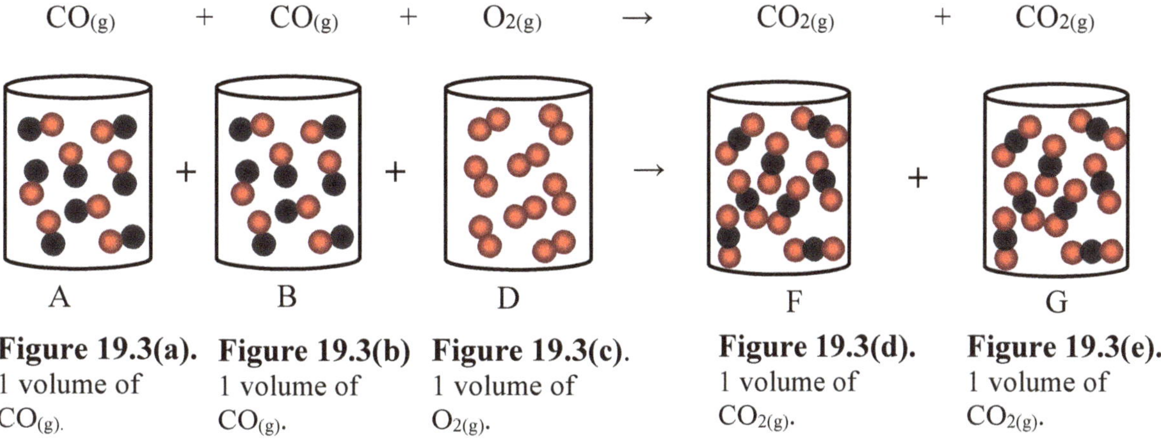

Figure 19.3(a). 1 volume of $CO_{(g)}$.

Figure 19.3(b). 1 volume of $CO_{(g)}$.

Figure 19.3(c). 1 volume of $O_{2(g)}$.

Figure 19.3(d). 1 volume of $CO_{2(g)}$.

Figure 19.3(e). 1 volume of $CO_{2(g)}$.

Rewriting these volumes, we have the following:

$$2\ CO_{(g)} \quad + \quad O_{2(g)} \quad \rightarrow \quad 2\ CO_{2(g)}$$

$$2\ \text{volumes} \quad + \quad 1\ \text{volume} \quad \rightarrow \quad 2\ \text{volumes}$$

Volume ratio: 2   :   1   :   2

The following are some other examples of Gay-Lussac's observations:

1.   Hydrogen   +   Chlorine $\rightarrow$ Hydrogen Chloride

$$H_{2(g)} \quad + \quad Cl_{2(g)} \quad \quad 2\ HCl$$

1 volume   **1 volume** $\rightarrow$ **2 volumes**

Volume ratio: **1**   :   **1**   :   **2**

2.   Nitrogen   +   Hydrogen $\rightarrow$ Ammonia

$$N_{2(g)} \quad + \quad 3\ H_{2(g)} \quad \rightarrow \quad 2\ NH_{3(g)}$$

1 volume   **3 volumes** $\rightarrow$ **2 volumes**

Volume ratio: **1**   :   **3**   :   **2**

Based on his findings, Gay-Lussac formulated his law of combining volumes, which states that, *in any chemical reaction involving gases measured at the same temperature and pressure, the volume of gaseous reactants and gaseous products are always in simple ratios of whole numbers.*

After Gay-Lussac formulated his law, it was challenged by **John Dalton** *who believed that gases were made up of single atoms, according to his atomic theory*. As such, it would be impossible for 1 volume of hydrogen reacting with 1 volume of chlorine to produce 2 volumes of hydrogen chloride. According to him, only 1 volume of hydrogen chloride should be produced. The following figures depict Dalton's contentions.

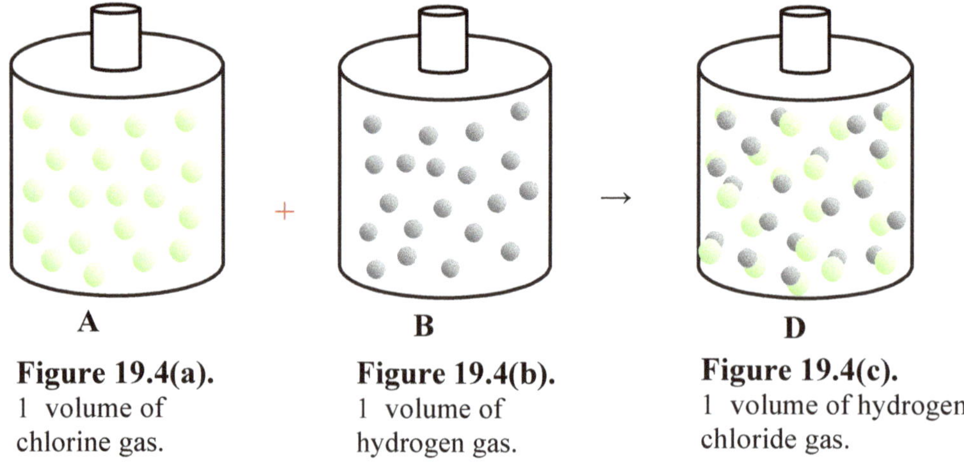

Figure 19.4(a).
1 volume of
chlorine gas.

Figure 19.4(b).
1 volume of
hydrogen gas.

Figure 19.4(c).
1 volume of hydrogen
chloride gas.

***Note that this was just what Dalton believed but was not necessarily correct.**

It was because of this understanding that Dalton assigned an incorrect relative atomic mass of 8 to oxygen. With subsequent accurate understanding, oxygen was reassigned its correct relative atomic mass of 16.

In 1811, while Gay-Lussac and Dalton were still in total disagreement, the Italian physicist Amedeo Avogadro (1776-1856) proposed a solution to this dilemma.

19.4 Avogadro's Hypotheses:

Avogadro proposed the following:
- *Equal volumes of gases measured at the same temperature and pressure contain equal number of molecules.*
- *These molecules, such as oxygen, chlorine, and hydrogen are divisible into "half-molecules." That is, each of these molecules must be composed of two atoms that split up into individual atoms before they react.* This is depicted in the following reaction:

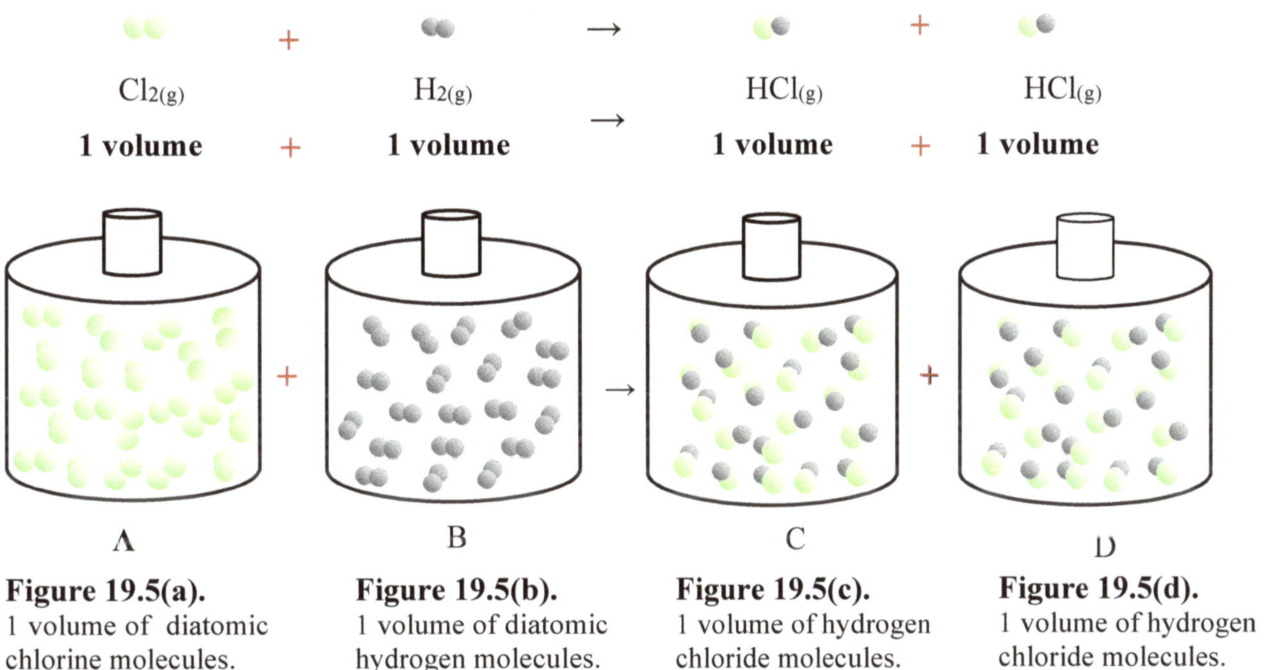

Figure 19.5(a).	**Figure 19.5(b).**	**Figure 19.5(c).**	**Figure 19.5(d).**
1 volume of diatomic chlorine molecules.	1 volume of diatomic hydrogen molecules.	1 volume of hydrogen chloride molecules.	1 volume of hydrogen chloride molecules.

19.5 Molar Volume

Standard Molar Volume

Based on Avogadro's *first hypothesis*, it can be deduced *that equal number of gaseous particles measured at the same temperature and pressures must occupy the same volume.* This means that **1 mole of any gas (6.02x10^{23} gas particles) at 0 °C and 101.3 kPa should occupy the same volume.**

Density
As can be recalled, density is defined as mass per unit volume.

$$density = \frac{mass}{volume}$$

$$\therefore volume = \frac{mass}{density}$$

The molar volumes of any gas at STP can thus be found by dividing its molar mass by its density. The following table shows how this is done for a few gases.

Table 19.3. Standard molar volumes for a few gases at STP

Gas	Density g/L	Molar mass g/mol	$V = \dfrac{Molar\ mass}{Density}$	Molar volume (All equal) L/mol	Number of particles/mol
O_2	1.429	32.00	$32.00\ g/mol\ /\ 1.429\ g/L$	22.39 L/mol	6.022×10^{23} *molecules*
N_2	1.251	28.01	$28.01\ g/mol\ /\ 1.251\ g/L$	22.39 L/mol	6.022×10^{23} *molecules*
H_2	0.0898	2.016	$2.016\ g/mol\ /\ 0.0898\ g/L$	22.43 L/mol	6.022×10^{23} *molecules*
Ne	0.9000	20.18	$20.18\ g/mol\ /\ 0.9000\ g/L$	22.42 L/mol	6.022×10^{23} *atoms*
CH_4	0.1760	16.04	$16.04\ g/mol\ /\ 0.7160\ g/L$	22.40 L/mol	6.022×10^{23} *molecules*
CO_2	1.977	44.01	$44.01\ g/mol\ /\ 1.964\ g/L$	22.41 L/mol	6.022×10^{23} *molecules*

Sparing the minor differences, the universally accepted value for the volume of **one mole of a gas at STP is 22.4 L.**

According to Avogadro's hypothesis, which states that equal volumes of gases measured at the same temperature contain equal number of particles, each of the molar volumes (22.4 L) in the previous table must contain the same number of particles, 6.022×10^{23} (Avogadro's number), *regardless of their molecular masses.* The following figures depict this concept for a few of the gases in the previous table under investigation.

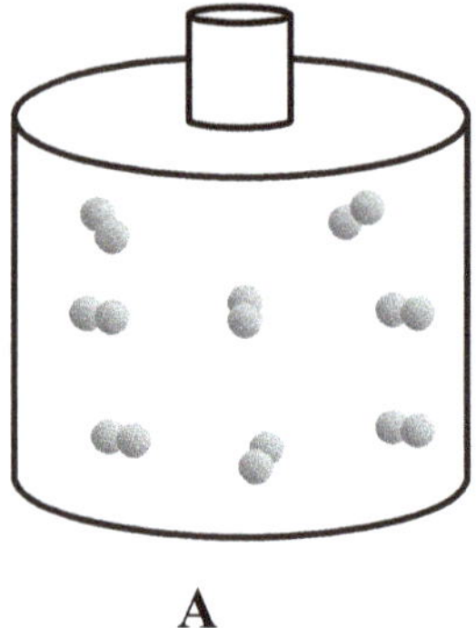

A

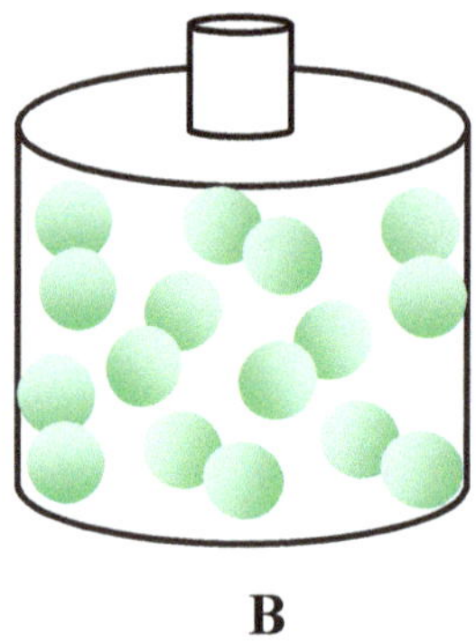

B

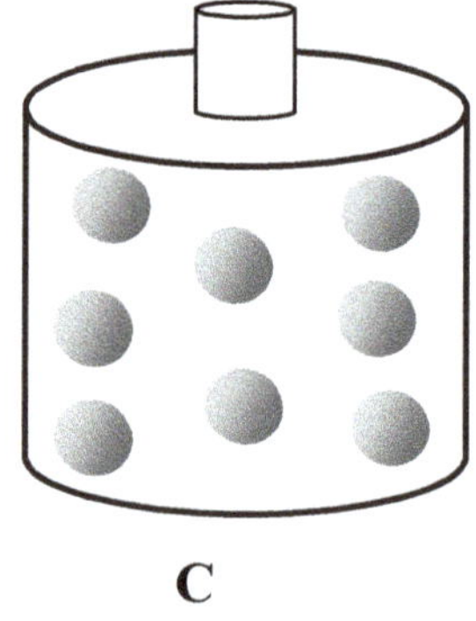

C

Figure 19.6(a). 1 mol $H_{2(g)}$ has a volume of 22.4 L at STP and contains 6.022×10^{23} molecules.

Figure 19.6(b). 1 mol $N_{2(g)}$ has a volume of 22.4 L at STP and contains 6.022×10^{23} molecules.

Figure 19.6(c). 1 mol $Ar_{(g)}$ has a volume of 22.4 L at STP and contains 6.022×10^{23} atoms.

Molar Volume at Standard Ambient Temperature and Pressure (SATP)

SATP values are measurements made at 25.0 °C and 100.0 kPa.

The universally accepted value for **the volume of one mole of a gas at SATP is 24.8 L.**
Explain why the measured molar volume at SATP is higher than that at STP.

Sample problem3:

What volume is occupied by 0.0100 mol of $O_{2(g)}$ at STP?

Solution:

Provided quantities:

$$Molar\ volume\ at\ STP = 22.4\ L$$

$$n_{O_{2(g)}} = 0.0100\ mol$$

Required quantity:

$$Volume\ of\ 0.0100\ mol\ O_{2(g)}\ at\ STP$$

$$V_{O_2} = 0.0100\ mol \times \frac{22.4\ L}{mol} = 0.224\ L$$

$$= 224\ mL$$

Sample Problem 4:

What amount of carbon dioxide, in moles, is required to produce a volume of 62.0 L at SATP?

Solution:

Provided quantities:

$$Volume\ of\ CO_{2(g)} = 62.0\ L$$

$$Molar\ volume\ at\ SATP = 24.8\ L$$

Required quantity:

$$\#\ mol\ of\ CO_{2(g)}$$

$$n_{CO_{2(g)}} = \frac{62.0\ L}{24.8\ L/mol}$$

$$= 2.50\ mol$$

Converting amount in mol to mass we have:

$$m = n \times M$$

$$m_{CO_{2(g)}} = 2.50\ mol \times \frac{44.01\ g}{mol}$$

$$= 1.10 \times 10^2\ g$$

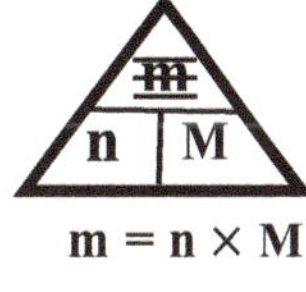

What volume of $CO_{2(g)}$ at SATP will be produced from the decomposition of 17.00 g of $NaHCO_{3(s)}$?

Solution:

To solve this problem, it is useful to follow the steps below:
Step 1. Write a balanced equation for the reaction:

(Given)   *(Required)*

$$2\ NaHCO_{3(s)} \longrightarrow Na_2CO_{3(s)} + H_2O_{(g)} + CO_{2(g)}$$

Step 2. Calculate the number of moles of $NaHCO_{3(s)}$ in the given sample:

$$n_{NaHCO_3} = \frac{17.00\ g}{85.01\ g/mol}$$

$$= 0.2000\ mol$$

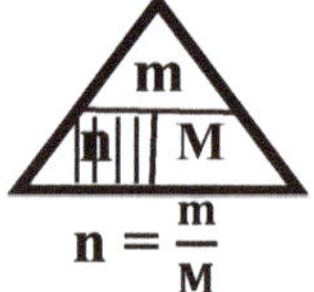

$$n = \frac{m}{M}$$

Step 3. Use the mole ratios in the balanced equation to find the number of moles of the required quantity:

$$\frac{Given}{required} = \frac{2\ mol\ NaHCO_3}{1\ mol\ CO_2} = \frac{0.2000\ mol}{X\ mol}$$

$$2X = 0.2000\ mol\ CO_2$$

$$X = 0.1000\ mol\ CO_2$$

Step 4. Find the volume occupied by 0.1000 mol $CO_{2(g)}$ at SATP:

$$V_{CO_2} = 0.1000\ mol \times 24.8\ L/mol$$

$$= 2.48\ L$$

Exercise 19.2

1. Determine the number of molecules in 33.6 L of $N_{2(g)}$ at STP. T/I

2. Find the volume of $H_{2(g)}$ at SATP that is produced from the decomposition of 3.40 g of $H_2O_{2(l)}$. The following equation shows how this decomposition happens: T/I
$$2\ H_2O_{2(l)} \longrightarrow 2\ H_2O_{(l)} + O_{2(g)}$$

3. The following equation shows how methane gas is burnt in oxygen: T/I
$$CH_{4(g)} + 2\ O_{2(g)} \longrightarrow 2\ H_2O_{(l)} + CO_{2(g)}$$
What volume of oxygen at SATP is required for complete combustion of 4.00 g of methane?

4. Ammonia ($NH_{3(g)}$) is synthesised by the reaction between nitrogen and hydrogen.
What volumes of nitrogen and hydrogen will be produced at SATP by the decomposition of 74.49 L of $NH_{3(g)}$ at SATP? T/I

5. A 4-L cylinder containing 5.0 moles of compressed $He_{(g)}$. How many 2-L balloons at SATP can be filled by the compressed $He_{(g)}$, assuming all the gas is emptied? T/I

6. Potassium chlorate decomposes according to the following equation:

$$2\,KClO_{3(s)} \rightarrow 2\,KCl_{(s)} + 3\,O_{2(g)}$$

If the only source of oxygen for the burning of carbon is from the decomposition of potassium chlorate, what mass of it must decompose to produce 124.0 L of oxygen at SATP? What mass of carbon can be completely burnt by this volume of oxygen? T/I

7. Sodium azide decomposes extremely quickly to inflate air bags according to the following equation:

$$2\,NaN_{3(s)} \rightarrow 2\,Na_{(s)} + 3\,N_{2(g)}$$

What mass of sodium azide must decompose to inflate a 70.0 L air bag at SATP? T/I

Chapter Content:
20.1 The ideal Gas Equation
20.2 Gas Stoichiometry

The Ideal Gas

An ideal gas does not exist. It is a theoretical gas that is assumed to have the following properties:
- The gas particles behave as small rigid spheres.
- The collisions between the gas particles are purely elastic; no attraction between themselves or with the walls of the container.
- The volume of the gas particles is very small compared to the space within which they are confined.
- The gas particles are in constant random motion.
- The gas particles do not attract or repel each other nor their container.
- They move in a perfect straight line.

Some real gases such as oxygen, hydrogen, nitrogen and noble gases conform to the ideal gas model reasonably well. Generally, these gases tend to behave like ideal gases at conditions of low pressures and high temperatures. At these two conditions, the gases are widely spaced apart, thereby reducing the chance for any deviations from the above assumptions. At other conditions, most gases do not demonstrate the properties of an ideal gas; they attract each other, their collisions are not purely elastic, and their molecules have their own volume and they do not move in straight lines.

Molar volumes of gases are normally measured at two sets of conditions: standard temperature and pressure (STP) – 0^0C and 101.325 kPa, and standard ambient temperature and pressure (SATP) – 25 ^{0}C and 100.0 kPa. The ideal gas equation, $PV = nRT$ allows gas volumes at different conditions of temperature and pressure other than at STP and SATP to be calculated.

Many chemical reactions produce gases. Gas stoichiometry is thus important in knowing what amounts of reactants to use in order to produce a specific volume of a gas at some conditions of temperature and pressure. For example, to fill a car airbag with a certain volume of nitrogen gas, a specific amount of sodium azide must be decomposed.

$$2 \text{ NaN}_{3(s)} \longrightarrow 2 \text{ Na}_{(s)} + 3 \text{ N}_{2(g)}$$

$$2 \text{ mol} \qquad\qquad\qquad 3 \text{ mol} = 74.4 \text{ L at SATP}$$

From the above stoichiometry, any specific volume of $N_{2(g)}$ at SATP can be generated by the decomposition of a calculated amount of sodium azide.

20.1 The Ideal Gas Law

In all the previous calculations involving real gases so far, it was assumed that they behaved like the ideal gases.

The Ideal Gas Equation

So far, the calculations for volumes of gases produced are restricted to conditions at STP and SATP. How could calculations be made for volumes of real gases produced at conditions of temperature and pressure other than those at STP and SATP? The ideal gas equation allows us to do this. It incorporates the variables of temperature, pressure, volume and the number of moles of real gases, with the assumption that they behave like ideal gases.

- According to Boyle's law: $V \propto \dfrac{1}{p}$

- According to Charles's law: $\quad V \propto T \quad$ (*Kelvin temperature*)
- From common experience: $\quad V \propto n \quad$ (*n = # of mol of gas particles*)
  (For example, inflating a balloon)

Combining these equations, we get: $\qquad V \propto \dfrac{nT}{p}$

Removing the proportionality sign $(\propto)$ and replacing it with a constant, R we get:

The ideal gas equation $\quad =$

$$V = \frac{nRT}{p}$$
$$or$$
$$PV = nRT$$

The constant **R** is called the **gas constant**. Its units can be derived by making it the subject of the equation:

$$R \;=\; \frac{PV}{nT} = \frac{kPa \times L}{mol \times K}$$

Scientists calculated the value for R to be: $\quad \dfrac{8.31\,kPa.L}{mol.K}$

What volume would 2.00 moles of hydrogen gas occupy at 30.0 ºC and 110.0 kPa?

***Provided quantities*:**

   n = 2.00 mol

   p = 110.0 kPa

   T = 30.0 ºC = 303.0 K

***Required quantity*:**

   Volume: V

 Using the ideal gas equation, we get:

$$V = \frac{nRT}{P}$$

$$V = \frac{2.00\,mol \times \dfrac{8.31\,kPa.L}{mol.K} \times 303.0\,K}{110.0\,kPa}$$

$$V = 45.8\,L$$

Note that all temperatures must be in Kelvin.

What is the mass of 10.0 L of $CO_{2(g)}$ at pressure 100.0 kPa and temperature 20.0 °C?

Solution:

Provided quantities:
  V = 10.0 L
  P = 100.0 kPa
  T = 293.0 K

Required quantity:
  Mass of $CO_{2(g)}$

Before the mass of $CO_{2(g)}$ can be calculated, it is first necessary to find its number of moles. Using the ideal gas equation and making moles, **n**, its subject, we get:

$$n = \frac{PV}{RT}$$

$$n = \frac{100.0\,kPa \times 10.0\,L}{\dfrac{8.31\,kPa.L}{mol.K} \times 293.0\,K} = \frac{100.0\,kPa \times 10.0\,L}{} \times \frac{mol.K}{8.31\,kPa.L} \times \frac{1}{293.0\,K}$$

$$n_{CO_2} = 0.411\,mol$$

$$m_{CO_2} = n \times M$$

$$= 0.411\,mol \times 44.01\,g/mol$$

$$= 18.1\,g$$

20.2 Gas Stoichemistry

What volume of hydrogen gas is produced at 20.0 °C and 104.0 kPa by the reaction between 100.0 mL of 0.100 mol/L $HCl_{(aq)}$ and 2.40 g of $Mg_{(s)}$?

Solution:

This is a multi-step problem that involves finding the limiting reagent, then using it to find the number of moles of hydrogen produced and then changing this to a volume at those conditions of temperature and pressure stated. Problems like this have already been explained and solved in chapters 6 and 7.

Step 1. Write a balanced equation for the reaction:

$$2\,HCl_{(aq)} \;+\; Mg_{(S)} \;\longrightarrow\; MgCl_{2(aq)} \;+\; H_{2(g)}$$

Step 2. Find the number of moles of $Mg_{(s)}$ used

$$n_{Mg} = \frac{2.40\,g}{24.31\,g/mol}$$

$$= 0.099\,mol$$

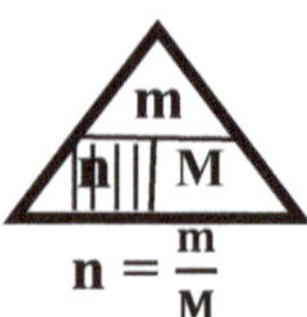

Step 3. Find the number of moles of $HCl_{(aq)}$ used:

Since this is an aqueous solution, to find the number of moles, the following equation is used:

$$n_{HCl_{(aq)}} = C \times V$$

$$= 0.100 \ mol \, / \, L \times 0.1000 \, L$$

$$= 0.0100 \ mol \quad \textit{(experimental value)}$$

$$n = C \times V$$

Step 4. Find the **limiting reagent:**

Choose any one of the above reagents to find out how much of the other one is required to react with it. Using $Mg_{(s)}$ (*now called the given reagent*) to find the number of moles of $HCl_{(aq)}$ (*now called the required reagent*),** we get:

$$\frac{Given}{Required} = \frac{1 \, mol}{2 \, mol} = \frac{0.099 \, mol \ Mg}{x \, mol \ HCl_{(aq)}}$$

$$X = 0.198 \ mol \ HCl_{(aq),} \ \textit{(theoretical value)}$$

When this ***theoretical value*** of $0.198 \ mol \ HCl_{(aq)}$ is compared with $0.0100 \ mol \ HCl_{(aq)},$ ***(experimental value)*** it be can seen that less $\mathbf{HCl_{(aq)}}$ was used to react completely with the amount of $Mg_{(s)}$ used, 0.099 mol. It can thus be concluded that the $\mathbf{HCl_{(aq)}}$ is the limiting reagent.

Step 5. Use the limiting reagent to find the number of moles of $H_{2(g)}$ produced. Since the limiting reagent is known, it is now considered the **given reagent** in this step:

$$\frac{Given}{Required} = \frac{2 \, mol}{1 \, mol} = \frac{0.0100 \, mol \ HCl_{(aq)}}{X \, mol \ H_{2(g)}}$$

$$2X = 0.0100 \ mol \ H_{2(g)}$$

$$\backslash X = 5.00 \times 10^{-3} \ mol \ H_{2(g)}$$

Step 6. Use the equation for the ideal gas to find the volume of $H_{2(g)}$ produced:

$$V = \frac{n \times R \times T}{P}$$

$$V = \frac{5.00 \times 10^{-3} \ mol \times \dfrac{8.31 \ kPa. \ L}{mol.K} \times 293.0K}{104.0 \ kPa}$$

$$= 0.117 \ L$$

$$= 117 \ mL$$

1. Sodium reacts with water to produce hydrogen gas according to the following equation:

   $$2 \, Na_{(s)} \ + \ 2 \, H_2O_{(l)} \ \rightarrow \ 2 \, NaOH_{(aq)} + \ H_{2(g)}$$

   What volume of $H_{2(g)}$ is produced at 30 °C and 100 kPa if 1.15 g of $Na_{(s)}$ reacts with excess water? T/I

2. Sodium hydrogen carbonate reacts with hydrochloric acid according to the following equation:

   $$NaHCO_{3(s)} + HCl_{(aq)} \ \rightarrow \ NaCl_{(aq)} + H_2O_{(l)} + \ CO_{2(g)}$$

   What volume of $CO_{2(g)}$ is produced at 15 °C and 100 kPa if 3.40 g $NaHCO_{3(s)}$ is allowed to react with 200.0 mL of 0.100 mol/L $HCl_{(aq)}$? T/I

3. Potassium chlorate decomposes when heated to produce oxygen gas according to the following equation:

$$2 \, KClO_{3(s)} \rightarrow 2 \, KCl_{(s)} + 3 \, O_{2(g)}$$

What mass of $KClO_{3(s)}$ must be decomposed to produce 50.0 L of $O_{2(g)}$ at 35 °C and 110.0 kPa? T/I

4. Magnesium reacts with hydrochloric acid according to the following balanced equation:

$$Mg_{(s)} + 2 \, HCl_{(aq)} \rightarrow MgCl_{2(aq)} + H_{2(g)}$$

What volume of $H_{2(g)}$ is produced at 40 °C and 105.0 kPa if 1.20 g of $Mg_{(s)}$ reacts with 400.0 mL of 1.00 mol/L $HCl_{(aq)}$? T/I

5. What volume of $H_{2(g)}$ is produced at 30 °C and 100.0 kPa between the reaction of 300.0 mL of 1.0 mol/L $H_2SO_{4(aq)}$ and 2.4 g of $Mg_{(s)}$. T/I

6. What mass of $Zn_{(s)}$ must react with 200.0 mL of 0.10 mol/L $H_2SO_{4(aq)}$ to produce 248.0 mL of $H_{2(g)}$ is produced at 25 °C and 100.0 kPa. T/I

7. 50 mL of 0.40 mol/L of a certain acid was reacted with excess magnesium and 258.4 mL of hydrogen was produced at 38 °C and 100 kPa. Was the acid used $HCl_{(aq)}$ or $H_2SO_{4(aq)}$? T/I

CHAPTER 21
Atmospheric Pollution

Chapter Content:

Our planet Earth is endowed with certain distinct features that make it the only known habitable one known to scientists, as of now. Tilted on its axis at an angle of 23.5 ^{0}C and located at a distance neither too close or far away from the sun, it enjoys a moderate average temperature that is supportive of various life forms. The tilting of its axis produces seasonal fluctuations that make possible the proliferation of living organisms over all parts of the Earth, and without which some parts (poles) would be in perpetual darkness. The two extremes of temperatures caused by Winter and Summer (-65 ^{0}C and 45 ^{0}C respectively), are not too severe to prevent organisms from adapting and thriving in both terrestrial or aquatic ecosystems. Of vital significance is the unique composition of our atmosphere. The presence of life-sustaining oxygen and carbon dioxide is what distinguishes Earth from the other planets; without oxygen there would be no aerobic cellular respiration for energy production, and without carbon dioxide there would be no photosynthesis for food production. Also the presence of carbon dioxide gas a helps to keep the Earth warm. The presence of a thin layer of ozone (3 millimeters thick, if compressed) at above 15-25 kilometers above the atmosphere blocks most of the dangerous UV B from reaching the Earth's surface. The rotation of the Earth not only causes day and night but also produces a strong magnetic field that extends outwards to about 65 kilometers, where it traps dangerous solar winds generated by the sun, preventing them from reaching the Earth.

Human beings and other organisms have lived and survived on the Earth for thousands of years. There has always been a harmonious balance between the activities of humans and the cycles of nature. However, as the human population grows, the need for more space and resources becomes important for survival. Humans have thus resorted to more deforestation and subsequent burning to clear space for more robust agriculture and habitation. The need for more transportation has seen a rapid proliferation of motor vehicles and aircrafts etc., that work by the combustion of fossil fuels. Industrial and technological revolutions have resulted in the manufacture of various types of new products and chemical compounds. Many of these inadvertently enter our environment, thereby polluting it.

The following sections will examine how human human activities are having profound destructive effects on our environment. These effects include global warming, acid precipitation, gound ozone and ozone layer depletion. There is growing evidence that global warming presents the most dangerous existential threat that is causing climates changes with catastrophic consequences. To save our planet from these ecological disasters, everyone needs to make a concerted effort in order to restore our environments to states that are life-sustainable.

21.1   Formation of the Atmosphere

Scientists believe that after the Earth was formed 4.5 billion years ago, the first atmosphere, comprising mainly of hydrogen and helium, was produced from violent volcanic eruptions within the Earth. These gases, being light and energetic, eventually disappeared into space. These were replaced by methane, ammonia, carbon dioxide and water vapour, also produced from volcanic eruptions. With the subsequent cooling of the Earth, the water vapour condensed forming oceans and other bodies of water. With the advent of plant life and water, most of the carbon disappeared by dissolving in the surface water and by the process of phosynthesis. Oxygen concentration gain ascendance after being produced as a byproduct of photosynthesis. All of the ammonia molecules, under the influence of sunlight, decomposed into nitrogen and hydrogen. The latter, being very light, escaped into space, but nitrogen, being much heavier, stayed behind and became the predominant gas and has remained so to this day. It is also believed that the ozone layer was formed from oxygen molecules reacting together with energy from sunlight. It is believed that all of the processes took billions of years to gradually happen.

21.2   Composition of the Atmosphere

Table 21.1. Percentage composition of the atmosphere by volume.

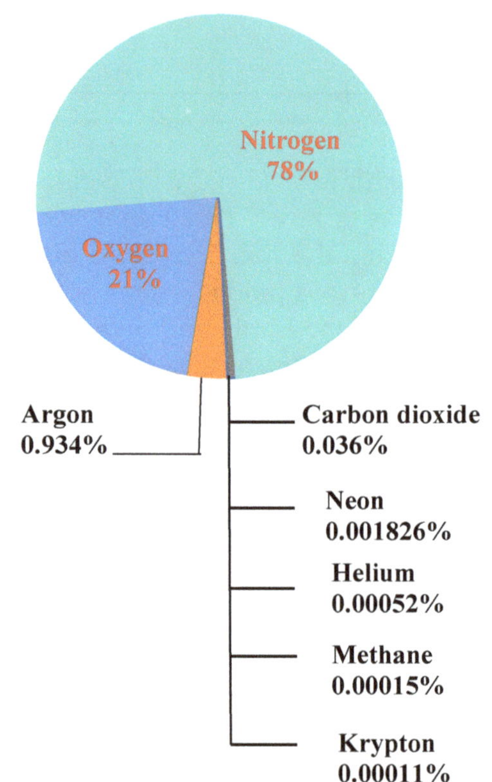

Fig. 21.1. Pie chart showing percentage composition of the atmosphere by volume.

Gases	Percentage by Volume	Molecular Structure
Nitrogen (N_2)	78.08	
Oxygen (O_2)	20.95	
Argon (Ar)	0.934	
Carbon dioxide(CO_2)	0.036	
Neon (Ne)	0.00182	
Helium (He)	0.00052	
Methane (CH_4)	0.00015	
Krypton (Kr)	0.00011	

21.3   The Nitrogen Cycle

The element, nitrogen is used to synthesize essential organic compounds such as proteins and nucleic acids in living organisms; without these, living organisms would cease to exist. Nitrogen gas exists abundantly in the atmosphere in the form of diatomic molecules. These, being triple bonded, are very stable are thus not as reactive as other gasses. Nitrogen gas would only react with other elements in conditions that provide adequate energy to zap their molecules. Examples of these are lightning, and the Haber-Bosch process that uses nitrogen to manufacture ammonia. Another natural way that gaseous nitrogen is made to react to produce other nitrogen-containing inorganic compounds, is by certain bacteria found in the root nodules of leguminous plants. The bacteria use some of these compounds for their own growth and reproduction, while the host plant use the rest to manufacture their proteins and nucleic acids. These plant proteins and their DNA eventually get assimilated into animal proteins and DNA after they are eaten by animals. Animals produce nitrogenous wastes. and eventually die along with plants, and their bodies eventually decompose.The following cycle illustrates how atmospheric nitrogen and that found in animal wastes and dead organisms are recycled in nature.

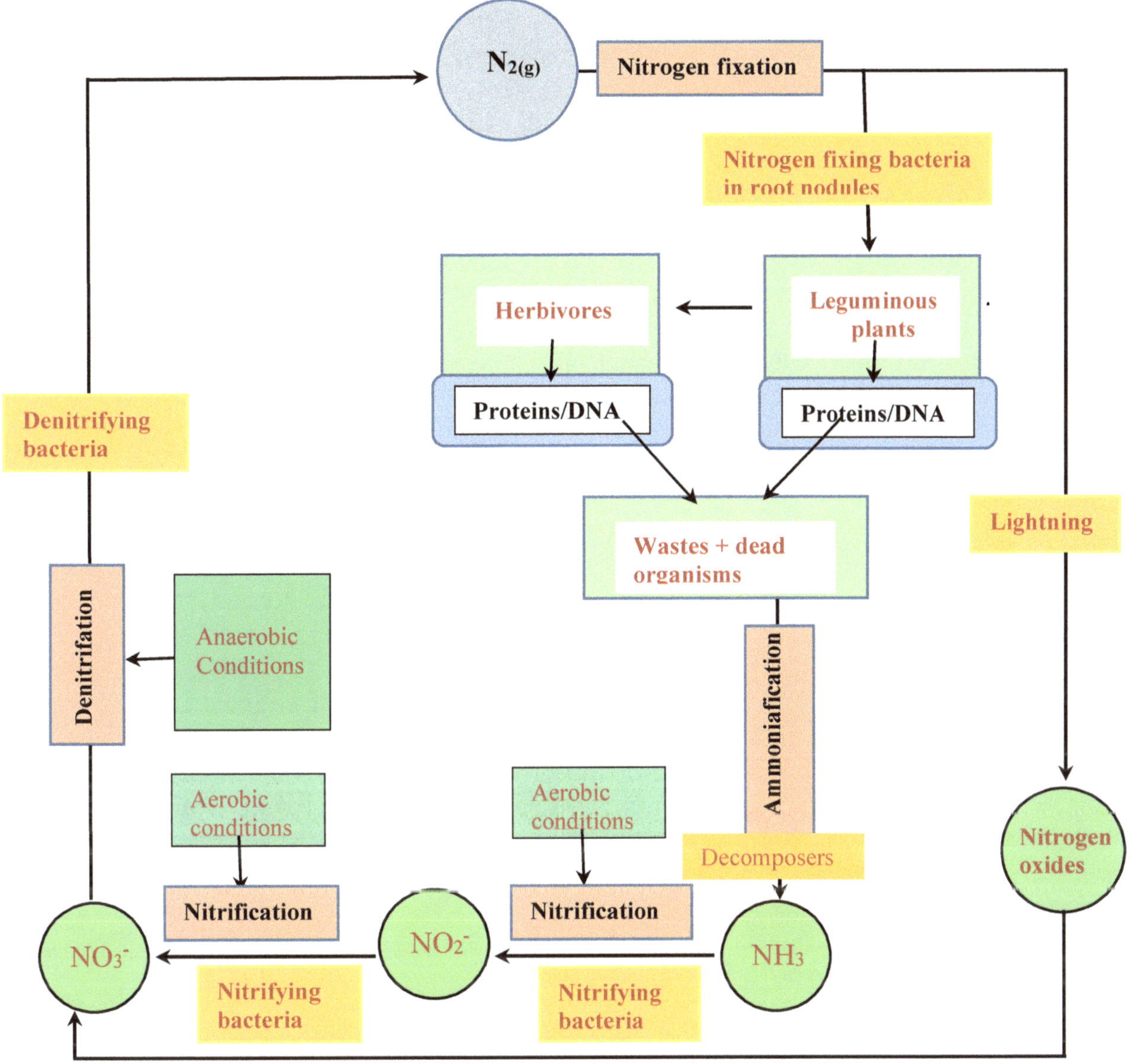

Fig. 21.2. The Nitrogen Cycle.

For the sake of clarity, the nitrogen cycle will be categorized into a number of steps.

- **Nitrogen Fixation:** This happens in two ways:
  *Flashes of lightning with their high energy, zap the nitrogen molecules allowing the atoms produced to react with oxygen in the air, forming oxides of nitrogen. These dissolve in rainwater and eventually reach the soil, where they are changed into nitrate ions.
  *Nitrogen-fixing bacteria present in the root nodules of leguminous plants (peas, beans, clover), change nitrogen in air spaces in the soil, into nitrates ions. The bacteria use some of these ions to make their own proteins/DNA and the rest is absorbed by the plants to produce theirs.
- **Assimilation:** When these plants or their produce are eaten by animals, their proteins and DNA are assimilated into animal types, for their growth and reproduction.
- **Decomposition:** Dead animals and plants, along with their wastes (urea and uric acid), are acted upon by decomposers (bacteria and fungi) to gain nutrients for their growth and reproduction. As this happens, minerals and some compounds from the dead organism and their waste, are returned to the soil, while others escape into the air.
  One nitrogenous compound that is produced is ammonia by a process called *ammoniafication*.

➢ **Nitrification:** In the soil, the ammonia produced is then converted into nitrate ions by a series of aerobic nitrifying bacteria into nitrate ions, some of which are used by the plants. This process is called **nitrification**. These bacteria are called aerobic, since they need molecular oxygen for their activities.

➢ **Denitrification:** If conditions become anaerobic, such as in water-logged soil, anaerobic bacteria (these don't need molecular oxygen from air, but use the oxygen bonded to the nitrate ion for their activities) convert any nitrate ions present into nitrogen molecules that return to the air.

Left all by itself, there would be a nice balance between the amount of nitrogen taken from the air and that returning, without causing any negative ecological consequences. However, with the use of nitrogenous and phosphate fertilizers for greater food production, the cycle has been disrupted with serious ecological impacts. These fertilizers find their way into oceans and other bodies of water from runoffs from agricultural lands, where they produce algal blooms with subsequent eutrophication of water. This loss of dissolved oxygen has seriously affected marine ecosystems, creating thousands of square kilometers of dead zones.

Exercise 21.1

1. Two nitrogen atoms are triply bonded to form a molecule of nitrogen. How might the percentage composition of nitrogen in the atmosphere change if the atoms were doubly bonded? K/U

2. At the end of a growing season, the lots of nutrients are still locked up in the leaves, roots and stems of plants. Instead of burning or discarding these as wastes, why would it be more beneficial to the soil if these are allowed to decompose in it? T/I A

3. During the denitrification process, **anaerobic** bacteria utilize oxygen bonded to nitrate ions for their respiration. In so doing, the bacteria decompose these valuable ions present in the soil producing nitrogen gas. What must farmers do to ensure that very little amount of nitrate is lost from the soil? A T/I

4. Fish farmers use nitrogenous fertilizers in water to promote the growth of algae that the fish feed on. Why must care be taken not to over fertilize the water? T/I A

21.4 The Carbon Cycle

Carbon is an essential part of most of the organic compounds in our bodies. It is an essential part of carbohydrates, fats, proteins, vitamins and nucleic acids. It is also found as coal, in hydrocarbons and in other types of inorganic compounds found in nature. The following table summarizes the different ways that carbon is recycled in nature; mostly by the removal of $CO_{2(g)}$ out and into the atmosphere.

Table.21.2. The ways in which carbon is taken in and out of the air and Earth.

Ways that $CO_{2(g)}$ enters atmosphere	Ways that $CO_{2(g)}$ is removed from atmosphere
Cellular respiration by decomposers and by both aquatic and terrestrial plants and animals $C_6H_{12}O_6 + 6\,O_2 \longrightarrow 6\,CO_2 + 6\,H_2O + E$	Photosynthesis: $6\,CO_2 + 6\,H_2O \longrightarrow C_6H_{12}O_6 + 6\,O_2$
Combustion of coal and hydrocarbons $C + O_{2(g)} \longrightarrow CO_{2(aq)} + E$ $CH_4 + 2\,O_2 \longrightarrow CO_2 + 2\,H_2O + E$	Diffusion: $CO_{2(g)} + H_2O \longrightarrow CO_{2(aq)}$
	Synthesis of calcium and other compounds $CO_3^{2-}{}_{(aq)} + Ca^{2+} \longrightarrow CaCO_3$ Fossilization: Dead organisms $\longrightarrow$ fossils

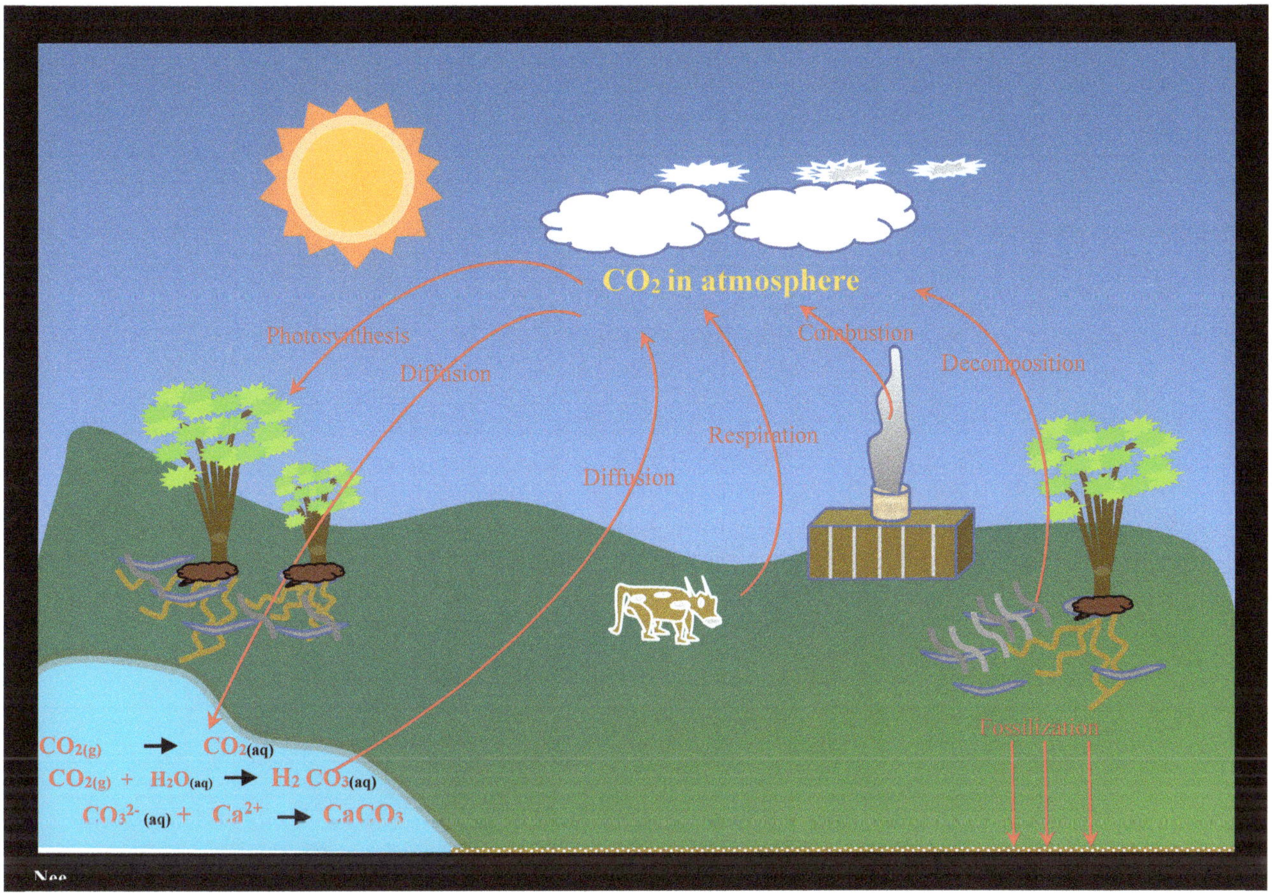

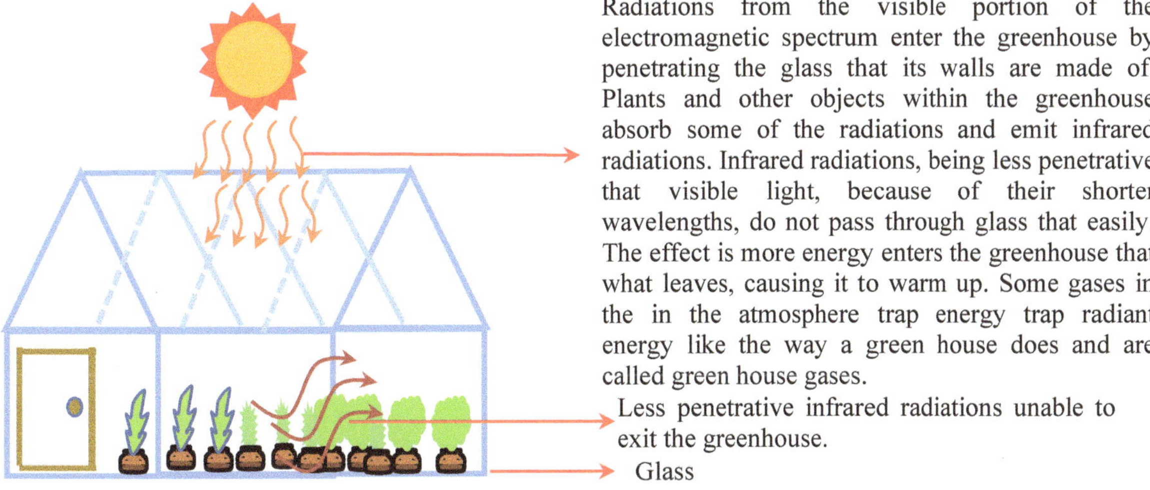

21.5 The Greenhouse

The greenhouse is utilized by people who operate nurseries to germinate seeds and to grow plants starting in early Spring when the ambient temperature is not conducive to doing this. Use of a greenhouse makes this possible; its feature enables its inside to attain a suitable higher temperature than the colder outside. The following figure models a greenhouse and provides an explanation of how it becomes warm even when the outside is cold.

Radiations from the visible portion of the electromagnetic spectrum enter the greenhouse by penetrating the glass that its walls are made of. Plants and other objects within the greenhouse absorb some of the radiations and emit infrared radiations. Infrared radiations, being less penetrative that visible light, because of their shorter wavelengths, do not pass through glass that easily. The effect is more energy enters the greenhouse that what leaves, causing it to warm up. Some gases in the in the atmosphere trap energy trap radiant energy like the way a green house does and are called green house gases.

Less penetrative infrared radiations unable to exit the greenhouse.

Glass

Fig.21.5. Model of a green house.

21.6   Greenhouse Gases

A number of gases are capable of trapping long wave solar energy in the atmosphere, but the main ones are methane, oxides of nitrogen, water vapour and carbon dioxide, with water vapour doing the bulk of it. Solar energy is trapped by these molecules when their bonds undergo vibrations; the C-H bonds in methane, the N-O and bonds in oxides of nitrogen, the O-H bonds in water and the $C=O$ bonds in carbon dioxide. Other molecules such as those of nitrogen and oxygen are not capable of doing this. As the bonds stop vibrating, their stored energy is released as heat in the form of infrared radiations that keeps the atmosphere warm; without these gases, the Earth and everything on it would freeze. However, because of a constant rise in the concentrations of some of these gases; especially carbon dioxide from the burning of coal and fossil fuels, the has been a rise the average temperature of the atmosphere. This phenomenon is called **global warming** which is trending towards catastrophic ecological consequences if left unchecked.

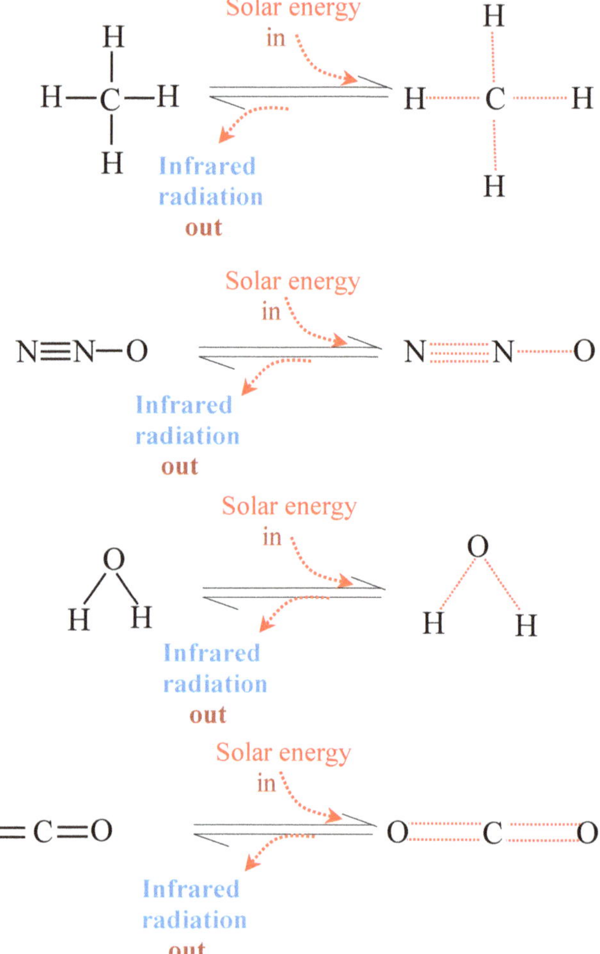

Fig.21.6. Greenhouse gases absorbing solar radiations and emitting infrared radiations.

21.7   Global Warming

Evidence

The unprecedented ecological changes that are taking place can only point to one cause: global warming. The following are some of the alarming changes that are most evident:

- Melting of the polar icecaps and shrinking of glaciers
- Rising of sea levels
- More violent hurricanes and storms
- Lowering of the pH in oceans
- More frequent destructive wildfires

Ecological Impacts

The melting of the polar icecaps causes rising of sea levels that causes more floods in low-lying coastal countries. This year, in my native country of Guyana, which lies ten feet below sea level, there has been unprecedented wide-scale flooding. Shrinking of the polar icecaps and glaciers represent a loss of **albedo effect** (**ability to reflect solar energy back into space**); the surfaces of the Earth and other reflecting ones like clouds reflect energy back into the atmosphere and space to help maintain energy balance of the globe. Less reflection of energy means more retention of heat on the Earth. The survival of some species such as the penguins and polar bears that habitat in these regions is threatened.

Along with other acidic gases, carbon dioxide has caused an increase in acidity of the oceans. There is growing evidence that coral reefs are getting smaller, and this is directly linked to the lowering of the pH of the oceans. The calcium carbonate, that constitute the bulk of coral reefs, reacts with acids in the water and gets dissolved in the process. The disappearance of these reefs represents a loss of precious habitats for a countless number of species of aquatic organisms.

Over the years, the frequency and ferocity of wildfires has increased with devastating consequences. In the year 2020, there has been 58 950 wildfires compared to 50 477 in 2019. Wildfires of 2020 in Australia caused the death of an estimated three billion animals and the destruction of a countless number of habitats. Last year, wildfires destroyed the entire village of Lytton in British Colombia. Record soaring heat waves there were responsible for over 1000 deaths. A few weeks ago they wreaked hovac in Alberta, Canada. Currently, unprecented historic wild fires are doing the same, but on larger scales in various provinces in Canada again. The enormous amount of smoke produced is causing dangerously high poor air quality in both Canada and neighbouring United States.

Storms, hurricanes and tornadoes have all become more powerful; the excess energy for which is provided by a warmer globe.

The Kyoto Protocol

The Montreal Protocol was signed to limit the use ozone layer depleting gases, CFCs, and so did the international community on 11 December 1997, in Kyoto, Japan to reduce the emission of greenhouse gases (GHG). Canada, along with 160 other countries signed on to this agreement. One of the objectives was to reduce the emission, measured at the year 1990, by 5% in the years of 2008-2012. To achieve this set target, a greater responsibility was placed on the more developed industrialized countries to do more in this respect, since they were more responsible for GHG emissions. Different nations sought to cut emissions in their own ways; either by reducing emission in their own countries or indirectly by assisting other less developed nations reduce their emissions in some tangible ways. For example, my native country of Guyana, entered an agreement with Norway in 2010, in which it will receive $250 million if keeps deforestation at its minimum for at least five years. Other developing countries are paid these 'carbon credits' by the richer developed countries for planting more trees. Some countries such as Pakistan and Saudi Arabia have recently, on their own volition, undertaken a 'tsunami of trees' project, vowing to plant at least 5 billion trees in their respective countries. More on the Kyoto Protocol can obtained from the following website: https://unfccc.int/kyoto_protocol#:~:text=In%20short%2C%20the%20Kyoto%20Protocol,accordance%20with%20agreed%20individual%20targets.

21.8   Acid Rain Formation

Lakes in North America are less productive with clearer waters, forests around the world are gradually disappearing, coral reefs are receding, soils in limstone regions are losing their topography, more people in industrialzed countries are dying of cardiovascular diseases, and major stuctural and heritage buildings throughout the world are weakened or defaced; the cause of which is acid-rain pollution. Acid rain pollution refers to the precipitation the reaches the ground that contains acids, made in air by water reacting with oxides of nitrogen and suplhur dioxide. Oxides of nitrogen and sulphur dioxide, produced from the burning of coal and fossils fuels, react with water in the air in the form of rain, hail or snow to form nitric acid and sulphuric acid.

Lower pH kills aquatic plants such as algae which form the basis of aquatic food chains; without these whole ecosystems are destroyed. Acid rain destroys terrestrials plants in two ways; direct bleaching of their exposed structures and by leaching some essential nutrients deep down into the soil, depriving the roots of the plants from absorbing them for their metabolism and growth. Acids dissolve the limestone in present in soil and coral reefs by reacting with them; the same can be said for structures that are made of marble. The destuction of weight-bearing metallic structures such as bridges and other buildings pose different threats, in that these can suddenly collapse, killing people. Some acidic particles (NO_2) can react with others such as ground level ozone in the air, in **sygergenic ways** to produce carcinogenic and other more destructive compounds. The happens more often when there is an atmospheric inversion (cold air is trapped beneath warm air for a prolonged period of time) which allows for the formation of smog in which various compounds present are allowed to mix and react with solar energy to form these harmful sustances. Asian countries (China, India, Pakistan and Bangladesh) that are densely populated have a greater percentage of people dying of atmospheric pollution related illnesses than elsewhere in the world.

Exercise 21.1

1. Of the main greenhouse gases, which one is most effective at absorbing solar energy? K/U
2. Which greenhouse gas increase is of most concern to us? A
3. Refering to the carbon cycle, which process do humans have most control over, in terms of limiting global warming? K/U
4. Photosynthesis is one natural process by carbon dioxide is removed from the air, how could the international community make used of this fact to offset global warming? K/U A
5. People living in poor developing countries practice the slash/burn process to clear away forested areas for crop farming. This practice is sometimes described as a double whammy. Explain. What role could the more developed nations play to empower these people to reduce this practice? T/I A
6. Create a list of things that you and others can personally do to limit global warming. K/U A
7. Identify the three main gases that produce acid rain and write balanced equations for their reactions with water. K/U
8. Why do farmers periodically sprinkle pellets of calcium carbonate on their soil? A
9. Why does burning of fossil fuels, which are hydrocarbons, release acidic gases? K/U
10. What better legislations could be put in place to ensure the makers of motor vehicles, owners of factory and smelter comply more with cleaner air policies? A

21.9 Layers of the Atmosphere

The layers of the atmosphere perform a number of activities that make life possible on Earth. Apart from providing life-sustaining gases such as carbon dioxide and oxygen, they regulate weather patterns and protect us from harmful solar radiation coming from outer space.

If a vertical analysis of the atmosphere is made, in terms of its composition, pressure and temperature, starting from sea level, a number of layers can be identified. These are illustrated by the figure below.

Table.21.3. Features of the atmospheric layers

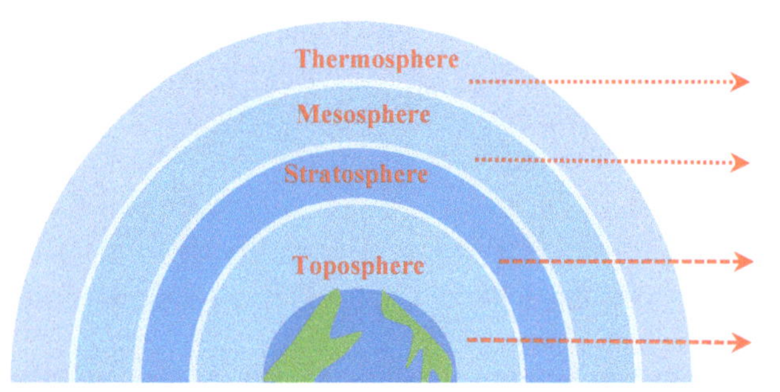

Fig.21.7. Layers of the atmosphere

Distance above sea level (Km) and accompanying pressure kPa	Temperature range ^{0}C
120 kM with 1.00×10^{-4} kPa	-85.5 ^{0}C – 1200 ^{0}C
80 kM with 0.10 kPa	-2.5 ^{0}C - -85.5 ^{0}C
50 kM with 0.10 kPa	-67 ^{0}C – 2.5 ^{0}C
10 kM with 10 kPa	-67 ^{0}C – 35 ^{0}C

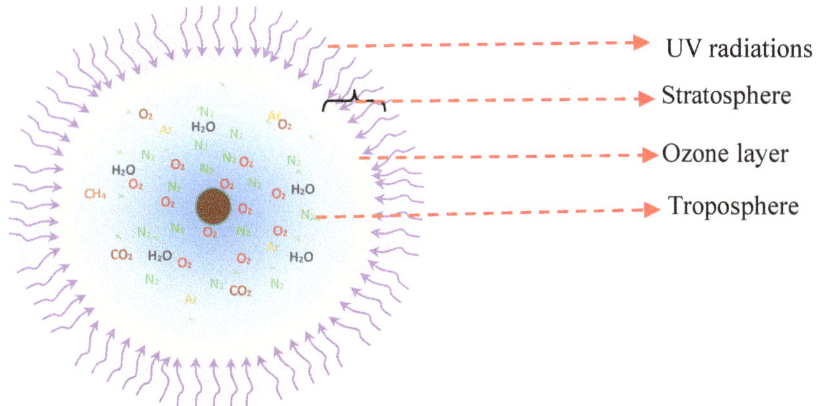

Fig.21.8. Stratosphere being shielded from UV by ozone layer within the stratosphere.

The troposphere (means change) is where we live, and it contains 75% of the gases present in the atmosphere. It has 99 % of all water present in the atmosphere. The heat trapped by the Earth (land and water) from the sun is radiated back into the troposphere and warms it up. This is where mixing of warm and cool air happens; warm air rises and cooler denser air sinks. These movements help distribute energy over the Earth and produce weather patterns. Hot air balloons operate in the upper part of this layer to monitor weather changes. As one ascends the troposphere the air becomes much thinner at extremely low pressure. **The tropopause** is the transition region between the troposphere and the stratosphere.

The stratosphere (means layer**)** experiences some degree of warming and since its warmer air sits on top of denser colder tropospheric air; no mixing happens. Very little disturbance in weather pattern happens this area; pilots make use of this stable feature to fly. In this layer is found the **ozone layer** (15 – 30 km). The ozone layer is important since it blocks UVB from reaching living organisms on the Earth (Fig.21.8). The energy absorbed is what raises the temperature of the stratosphere from -67 ^{0}C – 2.5 ^{0}C. **The stratopause** is the transition region between the the **troposphere** and the **mesosphere.**

The mesosphere (means middle) experiences a decrease in temperature with altitude as there is little or no greenhouse gases present to absorb solar energy. Life would be impossible there; no air to breathe, too low a pressure, and direct exposure to UV. Its only source of energy is what reaches it from the troposphere. Meteorites burn up in this region before reaching below. **The mesopause** is the transition region between the **mesosphere** and the **thermosphere**.

Thermosphere (means heat) has oxygen, helium and hydrogen gases. Much of the solar energy is absorbed in this region before the rest reaches the Earth's surface below.

21.10 Ozone Layer Depletion

The Ozone Cycle

Located as a layer within the stratosphere, just above the troposphere and extending out to about 20 km, a layer of ozone is located. There, it intercepts and blocks very energetic harmful ultraviolet radiations, UVC and most of UVB, from reaching the Earth, limiting their destructiveness. In absorbing these radiations, the ozone molecules get decomposed and then go through a cycle in which they are reformed. This cycle is illustrated as follows:

Initially, an oxygen molecule absorbs UV and is decomposed as follows:

(a) $O_{2(g)} \xrightarrow{\text{UV}} O_{(g)} + O_{(g)}$

Each of the activated oxygen atom then reacts with an oxygen molecule to form a molecule of ozone: **(b)** $O_{2(g)} + O_{(g)} \longrightarrow O_{3(g)}$

The ozone molecules, once created, intercept UV by absorbing their energy but in the process get decomposed; each molecule breaks down to produce a molecule of oxygen and an active oxygen atom. The energy absorbed in this layer is what is responsible for raising the temperature of the stratosphere from -65 ^{0}C to $-2.5\,^0$C: **(c)**

The products of the previous reaction, **(c)** recombine to reform an ozone molecule again. The last two processes, **(c) and (d),** keep repeating and this is how the ozone layer is maintained.

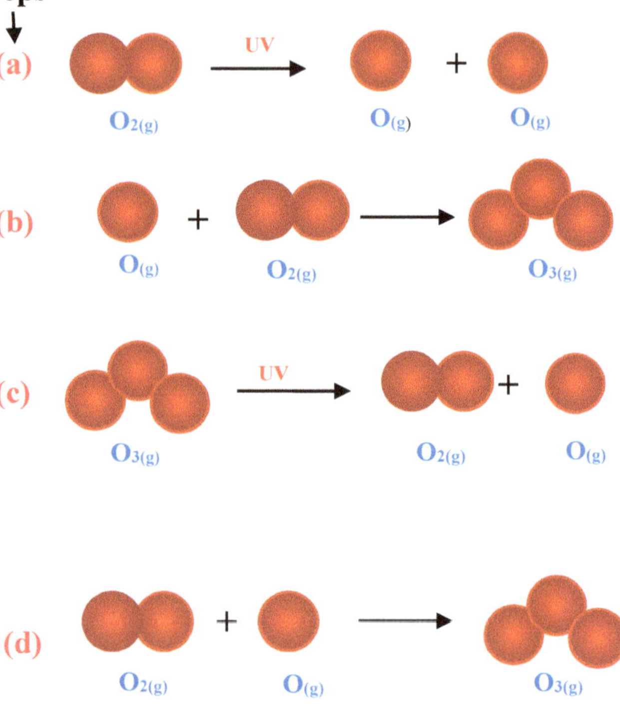

Fig.21.8. The ozone cycle.

CFCs and Ozone Depletion

CFCs are a group of compounds collectively known as **chlorofluorocarbons.** These are produced when one or more hydrogen atoms in a hydrocarbon molecule is replaced by either chlorine or fluorine atoms. In an effort to replace toxic and flammable compounds used as refrigerants and insulators of foam mattresses, Thomas Midgley of General Motors invented the first of the CFCs, Freon (CCl_2F_2). Because of its nontoxicity and efficiency, its use became widespread; millions of tons were in use by the early 1970s. What people did not realize until 1970s, was that CFCs being very stable, were accumulating in the troposphere and gradually finding their way into the ozone layer, gradually destroying some parts of it.

The first to raise concerns about where all the released Freon might have ended up were University of California chemists, F. Sherwood Rowland and Mario Molina. They were the first to speculate and investigate that the massive amounts of CFCs released ended up unchanged into the stratosphere and was having an effect there. In 1985, a hole in the ozone layer in the Antarctic was discovered by a team of three British scientists; they found that the ozone was disappearing at a rate of 1% per day. Along with one of these three British scientists, Paul Crutzen, F. Sherwood Rowland and Mario Molina were awarded the Nobel Prize in 1995 for their outstanding work.

Steps

(a) When a molecule of CFC such as Freon encounters ultraviolet radiations, it decomposes with one of its chlorine atoms breaking away.

(b) This free chlorine atom can react with other molecules such as ozone. When it reacts with a molecule of ozone, it breaks it into molecular oxygen and itself forming a chloro-oxygen radical (ClO).

(c) The ClO radical formed can react with an oxygen atom and get freed.

The free chlorine atom can go on to destroy as many as 100 000 ozone molecules.

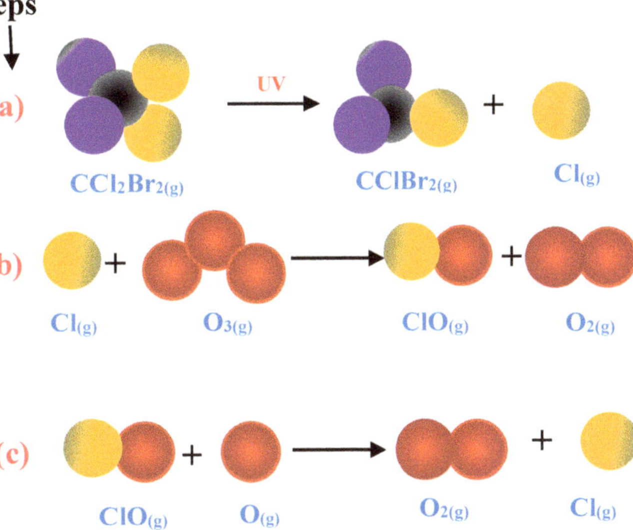

Fig.21.9. The depletion of the ozone layer reactions.

The Montreal Protocol

In what could be described as one of the most unprecedented, concerted efforts, in September 16, 1987, an international agreement was signed in Montreal, mandating countries of the world to stop the production and import of ozone depleting substances and to phase out those in use in a timely way. By phasing out these compounds and replacing them with those that pose no harm to the ozone, it was hoped that the hole in the ozone layer will be repaired by the year 2050. The new safer compounds to replace CFCs were **hydrofluorocarbons** HFCs. The use of these necessitated the replacement of all existing structures that once used CFCs, since they were not compatible for HFCs usage. This endeavor was not feasible for developing countries, as it proved to be too costly. The protocol was ratified for these countries to have a more gradual phasing out process; developed countries were to adhere to more stringent regulations. Also, to ensure that this became a global success, under an international treaty, a Multilateral Fund was created into which billions of dollars were donated by the rich countries of the world to support these projects in developing countries. Every three years, this fund is to be replenished as progress is monitored across the world. More information about the Montreal Protocol can be obtained from the following

website: 1https://www.environment.gov.au/protection/ozone/montreal-protocol.

As opposed to stratospheric level ozone that undergoes a cycle of decomposition and reformation, and which protects us from dangerous UV, ground-level ozone is created in the troposphere where it poses severe risks to human health and harm to vegetation. Ozone being a relatively unstable molecule, can easily react with other compounds, producing chemical changes. It does this in the tissues of living organisms causing harm to them. In humans, it irritates eyes and throat, triggers asthmatic attacks, aggravates pre-existing conditions such as bronchitis and emphysema, causes tightness of chest and shortness of breath. Prolong exposure to ozone may even cause scaring of lung tissues. In plants, it enters the leaves through their stomata, oxidizing tissues there. This results in loss of crops and damage to ecosystems around the globe.

Ground-level ozone is formed in a photochemical reaction between the various oxides of nitrogen (NOx) and volatile organic compounds (VOCs) in the presence of sunlight. The various oxides of nitrogen; $N_2O_{(g)}$, $NO_{(g)}$, $NO_{2(g)}$, and $N_2O_{4(g)}$ are produced by human activities associated in transportation, power plants and chemical plants. VOCs enter the air from sources they escape; gas stations, paint products, petrochemical, synthetic materials, and waste disposals. Once produced, these pollutants can travel freely to other distant places over the globe, where they react with sunlight to produce ozone. The latter process is helpful, in that, it lessens the concentration of ground level ozone in some places. It may worsen some situations if the pollutants go places where these pollutants are already present.

The formation of ozone begins first with the formation of nitric oxide. This gas is produced when fuels are burnt at a very high temperature, such as in the engine of motor vehicles. Such high temperatures allow nitrogen, present as impurity in fuels, to react with oxygen to form nitric oxide:

$$N_{2(g)} \;+\; O_{2(g)} \longrightarrow 2\,NO_{(g)}$$
$$2\,NO_{(g)} \;+\; O_{2(g)} \longrightarrow 2\,NO_{2(g)}$$

Once formed, it spontaneously reacts with oxygen in the air to form nitrogen dioxide. Nitrogen dioxide is a reddish-brown gas that reacts with water to form nitric acid. It is what imparts the color to smog that forms during air inversions. If inhaled, it irritates the lungs and triggers asthmatic attacks. In the presence of visible light, nitrogen dioxide is photolyzed according to the following equation:

$$NO_{2(g)} \xrightarrow{\text{Sunlight}} NO_{(g)} + O_{(g)}$$

The oxygen atom produced, being an activated species, reacts with a molecule of oxygen to form ozone:

$$O_{(g)} \;+\; O_{2(g)} \longrightarrow O_{3(g)}$$

The reaction of VOCs with oxides of nitrogen in forming ozone is a bit more intriguing and complicated, and as such will not be discussed here.

Exercise 21.2

1. Some of the ways of managing ground level ozone are by doing the following:

-   Driving less in the afternoons when it is hot and sunny,

-   Avoid refueling your vehicles when it is hot and sunny,

-   Avoid idling your vehicles when it is hot and sunny,

-   Avoid activities that use fires when it is hot and sunny,

Use your knowledge of how ground level ozone is formed to justify these steps list above.

CHAPTER REVIEW GAS MIXTURES (18-20)

Matching

Match each term in the table below with the correct statements that follow

A	Dalton	G	Gay-Lussac
B	Charles	H	Kelvin
C	$P_1V_1 = P_2V_2$	I	100 kPa and 25°C
D	Boyle	J	$V = kT$
E	$P = kT$	K	$PV = nRT$
F	Avogadro	L	24.8 L

1.	He formulated the law of partial pressures for gases.
2.	He extrapolated graphs of volume versus temperature for real gases to find absolute zero.
3.	He discovered that volume of a fixed mass of gas varies directly with temperature changes on the gas.
4.	He hypothesized that equal volumes of gases at the same pressure and temperature contain the same number of particles.
5.	He formulated the law of combining volumes for gaseous reactants and products during reactions of gases.
6.	He formulated law that the volume of a gas is inversely proportional to the pressure acting upon it.
7.	Boyle's law equation.
8.	Standard ambient pressure and temperature.
9.	Molar volume at SATP.
10.	Charles' law equation.
11.	The pressure law equation.
12.	The ideal gas equation.

True or False

Read each of the following statements and then decide if it is *True* or *False*

1.	The pressure of a gas in a container is due to the bombardments of the gas particles against its walls.
2.	The particles in an ideal gas attract each other and the walls of the container they bombard.
3.	As the temperature of a gas is raised so does the average kinetic energy of its particles.
4.	The higher we ascend into the atmosphere, the greater the atmospheric pressure becomes.
5.	For a fixed mass of gas at constant volume, its pressure decreases as its temperature is

	raised.
6.	Increasing the number of gas particles in a container at constant temperature, raises its pressure.
7.	The volume of a fixed mass of gas at constant pressure, will increase if its temperature is raised.
8.	On top of a high mountain, water may boil as low as 70 °C
9.	The pressure of a fixed mass of gas of fixed volume, will increase if its temperature is lowered.
10.	The total pressure of a mixture of non-reacting gases is due to the sum of the partial pressures of each gas making up the mixture.
11.	One mole of $H_{2(g)}$ occupies the same volume as one mole $CO_{2(g)}$ at SATP.
12.	16.0 g of $O_{2(g)}$ will have the same volume as 1.0 g $H_{2(g)}$ at SATP.

Multiple Choice

Choose the letter that best answers questions.

1. Which of the following are **true** statements about real gases?

a	They are highly compressible
b	Their particles attract each other upon collision
c	They condense when they are cooled to low temperatures
d	The force of attraction between their particles is very weak
e	All of the above

2. Which of the following statements is **false** about ideal gases?

a	Their particles have negligible volumes
b	Their particles are not attracted to the walls of their container
c	They do not condense when they are cooled to low temperatures
d	There is no force of attraction between their particles.
e	none of the above

3. The pressure of the gas in an open container at sea level is

a	the same as that of the atmosphere
b	different from that of the atmosphere
c	approximately 101 kPa
d	a and c only
e	none of the above

4.

	Atmospheric pressure is caused by
a	the planets
b	the force of gravity on the air particles above the Earth's surface
c	the various things on the Earth's surface
d	the wind blowing over the Earth's surface
e	none of the above

5. If a mixture of hydrogen gas and chlorine gas react in a sealed container, there will be

a	a decrease of pressure
b	no difference in pressure when the products cool down
c	an initial increase in pressure
d	all of the above
e	b and c

6. The pressure of a gas in a container at a fixed temperature can be increased by

a	reducing it volume	*d*	reducing the number of particles
b	increasing its volume	*e*	a and c only
c	increasing the number of particles		

7. A fixed mass of gas has a volume of 2.0 L at 100 kPa. Its volume at 200 kPa will be

a	2.0 L	*d*	4.0 L
b	1.0 L	*e*	none of the above
c	20.0 L		

8. A fixed mass of gas has a volume of 2.0 L at 100 kPa. At what pressure will its volume be 0.5 L?

a	200 kPa	*d*	400 kPa
b	300 kPa	*e*	none of the above
c	50 kPa		

9. The pressure of a fixed mass of gas at constant volume increases with temperature because its

a	molecules break up	*d*	molecules collide more frequently
b	molecules gain more energy	*e*	all except a
c	molecules move more forcefully		

10.

A rigid container has a mixture of $N_{2(g)}$, $CO_{2(g)}$ and $O_{2(g)}$ in a 12: 5: 3 mole ratios respectively. If the total pressure in the container is 100 kPa, the partial pressure of $N_{2(g)}$ must be

a	60 kPa	*d*	12 kPa
b	5 kPa	*e*	25 kPa
c	15 kPa		

11. A rigid container has 10 L of $NH_{3(g)}$ at SATP. What volume of $H_{2(g)}$ can be collected at SATP if all the $NH_{3(g)}$ is decomposed?

a	10 L	*d*	15 L
b	20 L	*e*	none of the above
c	30 L		

12. The pressure of oxygen gas collected above water at a certain temperature is 101 kPa. If the vapour pressure of water at that temperature is 4.0 kPa, the pressure of dry oxygen gas will be

a	101kPa	*d*	115 kPa
b	97 kPa	*e*	none of the above
c	4 kPa		

13. Sulphur dioxide and oxygen react according to the following equation:

$$2\ SO_{2(g)} + O_{2(g)} \rightarrow 2\ SO_{3(g)}$$

Which of the following ratios represents the correct volumes of reactants and products?

a	1: 2: 1	*d*	2: 1: 1
b	1: 1: 1	*e*	none of the above
c	2: 1: 2		

14. If one mole of any gas at STP occupies 22.4 L, the volume occupied by 1.01 g $H_{2(g)}$ will be

a	1.01 L	*d*	2.02 L
b	11.2 L	*e*	none of the above
c	22.8 L		

15. At STP 8.0 g of $O_{2(g)}$ and 7.0 g of $N_{2(g)}$ will occupy

a	the same volume	*d*	11.2 L each
b	different volumes	*e*	a and c only
c	5.6 L each		

16. The number of molecules in 6.2 L of $CO_{2(g)}$ at SATP is

a	1.50×10^{23}	*d*	6.02×10^{23}
b	3.01×10^{23}	*e*	none of the above
c	1.50×10^{24}		

17.

The following equation represents the reaction between magnesium and sulphuric acid: $Mg_{(s)} + H_2SO_{4(aq)} \rightarrow MgSO_{4(aq)} + H_{2(g)}$. If 2.0 g of $Mg_{(s)}$ reacts with excess $H_2SO_{4(aq)}$, what volume of $H_{2(g)}$ at STP will be produced?

a	0.92 L	d	3.68 L
b	1.84 L	e	none of the above
c	44.8 L		

18.

The following equation represents the reaction between hydrochloric acid and sodium carbonate: $2\,HCl_{(aq)} + Na_2CO_{3(s)} \rightarrow 2\,NaCl_{(aq)} + CO_{2(g)} + H_2O_{(l)}$. What mass of $Na_2CO_{3(s)}$ is required to react with excess $HCl_{(aq)}$ to produce 12.4 L of $CO_{2(g)}$ at SATP?

a	105.99 g	d	211.98 g
b	52.99 g	e	none of the above
c	26.50 g		

19.

If a rigid 20 L container has 64.16 g $CH_{4(g)}$, what is the pressure of the gas if the cylinder is kept at a temperature of 30 $^{\circ}$C?

a	101 kPa	d	503.6 kPa
b	100 kPa	e	none of the above
c	250 kPa		

20.

Which of the following gases cause destruction of the ozone layer?

a	CO_2	d	CO
b	CFC's	e	CH_4
c	SO_2		

21.

Decomposition of potassium chlorate produces oxygen as follows:

$$2\,KClO_{3(s)} \rightarrow 2\,KCl_{(s)} + 3\,O_{2(g)}$$

What mass of $KClO_{3(s)}$ must be decomposed to produce 50 L of $O_{2(g)}$ at 250 $^{\circ}$C and 120 kP?

a	112.75 g	d	245.1 g
b	122.55 g	e	none of the above
c	81.7 g		

22.

Which of the following gases **does not** cause global warming?

a	CO_2	d	CO
b	NO	e	CH_4
c	$H_2O_{(g)}$		

23.

Which of the following gases causes blood poisoning?

a	CO_2	*d*	SO_2
b	CFC's	*e*	CH_4
c	CO		

24.

Which of the following gases **would not cause** acid rain?

a	NO_2	*d*	SO_2
b	CO	*e*	SO_3
c	CO_2		

25.

Which of the following gases is found in rotten eggs?

a	NO_2	*d*	H_2S
b	CH_4	*e*	SO_2
c	CO_2		

26.

This greenhouse gas is found in cow's flatulence?

a	NO_2	*d*	CO_2
b	CH_4	*e*	H_2O
c	SO_2		

UNIT:1   Matter

CHAPTER 1: (Classification of matter)

Exercise1.1
1. (a) liquid   (b) solid   (c) gas   (d) liquid   (e) solid
2. (a) sublimation (b) evaporation (c) melting (d) condensation (e) freezing
3. (a) physical change   (b) physical property   (c) chemical property
   (d) chemical change   (e) chemical property   (f) physical change
   (g) physical property
4. (a) pure   (b) heterogeneous mixture   (c) homogeneous mixture
   (d) heterogeneous mixture   (e) homogeneous mixture   (f) pure
5. (a) released   (b) released   (c) absorbed (d) released   (e) absorbed
   (f) absorbed
6. (a) pure   (b) quantitative   (c) weakest (d) qualitative   (e) impure
   (f) impure (g) pure
7. (a) false (b) false (c) true   (d) false   (e) false   (f) true   (g) true

CHAPTER 2 : (Atomic theories)

Exercise 2.1
1. (a) Chadwick (b) Thomson (c) Goldstein (d) Rutherford (e) Bohr (f) Planck
2. (1) central (2) electrons (3) protons   (4) neutrons   (5) shells (6) energy levels
   (7) force   (8) positively (9) ground   (10) quantum (11) jump (12) excited
   (13) released   (14) light   (15) line

CHAPTER 3: (Isotopes)

Exercise 3.1

Answer to graph on Iodine
1.   8.07 days   **2.** 24.2 days   **3.** 32.3 days   **4.** The first **5.** 16.1 days

Exercise 3.2
1. (a) $^{14}_{6}C$ $\rightarrow$ $^{14}_{7}C$ + $^{0}_{-1}e$ - beta decay
   (b) $^{31}_{15}P$ $\rightarrow$ $^{28}_{14}Si$ + $^{3}_{1}H$ - nuclear fission
   (c) $^{90}_{38}Sr$ $\rightarrow$ $^{90}_{39}Sr$ + $^{0}_{-1}e$ - beta decay
   (d) $^{73}_{31}Ga$ $\rightarrow$ $^{0}_{-1}e$ + $^{73}_{32}Ga$ - beta decay
   (e) $^{231}_{90}Th + ^{1}_{0}n \rightarrow ^{232}_{90}Th$ - neutron enrichment
   (f) $^{12}_{5}B$ $\rightarrow$ $^{7}_{3}Li + ^{1}_{0}n + ^{4}_{2}He$ - nuclear fission
   (g) $^{226}_{88}Ra$ $\rightarrow$ $^{4}_{2}He$ + $^{222}_{86}Rn$ - nuclear fission

Exercise 3.3
   Answer to graph on Antibiotic:
1. (a)   160 days   (b) 80 days   (c) 140 g   **2.** Short half-life. Since an acute illness lasts for a relatively short time, an antibiotic with a short is the better choice as it would disappear from the body quickly with minimum side effects. **3.** 28.09, **4.** K-41
   5. 0.014 Bq

Chapter 4: (Trends in the periodic table)

Exercise 4.1

1. a) nuclear charge b) number of shells c) degree of shielding

2. It decreases down the group. More shells, more shielding of outer electrons, less force of attraction on outer electron(s).

3. Second electron experiences less electron-electron repulsion than the first.

4. When the fluoride atom gains an electron, there is now more electron-electron repulsion that causes expansion of the valence shell.

5. Magnesium has a smaller atomic radius because it has a greater nuclear charge than sodium; the force of attraction between the nucleus of magnesium and its outer shell electrons is greater than the force of attraction between the nucleus of sodium and its outer electrons. It is thus more difficult to remove a valence electron from the magnesium atom than it from a sodium atom.

Chapter 5: (Chemical Bonding)

Exercise 5.1

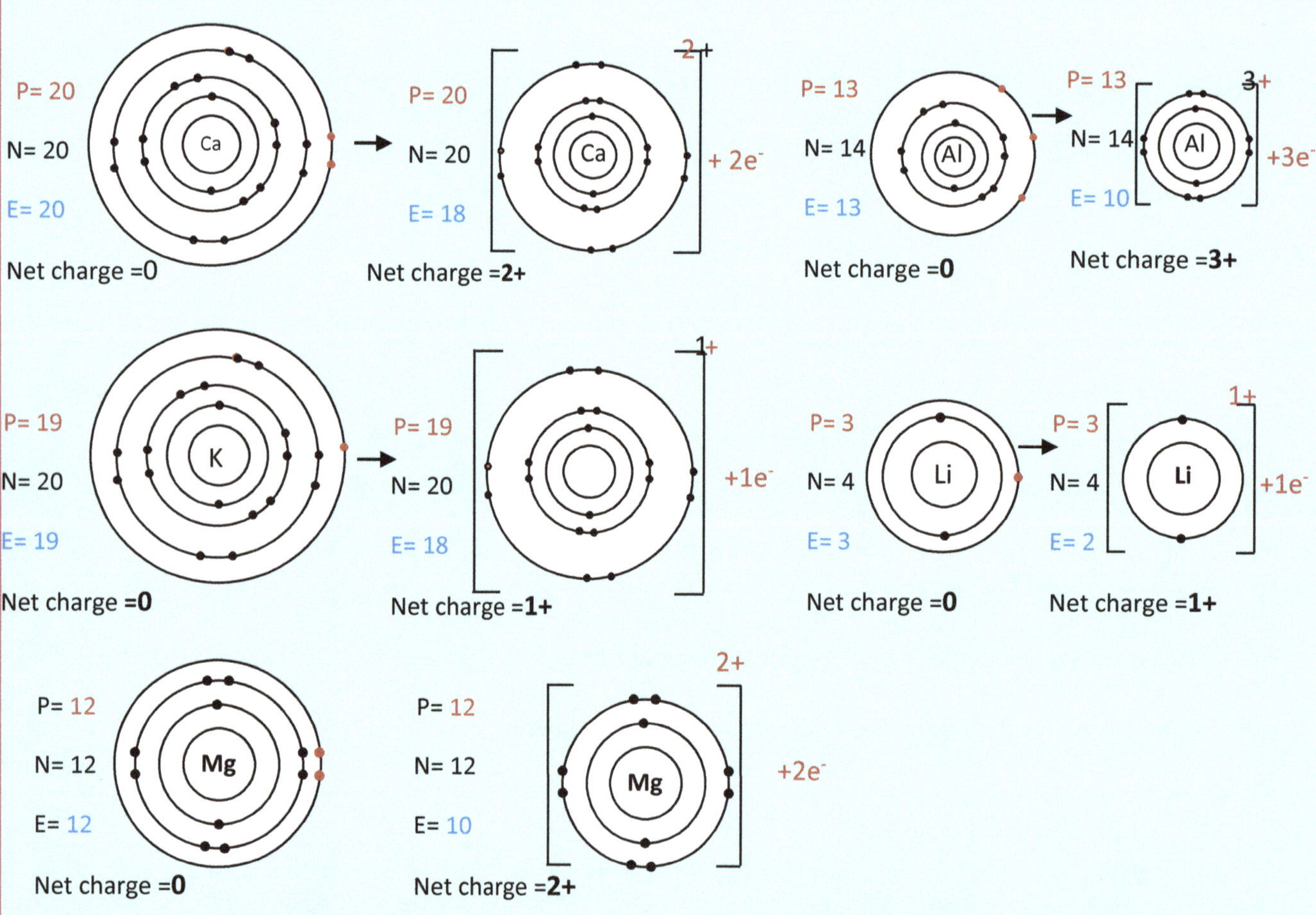

Exercise 5.2

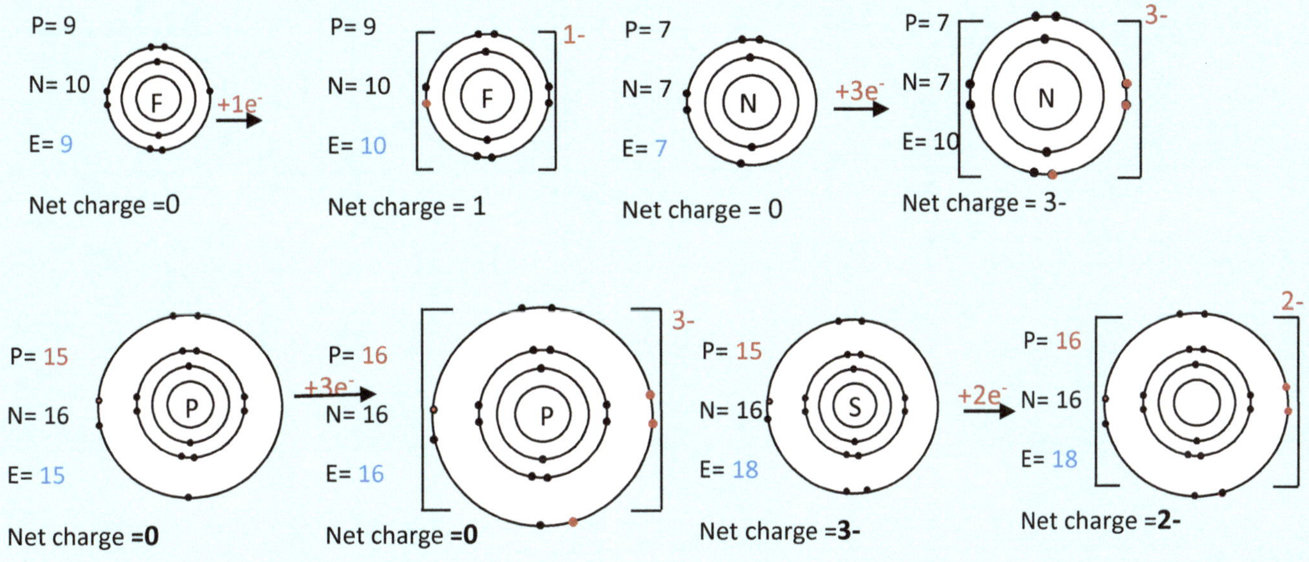

P= 9
N= 10
E= 9
F
+1e-
Net charge =0
P= 9
N= 10
E= 10
F
1-
Net charge = 1
P= 7
N= 7
E= 7
N
+3e-
Net charge = 0
P= 7
N= 7
E= 10
N
3-
Net charge = 3-
P= 15
N= 16
E= 15
P
+3e-
Net charge =0
P= 16
N= 16
E= 16
P
3-
Net charge =0
P= 15
N= 16
E= 18
S
+2e-
Net charge =3-
P= 16
N= 16
E= 18
2-
Net charge =2-

Exercise 5.3

Mg
Cl
Cl
1-
Cl
2+
Mg
1-
Cl

$MgCl_2$

Li
+
F
1+
Li
1-
F

LiF

Ca
+
O
2+
Ca
2-
O

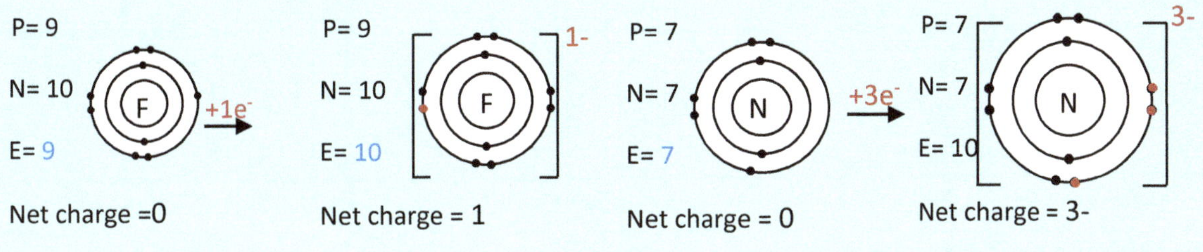

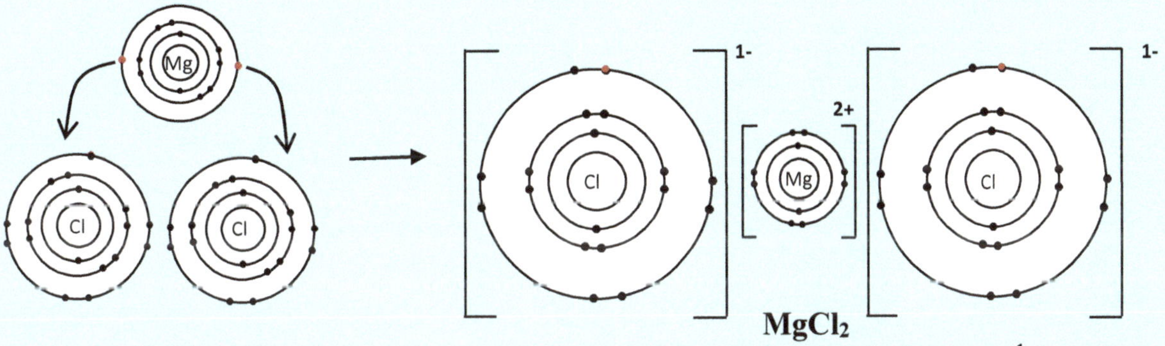

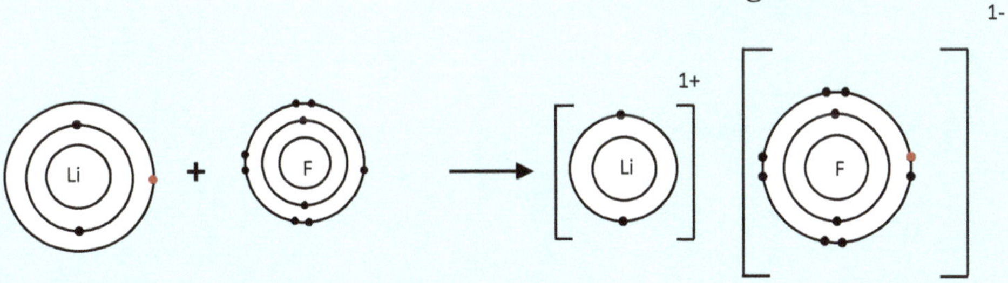

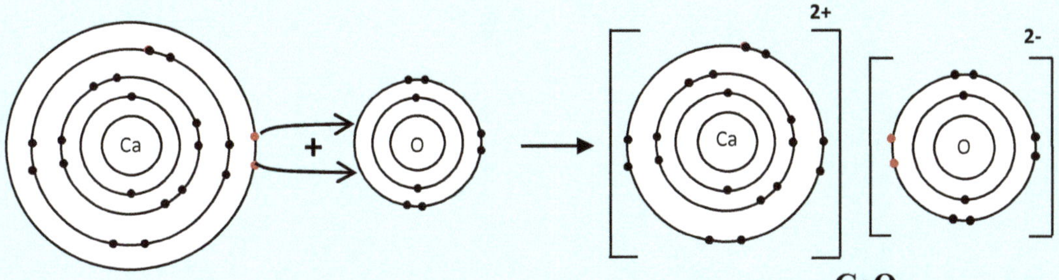

CaO

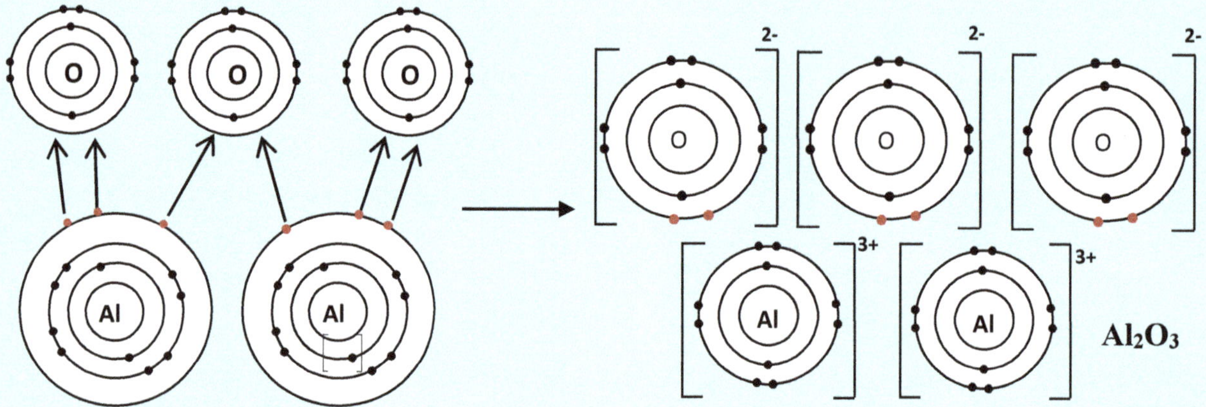

Al₂O₃

Exercise 5.4

LiBr

Na₂S

CaI₂

MgO

Al₂O₃

AlF₃

Exercise 5.5
Answers

(a)   (b)   (c)

(d)   (e)   (f)

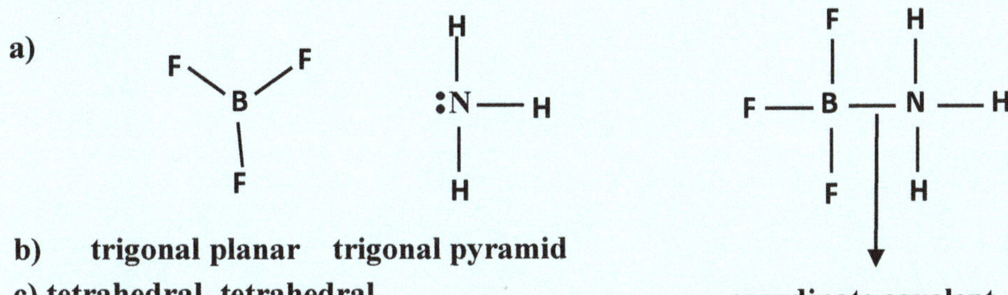

- Please ignore the electrons at the end of the lines.

Chapter 6:(**Molecular geometry**)

Exercise 6.1
(a) polar   (b) polar (c) polar (d) polar (e) polar (f) polar

Exercise 6.2

(a) **polar**   (b) **polar**   (c)   **polar**   (d) **polar**

Exercise 6.4

a)

b)   **trigonal planar   trigonal pyramid**

c) **tetrahedral, tetrahedral**

d) **coordinate covalent bond**

coordinate covalent

Exercise 6.3

Table 6.3: Complete the following table

Compounds	Lewis dot Structure	A_xX_y Type	Electron Arrangement	Molecular Geometry	Bond Angle	Polarity
SiH_4		AX_4	tetrahedral	tetrahedron	109.5°	non-polar
PCl_3		AX_3E	tetrahedral	Trigonal pyramid	107°	polar
H_2S		AX_2E_2	tetrahedral	bent	105°	polar
$BeCl_2$		AX_2	linear	linear	180°	non-polar
AlF_3		AX_3	trigonal planar	trigonal planar	120°	non-polar
SO_2		AX_2E	trigonal planar	bent	120°	polar
NO_3^-		AX_3	trigonal planar	trigonal planar	120°	polar

1. (b) CH_4 (d) C_2H_6 (c) C_3H_8 (a) C_4H_{10}

2. (a) Br_2 (b) ICl

1. 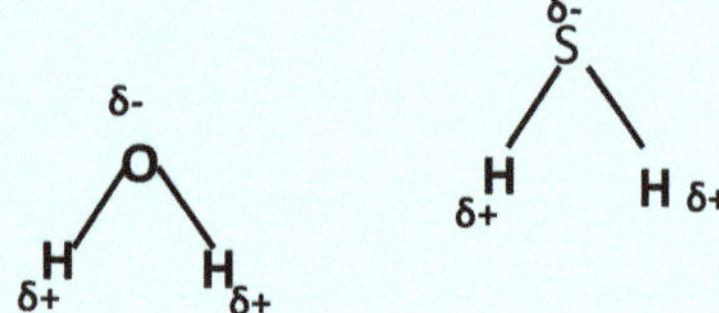

   Both have the same shape with polar bonds

   Both molecules are polar

   The O-H bond is more polar than the S-H bond

   The intermolecular force between H_2O molecules is greater than that between H_2S molecules.

2. Since they have the same number of electrons, their London dispersion forces is the same. There is no electronegativity difference between the Br and Br atoms so the molecules are non-polar. The electronegativity difference between the I and Cl atoms is 0.5, making the bond as well as the molecules polar. There is thus greater intermolecular between ICl molecules than Br_2 molecules and for this reason I-Cl has a higher boiling point than Br_2.

3. 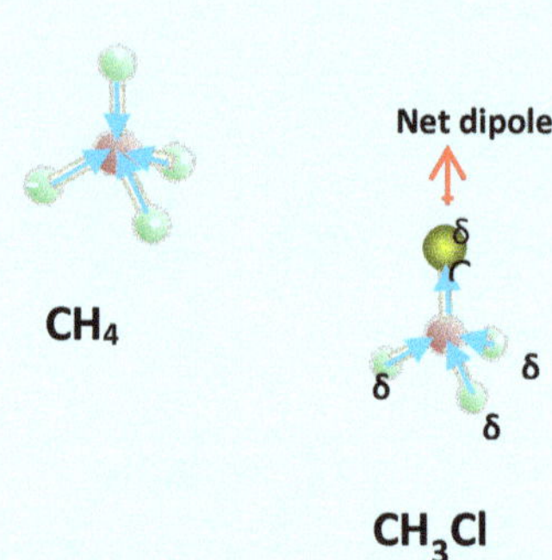

   Both of these molecules are tetrahedrons. The CH_4 molecule is non-polar because of its symmetry even though the C-H bonds are polar. The C-Cl bond is much more polar than the three other C-H bonds. Because of this the molecule is polar with a net dipole. The CH_3Cl compound would have stronger intermolecular force than CH_4; CH_3Cl having both L.D.F and dipole-dipole forces while CH_4 having only L.D.F.

4. a) 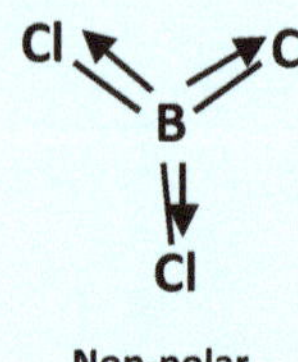   b) 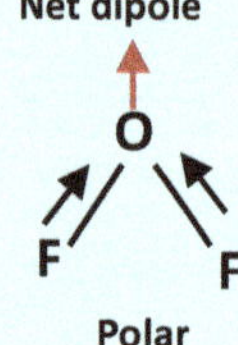   c) I⟵Be⟶I Non-polar   d) 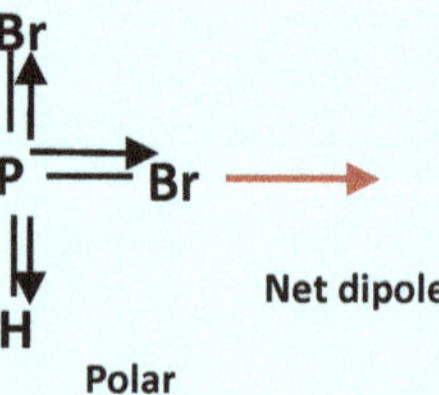

1. a) 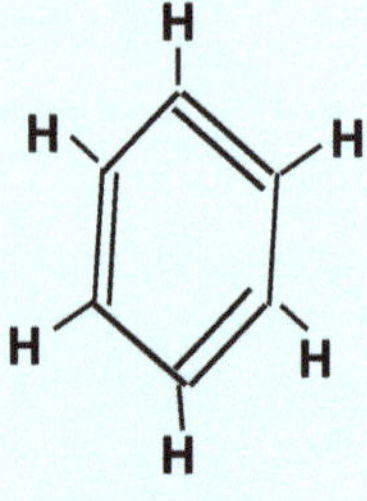

   No effect

   b) Br—Br

   No effect

   c) H–C(–H)(–H)–C(=O)–O–H

   attracted

   d) H–C(–H)(–H)–O–H

   attracted

Recognizing Chemical Reactions

Exercise 7.1
1.   Evolution of $H_{2(g)}$
2.   Formation of solid $Ag_{(s)}SO4_{(s)}$
3.   Evolution of $O_{2(g)}$
4.   Evolution of $O_{2(g)}$
5.   Energy change
6.   Evolution of $HCl_{(g)}$ and $NH_{3(g)}$ or disappearance of $NH_4Cl_{(s)}$, a solid
7.   Formation of solid $AgCl_{(s)}$
8.   Formation of solid $CaCO_{3(s)}$
9.   Formation of solid $Fe(OH)_{3(s)}$
10. Heat production

Exercise 7.2

For the following word equations, write the chemical formulas of the reactants and products and then balance them:

a) sodium hydroxide + sulphuric acid $\rightarrow$ sodium sulphate + water
$2NaOH_{(aq)}$ + $H_2SO_{4(aq)}$ $\rightarrow$ $Na_2SO_{4(aq)}$ + $2H_2O_{(l)}$

b) calcium hydroxide + nitric acid $\rightarrow$ calcium nitrate + water
$Ca(OH)_{2(aq)}$ + $2HNO_{3(aq)}$ $\rightarrow$ $Ca(NO_3)_{2(aq)}$ + $2H_2O_{(l)}$

c) hydrochloric acid + magnesium hydroxide $\rightarrow$ magnesium chloride + water
$2HCl_{(aq)}$ + $Mg(OH)_{2(aq)}$ $\rightarrow$ $MgCl_{2(aq)}$ + $2H_2O_{(l)}$

d) sodium + water $\rightarrow$ sodium hydroxide + hydrogen (gas)
$2Na_{(s)}$ + $2H_2O_{(l)}$ $\rightarrow$ $2NaOH_{(aq)}$ + $H_{2(g)}$

e) potassium + oxygen $\rightarrow$ potassium oxide
$4K_{(s)}$ + $O_{2(g)}$ $\rightarrow$ $2K_2O_{(s)}$

f) zinc + sulphuric acid $\rightarrow$ zinc sulphate + hydrogen (gas)
$Zn_{(s)}$ + $H_2SO_{4(aq)}$ $\rightarrow$ $ZnSO_{4(aq)}$ + $H_{2(g)}$

g) magnesium + nitric acid $\rightarrow$ magnesium nitrate + hydrogen (gas)
$Mg_{(s)}$ + $2HNO_{3(aq)}$ $\rightarrow$ $Mg(NO_3)_{2(aq)}$ + $H_{2(g)}$

h) copper + oxygen $\rightarrow$ copper (11) oxide
$2Cu_{(s)}$ + $O_{2(g)}$ $\rightarrow$ $2CuO_{(s)}$

i) sodium carbonate + hydrochloric acid $\rightarrow$ sodium chloride + water + carbon dioxide
$Na_2CO_{3(aq)}$ + $2HCl_{(aq)}$ $\rightarrow$ $2NaCl_{(aq)}$ + $H_2O_{(l)}$ + $CO_{2(g)}$

Exercise 7.3
1. $4 P_{(s)}$ + $3 O_{2(g)}$ $\rightarrow$ $2P_2O_{3(s)}$
2. $Zn_{(s)}$ + $2 HCl_{(aq)}$ $\rightarrow$ $ZnCl_{2(aq)}$ + $H_{2(g)}$
3. $2 C_{(s)}$ + $O_{2(g)}$ $\rightarrow$ $2 CO_{(g)}$
4. $Na_2CO_{3(aq)}$ + $2 HCl_{(aq)}$ $\rightarrow$ $2 NaCl_{(aq)}$ + $CO_{2(g)}$ + $H_2O_{(l)}$
5. $2 KNO_{3(s)}$ $\rightarrow$ $2 KNO_{2(s)}$ + $O_{2(g)}$
6. $2 KClO_{3(s)}$ $\rightarrow$ $2 KCl_{(s)}$ + $3O_{2(g)}$
7. $2 H_2O_{2(l)}$ $\rightarrow$ $2 H_2O_{(l)}$ + $O_{2(g)}$
8. $2 Al_{(s)}$ + $3 CuCl_{2(aq)}$ $\rightarrow$ $2 AlCl_{3(aq)}$ + $3 Cu_{(s)}$
9. $Ca(OH)_{2(aq)}$ + $2 HNO_{3(aq)}$ $\rightarrow$ $Ca(NO_3)_{2(aq)}$ + $2 H_2O_{(l)}$
10. $2 AgNO_{3(aq)}$ + $BaCl_{2(aq)}$ $\rightarrow$ $2 AgCl_{(s)}$ + $Ba(NO_3)_{2(aq)}$
12. $Na_2CO_{3(aq)}$ + $CaCl_{2(aq)}$ $\rightarrow$ $CaCO_{3(s)}$ + $2 NaCl_{(aq)}$

13.	$2\,CO_{(g)}$	+	$O_{2(g)}$	$\rightarrow$	$2\,CO_{2(g)}$		
14.	$Fe_2O_{3(s)}$	+	$3\,CO_{(g)}$	$\rightarrow$	$2\,Fe_{(s)}$	+	$3\,CO_{2(g)}$
15.	Fe_2O_3	+	$6\,H_{2(g)}$	$\rightarrow$	$2\,Fe_{(s)}$	+	$3\,H_2O_{(l)}$
18.	$2\,Al(OH)_{3(s)}$	+	$3\,H_2SO_{4(aq)}$	$\rightarrow$	$Al_2(SO_4)_{3(aq)}$	+	$6\,H_2O_{(l)}$
19.	$CO_{2(g)}$	+	$2\,NH_{3(g)}$	$\rightarrow$	$CO(NH_2)_{2(aq)}$	+	$H_2O_{(l)}$

Types of Chemical Reactions

Exercise 7.4

1. For the following reactions, determine if they are synthesis or decomposition.

(a) $3H_{2(g)} + N_{2(g)} \rightarrow 2NH_{3(g}$ Synthesis
(b) $2HgO_{(s)} \rightarrow 2Hg + O_{2(g)}$ Decomposition
(c) $H_2CO_{3(l)} \rightarrow H_2O_{(l)} + CO_{2(g)}$ Decomposition
(d) $CaO_{(s)} + H_2O_{(l)} \rightarrow Ca(OH)_{2(s)}$ Synthesis
(e) $2NaHCO_{3(s)} \rightarrow Na_2CO_{3(s)} + CO_{2(g)} + H_2O_{(l)}$ Decomposition

2. The following are synthesis reactions. Complete the equations by first predicting the product, and then balancing the equation.

(a) $2Ba_{(s)} + O_{2(g)} \rightarrow 2BaO_{(s)}$ (d) $Na_2O_{(s)} + H_2O_{(l)} \rightarrow 2NaOH_{(aq)}$
(b) $2H_{2(g)} + O_{2(g)} \rightarrow 2\,H_2O_{(g)}$ (e) $SO_{3(g)} + H_2O_{(l)} \rightarrow H_2SO_{4(aq)}$
(c) $4Fe_{(s)} + 3O_{2(g)} \rightarrow 2Fe_2O_{3(s)}$ (f) $2P_{(s)} + 3Cl_{2(g)} \rightarrow 2PCl_{3(g)}$

3. The following are decomposition reactions. Complete the equations by first predicting the product, and then balancing the equation.

(a) $2Ag_2O_{(s)} \rightarrow 4Ag_{(s)} + O_{2(g)}$ (d) $CaCO_{3(s)} \rightarrow CaO_{(s)} + CO_{2(g)}$
(b) $2H_2O_{2(l)} \rightarrow 2\,H_2O_{(l)} + O_{2(g)}$ (e) $2AlBr_{3(s)} \rightarrow 2Al_{(s)} + 3Br_{2(g)}$
(c) $K_2S_{(s)} \rightarrow 2K_{(s)} + S_{(s)}$ (f) $2CuO_{(s)} \rightarrow 2Cu_{(s)} + O_{2(g)}$

Balancing Single displacement Reactions

Exercise 7.5

1. Balance the following single displacement reactions

(a) $Cu_{(s)} + 2AgNO_{3(aq)} \rightarrow Cu(NO_3)_{2(aq)} + Ag_{(s)}$
(b) $Cl_{2(g)} + 2NaBr_{(aq)} \rightarrow NaCl_{(aq)} + Br_{2(l)}$
(c) $2Fe_{(s)} + 3SnCl_{2(aq)} \rightarrow 2FeCl_{3(aq)} + 3Sn_{(s)}$
(d) $Mg_{(s)} + 2HClO_{3(aq)} \rightarrow Mg(ClO_3)_{2(aq)} + H_{2(g)}$
(e) $Zn_{(s)} + 2HI_{(aq)} \rightarrow ZnI_{2(aq)} + H_{2(g}$

2. The following are single displacement reactions. Complete the following equations by first predicting the products, and then balancing the equation.

(a) $Zn_{(s)} + 2HBrO_{3(aq)} \rightarrow Zn(BrO_3)_{2(aq)} + H_{2(g)}$
(b) $Br_{2(g)} + 2NaI_{(aq)} \rightarrow I_{2(s)} + 2NaBr_{(aq)}$
(c) $Cu_{(s)} + 2AgNO_{3(aq)} \rightarrow 2Ag_{(s)} + Cu(NO_3)_{2(aq)}$
(d) $4Al_{(s)} + 6H_2SO_{4(aq)} \rightarrow 2Al_2(SO_4)_{3(aq)} + 3H_{2(g)}$
(e) $Ca_{(s)} + 2H_2O_{(l)} \rightarrow Ca(OH)_{2(aq)} + H_{2(g)}$
(f) $F_2O_{3(s)} + 3CO_{(g)} \rightarrow 2\,Fe_{(s)} + 3CO_{2(g)}$

3. Use the reactivity series to determine whether or not the following single reactions can happen. If a reaction is possible, complete the equation by first predicting the products, and then balancing the equation. If a reaction cannot happen just write 'NR'.

(a) $Cu_{(s)}$ + $SnCl_{2(aq)}$ $\rightarrow$ NR

(b) $Cl_{2(g)}$ + $2NaI_{(aq)}$ $\rightarrow$ $I_{2(s)}$ + $2NaCl_{(aq)}$

(c) $Ag_{(s)}$ + $Cu(NO_3)_{2(aq)}$ $\rightarrow$ NR

(d) $Zn_{(s)}$ + $H_2O_{(l)}$ $\rightarrow$ NR

(e) $Na_{(s)}$ + $2H_2O_{(l)}$ $\rightarrow$ $2NaOH_{(aq)}$ + $H_{2(g)}$

(f) $Au_{(s)}$ + $H_2SO_{4(aq)}$ $\rightarrow$ NR

(g) $Br_{2(g)}$ + $NaCl_{(aq)}$ $\rightarrow$ NR

(h) $2Al_{(s)}$ + $3SnSO_{4(aq)}$ $\rightarrow$ $Al_2(SO_4)_{3(aq)}$ + $3Sn_{(s)}$

Exercise 7.6

a) $Mg_{(s)}$ + $CuSO_{4(aq)}$ $\longrightarrow$ $MgSO_{4(aq)}$ + $Cu_{(s)}$

b) $Cu_{(s)}$ + $ZnSO_{4(aq)}$ $\longrightarrow$ NR

c) $Al_{(s)}$ + $Mg(NO_3)_{2(aq)}$ $\longrightarrow$ NR

d) $Fe_{(s)}$ + $CuSO_{4(aq)}$ $\longrightarrow$ $FeSO_{4(aq)}$ + $Cu_{(s)}$

e) $3Zn_{(s)}$ + $2Fe(NO_3)_{3(aq)}$ $\longrightarrow$ $3Zn(NO_3)_{2(aq)}$ + $2Fe$

Double Displacement Reaction

Exercise 7.8

For the following pairs of aqueous ionic compounds, complete the skeleton equations. If a double displacement reaction occurs, balance the equation. If no reaction occurs write NR.

(a) $2NaOH_{(aq)}$ + $H_2SO_{4(aq)}$ $\rightarrow$ $Na_2SO_{4(aq)}$ + $2H_2O_{(l)}$

(b) $2KI_{(aq)}$ + $Pb(NO_3)_{(aq)}$ $\rightarrow$ $2KNO_{3(aq)}$ + $PbI_{2(s)}$

(c) $MgCl_{2(aq)}$ + $NaNO_{3(aq)}$ $\rightarrow$ NR

(d) $NH_4Cl_{(aq)}$ + $KOH_{(aq)}$ $\rightarrow$ $KCl_{(aq)}$ + $NH_{3(g)}$ + $H_2O_{(l)}$

(e) $H_2SO_{4(aq)}$ + $2CsOH_{(aq)}$ $\rightarrow$ $Cs_2SO_{4(aq)}$ + $2H_2O_{(l)}$

(f) $CaS_{(aq)}$ + $2HCl_{(aq)}$ $\rightarrow$ $CaCl_{2(sq)}$ + $H_2S_{(g)}$

(g) $ZnBr_{2(aq)}$ + $HNO_{3(aq)}$ $\rightarrow$ NR

(h) $K_2CO_{3(aq)}$ + $2HCl_{(aq)}$ $\rightarrow$ $2KCl_{(aq)}$ + $CO_{2(g)}$ + $H_2O_{(l)}$

Reagents	HCl$_{(aq)}$ $\underline{H^+ \mid Cl^-}$ ions	K$_2$SO$_{4(aq)}$ $\underline{K^+ \mid SO_4^{2-}}$ ions	NaOH$_{(aq)}$ $\underline{Na^+ \mid OH^-}$ ions	ZnI$_{2(aq)}$ $\underline{Zn^{2+} \mid I^-}$ ions	Li$_3$PO$_{4(aq)}$ $\underline{Li^+ \mid PO_4^{3-}}$ ions	K$_2$S$_{(aq)}$ $\underline{K^+ \mid S^{2-}}$ ions
CuCl$_{2(aq)}$ $\underline{Cu^{2+} \mid Cl^-}$ ions	NR	NR	NR	NR	Cu$_3$(PO$_4$)$_{2(s)}$ + LiCl$_{(aq)}$ √	√ CuS$_{(s)}$ +KCl$_{(aq)}$
Ba(NO$_3$)$_{2(aq)}$ $\underline{Ba^{2+} \mid NO_3^-}$ ions	NR	BaSO$_{4(s)}$ + KNO$_{3(aq)}$ √	NR	NR	Ba$_3$(PO$_4$)$_{2(s)}$ + LiNO$_{3(aq)}$ √	√ BaS$_{(s)}$ + KNO$_{3(aq)}$
AgNO$_{3(aq)}$ $\underline{Ag^+ \mid NO_3^-}$ ions	AgCl$_{(s)}$ +HNO$_{3(aq)}$ √	√ Ag$_2$SO$_{4(s)}$ + KNO$_{3(aq)}$	AgOH$_{(s)}$ + NaNO$_{3(aq)}$ √	√ AgI$_{(s)}$ +Zn(NO$_3$)$_{2(aq)}$	Ag$_3$PO$_{4(s)}$ + LiNO$_{3(aq)}$ √	√ Ag$_2$S$_{(s)}$ + KNO$_{3(aq)}$
HNO$_{3(aq)}$ $\underline{H^+ \mid NO_3^-}$ ions	NR	NR	H$_2$O$_{(l)}$ + NaNO$_{3(aq)}$ + heat √	NR	NR	H$_2$S$_{(g)}$ + KNO$_{3(aq)}$ √
Al(NO$_3$)$_{3(aq)}$ $\underline{Al^{3+} \mid NO_3^-}$ ions	NR	NR	Al(OH)$_{3(s)}$ + NaNO$_{3(aq)}$ √	NR	Al$_3$PO$_{4(s)}$ +LiNO$_{3(aq)}$ √	√ Al$_2$S$_{3(s)}$ + KNO$_{3(aq)}$
Na$_2$CO$_{3(aq)}$ $\underline{Na^+ \mid CO_3^{2-}}$ ions	H$_2$O$_{(l)}$ + CO$_{2(g)}$ + NaCl$_{(aq)}$ √	NR	NR	ZnCO$_{3(s)}$ +NaI$_{(aq)}$ √	NR	NR

Answer:

Combustion of Acetylene.

1. Test tube 1.   2. Test tube 4.   3. Test tube 4.   4. Test tubes 1, 2 and 3.

5. Test tube 4 produced the most heat and no soot.

Test tubes 1, 2 and 3 produced less energy and soot when compared to test tube 4.

6. Explanations

Test tube 1. Oxygen is only at the mouth of the test tube so combustion takes place only there and the flame was quickly extinguished. What amount of combustion took place was incomplete as soot was formed. The flame was yellow because of the presence of soot.

Test tube 2. Oxygen is present throughout the test tube so combustion took places throughout the test tube. However, the amount of oxygen present was insufficient for complete combustion. As a result of this soot was present and the flame was yellow.

Test tube 3. There was greater oxygen to fuel ratio, so combustion was more complete than in the former cases. However, the amount of oxygen was still not adequate for complete combustion as trace of soot was present.

Test tube 4. The oxygen to fuel ratio was adequate so complete combustion took. This is evident by the large amount of energy and blue flame produced and the absence of soot.

7. The more oxygen was present the greater the degree of combustion.

8. It is important to tune up your vehicle for the following reasons:

- To supply enough oxygen for the fuel to be burnt efficiently

- To ensure that the fuel is ignited in all the cylinders of the engine

- To avoid incomplete combustion that leads to energy loss and poor air quality.

Chapter 8:

Writing Names and Formulas for Binary Ionic

Compounds Exercise 8.1

Compound	Formula	Compound	Fornula
(a) magnesium fluoride	MgF_2	(k) aluminum nitride	AlN
(b) lead(II) oxide	PbO	(l) tin(IV) bromide	$SnBr_2$
© potassium sulphide	K_2S	(m) iron(II) chloride	$FeCl_2$
(d) zinc iodide	ZnI_2	(n) strontium nitride	Sr_3N_2
(e) iron(III) sulphide	Fe_2S_3	(o) sodium posphide	Na_3P
(f) calcium bromide	CaBr	(p) aluminum oxide	Al_2O_3
(g) barium chloride	$BaCl_2$	(q) cesium bromide	CsBr
(h) copper(II) oxide	CuO	(r) rubidium nitride	Rb_3N
(i) silver iodide	AgI	(s) gold(I) chloride	AuCl
(j) lead(IV) oxide	PbO_2	(t) mercury(II) oxide	HgO

Exercise 8.2

(a) Li_2O lithium oxide
  mercury(II) oxide
(b) CaO   calciumoxide
(c) $AlCl_3$ aluminum chloride
(d) MgS   magnesium sulphide
(e) Na_2S sodium sulphide
(f) FeO   iron(II) oxide
(g) $CuCl_2$ copper(II) chloride
(h) Na_3N sodium nitride

(i) HgO
(j) Ag_2S silver sulphide
(k) $PbCl_4$ lead(IV) chloride
(l) BaI_2 barium iodide
(m) Cs_2O cesium oxide
(n) $SnBr_4$ tin(IV) bromide
(o) KF   potassium fluoride
(p) $CoCl_2$ cobalt(II) chloride

Exercise 8.3

(a) carbon tetrachloride CCl_4
(b) phosphorus trichloride PCl_3
(c) dinitrogen pentoxide N_2O_5
(d) sulphur hexachloride SCl_6
(e) nitrogen dioxide NO_2
(f) diphosphorus trioxide P_2O_3
(g) sulphur trioxide SO_3
(h) ammonia NH_3

Exercise 8.4

(a) SBr_6 sulphur hexabromide
(b) CS_2 carbon disulphide
(c) SiO_2 silicon dioxide
(d) CO   carbon monoxide
(e) CF_4 carbon tetrafluoride
(f) P_2O_5 diphosphorus pentoxide
(g) $HCl_{(g)}$ hydrogen chloride
(h) N_2O_4 dinitrogen tetroxide

Write the IUPAC name for each of the following binary compounds.
(a) AuCl gold(I) chloride (f) SnO_2 tin(IV) oxide   k) zinc nitride p) mercury(1) oxide
(b) BaO barium oxide   (g) NCl_3 nitrogen trichloride l)gold(III) chloride q) dinitrogen mon
(c) PbI_2 lead(II) iodide   (h) K_3N potassium nitride m)lead(II) nitride r mercury(I)iodide
(d) ZnO zinc oxide   (i) CuO copper(II) oxide   n) ozone r) hydrogen iodide
(e) $HBr_{(g)}$ hydrogen bromide (j)Ag_2S silver sulphide   m) silver nitride t) zinc iodide

Chapter 9 Atomic structure
Exercise 9.2

Atomic Notation	Number of Protons	Number of Neutrons	Number of Electrons	Mass Number
$^{7}_{3}Li$	3	4	3	7
$^{23}_{11}Na$	11	12	11	23
$^{40}_{20}Ca$	20	20	20	40
$^{27}_{13}Al$	13	14	13	27
$^{39}_{19}K$	19	20	19	39
$^{31}_{15}P$	15	16	15	31

Exercise 9.1
1. (a) true   (b) false   (c) false   (d) true
(f)   True   (g) true
2. (a) The number of protons   (b) 6   (c) 6
   (d) $^{12}_{6}C$ has 6 $^{13}_{6}C$ has 7 $^{14}_{6}C$ has 8 neutrons each respectively.
   (e) The number of neutrons that they have.

Quantities in Chemical formulas
Answers 9.3

1. (a) Na = 22.99g/mol (b) Ca = 40.08 g/mol   (c) Cl = 35.45 g/mol   (d) Zn = 65.3g/mol

2. One mole or 6.02×10^{23} atoms

3. Zn>, Ca>, Cl>Na

4. C

5. C-14

Exercise 9.4

   (a)   200.4 g   (b) 97.75 g   (c) 3.03 g   (d) 141.8 g
Exercise 9.5

   (a) 2.00 mol   (b) 3.000 mol   (c) 1.500 mol   (d)   3.000 mol

Exercise 9.6

   a) 3.61×10^{24} atoms   (b) 1.51×10^{24} atoms   (c) 2.41×10^{24} atoms
   (d) 1.81×10^{24} atoms   (e) 3.01×10^{23} atoms

Exercise 9.7

(a) 1.51×10^{23} atoms (b) 3.01×10^{23} atoms (c) 1.81×10^{24} atoms (d) 1.81×10^{24} atoms

Exercise 9.8

a) 0.050 mol   (b) 3.00 mol   (c) 0.250 mol   (d) 0.100 mol

Exercise 9.9

a) 216 g   (b) 6.54 g   (c) 53.2g   (d) 0.480 g

Exercise 9.10 $CaCl_2$ and $Ca(NO_3)_2$ are formula units and CO_2 and PCl_3 are molecules.

Exercise 9.11

a) 74.10 g   (b) 56.11 g   (c) 159.70 g   (d) 294.20 g   (e) 180.18 g
(f) 259.37 g   (g) 342.14 g   (h) 28.02 g   (i) 44.01 g   (j) 148.37 g
(k) 389.88 g   (l) 162.50 g   (m) 249.71 g   (n) 125.60 g   (o) 34.02 g
(p) 106.44 g   (q) 261.35 g   (r) 282.77

Exercise 10.1

a) K = 38.67%   b) N = 35.0%   c) N = 21.2%   d) NH_4NO_3
   N = 13.86%   O = 59.94%,   H = 6.10%   e) i. 4.0 g
   O = 47.47%   H = 5.06%   S = 24.28%   ii. 32.0 g

Exercise 10.2

1.a) $C_6H_{12}O_6$ 2. C_4H_{10} 3. $C_4H_5N_4O_2$ 4. $C_6H_8O_6$, 2936 days
  b) 11.10 mol

5. C = 84.76%   6. C = 51.67%   7. $C_3H_6O_3$ 8. $C_{14}H_{18}N_2O_5$,   9. $C_{17}H_{19}NO_3$
   H = 11.20%   H = 4.35%,   5887 packs.
   O = 4.04%   N = 22.22%,
            O = 21.75%   10. $C_{22}H_{30}N_6O_4S$ 11. $C_{33}H_{35}FN_2O_5$ 12. $C_8H_{10}N_4O_2$
            110300 packs
                   13. C_5H_{10}

CHAPTER 11: QUANTITIES IN CHEMICAL REACTIONS
Exercise 11.1

(a) $2HCl_{(aq)} + Na_2CO_{3(aq)} \rightarrow 2NaCl_{(aq)} + H_2O_{(l)} + CO_{2(g)}$

    4 mol   **2 mol**   **4 mol**   **2 mol**   **2 mol**

(b) $2HCl_{(aq)} + Na_2CO_{3(aq)} \rightarrow 2NaCl_{(aq)} + H_2O_{(l)} + CO_{2(g)}$

    1 mol   **½ mol**   **1 mol**   **½ mol**   **½ mol**

(c) $H_2SO_{4(aq)} + 2NaOH_{(aq)} \rightarrow 2NaCl_{(aq)} + 2H_2O_{(l)}$

    ½ mol   **1 mol**   **1 mol**   **1 mol**

(d) $H_2SO_{4(aq)} + 2NaOH_{(aq)} \rightarrow 2NaCl_{(aq)} + 2H_2O_{(l)}$

    ¼mol   **½ mol**   **½ mol**   **½ mol**

(e) $Ba(OH)_{2(aq)} + 2HNO_{3(aq)} \rightarrow Ba(NO_3)_{2(aq)} + 2H_2O_{(l)}$

    2 mol   **4 mol**   **2 mol**   **4 mol**

(f) $Ba(OH)_{2(aq)} + 2HNO_{3(aq)} \rightarrow Ba(NO_3)_{2(aq)} + 2H_2O_{(l)}$

    1/3 mol   **2/3 mol**   **1/3 mol**   **2/3 mol**

(g) $2BrO_3^-{}_{(aq)}$ + $5HSO_3^-{}_{(aq)}$ → $Br_{2(aq)}$ + $5SO_4^{2-}{}_{(aq)}$ + $H_2O_{(l)}$ + $3H^+{}_{(aq)}$
     3 mol   **7.5 mol**   **1.5 mol**   **7.5 mol**   **1.5 mol**   **4.5 mol**

(h) $2BrO_3^-{}_{(aq)}$ + $5HSO_3^-{}_{(aq)}$ → $Br_{2(aq)}$ + $5SO_4^{2-}{}_{(aq)}$ + $H_2O_{(l)}$ + $3H^+{}_{(aq)}$
     0.8 mol   **2 mol**   **0.4 mol**   **2 mol**   **0.4 mol**   **1.2 mol**

(i) $3NO_{2(g)}$ + $H_2O_{(l)}$ → $2HNO_{3(aq)}$ + $NO_{(g)}$
     9 mol   **3 mol**   **6 mol**   **3 mol**

(j) $3NO_{2(g)}$ + $H_2O_{(l)}$ → $2HNO_{3(aq)}$ + $NO_{(g)}$
     4 ½ mol   **1 ½ mol**   **3 mol**   **1 ½ mol**

Exercise 11.2

a) $BaCl_2$ = 139.57 g   (b) $CuSO_4$ = 15.96 g   © Mg = 2.43 g
                (d) $Al(OH)_3$ = 13.00 g   (e) H_2SO_4 = 24.52 g

Exercise 11.3

(a) $H_2SO_{4(aq)}$ + $2NaOH_{(aq)}$ → $2NaCl_{(aq)}$ + $2H_2O_{(l)}$
     20.00 g   *excess*   *23.84 g*   *7.35 g*

(b) $Ba(OH)_{2(aq)}$ + $2HNO_{3(aq)}$ → $Ba(NO_3)_{2(aq)}$ + $2H_2O_{(l)}$
     excess   *15.75 g*   *32.6 g*   *4.15 g*

(c) $3NO_{2(g)}$ + $H_2O_{(l)}$ → $2HNO_{3(aq)}$ + $NO_{(g)}$
     23.00 g   *excess*   *21.00 g*   *5.00 g*

(d) $CaCO_{3(s)}$ + $2HCl_{(aq)}$ → $CaCl_{2(aq)}$ + $H_2O_{(l)}$ + $CO_{2(g)}$
     excess   *9.12 g*   *13.87 g*   *2.25 g*   *5.50 g*

Exercise 11.4

1) 0.01 g $H_{2(g)}$ **(2)** 4.88 g $Cu(OH)_{2(s)}$ **(3)** 5.50 g $CO_{2(g)}$ and 4.51 g

(4) 14.33 g $AgCl_{(s)}$, 84.44%   **(5)** 2.33 g $BaSO_{4(s)}$, 90.13%   **(6)** 6.22 g $PbI_{2(s)}$

(7) a) 12.97 g b) 92.60% c) 3.05 g **(8)** a) 30.75 g , b) $C_{17}H_{35}COOH$ c) Oily **(9)** 97.78%

CHAPTER 12: Nature and Properties of Solution

Exercise 12.1

1. $ZnCl_{2(s)}$ + $H_2O_{(l)}$ → $Zn^{2+}{}_{(aq)}$ + $2Cl^-{}_{(aq)}$

2. $AlCl_{3(s)}$ + $H_2O_{(l)}$ → $Al^{3+}{}_{(aq)}$ + $3Cl^-{}_{(aq)}$

3. $Al_2(SO_4)_{3(s)}$ + $H_2O_{(l)}$ → $2Al^{3+}{}_{(aq)}$ + $3SO_{4(aq)}$

4. $Ca(HCO_3)_{2(s)}$ + $H_2O_{(l)}$ → $Ca^{2+}{}_{(aq)}$ + $2HCO_3^-{}_{(aq)}$

5. $Mg(BrO_3)_{2(s)}$ + $H_2O_{(l)}$ → $Mg^{2+}{}_{(aq)}$ + $2BrO_3^-{}_{(aq)}$

6. $K_2CO_{3(s)}$ + $H_2O_{(l)}$ → $2K^+{}_{(aq)}$ + $CO_3^{2-}{}_{(aq)}$

7. $FeBr_{3(s)}$ + $H_2O_{(l)}$ → $Fe^{3+}{}_{(aq)}$ + $3Br^-{}_{(aq)}$

Exercise 12.2

(c) homogeneous   (b) homogeneous   (a) heterogeneous   (d) heterogeneous
(h) homogeneous   (g) homogeneous   (f) heterogeneous   (e) heterogeneous
(l) homogeneous   (k) heterogeneous   (j) homogeneous   (i) heterogeneous

Exercise 13.1

e) 1.0 mol/L   c) 1.25 mol/L   d) 1.33 mol/L   a) 0.40 mol/L

(e) 0.20 mol/L   f) 0.010 mol/L   b)

Exercise 13.2

a) 0.500 mol/L   b) 0.153 mol/L   c) 2.00 mol/L   d) 1.25 mol/L   e) 5.0×10^{-3} mol/L   f) 0.0200 mol/L

Exercise 13.3

a) 7.85 g   b) 23.4 g   c) 3.64 g   d) 3.95 g   e) 26.4 g

Exercise 13.4

a) 11.8 g $CaCl_2 \cdot 2H_2O_{(s)}$ b) 25.8 g $Na_2SO_4 \cdot 10H_2O_{(s)}$

Exercise 13.5

a) $NaOH_{(aq)} \rightarrow Na^+_{(aq)} + OH^-_{(aq)}$

1 mol/L   1 mol/L   1 mol/L

b) $ZnSO_{4(aq)} \rightarrow Zn^{2+}_{(aq)} + SO_4^{2-}_{(aq)}$

1 mol/L   1 mol/L   1 mol/L

c) $CaCl_{2(aq)} \rightarrow Ca^{2+}_{(aq)} + 2Cl^-_{(aq)}$

1 mol/L   1 mol/L   2 mol/L

d) $Mg(OH)_{2(aq)} \rightarrow Mg^{2+}_{(aq)} + 2(OH^-)_{(aq)}$

1 mol/L   1 mol/L   2 mol/L

f) $FeCl_{3(aq)} \rightarrow Fe^{3+}_{(aq)} + 3Cl^-_{(aq)}$

1 mol/L   1 mol/L   3 mol/L

g) $Al_2(SO_4)_{3(aq)} \rightarrow 2Al^{3+}_{(aq)} + 3SO_4^{2-}_{(aq)}$

1 mol/L   2 mol/L   3 mol/L

Exercise 13.6

a) $HCl_{(aq)} \rightarrow H^+_{(aq)} + Cl^-_{(aq)}$

1 mol/L   1 mol/L   1 mol/L

b) $HNO_{3(aq)} \rightarrow H^+_{(aq)} + NO_3^-_{(aq)}$

1 mol/L   1 mol/L   1 mol/L

c) $HI_{(aq)} \rightarrow H^+_{(aq)} + I^-_{(aq)}$

1 mol/L   1 mol/L   1 mol/L

d) $HCN_{(aq)} \rightarrow H^+_{(aq)} + CN^-_{(aq)}$

1 mol/L   1 mol/L   1 mol/L

e) $HClO_{3(aq)} \rightarrow H^+_{(aq)} + ClO_3^-_{(aq)}$

1 mol/L   1 mol/L   1 mol/L

Exercise 13.7

a) $[Cu^{2+}_{(aq)}] = 0.0250$ mol/L $[SO_4^{2-}_{(aq)}] = 0.0250$ mol/L
b) $[Na^+_{(aq)}] = 0.05$ mol/L $[SO_4^{2-}_{(aq)}] = 0.025$ mol/L
c) $[Ca^{2+}_{(aq)}] = 0.200$ mol/L $[Cl^-_{(aq)}] = 0.400$ mol/L
d) $[Ba^{2+}_{(aq)}] = 0.050$ mol/L $[OH^-_{(aq)}] = 0.100$ mol/L

Exercise 13.8

a) 0.025 mol/L
b) 0.10 mol/L
c) 0.020 L or 2.0×10^1 mL
d) 0.20 L or 2.0×10^2 mL
e) (i) 0.125 mol/L $H^+_{(aq)}$ (ii) 0.0625 mol/L $SO_4^{2}_{(aq)}$
f) (i) 0.252 mol/L $Mg(OH)_{2(aq)}$
    (ii) 0.050 mol/L $Mg^{2+}_{(aq)}$
    (iii) 0.10 mol/L $OH^-_{(aq)}$

g) 10.6 g
h) 1.82 g $HCl_{(aq)}$

Exercise 13.9

1. (a) 5.0×10^2 mL (b) 524 g (c) 8.74 mol
2. 97.5 % v/v
3. 1.2×10^2 g
4. 6.60 g
5. 1.39 mol

Exercise 13.11
1. 3.00×10^6 ppb
2. 15 ppm
3. 0.70 g

Exercise 13.12
1. **10.32 mol/L**
2. **8.70 mol/L**
3. **10.39 mol/L**

CHAPTER 14: Factors That Affect Solubility

Activity X

Analysis 1.

1. Carbon dioxide **2.** Carbon dioxide **3.** No

4. Gases dissolve less in warm solvents than cold solvents. In this case as the soda was heated up by the hot water, the dissolved carbon dioxide gas molecules gained more kinetic energy and escape at a greater rate from the soda. This was evident by the vigorous effervescence produced. The greater volume of carbon dioxide produced at a faster rate was able to extinguish the burning candle faster than in the cold soda where the carbon dioxide gas was escaping at a much slower rate.

5. One of the effects of thermal pollution in bodies of water is the lack of dissolved gases such as oxygen and carbon dioxide essential to support life forms in aquatic ecosystems.

6. In the warm can of soda less gas dissolves in the solution because of the higher temperature so more gas particles occupy the space above the solution creating a higher pressure there than in the cold can. Because of this higher pressure the gas makes a louder sound as it tries to escape than in the cold can.

Activity Y

Analysis 2.

1. Unsaturated **2.** Saturated **3.** As the temperature increases the solubility increases.

Conclusion: The solubility of KNO_3 increases with temperature rise.

Analysis 3.

1. KI **2.** 72.0°C **3.** 45 g **4.** $SO_{2(g)}$ and $NH_{3(g)}$

5. NH_3 is more polar SO_2 than and also it forms hydrogen bonds with water. It is thus more attracted to water molecules than SO_2.

6. 73.0 g.

7. No, because the exact volume of the solution is unknown. This will have to be measured and provided after the solute is dissolved in the 100.0 mL of water.

Exercise 14.1

1. Lowering temperatures cause less solubility and therefore eventual crystallization
2. Gently heat the jar in a water bath to temperature less than 40°C. Do not over heat.

3. Particles of wax and pollens etc, in raw honey act as nuclei for crystallization not present in treated honey.
4. Nothing changes except that the sugars have crystallized out of solution.
5. It could be could be found in the liquid part since it has a higher solubility than glucose which must have crystallized out of solution before.

CHAPTER 15: Soap and Detergents

Part A: Table 15.1 **Observation After Shaking**

1. Distilled water + soap	A lot of lather was readily formed
2. Distilled water + soap + $MgCl_{2(aq)}$	A creamy precipitate was formed. Only a small amount of lather was formed
3. Distilled water + soap + $CaCl_{2(aq)}$	A creamy precipitate was formed. Only a small amount of lather was formed
4. Distilled water + soap + $NaCl_{(aq)}$	A lot of lather was readily formed

Analysis 1.

1. Test tubes 1 and 4

2. Test tubes 2 and 4. Because the $Mg^{2+}_{(aq)}$ and $Ca^{2+}_{(aq)}$ ions present reacted with the soap molecules.

3. Test tubes 2 and 4   **4.** test tubes 2 and 4

Conclusion

1. $Mg^{2+}_{(aq)}$ and $Ca^{2+}_{(aq)}$ ions are responsible for the hardness of water.

$$Mg^{2+}_{(aq)} + 2\ C_{17}H_{35}COO^{-}_{(aq)} \longrightarrow Mg(C_{17}H_{35}COO)_{2(s)}$$

$$Ca^{2+}_{(aq)} + 2\ C_{17}H_{35}COO^{-}_{(aq)} \longrightarrow Ca(C_{17}H_{35}COO)_{2(s)}$$

Part B:
Table 15.2

Water samples	Observation Before Shaking
1 Distilled water + soap + $Na_2CO_{3(aq)}$	No visible change took place
2 Distilled water + $Na_2CO_{3(aq)}$ + soap + $MgCl_{2(aq)}$	A creamy precipitate was formed
3 Distilled water + $Na_2CO_{3(aq)}$ + soap + $CaCl_{2(aq)}$	A creamy precipitate was formed
4 Distilled water + $Na_2CO_{3(aq)}$ + soap + $NaCl_{(aq)}$	No visible change took place

1. (a) Test tubes 2 and 3.

    (b) $Mg^{2+}_{(aq)} + CO_3^{2-}_{(aq)} \rightarrow MgCO_{3(s)}$

    $Ca^{2+}_{(aq)} + CO_3^{2-}_{(aq)} \rightarrow CaCO_{3(s)}$

Table15.3

Water samples	Observation After Shaking
1. Distilled water + soap + $Na_2CO_{3(aq)}$	A lot of lather was readily formed
2. Distilled water + $Na_2CO_{3(aq)}$ + soap + $MgCl_{2(aq)}$	A lot of lather was readily formed Precipitate remains
3. Distilled water + $Na_2CO_{3(aq)}$ + soap + $CaCl_{2(aq)}$	A lot of lather was readily formed Precipitate remains
4. Distilled water + $Na_2CO_{3(aq)}$ + soap + $NaCl_{(aq)}$	A lot of lather was readily formed

Analysis 3.

1. Yes. Test tubes 2 and 4 produced significantly more lather than in procedure A.

2. No

3. The water was softened because the $CO_3^{2-}_{(aq)}$ ions removed the $Mg^{2+}_{(aq)}$ and $Ca^{2+}_{(aq)}$ ions from the water in the form of precipitates.

Table15.3

1. Distilled water + detergent	A lot of lather was readily formed
2. Distilled water + detergent + $MgCl_{2(aq)}$	No precipitate but lots of lather formed
3. Distilled water + detergent + $CaCl_{2(aq)}$	No precipitate but lots of lather formed
4. Distilled water + detergent + $NaCl_{(aq)}$	A lot of lather was readily formed

Analysis 4

1. No   2. No   3. No

Conclusion: Detergents work well in hard water because its molecules do not form precipitates with $Mg^{2+}_{(aq)}$ and $Ca^{2+}_{(aq)}$ ions.

1. a) The water is not well sterilized. This can be rectified by adding more sanitizer.

   High nitrates means that there was not enough time spent by the effluent in the anoxic zone for nitrifying bacteria to act. To solve this problem, allow the effluent to spend more time in the anoxic zone.

   b. There would be inadequate oxidation of ammonia to nitrates. The water ensuing water will have unchanged ammonia.

   c. The dissolved oxygen gets low. It will cost more to replenish the oxygen by the pumping process.

   d. There would not be enough bugs for the breakdown of the sewage.

   O turbidity, E. *Coli* and low oxygen are typical of runoff from feeding lots etc. The low oxygen is due to the high BOD of the water.

   b) Source B is probably from an uncontaminated source such treated water or spring water.

   3. a) Source A is unsafe.

CHAPTER 17: Acids and Bases
ACID-BASE THEORIES

Exercise17.1

a) $HCO_3^-{}_{(aq)}$ and $CO_3^{2-}{}_{(aq)}$, $HS^-{}_{(aq)}$ and $S^{2-}{}_{(aq)}$

b) $HC_2H_3O_2$ and $C_2H_3O_2^-{}_{(aq)}$, $HCO_3^-{}_{(aq)}$ and $CO_3^{2-}{}_{(aq)}$

c) $H_3PO_4{}_{(aq)}$ and $H_2PO_4^-{}_{(aq)}$, $HClO_{(aq)}$ and $OCl^-{}_{(aq)}$

d) $HSO_4^-{}_{(aq)}$ and $SO_4^{2-}{}_{(aq)}$, $H_2PO_4^-{}_{(aq)}$ and $HPO_4^{2-}{}_{(aq)}$

Exercise17.3

1. **a) weak acid**
   b) close to 7
   c) $HC_6H_7O_6{}_{(aq)} + H_2O_{(l)} \rightleftharpoons C_6H_7O_6^-{}_{(aq)} + H_3O^+{}_{(aq)}$
   d) ($H_3O^+{}_{(aq)}$ and $H_2O_{(l)}$) and ($HC_6H_7O_6{}_{(aq)}$ and $C_6H_7O_6^-{}_{(aq)}$)
   e) 1 mol, $HC_6H_7O_6{}_{(aq}$ + NaOH $\rightarrow$ $NaC_6H_7O_6$ + $H_2O_{(l)}$

2.
   a) Strong base.. Complete ionization with production of $OH^-{}_{(aq)}$ ions.
   b) Good conduction. Complete ionization means lot of ion for conduction
   c) Well above 7
   d) i) KOH + $H_2SO_4{}_{(aq)} \rightarrow$ $KHSO_4{}_{(aq)}$ + $H_2O_{(l)}$
      ii) The solution is acidic

3. a) A weak base
   b) $NH_3{}_{(aq)}$
   c) $NH_3{}_{(aq)} + + H_2O_{(l)} \rightleftharpoons H_3O^+{}_{(aq)} + NH_4^+{}_{(aq)}$

Exercise17.4

Complete and balance the following equations:

1. $2HCl_{(aq)}$ + $Ca(OH)_2{}_{(aq)} \rightarrow$ $CaCl_2{}_{(aq)}$ + $2H_2O_{(l)}$

2. $H_2SO_4{}_{(aq)}$ + $2KOH_{(aq)}$ $\rightarrow$ $K_2SO_4{}_{(aq)}$ + $2H_2O_{(l)}$

3. $Mg(OH)_{2(aq)} + H_2SO_{4(aq)} \rightarrow MgSO_{4(aq)} + 2H_2O_{(l)}$

4. $H_3PO_{4(aq)} + 3\,NaOH_{(aq)} \rightarrow Na_3PO_{4(aq)} + 3H_2O_{(l)}$

5. $CH_3COOH_{(aq)} + KOH_{(aq)} \rightarrow CH_3COOK_{(aq} + H_2O_{(l)}$

6. $2H_3PO_{4(aq)} + 3Ba(OH)_{2(aq)} \rightarrow Ba_3(PO_4)_{2(aq)} + 6H_2O_{(l)}$

7. $2LiOH_{(aq)} + H_2SO_{4(aq)} \rightarrow Li_2SO_{4\,(aq)} + 2H_2O_{(l)}$

8. $H_2SO_{3(aq)} + Ca(OH)_{2(aq)} \rightarrow Ca\,SO_{3(aq)} + 2H_2O_{(l)d}$

9. $2HClO_{3(aq)} + Mg(OH)_{2(aq)} \rightarrow Mg(ClO_3)_{2(aq)} + 2H_2O_{(l)}$

10. $3HBrO_{4(aq)} + Al(OH)_3 \rightarrow Al(BrO_3)_{3(aq)} + 3H_2O_{(l)}$

Exercise 17.5

1. 20 mL   2. 500 mL   3. 5.83 g   4. 2 mL   5. 0.125 mol/L 6. 0.50 mol/L

7. 400 mL   8. 0.36 g Mg   9. 0.84 g $NaHCO_3$ 11. 0.043 mol/L 10. a) 0.4 g, b) 0.53g, c) 0.79g
                                                                 d) B

ACID BASE TITRATION

Exercise 17.6

d) 0.05 mol/L   c) 0.5 mol/L b) 0.037 mol/L   a) i) 0.066 mol/L
                                                                 ii) 3.5 g $Na_2CO_{3(s)}$

pH OF A SOLUTION

Exercise 17.7

a) 3   c) 1.92   b) 3.7

Exercise 17.8

a) 1   b) 0.6   c) 1.6   d) 1.6   e) 1.2

Exercise 17.9

a) 1×10^{-2} mol/L   b) 1×10^{-4} mol/L   c) 3.98×10^{-5} mol/L   d) 1×10^{-3} mol/L e) 1×10^{-5} mol/L

Exercise 17.11

a) 1×10^{-3} mol/L   b) 1×10^{-5} mol/L   c) 10 L   d) (i) pH = 1.3 (ii) pH = 1.6 (iii) 0.3

e) (i) pOH = 12.74 (ii) pH = 1.26   f) 0.08 g $NaOH_{(s)}$ g) pH = 13   h) pH = 1.47

i) 0.08 g $NaOH_{(s)}$ j(i) No, there was excess acid, (ii) 2.65 g $Na_2CO_{3(s)}$ k) 12.7

Exercise 17.12

Write the formulas for the following compounds.

(a) magnesium bromate $Mg(BrO_3)_2$ (h) ammonium sulphate $(NH_4)_2SO_4$
(b) zinc iodite $Zn(IO_2)_2$ (i) potassium cyanide KCN
(c) lead(II) sulphite $PbSO_3$ (j) sodium periodate $NaIO_4$

(d) ferric chlorate $Fe(ClO_3)_3$ (k) plumbic nitrate $Pb(NO_3)_4$
(e) tin(IV) carbonate $Sn(CO_3)_2$ (l) potassium hydrogensulphate $KHSO_4$
(f) cupric acetate $Cu(CH_3COO)_2$ (m) lithium phosphate Li_3PO_4
(g) calcium hypoiodite $Ca(IO)_2$ (n) calcium nitrite $Ca(NO_2)_2$

Exercise 17.13

Write the classical and or IUPAC names for the following compounds.

(a) $AgIO_2$ silver iodite   (j) K_3PO_4 potassium phosphate
(b) $MgCO_3$ magnesium carbonate   (k) $Be(CN)_2$ beryllium cyanide
(c) $Ca(HSO_4)_2$ calcium hydrogen sulphate   (l) $Ba(HCO_3)_2$
(d) $Na_2SO_4.7H_2O$ sodium sulphate heptahydrate   (m) $SrSO_3$ strunctium sulphite
(e) $HBrO_{3(aq)}$ aqueous hydrogen bromate   (n) $KClO$ potassium hypochlorite
(f) $Sn(NO_3)_2$ tin(II) nitrate   (o) $KMnO_4$ potassium permanganate
(g) $(NH_4)_2SO_4$ ammonium sulphate   (p) $HClO_{4(aq)}$ aqueous hydrogen perchlorate
(h) $Pb(NO_2)_2$ lead(II) nitrite   (q) $Fe(IO_3)_3$ iron(III) iodate
(i) $Mg(CH_3COO)_2$ magnesium acetate   (r) $Zn_3(PO_3)_2$ zinc phosphite

CHAPTER 18: Gases

Exercise 18. 1

1. (a) It will expand (b) $8.0x10^1$ kPa   2. $7.0x10^1$ balloons with 5.0 L of gas remaining in the cylinder
   3. 0.50 L4. 135 kPa

Exercise 18.2

a) (i) 283 K   (ii) 373 K   b) (i) -198°C (ii) -153 °C

Exercise 18.3

a) 4.9 L   b) 21 L   d) 3.56 °C
c) 56.6 °C

Exercise 18.4

a) 126 kPa   b) 610 °C   c) 92.4 kPa   d) 248 kPa

Exercise 18.5

a) 46 balloons with 10 L remaining in the cylinder.   b) 5.5 L   c) 237 kPa   d) $1.80x10^2$ mL   e) 86.1 °C   f) − 102.6 °C   g. 115.02 kPa

Exercise 19.1

2) 1 atm.   1)

Gases	Inhaled air	Exhaled air
O_2	> pp	< pp
CO_2	< pp	> pp
H_2O	< pp	> pp

Exercise 19.2

4) $9.03 x10^{23}$ molecules   3) 1.24 L   1) 12.4 L   2) 37.2 L $N_{2(g)}$ and 112 L $H_{2(g)}$.

5) 62 balloons   6) 408 g $KClO_3$ and 60.1 g C   7) 122 g

Exercise 20.1

5) 478 mL   4) 263 g   3) 1.24 L   2) 629 mL   1) 2.52 L   6) 0.654 g

7) $HCl_{(aq)}$

UNIT1: Matter

CHAPTER 1-4

True/False

1.) T   2.) F   3.) F   4.) T   5.) F   6.) F   7.) F   8.) F

9.) T   10.) F   11.) T

Multiple Choice

1.) E   2.) C   3.) E   4.) A   5.) B   6.) B   7.) C   8.) A

9.) E   10.) A   11.) C   12.) E   13.) E   14.) C   15.) E   16.) D

17.) E   1 8.) C   19.) C   20.) D   21.) D   22.) C

Matching :

1.) H   2.) B   3.) F   4.) C   5.) I   6.) J   7.) D   8.) L

9.) E   10.) M   11.) A   12.) G

CHAPTER 5-6

True/False

1.) T   2.) F   3.) T   4.) T   5.) F   6.) F   7.) F   8.) F

9.) T   10.) F   11.) T

Multiple Choice

1.) D   2.) C   3.) C   4.) D   5.) E   6.) B   7.) B   8.) A

9.) A   10.) B   11.) C   12.) D   13.) B   14.) B   15.) C   16.) C

17.) E   1 8.) E   1 9.) E   20.) B

Matching :

1.) G   2.) K   3.) E   4.) L   5.) H   6.) B   7.) A   8.) F

9.) D   10.) J   11.) I   12.) C

CHAPTER 7-8

True/False

1.) F   2.) T   3.) T   4.) T   5.) T   6.) T   7.) T   8.) T

9.) T   10.) T   11.) T   12.) T   13.) T   14.) T   15.) T

Multiple Choice

1.) E   2.) E   3.) C   4.) E   5.) C   6.) E   7.) D   8.) D   9.) D   10.) C   11.) E

12.) A   13.) B   14.) D   15.) E   16.) C   17.) E   18.) E   19.) C   20.) D   21.) C   22.) A

23.) A   24.) E   25.) E

Matching :

1.) K   2.) G   3.) E   4.) F   5.) J   6.) L   7.) A   8.) D

9.) I   10.) B   11.) C   12.) H

UNIT 2: Quantities In Chemical Reactions

CHAPTER 9-11

True/False

1.) T   2.) T   3.) T   4.) F   5.) F   6.) T   7.) T   8.) F

9.) T   10.) T   11.) F

Multiple Choice

1.) B   2.) C   3.) E   4.) C   5.) A   6.) B   7.) B   8.) A

9.) C   10.) D   11.) C   12.) C   13.) D   14.) B   15.) B   16.) E

17.) B   1 8.) B   1 9.) E   20.) C   21.) D   22.) A   23.) C   24.) E

25.) B   26.) A   27.) C   28.) A   29.) D

Matching :

1.) K   2.) D   3.) A   4.) J   5.) L   6.) F   7.) H   8.) G

9.) I   10.) E   11.) C   12.) B

UNIT 3: Solutions And Solubility

CHAPTER 12-15

True/False

1.) T   2.) F   3.) T   4.) T   5.) T   6.) F   7.) T   8.) F

9.) F   10.) F   11.) F   12.) T   13.) T   14.) F

Multiple Choice

1.) C   2.) E   3.) E   4.) C   5.) C   6.) A   7.) E   8.) A

9.) B   10.) D   11.) D   12.) E   13.) A   14.) B   15.) B   16.) B

17.) E   18.) D

Matching :

1.) G   2.) A   3.) I   4.) B   5.) H   6.) C   7.) D   8.) J

9.) E   10.) K   11.) F   12.) L

CHAPTER 16

True/False

1.) F   2.) T   3.) T   4.) F   5.) T   6.) F   7.) F   8.) T

9.) T   10.) F

Multiple Choice

1.) E   2.) D   3.) E   4.) B   5.) E   6.) E   7.) E   8.) D

9.) C   10.) E   11.) C   12.) E   13.) C   14.) B   15.) E

Matching :

1.) J   2.) A   3.) F   4.) K   5.) E   6.) L   7.) G   8.) H

9.) B   10.) C   11.) D   12.) J

CHAPTER 17

True/False

1.) F   2.) F   3.) T   4.) T   5.) F   6.) F   7.) T   8.) F

9.) T   10.) F   11.) T   12.) T   13.) T   14.) T

Multiple Choice

1.) D   2.) A   3.) C   4.) E   5.) E   6.) B   7.) E   8.) A

9.) E   10.) E   11.) E   12.) B   13.) D   14.) E   15.) D   16.) E

17.) E   18.) A   19.) B   20.) D   21.) E

Matching :

1.) L   2.) H   3.) J   4.) B   5.) I   6.) C   7.) K   8.) D

9.) G   10.) E   11.) N   12.) A   13.) M   14.) F

UNIT 4: Gases

CHAPTER 18-20

True/False

1.) T   2.) F   3.) T   4.) F   5.) F   6.) T   7.) T   8.) T

9.) F   10.) T   11.) T   12.) T

Multiple Choice:

1.) E   2.) E   3.) D   4.) B   5.) E   6.) E   7.) B   8.) D

9.) E   10.) A   11.) D   12.) B   13.) C   14.) B   15.) E   16.) A

17.) D   18.) B   19.) D   20.) E   21.) A   22.) D   23.) C   24.) B   25.) D   26.) B

Matching :

1.) A   2.) H   3.) B   4.) F   5.) G   6.) D   7.) C   8.) I

9.) L   10.) J   11.) E   12.) K

9.) E   10.) K   11.) B   12.) I

Table of Common Radicals

Chlorate	chlorite	hypochlorite	perchlorate
ClO_3^-	ClO_2^-	ClO^-	ClO_4^-
bromate	bromite	hypobromite	perbromate
BrO_3^-	BrO_2^-	BrO^-	BrO_4^-
iodate	iodite	hypoiodite	periodate
IO_3^-	IO_2^-	IO^-	IO_4^-
fluorate	fluorite	hypofluorite	perfluorate
FO_3^-	FO_2^-	FO^-	FO_4^-
nitrate	nitrite		
NO_3^-	NO_2^-		
carbonate	carbonite	hydrogen carbonate	
CO_3^{2-}	CO_2^{2-}	HCO_3^-	
sulphate	sulphite	hydrogen suplhate	persulphate
SO_4^{2-}	SO_3^{2-}	HSO_4^-	SO_5^{2-}
manganate	manganite		permanganate
MnO_3^-	MnO_2^-		MnO_4^-
phosphate	phosphite	hydrogen phosphate	dihydrogen phosphate
PO_4^{3-}	PO_3^{3-}	HPO_4^{2-}	$H_2PO_4^-$

Table of Acids

sulphuric	H_2SO_4
sulphurous	H_2SO_3
carbonic	H_2CO_3
carbonous	H_2CO_2
phosphoric	H_3PO_4
phosphorous	H_3PO_3
nitric	HNO_3
nitrous	HNO_2
chloric	$HClO_3$
chlorous	$HClO_2$
bromic	$HBrO_3$
bromous	$HBrO_2$
iodic	HIO_3
iodous	HIO_2
acetic	CH_3COOH

Table of radicals and unusual ions

cyanide	CN^-
hydroxide	OH^-
dichromate	$Cr_2O_7^{2-}$
chromate	CrO_4^{2-}
thiosulphate	$S_2O_3^{2-}$
acetate	$C_2H_3O_2^-$
ammonium	NH_4^+
oxalate	$C_2O_4^{2-}$
peroxide	O_2^{2-}
hydride	H^-
manganate	MnO^-
Hydrogen carbonate	HCO_3^-

APPENDIX : Table of Elements, atomic number, electron configurations and atomic mass and valence

Table of Atomic masses, atomic numbers, mass number and electron configurations

Elements	Symbol	Atomic mass	Atomic number	Electron configuration	Elements	Symbol	Atomic mass
Hydrogen	H	1.01	1	1	Barium	Ba	137.33
Helium	He	4.00	2	2	Cobalt	Co	58.93
Lithium	Li	6.94	3	2, 1	Copper	Cu	63.54
Boron	B	10.81	5	2, 3	Gold	Au	196.96
Carbon	C	12.01	6	2,4	Iodine	I	126.90
Nitrogen	N	14.01	7	2,5	Iron	Fe	55.84
Oxygen	O	15.99	8	2,6	Krypton	Kr	83.80
Fluorine	F	18.99	9	2,7	Lead	Pb	207.20
Neon	Ne	20.18	10	2,8	Manganese	Mn	54.93
Sodium	Na	22.99	11	2,8,1	Mercury	Hg	200.59
Magnesium	Mg	24.30	12	2,8,2	Nickel	Ni	58.69
Aluminum	Al	26.98	13	2,8,3	Platinum	Pt	195.08
Silicon	Si	28.08	14	2,8,4	Plutonium	Pu	244
Phosphorus	P	30.97	15	2,8,5	Polonium	Po	209
Sulphur	S	32.06	16	2,8,6	Radium	Ra	226
Chlorine	Cl	35.45	17	2,8,7	Silver	Ag	107.86
Argon	Ar	39.94	18	2,8,8	Strontium	St	87.62
Potassium	K	39.09	19	2,8,8,1	Tin	Sn	118.71
Calcium	Ca	40.08	20	2,8,8,2	Zinc	Zn	65.39

Tables of valences

Elements	Valence
Hydrogen	1+, 1-
Lithium	1+
Sodium	1+
Potassium	1+
Rubidium	1+
Cesium	1+
Francium	1+
Beryllium	2+
Magnesium	2+
Calcium	2+
Strontium	2+
Barium	2+
Boron	3+
Aluminum	3+

Elements	Valence
Scandium	3+
Titanium	3+, 4+
Chromium	2+, 3+
Manganese	2+, 4+
Iron	2+, 3+
Cobalt	2+, 3+
Nickel	2+, 3+
Copper	1+, 2+
Zinc	2+
Silver	1+
Tin	2+, 4+
Platinum	2+, 4+
Gold	1+, 3+
Mercury	1+, 2+

Elements	Valence
Boron	3+
Carbon	4+, 0, 4-
Silicon	
Nitrogen	3-
Phosphorus	3-
Oxygen	2-
Suplhur	2-
Fluorine	1-
Chlorine	1-
Bromine	1-
Iodine	1-
Helium	0
Neon	0
Argon	0

Glossary

absolute zero: the lowest possible temperature that can be attained (-273 ^{0}C)

acid: Arrhenius' theory (a compound that ionizes in water to produce $H^+_{(aq)}$ ions) or according to the Bronsted-Lowry theory, one that is a proton donor.

acid precipitation: water in the form of rain, hail or snow that has elevated amounts of acids due the absorption acidic gases

activity series: a list of elements based on the order of their relative reactivity with each other; the most reactive one is at the top

actual yield: in chemistry, it is the actual amount of product that is produced at completion of a chemical reaction

activated sludge: a type of waste matter in the sewage treatment plant that still has BOD and bugs (decomposers)

agricultural runoff: wastes from feeding lots and farms that get washed away by running surface water and eventually enter bodies of water, causing pollution there

algal bloom: a rapid exponential growth and subsequent death of algae due to an excess of any nutrients (nitrates, phosphates) in water

alpha decay: the nuclear reaction in which alpha particles are produced

alpha particle: a helium atom without it electrons; two protons and two neutrons

amphotheric: a substance that can behave as both an acid or base

anion: ions in an electrolyte that migrates to the anode

anode: the electrode that is positively charged

aqueous solution: a solution that has dissolved solute(s), where water is the solvent

atmospheric pressure: pressure due to the atmosphere on the Earth's surface or on objects. At sea level its average value is 101.0 kPa at 0 ^{0}C.

atom: the smallest unit of an element that has all the characteristics of that element

atomic mass: this includes sum of the total masses of all the electrons, protons and neutrons that an atom has. Other definition has it as its mass relative to one atom of carbon-12.

average atomic mass: the average of the the atomic masses of all the isotopes in a natural sample of an element

atomic mass unit: the unit of this 1u and is defined as the mass of 1/12 that of the carbon-12 atom.

atomic number: symbolized by the letter, Z, it is the number of protons found in nucleus of every atom of a particular element

atomic radius: the distance from the atom's nucleus to the outermost edge or half the distance between the nuclei of two atoms just touching each other. This is indicative of the size of the atom

Avogadro's constant: the number of entities found in one mole of anything: 6.022×10^{23} and is represented as N_A

Avogadro's hypothesis: equal volumes of gasses at the same temperature and pressure contain equal number of particles

Balmer's series: emission spectrum in the visible portion of the electromagnetic spectrum; seen when electrons are bumped up and return to energy level, n= 2

base: Arrhenius' theory (a compound that ionizes in water to produce $OH^-_{(aq)}$ ions) or according to the Bronsted-Lowry theory, one that is a proton acceptor

beta decay: the nuclear reaction in which beta particles are produced

beta particle: a fast-moving energetic electron that is emitted during a nuclear reaction

binary acid: a compound with hydrogen and a non-metal that ionizes in water to produce $H^+_{(aq)}$ ions

binary ionic compound: a compound that is made up only two elements; a metal and a non-metal

binary molecular compound: a compound that is made up only two non-metallic elements

bonding capacity: the number bonds an atom makes when it bonds with other elements

biological oxygen demand, BOD: the amount of dissolved oxygen that bacteria and other microbes use up when they decompose organic materials present in water. This is indicative of the amount of the amount of biodegradable matter present in water

Boyle's law: the volume of a fixed mass of gas at a certain temperature is inversely proportional to its pressure; volume decreases as pressure increases.

bug: commonly refers to all decomposing micro-organisms precent

catalyst: a substance that alters the rate of a reaction without itself being changed

catalytic converter: the device that contains a number of catalysts which convert harmful gases emanating from the engines of automobiles, into harmless ones. It is normally placed in the exhaust system

cations: positively charges aqueous ions in an electrolyte that migrates to the cathode

cathode: the electrode that is negatively charged

cathode ray tube: a device that is used to produce cathode rays

cathode rays: fast moving electrons emanating from the cathode in a cathode ray tube

cellular respiration: the process by which carbohydrates or other organic molecules are oxidized in cells with the use of oxygen

Charles' law: this states that the volume of a fixed mass of gas at a constant pressure, is directly proportional to its temperature; as its temperature increases its volume increases

chemical property: this describes how an element behaves when it is in contact with other substances

chemical change: one in which the original substance(s) is change to one or more new substances, having different chemical and physical properties from the original

compound: this is produced when atoms of two or elements combine chemically

combustion: process by which fuels are burnt in oxygen to produce energy and other substances; notably oxides

complete combustion: reaction of an element or compound where all products have attained fullest oxidation states

concentration: a measure of the amount of solute per unit volume

condensation: change of state from gas to liquid

conjugate acid-base pair: a pair of compounds where one is an acid, and the other one, a base; what is left if an $H^+_{(aq)}$ is removed from the **same** acid. For example, $CH_3COOH_{(aq)}$ and $CH_3COO^-_{(aq)}$

coordinate covalent bond: a covalent bond that is formed when the shared pair of electrons is provided by one of the bonded atoms

Coulomb's law: the force between two charges is directly proportional to their magnitude but inversely proportional to the distance between them

covalent bond: a bond formed between two non-metals by the sharing of their electrons

crystal lattice: the unique regular arrangement of the atoms, molecules, ions in elements and compounds.

critical mass: the minimum mass of radio- active atoms that must be present for a chain reaction to happen

Dalton's law of partial pressure: the total pressure of a mixture of non-reacting gases is the sum of the partial pressures of the individual gases.

decomposition reaction: a chemical in which a single compound is broken down into smaller molecules

electron cloud: a region of space about the nucleus where the electrons move

decomposers: microorganisms that break down dead organic materials

density: mass per unit volume of a substance

detergent: a surfactant that cleanses fabrics etc. It is not affected by hard water

diffracting grating: an optical device that diffracts light into it various component colours

dissociation: the procession by which the ions in an ionic compound separate when it dissolves in water or in its molten state

dipole moment: This is a measure of the degree of charge separation in a covalent bond

dipole-dipole forces: the force of attractions between opposite charges on neighboring polar molecules

double displacement reaction: a reaction between aqueous solutions of two ionic compounds where two new compounds are formed; cations in the two compounds displace each other

effective nuclear charge: the force of attraction that any electron feels from the nuclear charge

electrolytes: compounds in their aqueous form or molten state that can conduct electricity

electromagnetic spectrum: the full spectrum of all electromagnetic radiations

electron affinity: the energy change that occurs when an electron is added to atom in its gaseous state to form a negative ion

electron cloud: a region of space about the nucleus of an atom where there is high probability of finding the electrons

electronegativity: a measure of the ability of an atom in a molecule to the draw bonding pair electrons to itself.

electrostatic force: the attractive force between a positive and a negatively objects or the repulsive force between them if they have the same charge

elements: A substance that cannot be broken down into simpler substances by chemical means; has atoms with the same atomic number

emission spectrum: this is what is produced when electrons in an atom are bumped up into higher energy levels and they subsequently return to their ground states

empirical formula: the smallest whole-number ratio of the atoms in a compound.

equivalence point: the point in a titration when the required number of moles of the titrant (eg. a base) is added to the Erlenmeyer (with the acid) to completely neutralize it

excited atom: state of the atom that results when an electron of the atom is bumped from its ground state to a higher energy level

excess reagent: the reagent that is left over after a reaction is complete; more of it was used that was required for complete reaction with the other reagent(s)

flame test: a method used to identify the metal in an ionic compound by observing the colour of the flame produced when the compound is heated in a Bunsen flame

floc: aggregates of bugs and other colloidal materials in the activated sludge in the sewage treatment plant

fission reaction: a nuclear reaction in which a large unstable nucleus of isotope splits into two smaller nuclei with the release of a tremendous amount of energy; used for making nuclear bombs and the production of nuclear energy

freezing: the change of state from liquid to solid

fusion reaction: a nuclear reaction in which two smaller nuclei combine to form a larger nucleus with the release of a very large amount of energy

gamma radiation: high-energy electromagnetic radiation resulting from radio active decay

ground state: the state of the atom in which all of its electrons are in their lowest possible energy levels

Gay-Lussac's law: in any chemical reaction involving gases measured at the same temperature and pressure, the volume of gaseous reactants and gaseous products are always in simple ratios of whole numbers.

global warming: the gradual rise in the average temperature of the Earth's atmosphere; causes catastrophic climate changes

greenhouse effect: the effects of $CO_{2(g)}$, $CH_{4(g)}$ oxides of nitrogen and water vapor in the atmosphere, where they trap heat the way the glass in greenhouse does

half-life: the time it takes for half of a sample of a radioactive to decay

halogens: elements in group 17 of the periodic table

hard water: water that forms a scum with soap; caused by the presence of $Ca^{2+}_{(aq)}$ and $Mg^{2+}_{(aq)}$ ions

heterogeneous mixture: a mixture in which there is more than one phase can be seen.

hetero-nuclear diatomic molecule: a diatomic molecule where the two atoms are different

homogeneous mixture: sometimes called a solution, it is a mixture where only one phase can be observed

homo-nuclear diatomic molecule: a diatomic molecule where the two atoms are identical

hydration energy: energy produced when ions are hydrated during the solution process

hydrogen bonding: a bond that takes place between a lone pair of electrons on an atom (usually only N, O and F) of one molecule, and a positive charge on the hydrogen atom of a neighbouring molecule

hydrophobic: water-repelling

hydrophilic: water-loving

hydroxide ion: these cause solutions to be basic

hydrate: a compound that incorporates molecules of water into its crystals

hydronium ion: this formed when a $H^+_{(aq)}$ bond with a water molecule; $H_3O^+_{(aq)}$.

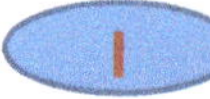

ideal gas: an imaginary gas that has no intermolecular attractions, no size and having with perfect elastic collisions.

ideal gas law: $PV = nRT$

impure substance: this has a mixture of two or more pure substances

indicators: these change colour in acids and bases

ionic bond: this formed between metals and non-metals in ionic compounds

ionic equation: an equation that shows all the possible aqueous ions of the reactants and products in a chemical reaction

intermolecular force: force of attraction between molecular compounds

ion exchanger: a device that is used to cleanse hard water of $Ca^{2+}_{(aq)}$ and $Mg^{2+}_{(aq)}$ ions by exchanging them with Na^+ ions laced on a resin.

ionization energy: This minimum amount of energy that is required to knock one of the outermost electrons from a gaseous element out of its orbit.

ionic product for water:

$$K_w = [H^+_{(aq)}] \times [OH^-_{(aq)}] = 1.0 \times 10^{-14}$$

ionic radius: distance from the centre of an ion to its outermost orbital; this indicative of the size of the ion

isotopes: atoms of the same elements having the same number of protons but different number of neutrons

isotopic abundance: the percentage of one isotope in random sample of isotopes of the same element

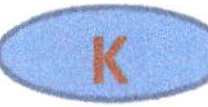

Kelvin temperature scale: this scale begins with -273^0C as 0 Kelvin and having every degree equaling that of the Celsius scale in terms of magnitude

law of conservation of mass: after a chemical change has taken place, the mass of the products equals that of the reactants; matter is neither created nor destroyed

law of definite proportion: In any specific compound, the elements are always present in definite proportions by mass.

Lewis diagrams: these use electrons dots as valence electrons to show how ionic and molecular compounds are formed; dashes are used to represent bonded pairs and dots are used to represent non-bonded pairs

line spectrum: this produced when the electrons of atoms are bumped up to higher energy levels and subsequently return to their ground state; line produced is of a specific wavelength and energy.

Limiting reagent: this is the reactant that is complete used up during a chemical change

London dispersion force: an intermolecular force between non-polar molecules

mass number: the sum of the number of neutrons and protons in the nucleus of an atom; represented by A and it is expressed in atomic mass unit, u

melting: change of state from solid to liquid

mixture: this results from the bringing together two or more pure substances

micelles: a sherical aggregate of soap or detergent moles in water, normally with their hydrophobic tails on the inside and their hydrophilic ends on the outside

molecular geometry: the three-dimentional shape of a molecule

mole: this contains Avogadro's number of entities; atoms, ions, molecules, etc

molar volume: the volume that one mole of any gas would occupy at a specific pressure and temperature; at STP = 22.4 L and at SATP = 24.8 L

neutralization reaction: a reaction between and acid and base to produce a salt and water

neutron: the subatomic particle in the nucleus that has no charge and has a mass of 1u

nomenclature: a system of naming chemical compounds

nitrifying bacteria: these degrade organic matter using oxygen from nitrate ions

nuclear reactions: these include radioactive decay, fission and fusion

orbital shape: the region about the nucleus where there is a high probability of finding the electrons

orbital energy levels: the way in which the energy of the orbital in every shell is depicted

oxyacids: binary acids that have incorporated oxygen atoms in their molecules

Pascal: a unit of atmospheric pressure

percentage composition: the composition of each element in a compound, expressed as a percentage of the total mass of the compound

percentage yield: this is calculated by dividing the mass of the actual yield by what was expected, and then multiplying this value by 100%

photon: an electromagnetic radiation of a specific quantum of energy

pH: a measure of the hydrogen ion concentration in a solution = -log $[H^+_{(aq)}]$

polar molecule: a molecule that has uneven distribution of the shared electron cloud that results in it having a small positive charge on one end and a small positive charge on the other

polarizability: the tendency of the electron cloud to shift within a molecule; this increases with the number of electrons

polyatomic compounds: these have more than two elements in their compounds

phosphor: a substance that glows when bombarded with electrons; zinc sulphide

physical property: one that is observable and measurable and may help to distinguish one substance from another

physical change: a change in physical property of a substance without affecting its composition; can be reversed

pOH: a measure of the hydroxide ion concentration in a solution = -log $[OH^-_{(aq)}]$

pure substance: this only one type of particle

quantum: a package of electromagnetic energy represented as, E = hf

quantum mechanics: this theory envisions electrons as having wave-like probability distribution with specific amount of energy

radioactive decay: the decomposition of an atomic nucleus with the release of radiations; beta, gamma and alpha rays

resultant force: the result of all the forces acting on an object in the same or from different directions

saturated solution: one that cannot dissolve anymore solute at particular temperature

shells: orbital or energy levels

single displacement reaction: a reaction in which, if an element if placed in a compound it will displace a less reactive one present in the compound

solubility: the total amount of solute that can dissolve in fixed volume of solvent at a particular temperature

solute: the substance that is dissolved by the solvent to form the solution

solvent: the substance that dissolves the solute

solution: a homogeneous mixture of the solute and solvent

standard solution: a solution of known concentration in a titration process

stoichiometry: The study of the relationships between the amounts of reactants used and products formed in chemical reactions

strong acid: one that ionizes completely in water to produce $H^+_{(aq)}$ ions.

strong base: one that ionizes completely in water to produce $OH^-_{(aq)}$ ions.

supersaturated solution: a solution that dissolves more solute than it normally does at a particular temperature

sublimation: a change of state from a solid directly to a gas, or from a solid to a gas

synthesis reaction: a reaction in which two or more simpler substances chemically combine to for a more complex substance (compound)

UV: a type of electromagnetic radiation

urea: a nitrogenous waste emanating from within the bodies of living organism

vapour pressure: the partial pressure due to water vapour in a mixture of gas at a particular temperature

Van der Waals forces: forces of attraction between molecules; London dispersion forces and dipole-diploe forces

water softening: the process of removing $Ca^{2+}_{(aq)}$ and $Mg^{2+}_{(aq)}$ from water.

weak acid: an acid which ionizes only partially to produce $H^+_{(aq)}$ ions.

weak base: an acid which ionizes only partially to produce $OH^-_{(aq)}$ ions.

weak base: an acid which ionizes only partially to produce $OH^-_{(aq)}$ ions.

x-rays: very energetic electromagnetic radiation; very penetrative in nature

A

Absolute zero, **329**
Acceleration due to gravity, **322**, **344**
Acids, **269**
    Arrhenius' theory, **264, 266**
    Brönsted-Lowry's theory, **266-267**
    conjugate acid-base, **267-268**
    comparison of weak/strong, **268-271, 273**
    monoprotic vs polyprotic, **271-272**
Acid-base neutralization, **273-274**
    stoichiometry, **173-174, 279**
Acid-base titration, **281**
    aliquot, **281**
    end point, **282**
    indicators, **281**
    standard solution, **281**
    titrant, **281**
Acid nomenclature, **312-314**
    Binary acids, **309**
    oxyacids, **309-311**
Acid rain, **86, 92-93, 109, 257, 359**
Activated sludge, **251--252**
Activity series, **95-97**
    -metals, **95-97**
    -halogens, **97-98**
Agricultural runoff, **256**
Airbags, **338, 348**
Algal bloom, **251, 257, 356**
Alpha particle, **12-13, 21-22**
Anode, **10-11, 265**
Anions, **265-269**
Aqueous solution, **201, 204-206**
Arrhenius' theory, **264, 266, 269**
Aristotle **8-9**
Atmospheric pressure, **321-324**
    demonstration of, **322**
    units, **324**
Atomic notation, **135, 138**
Atomic theories, **8-16**
    -timeline of, **8**
    -Aristotle, **9**
    -Dalton, **9**
    -Democritus, **9**
    -Thomson, **10**
    -Goldstein, **10-11**
    Rutherford, **12-13**
    Chadwick, **13**

    Bohr, **13-16**
Atomic mass, **135, 137-138**
Atomic mass unit, **136**
Atomic radius (definition), **26**
Atomic radius (trends across), **27-28, 32**
Atomic structure, **135**
Auto-ionization of water, **288**
Average atomic mass, **138, 134, 146**
Avogadro's constant, **141-146**
Avogadro's hypotheses, **343**

B

Bacteria (E.Coli), **249**
Balancing equations, **89-90**
Balmer's series, **15-16**
Bases, **266, 271-273**
Beta decay, **21**
Binary acids, **309**
Binary ionic compound, **119, 122-123**
Binary molecular comp. **124**
Biodegradation activity, **261-263**
BOD, **257**
Bohr N, **13-16**
Bond angles, **67, 69**
Bonding capacity, **53, 54**
Bohr-Rutherford diagrams (ions)**, 42-48**
Bohr-Rutherford diagrams (comp), **48**
Boyle's law, **325-326**
Bronsted-Lowry theory, **266-267, 270**
Bugs, **251**
Burette, **281-28**2

C

Canadian guidelines for safe water, **252**
Carbon cycle, **356-360**
Carbon monoxide, **109-111**
Catalytic converter, **112-113**
Cations, **265-266**
Cathode, **265**
Cathode ray tube, **10-11**
Cathode rays, **10-11**
Cellular respiration, **86, 109, 257-258**
Chadwick J, **8, 13**
Charles's law, **328, 330**

Bibliography

Bruckman, H.J., & Cruickshanks, A. (1988). *Understanding Chemistry*. Toronto: John Wiley & Sons Canada, Limited.

Ebbing, D.D. (1996). *General Chemistry*. 5th Edition. Wrighton, M.S. (Ed.) Toronto: Houghton Mifflin Company.

Jenkins, F., van Kessel, H., Davies, L., Lantz, O., Thomas, P., & Tompkins, D. (2000). *Chemistry 11*. Toronto: Nelson Thomson Learning.

Mustoe, F., Jansen, M.P., Doram, T., Ivanco, J., Clancy, C., & Ghazariansteja, A. (2001). *Chemistry 11*. Toronto: McGraw-Hill Ryerson.

Rayner-Canham, G., Damji, S., & Goering-Boone, U. (2001). *Chemistry 11*. Toronto: Addison Wesley.

Reger, D.L., Goode, S.R., & Mercer, E.E. (1993). *Chemistry: Principles & Practice*. Toronto: Saunders College Publishing.

Webber, H.D., Billings, G.R., & Hill, R.A. (1970). *Chemistry: A Search for Understanding*. Toronto: Holt,Rineheart and Winston of Canada, Limited.

Whitman, R.L., Zinck, E.E., & Nalepa, R.A. (1982). *Chemistry Today*. 2nd Edition. Scarborough, Ontario: Prentice-Hall Canada, Inc.

Zumdahl, S.S. (1997). *Chemistry*. 4th Edition. Stratton R. (Ed.) New York: Houghton Mifflin Company.

Internet Resources Consulted

Atomic Archive. (1998). *Nuclear Fusion*. Retrieved from
http:/www.atomicarchive.com/Fusion/Fusion1.shtml

Bennett, L. (2012). *Naming Acids and Bases*. Retrieved from
http//edtech2.boisestate.edu/lindabennett1/502/Compounds%20and%20Naming/naming%20acid%20base.html

Bennett, L. (2012). *Types of Radioactive Decay*. Retrieved from
http://edtech2boisestate.edu/lindabennett/502/Nuclear%20Chemistry/types%20of%20decay.html

Bishop, M. (2013). *Oxyacid Nomenclature*. Retrieved from
http:/www.preparatorychemistry.com/bishop_oxyacid_nomenclature.html

Bishop, M. (2013). *Supersaturated Solutions*. Retrieved from
http://preparatorychemistry.com/Bishop_supersaturated.html

Blaber, M. (1996). *Basic Concepts of Chemical Bonding*. Retrieved from
http://chemwiki.ucdavis.edu/Physical_Chemistry/Kinetics/Modeling_Reaction_Kinetics/Reaction_Profile

Blaber, M. (1996) *Molecular Geometry and Bonding Theories*. Retrieved from
http://www.mikeblaber.org/oldwine/chm1045/notes/Geometry/Covalent/Geom04.html

Buescher, L. (2004). *Atomic Structure Timeline*. Retrieved from http://atomictimeline.net/index.php

Campbell, H.A. (1998). *Avogadro's Law – What is it?* Retrieved from
http://www.chemistry.co.nz/avogadro.html

Clark, J. (2000). *Electron Affinity*. Retrieved from http://www.chemguide.co.uk/atoms/properties/eas.html

Clark, J. (2000). *Electronegativity*. Retrieved from
http://www.chemguide.co.uk/atom/bonding/electroneg.html

Clark, J. (2010). *Energy Profiles for Simple Reactions*. Retrieved from
http://www.google.ca#q=Energy%20profile%20diagrams

Clark, J. (2010). *Ideal Gases and the Ideal Gas Law*. Retrieved from
http://www.chemguide.co.uk/physical/kt/idealgases.html

Clark, J. (2006). *The Atomic Hydrogen Emission Spectrum*. Retrieved from
http://www.chemguide.co.uk/atom/properties/spectrum.html
Clark, J. (2002). *Theories of Acids and Bases*. Retrieved from
http://www.chemguide.co.uk/physical/acidbaseeqia/theories.html
Colwell, C.H. (1997). *Famous Experiments: The Discovery of the Neutron*. Retrieved from
http://dev.physicslab.org/Document.aspx?doctype=3&filename=AtomicNuclear_ChadwickNeutron.xml
Complete Dictionary of Scientific Biograpy. (2008). *Eugen Goldstein*. Retrieved from
http://www.encyclopedia.com/doc/1G2-2830901680.html
DeLeon, N. (Year unknown). *Rutherford's Planetry Model of the Atom*. Retrieved from
http://www.jun.edu/~cpanhd/C101webnoted/modern-atomic-theory/rutherford-model.html
Educating Online (Year unknown). *Factors Affecting Solubility*. Retrieved from
http://www.solubilityofthings.com/basics/factors_affecting_solubility.php
Foro Nuclear. (2013). *What are Radioisotopes*? Retrieved from http://www.foronuclear.org/consultas-
en/ask-the-expert/what-are-radioisotopes
Helmenstine, A.M. (2014). *Bohr Model of the Atom*. Retrieved from
http://chemistry.about.com/od/atomicstructure/a/bohr-model.html
Helmenstine, A.M. (2014). *Chemistry of Hard and Soft Water*. Retrieved from
http://chemistry.about.com/cs/howthingswork/a/aa082403a.html
Helmenstine, A.M. (2014). *How Do Detergents Clean?* Retrieved from
http://chemistry.about.com/od/howthingswork/f/detergentfaq.html
Helmenstine, A.M. (2014). *How Soap Cleans*. Retrieved from
http://chemistry.about.com/od/cleanerchemistry/a/how-soap-cleans.html
Helmentine, A.M. (2014). *Ionization Energy*. Retrieved from
http://chemistry.about.com/od/periodicitytrends/a/ionization-energy.html
Helmentine, T. (2014). *What is Dalton's Law of Partial Pressures?* Retrieved from
http://chemistry.about.com/od/chemistryfaqs/f/What-Is-Daltons-Law-Of-Partial-Pressures.html
Helmentine, T. (2014). *What is the Formula for Gay-Lussac's Law?* Retrieved from
http://chemistry.about.com/od/chemistryfaqs/f/What-Is-The-Forumla-For-Gay-Lussacs-Law.html
Jones, A.Z. (2014). *Cathode Ray*. Retrieved from http://physics.about.com/od/glossary/g/cathoderay.html
Kent Chemistry. (Year Unknown). *Covalent Bond Polarity*. Retrieved from
http://www.kentchemistry.com/links/bonding/bondpolarity.html
Kent Chemistry. (Year Unknown). *Solubility Curves*. Retrieved from
http://www.kentchemistry.com/links/Kinetics/SolubilityCurves.html
Kent Chemistry. (Year Unknown). *Use of Radioisotopes*. Retrieved from
http://www.kentchemistry.com/links/Nuclear/radioisotopes.html
Ladon, L. (2001). *Ideal and Real Gas Laws*. Retrieved from
http://www.info.com/properties%20of%20ideal%20gases?cb=57&q_sl=*&cmp=320434&q_csr=l
Netting, R. (2003). *Feeling Pressured*. Retrieved from http://kids.earth.nasa.gov/archive/air_pressure.html
Ophardt, C.E. (2003). *Acid and Base Strength*. Retrieved from
http://www.elmhurst.edu/~chm/vcchembook/185strength.html
Rice University. (1999). *Discovery of Parts of the Atom: Electron and Nuclei*. Retrieved from
http://cnx.org/contents/7ecc38ba-f855-43c4-bad7-50c40b70556d@6
School of Chemistry, University of Bristol. (2002). *VSERP Theory*. Retrieved from
http://www.chemlabs.bris.ac.uk/outreach/resources/vsepr.html
Science Buddies. (2000). *Acids, Bases, and the pH*. Retrieved from http://www.sciencebuddies.org/science-
fair-projects/project_ideas/Chem_AcidsBasespH.html
Shapley, P. (2011). *Self Ionization of Water*. Retrieved from
http://butane.chem.uiuc.edu/pshapley/GenChem1/L19/3.html
Shodor. (1996). *Gas Laws*. Retrieved from http://www.shodor.org/unchem/advanced/gas.html

Silly Beagle Productions (2014). *Coulomb's Law*. Retrieved from
http://www.aplusshysics.com/courses/honors/estat/Coulomb.html

Szaflarski, D., Dean, R., & Dean, M. (Year Unknown). *Nuclear Stability and Radioactive Decay*. Retrieved from http://people.chem.duke.edu/jds/cruise_chem/nuclear/stability.html

University of Waterloo. (Year Unknown). *Enthalpies of Reactions*. Retrieved from
http://www.science.uwaterloo.ca/~cchieh/cact/c120/heatreac.html

Unknown Author. (1994). *A Timeline of Atomic Structure*. Retrieved from
http://www.barcodesinc.com/articles/timeline-on-atomic-structure.html

Unknown Author. (Year Unknown). *Rutherford's Gold Foil*. Retrieved from
http://chemteacher.chemeddl.org/services/chemteacher/index.php?option=com_content&view=article&id=74

Winter, M. (1993). *Isotopes of Carbon*. Retrieved from http://www.webelements.com/carbon/isotopes.html

Wolfram Research. (Year Unknown). *Ionization Energies of the elements*. Retrieved from
http://www.periodictable.com/Properties/A/IonizationEnergies.html

Yoder, C. (2014). *Vapour Pressure of Water from 0^oC to 100^0C*. Retrieved from
http://www.wiredchemist.com/chemistry/data/vapor-pressure

Text Credits

The author of this textbook would like to thank the providers of the sources mentioned in the bibliography, since their usefulness was important to the successful completion of this text. After meticulous scrutiny was made of this text, the following sources were identified from which specific copyright materials were retrieved and used. The author will be very willing to accept suggestions with regards to any omissions and will take all necessary steps to ensure any further appropriate accreditation.

Unit 1: Matter and Chemical Bonding
Chapter 2: p. 11 Adaptation of figure 2.7(b) from © Bruckman, H.J., & Cruickshanks, A. (1988). *Understanding Chemistry*. Toronto: John Wiley & Sons Canada, Limited.
Chapter 2: p.11 and 10 some information retrieved from © Bruckman, H.J., & Cruickshanks, A. (1988). *Understanding Chemistry*. Toronto: John Wiley & Sons Canada, Limited.
Chapter 3: p.20 first paragraph – information retrieved from © Jenkins, F., van Kessel, H., Davies, L., Lantz, O., Thomas, P., & Tompkins, D. (2000). *Chemistry 11*. Toronto: Nelson Thomson Learning.
Chapter 3: p.20 second paragraph – information retrieved from © Rayner-Canham, G., Damji, S., & Goering-Boone, U. (2001). *Chemistry 11*. Toronto: Addison Wesley.
Chapter 3: p.22 figure 3.3. Adapted from © Zumdahl, S.S. (1997). *Chemistry*. 4th Edition. Stratton R. (Ed.) New York: Houghton Mifflin Company.
Chapter 6: p.77 figure 6.31 Adapted from © Ebbing, D.D. (1996). *General Chemistry*. 5th Edition. Wrighton, M.S. (Ed.) Toronto: Houghton Mifflin Company.
Chapter 6: p.77 Some materials used from © Ebbing, D.D. (1996). *General Chemistry*. 5th Edition. Wrighton, M.S. (Ed.) Toronto: Houghton Mifflin Company.
Chapter 7: p.114, first paragraph and some materials used were obtained from ©
https://chem.libretexts.org/Bookshelves/Inorganic_Chemistry/Supplemental_Modules_and_Websites_(Inorganic_Chemistry)/Coordination_Chemistry/Complex_Ion_Chemistry/Origin_of_Color_in_Complex_Ions

Unit 2: Quantities in Chemical Reactions
Chapter 10: p.167 Table 10.2 Values retrieved from © Rayner-Canham, G., Damji, S., & Goering-Boone, U. (2001). *Chemistry 11*. Toronto: Addison Wesley.

Unit 3: Solutions and solubility
Chapter 13: p.223 Sample problem 9 adapted from © Rayner-Canham, G., Damji, S., & Goering-Boone, U. (2001). *Chemistry 11*. Toronto: Addison Wesley.

Unit 4: Gases
Chapter 18: p.326 Table 18.1. Table format adapted from © Jenkins, F., van Kessel, H., Davies, L., Lantz, O., Thomas, P., & Tompkins, D. (2000). *Chemistry 11*. Toronto: Nelson Thomson Learning.
Chapter 18: p.330 Table 18.2. Table format adapted from © Jenkins, F., van Kessel, H., Davies, L., Lantz, O., Thomas, P., & Tompkins, D. (2000). *Chemistry 11*. Toronto: Nelson Thomson Learning.
Chapter 19: p.341 Second paragraph top right adapted from Whitman, R.L., Zinck, E.E., & Nalepa, R.A. (1982). *Chemistry Today*. 2nd Edition. Scarborough, Ontario: Prentice-Hall Canada, Inc.
Chapter 19: p.343 some materials used from © Bruckman, H.J., & Cruickshanks, A. (1988). *Understanding Chemistry*. Toronto: John Wiley & Sons Canada, Limited.
Chapter 19: p.344 Table 19.3. Table format adapted from © Bruckman, H.J., & Cruickshanks, A. (1988). *Understanding Chemistry*. Toronto: John Wiley & Sons Canada, Limited.
Chapter 21: p.363 Avogadro's hypotheses taken from © Bruckman, H.J., & Cruickshanks, A. (1988). *Understanding Chemistry*. Toronto: John Wiley & Sons Canada, Limited.

About the Author

Abdul Jalil Shakur was born in the Republic of Guyana, the only English-speaking country in South America. He grew up in his native village of Meten Meer Zorg, situated on the west coast of Demerara. He holds a Hon. B.Sc. degree from the Universities of Guyana and Atabasca (Canada), as well as a B.Ed. from the University of Toronto. He is a Canadian trained science teacher.

Jalil acquired a wealth of experience of as a teacher of integrated Science, Biology, Chemistry and Physics, having taught these subjects to high school students during his 43-year illustrious, successful teaching career. He was a high school teacher at the Zeeburg Secondary School (Guyana) from 1974-1978 and from 1982-1989. He taught Physics at St. Mary's College in St. Lucia from 1989 until 1991 when he migrated to Canada. During those times he served as the **Head of the Science Department** at the Zeeburg Secondary School, and as Chemistry examiner with the **Caribbean Examination Council (CXC)**. In Canada, he was employed by Toronto District School Board where he taught integrated Science, Biology, Chemistry at the C.W. Jefferys C.I. from 1997 until he retired in 2017.

Jalil lives in Toronto with his wife, Lilatool, his daughters Shazeeda, Nazeera, Nafeesa, (teachers with the Toronto District School Board) and son, Dr. Yaseer, and twelve wonderful grandchildren.

Jalil is sport-loving person, who enjoys socializing with people and is a strong advocate for the oppressed, poor and needy.

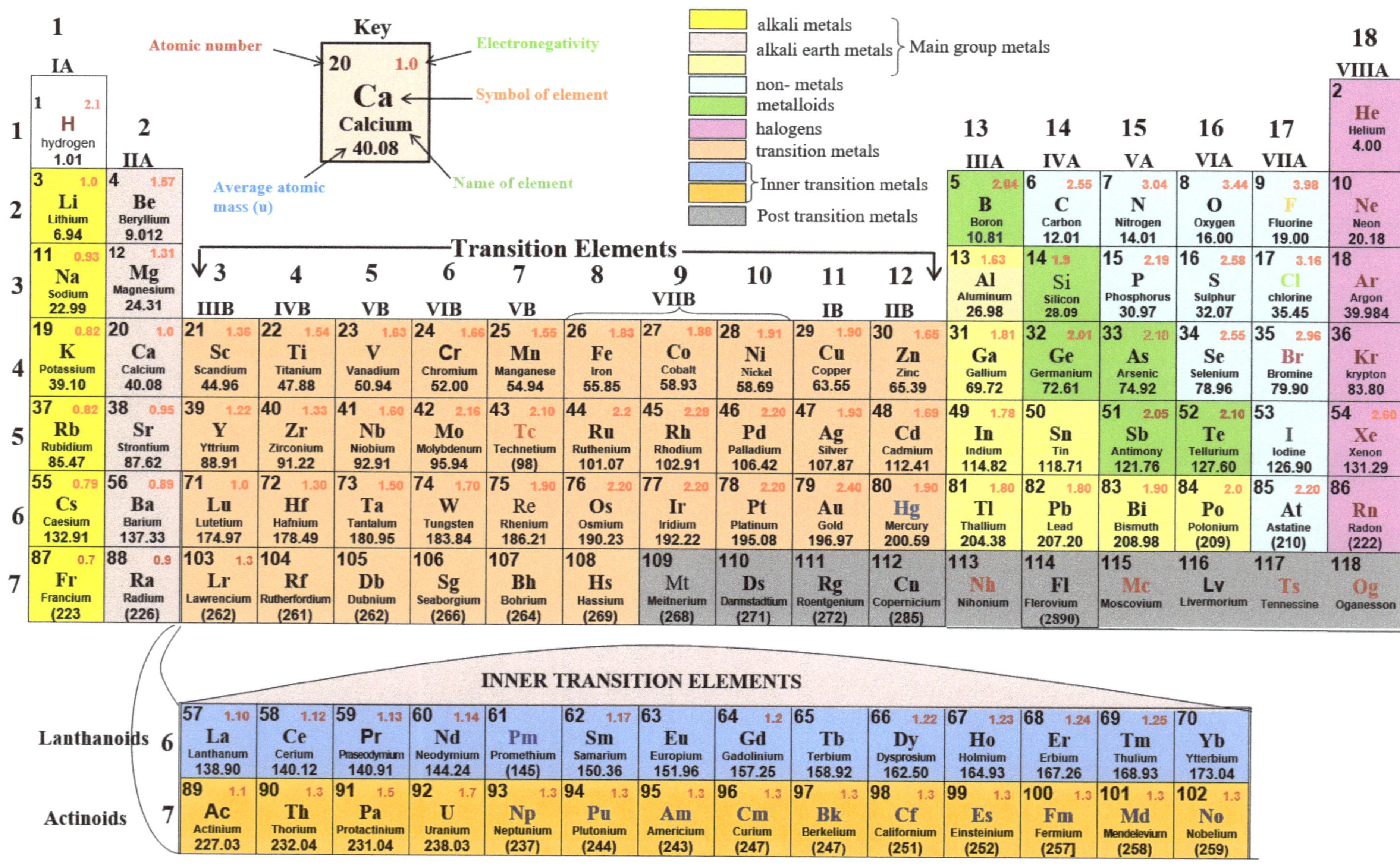

The Periodic Table of the Elements
Key
Atomic number
Electronegativity
20 1.0
Ca
Calcium
40.08
Symbol of element
Name of element
Average atomic mass (u)
alkali metals
alkali earth metals
non- metals
metalloids
halogens
transition metals
Inner transition metals
Post transition metals
Main group metals
Transition Elements
INNER TRANSITION ELEMENTS
Lanthanoids
Actinoids